Membrane molecular biology of neoplastic cells

Membrane molecular biology of neoplastic cells

Donald F.H. Wallach

with contributions by

Rupert Schmidt-Ullrich

1975

ELSEVIER SCIENTIFIC PUBLISHING COMPANY

AMSTERDAM – OXFORD – NEW YORK

ELSEVIER SCIENTIFIC PUBLISHING COMPANY
355 JAN VAN GALENSTRAAT
P.O. BOX.211, AMSTERDAM, THE NETHERLANDS

AMERICAN ELSEVIER PUBLISHING COMPANY, INC
52 VANDERBILT AVENUE
NEW YORK, NEW YORK 10017

ISBN: 0 444 41359 6

Printed in The Netherlands

To
my Parents
and
to
Fernando Bloedorn

Contents

Chapter 7 Membrane permeability *293*

D.F.H. WALLACH

Chapter 8 Cell contact and cell recognition *345*

D.F.H. WALLACH

Foreword

Like the bark of a tree or the skin of an animal, the plasma membrane is a highly specialized organ combining barrier functions that widely surpass the performance of the most closely guarded borders, with highly regulated bilateral exchange functions. Dr. Wallach's view that membranes are laterally segregated into domains of different composition and behavior provides a rational basis for this duality, characteristic for *all* cells. In higher organisms, cell differentiation adds a further, formidable superstructure of composition and function. The membrane becomes an array of specialized organelles essential to a wide variety of receptor transduction and signalling functions, involving both contactual, short-range and humorally mediated, long-range interactions. Differentiated membrane structures are beginning to be recognized as morphological, functional or antigenic units. The mapping of these entities is still in its earliest beginnings, although some cells, e.g. lymphocytes, already provide a certain amount of information, useful to the membranologist, the student of normal lymphocyte function and the lymphopathologist. A rapid increase of information can be expected with regard to other cell types.

The last few years have provided proof for the originally surprising concept that many proteins penetrate into or through the apolar cores of biomembranes in dynamic association with membrane lipids. Wallach's definition of the interacting units of a membrane lattice, which he calls protomers, as the intramembranous segments of membrane proteins associated with their boundary layer lipid is a further step towards a more concrete definition of this dynamic system. Studies on capping, redistribution, protein 'shedding' and ligand-induced 'shedding' have now become the favorite playtoys of membranologists, immunologists and experimental oncologists. Relatively simple experimental designs, including direct microscopial observation, have provided hitherto unparalleled insights into the interdependence or independence of different membrane-associated antigenic or receptor units, their mobilities or anchorages, their 'shedding', modulation, degradation and regeneration. Unexpected and as yet unexplained differences have been found in the behavior of different receptors on the same cell and of the same receptor on different cells. Suggestive differences between normal and neoplastic cells are beginning to emerge and are particularly interesting.

The thought that neoplasia is essentially a 'membrane disease' has been raised by Wallach at an early stage and with force. It is an exciting thought, because the cell membrane is an almost compulsory transfer point for the action of growth controlling signals. Changes in critically important membrane sites, or, as Wallach puts in, the alteration of concerted membrane behavior by oncogenic agents, may bring about the free-wheeling of the cell cycle recognized as neoplastic growth. Insertion of new, virally determined subunits or functional modification of existing units by the action of viral or non-viral agents may be responsible. The reality of such phenomena is strongly suggested by the documented existence of membrane changes in tumors, expressed as altered contact inhibition, anchorage independence, appearance of new and change or disappearance of old membrane antigens and changes in membrane permeability and transport. In addition to the notorious difficulty of distinguishing between causal and consequential phenomena, differences between normal and neoplastic cells must be interpreted with great caution, however, it is particularly important to scrutinize the representativeness of the normal cell since uniclonal tumor lines often express clonal markers present on a minority but not the majority of the control cell population. It is therefore of considerable significance that reproducible changes in membrane biochemistry are found to accompany the reversible transitions between transformed and non-transformed cellular phenotypes induced by shifts between permissive and non-permissive temperatures, in cells transformed by temperature-sensitive, oncogenic virus mutants.

Equally significant is the extensive evidence that membrane anomalies in neoplastic cells are not restricted to the plasma membrane and therewith the emergence of clues as to the metabolic mechanisms underlying both membrane-associated and other peculiarities of tumor cells.

All this strongly suggests that the biochemical and functional study of biomembranes has come to stay as one of the most important areas in tumor biology. The book of Wallach is an admirable review of the present status of the field, written by an outstanding 'membranologist', with a profound interest in tumor cell membranes from the outset and one who never loses sight, while analyzing the impressive wealth of factual information presented in this book, either of the biological problems or of the desperate need of this field to keep apart hard from soft facts, or well-founded hypotheses from premature generalizations. The triple capacity of the author as biochemist, physician and Scientific Director of a department devoted not only to tumor biology but also to tumor therapy accounts for the special interest and flavor of this book.

Stockholm, Sweden GEORGE KLEIN
September, 1975

Preface

O glücklich wer noch hoffen kann,
Aus diesem Meer des Irrtums aufzutauchen!
Was man nicht weiss, dass eben brauchte man,
Und was man weiss, kann man nicht brauchen.

GOETHE, (*Faust*, I, 1065)

The membrane biology of tumor cells has become a most active field of biomedicine. The area has excited the interest of scholars from numerous and diverse disciplines for many reasons, some relating to cancer as a disease and others to neoplasia as a fundamental biological process. Prominent among these reasons is the suspicion that the normal social interactions between tissue cells depend directly upon their surface membranes and that the invariable disruption of tissue homeostasis in cancer derives from cell surface deviations. Equally important is the realization that numerous metabolic functions commonly abnormal in tumor cells involve diverse intracellular membranes. Other sources of interest are the recognition that surface membranes can act as restriction sites in cancer-chemotherapy, as well as targets in immunotherapy, and the anticipation that the elucidation of tumor cell membrane anomalies will provide new therapeutic strategies. Finally, the remarkable maturation of general membrane technology and the great increase in understanding of fundamental membrane processes, now allows experimental and conceptual approaches to tumor membrane biology that were not accessible until recently.

In my efforts to achieve some understanding of the membrane anomalies of neoplastic cells and the relevance of these anomalies to malignancy, I have encountered some formidable obstacles that assuredly also trouble others in the field. Of these the polymorphism of tumor cells and the intricacies of membrane biology are features inherent in the biological problem. However, additional difficulties come from the fragmented character of the explosively growing literature, the oftentimes deficient cohesion between diverse research strategies and the inadequate integration of tumor membrane biology with basic membrane studies and experimentation on normal cells. Such factors, while understandable in view of the complexity of the field and of the social pressure to 'cure cancer', generate intellectual hindrances that appear as troublesome as the biological obstacles.

My purpose in writing this book is to integrate existing biochemical, biophysical and molecular biological information relevant to the field of tumor membrane biology, to relieve some of the communication problems in the area and to create some new perspectives. I have therefore attempted to present an objective, topical and rather comprehensive survey of membrane molecular biology *as it now applies to comparisons of normal and neoplastic* cells, integrating data and hypotheses from multiple sources and disciplines in a manner accessible to specialists in both tumor biology and membrane biology, as well as to a wide circle of graduate students and other scholars in diverse fields of biomedicine. It is my hope that the book will prove as helpful for the reader as it has been challenging to the writer.

This book could not have been completed without a great deal of support. I would therefore like to express my particular appreciation to R. Schmidt-Ullrich for his assistance in Chapters 2 and 6. In addition, I extend my sincere thanks to the following colleagues for their inspiration, constructive criticism and frequent permission to read important manuscripts prior to publication: J. Borysenko, H. Chen, L. Furcht, J.M. Graham, S.I. Hakomori, M. Hatanaka, A.A. Kandutsch, L. Kwock, P.S. Lin, A.L. Lehninger, C. McKhann, Jr., R. Mikkelsen, V. Najjar, F. Oosawa, C.A. Pasternak, J. Sheridan, S. Tevethia and R. Weinstein. I am particularly indebted to Kate Scheller and Tina Monroy for their invaluable help in preparing the manuscript and to the officers of ASP Biological and Medical Press, Jack Franklin in particular, for their continued interest and patient support.

Finally, I want to express my deep gratitude to my wife, both for her encouragement and for her patience in enduring the many evenings and weekends spent in the conception and writing of this book.

Boston, Mass.
May, 1975

DONALD F. HOELZL WALLACH

General introduction

The terms 'cancer' and 'malignant neoplasia' refer to a category of disease processes which may diverge markedly in detail, but which share critical features, including:

(1) Biologic autonomy.
(2) Cell proliferation.
(3) Asocial cell behavior.
(4) Invasiveness and metastasis.

As emphasized by Huxley [1], once an autonomous, neoplastic cell proliferates and gives rise to a tumor cell colony, a new, self-replicating species has been generated. Such a species has full capacity for further evolution through mutation and selection, but possesses certain unique properties.

First, it is universally *parasitic*. Outside the laboratory this parasitism is obligatory and, in nature, the parasite always originates from a host cell.

Second, each tumor species becomes extinct with the inevitable death of its host. Thus, normal evolution and divergence do not occur in nature, although, when tumor cells are maintained *in vitro*, or by grafting, the strain of the transplanted cells is a species equivalent. However, 'tumor species' cannot be rigorously classified, although approximate categorizations can serve heuristic or clinical purposes. Attempts at rigorous taxonomic classification fail because of the diversity and pleiomorphism characteristic of tumors. This diversity arises in part from:

(a) *Genetic heterogeneity*, including that involving the taxonomic specificity of the host, as well as intra-species variation.
(b) *Epigenetic heterogeneity*, including that due to viral and mitochondrial genomes, as well as the consequence of irreversible tissue differentiation within a given host (giving rise, e.g. to sarcomata vs. carcinomata).

(c) *Responsive heterogeneities*, reflecting the diverse reactions of various neoplastic cells to extrinsic stimuli.
(d) The phenotypic pleiomorphism characteristic of the neoplastic process *per se*.

Despite the obstacles facing all attempts at precise classification of tumors as distinct biologic entities, one must nevertheless enquire into possible mechanisms which can cause these species to emerge. This inquiry has preoccupied oncologists for centuries, yet, two-hundred years have passed since Pott demonstrated in 1775 that cancer in man can be caused by environmental agents [2] and sixty years since the existence of viruses carcinogenic to animals was proven [3, 4] and we still lack substantial concensus as to what processes in nature are critical in carcinogenesis. Indeed, the tumor biologist must grapple with two rather divergent concepts.

One concept invokes *genetic mutation and selection*. This hypothesis was first proposed by Boveri [5] and developed impressively in terms of new knowledge by Huxley [1] and Burnet [6]. As detailed by Burnet, a purely evolutionary hypothesis implies that cancers represent errors in the normal cell genome. Such genetic errors can occur randomly at both the germinal and somatic levels and at frequencies independent of normal environmental factors. Carcinogenic agents, such as chemicals and ionizing radiation, act to 'intensify' the processes responsible for spontaneous mutation and increase the probability of error in the replicative process. Burnet [6] does not dispute with evidence that certain viruses bear oncogenic information, but argues that viruses can also increase the probability of genetic error, that the consistent production of malignant disease offers no selective advantage to a virus and that most experimental, virus-induced tumors constitute laboratory artifacts.

Although not framed in precise genetic terms, the Warburg hypothesis, proposing that cancer derives from a heritable defect in mitochondrial material [7], also implies 'genetic error'.

Strictly evolutionist-selectionist hypotheses tend to consider viral oncogenesis as a relatively unimportant aspect of carcinogenesis *in nature*. This is in contrast to the '*protovirus*' *hypothesis*, first proposed in 1970 by Temin [8]. This concept merges, in molecular-biological terms, information derived from classical genetics with information from tumor virology. The *protovirus hypothesis* deals with possible mechanisms by which genetic material can be transferred between cells so as to 'amplify' existing genes and to generate *new* genetic information, *potentially present* in the somatic cell genome, but lacking *as such* in the germ line chromosome. When the postulated transfer of genetic material proceeds in an orderly way, it can serve as a genetic mechanism for normal development and differentiation. However, '*misevolution*' of protoviruses may generate the genetic information for cancer.

Temin proposes that *protoviruses* consists of specific regions of the cellular DNA, whose information can be transferred via RNA back to DNA. The postulation of a DNA → RNA → DNA sequence rests on the discovery of reverse transcription by virus-derived, RNA-dependent DNA polymerases [9,10]. Encoded in RNA, the information of the *protovirus* region can be transmitted between cells, and the new DNA generated therefrom by reverse transcriptase can be integrated into the DNA of the recipient cell. This integration could occur either at the same locus as that of the original *protovirus*, causing duplication, or at another site, generating, in conjunction with its neighboring DNA segments, a *new informational set*. This information can be duplicated and thus passed on to other chromosomes and to daughter cells. In principle, the postulated process of informational evolution can be repeated and diversified. Moreover, RNA generated on the novel template can lead to regeneration of the template as well as transfer of the novel information to other cells.

'Misevolution' of protoviruses could generate spontaneous cancers by *mutation* (as in mutation-selection hypothesis), by *integration of the protovirus at appropriate loci* or both. Carcinogenic chemicals, ionizing radiation and other carcinogens could act by increasing the probability of genetic error, increasing the chance of inappropriate integration of the *protovirus*, or both.

Several unique features of the *protovirus* hypothesis deserve comment here. *First*, the concept provides a plausible gene amplification mechanism; *second*, the hypothesis can account for the membrane pleiomorphism found in tumors, that is, if insertion of the '*protovirus*' occurs preferentially next to regions controlling membrane structure and/or functions.

Neither the classical evolution-selection hypothesis, nor the *protovirus* concept suggest that all carcinogenic events involve a common, underlying genetic mechanism. However, the '*oncogene*' *hypothesis*, first proposed in 1969 by Huebner and Todaro [11] does postulate such a process.

The *oncogene hypothesis* argues that the genomes of certain 'C-type' RNA viruses contain *virogenes*, necessary for the replication and perpetuation of the virus, as well as *oncogenes*, capable of causing neoplasia. The hypothesis further postulates that, early in phylogeny, such viruses gained access to the genome of vertebrates, that these genomes are now universally present in the cells of all vertebrates and are passed on to successive generations through the germ cells as part of the normal inheritance. The *virogenes and oncogenes* are envisaged to be subject to the same operational controls as normal genes, and can be repressed or depressed by host regulatory genes. However, various external agents, such as carcinogenic DNA viruses, carcinogenic chemicals and ionizing radion can depress the *oncogenes*. Spontaneous carcinogenesis is considered to arise from spontaneous depression of the *oncogenes*. Derepression of the *virogenes*, subsequent production of infectious virus and carcinogenesis by virus infection are assumed to be rare events in nature and relatively unimportant in cancer etiology. The Huebner–Todaro hypothesis further proposes that the genome of the 'C-type' RNA viruses provides critical genetic information for various *normal* developmental functions under *controlled* derepression (contrasted with *sustained* derepression in cancer).

A critical feature of the *oncogene* hypothesis is the postulate that all cancers result from the activation of a genetic potential present in all normal cells. This postulate appears unrealistic, since 'cancer' constitutes an almost unlimited series of biological entities. Neither the *mutation-selection* hypothesis, nor the *protovirus concept*, suffer from this weakness and both can account for a most important property of tumors induced by non-viral carcinogens, their *immunologic individuality*. This phenomenon was first demonstrated in 1943 by Gross [12] and rediscovered in 1957 by Prehn and Main [13]. The field has been comprehensively reviewed [14–18].

Tumors induced by chemical carcinogens, by ionizing radiation and by the implantation of inert plastic films, as well as spontaneous tumors, exhibit new, membrane-associated transplantation antigens, that are specific for *each* tumor, regardless of initiating agent. The antigenic diversity of non-viral tumors appears baffling, but, as argued by Burnet [6] can be explained in terms of mutation-selection hypotheses. Burnet [6] suggests that (a) the genetic loci determining the 'self'-specificity of cell surfaces are highly mutable; (b) the autochthonous parasitism unique to neoplasia, is invariably associated with new surface antigens and (c) the immunologic responses of higher animals, including the diversity of specific recognition processes, have evolved as a fundamental defense against autochthonous parasitism.

Burnet [6] develops his concept from the suggestion of Thomas [19] that 'adaptive immunity', arises either from the placental mother-fetus relation and/or in response to the vertebrate susceptibility of neoplasia. Burnet [6] reasons that ancestral lymphoid cells evolved recognition sites on their surfaces, allowing them to discriminate between 'self' and 'not-self', and to destroy the latter. In this he assumes: (a) a lability in the genes responsible for the surface detail of a cell (e.g. histocompatibility antigens) and (b) evolution of mechanisms to detect subtle deviation from 'self'. He further argues that, without the emergence of new histocompatibility antigens, i.e. tumor-specific antigens, and of populations of immunocytes capable of distinguishing these and initiating their elimination, neoplasia would be a contagious disease.

Burnet [6] reasons that the genetic loci determining the 'self'-specificity of cell surface proteins are as prone to diversity as the 'variable' segments of immunoglobulins, and provides evidence that this occurs at both the germinal and somatic levels. The frequency of detectable histocompatibility changes in mice (above 1% per generation) attests to the extent of diversification at the germinal level. Moreover, if early mammalian embryonal cells generally mutate as frequently as those of sheep, every cell in an adult animal should possess an antigen which is not common to every other cell. This would explain the otherwise baffling antigenic diversity in neoplastic cells, malignantly converted by agents other than oncogenic viruses.

Burnet argues that both immunoglobulins and histocompatibility antigens are plasma membrane derivatives and that antibody release reflects an 'over-production' of the 'recognition molecules' built into the plasma membrane of primitive immunocytes [6].

Virally-induced tumors also possess an immunologic individuality, but unlike chemically or physically induced tumors, each of which appears to be immunologically unique, tumors caused by a given virus exhibit transplantation antigens characteristic for that virus, and independent of the tissue or host involved. The virus-induced transplantation antigens are not unique to the neoplastic state but also appear during permissive, lytic viral infections. Because of this fact, presumably, *in vivo* infection of immunocompetent hosts by potentially oncogenic viruses does not generally lead to tumorigenesis; few cells survive the infection and become transformed and these are usually eliminated immunologically. In principle, therefore, viral carcinogenesis is a contagious disease which, in nature, occurs only following productive infection with oncogenic RNA viruses (DNA viruses induce tumors only in immunoincompetent hosts). Important in this context is the high incidence tumors in individuals with immunodeficiency diseases, or persons exposed to immunosuppressive therapy, and the decreased incidence of tumors following specific or non-specific activation of the immune response [14].

No hypothesis of cancer causation has yet dealt satisfactorily either with the fact that so many diverse agents can cause cancers, or the critical related issue, that cancer constitutes so many disparate entities. Moreover, present concepts of carcinogensis offer no substantial assistance to the cancer therapist and contribute substantially less to the problem of cancer control than information obtained by epidemiologists.

For these reasons alone, many investigators now focus their efforts on the definition and elucidation of the processes responsible for *malignancy* – accepting the likelihood that malignant behavior may be a manifestation of diverse processes due to different causes. *Malignancy* can be considered to derive from the conjoint operation of the following biological processes: *autonomy, cell proliferation, asocial cell behavior, invasiveness* and *metastasis.* Let us briefly define these processes:

1. Autonomy

All tumor cells are autonomous to some extent; i.e. their proliferation and function do not respond appropriately to the biologic control mechanisms normal for the host. However, autonomy may not be absolute. For example, certain tumors of endocrine origin can proliferate only in rather restricted hormonal environments and some tumors cannot survive in an immunocompetent host.

2. Cell proliferation

Unless neoplastic cells proliferate, they cannot give rise to a tumor. But *rapid* cell replication is not a necessary, unique or sufficient property to induce cancer. Thus the cells of numerous normal tissues (e.g. hematopoietic tissues, intestinal epithelium) normally proliferate more rapidly than cancer cells. Tissue repair involves very rapid cell replication and many benign tumors, e.g. leiomyomas, exhibit much greater growth rates than most cancers.

3. Asocial cell behavior

Cancer cells do not interact harmoniously and tumor tissue typically lacks the precise social organization of normal tissue. Tumors can be considered as 'organoids' [20], containing not only neoplastic cells, but also normal tissues, e.g. blood vessels, and supportive structures, albeit in arrangements of varying degrees of disorder. However, the tissue disorganization observed in tumors is not a necessary, sufficient or unique feature of cancer; it is also observed to some extent after tissue injury.

4. Invasiveness

Infiltration of tumor cells into adjacent normal tissues constitutes a critical feature of cancer. A non-invasive tumor, even if rapidly growing, is clinically benign, not malignant. Yet, some non-neoplastic cells can also manifest invasive behavior: lymphocytes and macrophages infiltrate normal tissues in carrying out their normal function. Moreover, hair-follicle cells during their growth cycle, the implanting mammalian embryo and mammary ducts in pregnancy exhibit distinct invasive behavior; however, such invasive processes are physiologically controlled and limited to certain stages of tissue development, while malignant cells invade in an uncontrolled, disordered, progressive and unrestrained manner.

5. Metastasis

Malignant tumors tend to disseminate neoplastic cells to distant parts of the hosts's body, where they establish secondary colonies characterized by the same abnormal cell behavior found in the 'primary'. Metastasis constitutes the major clinical feature of cancer, since, once metastasis has occurred, the disease cannot usually be eradicated by elimination of the 'primary' tumor. Metastasis does not necessarily require a large 'primary' tumor; indeed many patients die of metastatic disease originating from undiscovered 'primaries'.

Since cells are *immunologically foreign* to their hosts, one must ask how tumors can overcome this selective *in vivo* disadvantage. In particular, how can tumor cells that, during invasion and metastasis, venture into immunologically hostile zones, escape immune destruction? Very likely many such cells are eliminated, but there is evidence for a number of mechanisms by which tumor cells can avoid immune surveillance [14]. Two of these processes, antigenic modulation and surface-modification by tumor-cell proteases, appear to be membrane phenomena and will be discussed in this volume.

We have previously suggested [21–23] that cancer may be a '*membrane disease*' and that certain critical properties of malignant cells – those producing asocial cell behavior, invasiveness and metastasis – arise from defective membrane structure and/or function. This book was originally intended as a sequel to these earlier publications, incorporating relevant new data and reassessing earlier information. However, we have had to modify this simple goal because research during the past five years has produced a vast amount of multidisciplinary evidence, documenting the occurrence of a multitude of membrane anomalies in neoplasia. Moreover, many engaging arguments have been presented relating one or another membrane anomaly to the malignant process, and several international congresses have dealt exclusively with the membrane alterations in neoplasia [24–25].

This explosion of information has not yet yielded a proportional expansion of understanding, partly because the new knowledge derives from diverse, often widely separated disciplines and techniques, and partly because experimentation on tumors is not correlated adequately with the rapid progress in general membrane biology. We will therefore examine the membrane properties of neoplastic cells from the interdisciplinary point of view of modern membrane biology. We will assess the data as critically as is now feasible, including methodological detail where relevant. Our emphasis will be primarily on membrane *molecular* biology. In this, we cannot treat the very important area of *tumor immunology* in any detail because information in this area at the *molecular* level is insufficient, and the biologic and genetic aspects of the field are well reviewed elsewhere.

Our primary literature review extends through November, 1974 but important later contributions are either included in the text of are summarized in a general addendum.

The composition of the book is as follows: Chapter 1 deals with general aspects of membrane architecture. Chapters 2–6 treat diverse aspects of membrane composition and metabolism. Chapters 7–10 address the interactions of cells with their environments. Chapter 11 presents some major unifying concepts in the field and an assessment of its prospects.

References

[1] Huxley, J., *Biological Aspects of Cancer*, Harcourt, Brace, New York, (1958).
[2] Pott, P., reprinted in *Nat. Canc. Inst: Monogr. 10*, 8 (1963).
[3] Ellerman, V. and Bang, O., *Centralbl. Bakteriol. Parasitenk. 46*, 595 (1908).
[4] Rous, P., *J. Exp. Med., 12*, 696 (1910).
[5] Boveri, T., *The Origin of Malignant Tumors*, English translation by M. Boveri; Williams and Wilkins, Baltimore, Md. 1929.
[6] Burnet, M.F., in *Immunogenicity*, ed. F. Borek, North Holland, Amsterdam, 1972, Ch. 16.
[7] Warburg, O., *The Metabolism of Tumors*, Arnold Constable, London, 1930.
[8] Temin, H.M., *Perspect. Biol. Med. 14*, 11 (1971).
[9] Temin, H.M. and Mizutani, S., *Nature*, *226*, 1211 (1970).
[10] Baltimore, D., *Nature 226*, 1209 (1970).
[11] Huebner, R.J. and Todaro, G.J., *Proc. Natl. Acad. Sci. U.S.*, *64*, 1087 (1969).
[12] Gross, L., *Cancer Res.* 3, 326, 1943.
[13] Prehn, R.T. and Main, M.J., *J. Nat. Cancer Inst.* 18, 763 (1957).
[14] Oettken, H.F. and Hellström, K.E. in *Cancer Medicine*, J. Holland and E. Frei III, eds., Lea and Febiger, Philadelphia, 1973, p. 951.
[15] Baldwin, R.W., *Adv. Cancer Res.*, *18*, 1 (1973).
[16] Hersey, J.K., Spector, B.D. and Good, R.A., *Adv. Cancer Res.*, *18*, 211 (1973).
[17] Beers, R.F., Jr., Tilghman, R.C. and Bassett, E.G., in *The Role of Immunological Factors in Viral and Oncogenic Processes*, Johns Hopkins University Press, Baltimore, 1974.
[18] Tevethia, S.S. and Tevethia, M.J. in *A Comprehensive Treatise in Cancer*, ed. F. Becker, Plenum Press, New York, 1975, Vol. III, in press.
[19] Thomas, L., in *Cellular and Humoral Aspects of the Hypersensitive State*, H.S. Lawrence, ed., Cassell, London, 1959, p. 529.
[20] Tarin, D., in *Tissue Interaction in Carcinogenesis*, D. Tarin, ed., Academic Press, London–New York, 1972, p. xiii.
[21] Wallach, D.F.H., *Proc. Natl. Acad. Sci. U.S.*, *61*, 868 (1968).
[22] Wallach, D.F.H., *New Engl. J. Med.*, *280*, 761 (1969).
[23] Wallach, D.F.H., *Curr. Top. in Microbiol. Immunol.*, *47*, 152 (1969).
[24] Wallach, D.F.H., Knüferman, H. and Wunderlich, F., *FEBS Letters*, *33*, 275 (1973).
[25] *Membrane Transformations in Neoplasia*, eds. J. Schultz and R.E. Block, Academic Press, New York, 1974.
[26] *Cellular Membranes and Tumor Cell Behavior*, 28th Annual M.D. Anderson Symposium on Fundamental Cancer Research, 1975, ed. E.F. Walborg, Williams & Wilkins Publishers, in press.

1

Membrane structure and organization

1.1. The molecular assembly of biomembranes: general principles

1.1.1. Introduction

Biomembranes contain protein, lipid and carbohydrate. Membrane proteins constitute the major membrane macromolecules and generally make up 60%–70% of the membrane mass. Membrane carbohydrate is associated primarily with glycoproteins and to a lesser extent with glycolipids. We will deal with the individual categories of membrane lipids in Chapters 3–5, where we will also consider the anomalies of lipid composition and metabolism found in tumors. Similarly, we will address the altered protein composition and metabolism of tumors in Chapters 2 and 6. The emphasis in this chapter will be on the mechanisms involved in the assembly of membrane proteins and membrane lipids to form biomembranes.

1.1.2. Lipids

The biochemistry of membrane lipids has been often and well reviewed (e.g. [1,2]). The lipids found in biomembranes consist of three major classes, glycerophosphatides, sphingolipids and glycolipids, and steroids; the glycerophosphatides are the most abundant. The *glycerophosphatides* are derivatives of glycerol-3-phosphoric acid. Their hydrocarbon side chains are hydrophobic, whereas their polar 'head groups', containing the glycerol moiety, the phosphate groups and the choline, serine and ethanolamine or inositol residues, are water-soluble.

The phosphatides therefore orient at oil–water interfaces in such a way that the side chains are in the oil and the polar groups are in the water. At physiologic pH the polar groups may lack a net change as in phosphatidylcholine, be slightly anionic as in phosphatidylethanolamine, or bear a distinct net negative charge, as in phosphatidylserine and in phosphatidylionositol. *Acetal phosphatides* (plasmalogens) consist of glyceryl–phosphorylcholine (or -ethanolamine, or -serine) with one esterified fatty acid and one fatty acid in enol–ether linkage.

The *sphingolipids* contain sphingosine with a fatty acid in amide linkage with the sphingosine. Fatty acid amides of sphingosine are known as ceramides. In *sphingomyelin* a phosphate is in diester linkage between choline and the hydroxyl on the terminal carbon of ceramide. Sphingomyelin bears no net charge at neutral pH. Cerebrosides are glycosyl ceramides. The gangliosides are complex, sialic acid-containing derivatives of ceramide.

The third type of lipid observed in significant amounts in membranes comprises the *sterols*, particularly cholesterol. These compounds do not have charged polar groups.

Biomembranes differ extensively in the proportions of protein and lipid, as well as in lipid composition. The large lipid differences between membranes indubitably have some functional significance. It is, however, also possible that very diverse lipids may serve the same function, provided they participate suitably in the hydrophobic interactions required for membrane structure and function. The structures of diverse hemoglobins provide some useful analogies in this regard [3]. Although very similar in function, the hemoglobins of various species show extensive differences in amino acid composition and only nine out of more than 140 residues are the same in all the monomers. However, in *all* hemoglobins, 30 of the residues are always non-polar and lie in the proteins' hydrophobic cores. 'Core hydrophobicity', is therefore a critical structural parameter and this can be met by diverse apolar amino acids. One can thus anticipate that very different lipids might subserve very similar hydrophobic roles.

The proportions of lipids in the membranes of a given cell type may vary strikingly from one species to the next. This appears to be in part due to genetically determined differences among membrane proteins [4]. Finally, the fatty acid composition of membrane phosphatides can change with the functional state of a cell, cell age, culture conditions, etc.

1.1.3. Proteins

1.1.3.1. Introduction

The attention of many biologists tends to gravitate toward membrane lipids, although these substances are not the predominant constituents of most biomembranes and do not primarily direct membrane function. Indeed, much current research, applying elegant techniques and sophisticated models, deals with membrane lipids in preference to membrane proteins, and some of the most engaging membrane concepts, for example, the 'fluid-mosaic' model [5] derive largely from the known properties of membrane lipids.

These tendencies are readily understood. Membrane proteins are among the more intractable substances, while the lipids are small molecules which can be easily extracted, purified and characterized. Moreover, membrane lipids are readily accessible to convenient structural analysis and many of the lipids can be synthetized and derivatized. Also lipids lend themselves to the formation of simple model membranes, which can be readily examined, using a diversity of biophysical and biochemical and other techniques. Nevertheless, the critical issues in membrane biology, for example, membrane alterations in neoplasia, cannot be effectively approached without an understanding of the properties of membrane proteins and of the mechanisms of lipid protein interaction.

How does one define membrane proteins? Numerous variables must be considered even in the case of well-defined membrane domains isolated by optimal methods from specific cells in defined genetic, developmental and physiologic states. Membrane proteins are now often categorized as 'integral' or 'core' proteins and 'extrinsic' proteins, although this distinction is operational and somewhat artificial. 'Core' proteins are those that remain water-insoluble and associated with membrane lipids unless drastic solubilization methods are employed. Most authors would agree that membrane 'core' proteins are bound to their membranes by hydrophobic associations.

However, we cannot be certain that, what we define as a 'core' protein does not exist in a guise that may obscure recognition of its membrane connections. For example, the protein might exist in a soluble form, but in this form might exhibit a different configuration and/or functional state than when membrane-associated. Indeed, we now know that certain soluble proteins, for example, immunoglobulins, can serve as membrane proteins in lymphocytes and can exert a specific functional role in this fashion. Moreover, Langdon's recent, provocative data suggest that some protein moieties of human plasma lipoproteins may serve as major membrane constitutents of human erythrocytes [6].

Another obstacle to the classification of proteins as 'membrane proteins', derives from the circumstance that the study of membranes generally requires disruption of cells and application of various subcellular fractionation procedures [4]. These fractionation procedures introduce a number of complications. For example, (a) one cannot equate 'membrane' with organelle, and most of the proteins within various organelles (nuclei, mitochondria, lysosomes, etc.) are not membrane associated; (b) many membrane isolates are contaminated by proteins absorbed during fractionation, or contain soluble proteins, which are trapped when membrane vesicles form during cell disruption; (c) many cells contain lysomal peptidases and glycosidases which may alter the protein and/or glycoprotein structure during the purification of a membrane or the fractionation of its proteins; (d) certain proteins can exhibit a high affinity for the membranes of intact cells, but elute from the membranes during isolation; (e) the topologies of surface and intracellular membranes are not uniform and various membrane domains can differ in protein composition; (f) membrane protein composition depends upon cell variations due to physiologic state, cell cycle, age, etc.

Most membranes contain numerous proteins and the question of 'how many' depends on detection criteria. One of us has speculated on this topic in an immunologic context [7]: a smooth, spherical cell with a radius of 5×10^{-4} cm contains about 2×10^4 different proteins. Assuming that a typical membrane protein (mol wt. $\sim 10^5$) is a cylinder with a radius $\sim 2.5 \times 10^{-7}$ cm, that is, a cross sectional area of 2×10^{-13} cm^2, these 2×10^4 proteins will subtend an area of $\sim 4 \times 10^{-9}$ cm^2, i.e. $\sim 0.1\%$ of the 3×10^{-6} cm^2 surface area of the hypothetical cell. This computation suggests the astonishing possibility that the surface of each cell could bear, in addition to plasma membrane components proper, at least one copy of all protein components of the cell interior.

Even a simple list of known membrane-associated enzymes will involve hundreds of proteins; however, the contribution of each enzyme to the total membrane mass may be small. Thus, human erythrocytes contain less than 10^3 molecules of ouabain-sensitive ATPase per cell [8], a minute proportion of the total membrane protein. On the other hand, we know of instances where a biomembrane contains only one predominant protein. For example, rhodopsin accounts for nearly 90% of the protein mass in the membranes of retinal-rod outer segments [9].

Most membrane types, when analyzed by polyacrylamide gel electrophoresis in sodium dodecylsulfate (SDS-PAGE), exhibit considerable protein diversity (mol. wt. 15,000–300,000). However, within this diversity lie certain consistent patterns for a given membrane class. For example, SDS-PAGE of erythrocyte membranes from different species exhibit striking similarities [10]. Also, the plasma membranes of lymphoid cells reveal SDS-PAGE patterns different from those of erythrocyte membranes or from endoplasmic reticulum, but closely similar to each other [11]. We suspect, therefore, that proteins of a given membrane may possess certain 'architectural homologies' which permit their interaction with the lipids of that membrane, as well as with its other protein constituents. As proposed previously [12] such 'architectural homologies' of membrane proteins would derive from their genetically determined primary structuring and could explain: (a) nonspecific membrane affinity, arising from energetically favorable interactions with membrane lipids; (b) specific affinity of a given protein for a given membrane or membrane domain*.

* Certain membrane proteins are incorporated into or excluded from one type of membrane, but not another. For example, host membrane proteins do not occur in arbor virus membranes, although the virus buds from the plasma membrane.

1.1.3.2. Forces participating in lipid–protein interaction

The following forces may participate in lipid–protein associations [13]:

(a) *Electrostatic (Coulombic) forces.* These interactions will occur between the charges on membrane lipids, membrane proteins and membrane carbohydrates. Ideally, the interaction energy, W, between two ions of charge q, q' is given by:

$$W_e = (q \cdot q')/(Dr) \qquad (1.1)$$

where D is the effective dielectric constant of the medium separating the charges and r the distance between the charges. It is difficult to define the dielectric constant, but known values of D range from 80.4 for water to 2.6 for liquid linolenic acid. The energies of electrostatic attractions between two unit charges of opposite sign would therefore lie near 4 kcal/mole when $D \sim 15$ [13]. Electrostatic attraction plays a significant role in the binding of cytochrome c to acidic mitochondrial lipids.

(b) *Induction forces.* A negative charge (e.g. on a phospholipid phosphate) can polarize a nearby apolar group (e.g. a methylene on an amino acid residue). Ideally, the attraction energy, W_p, due to such a mechanism is given by:

$$W_p = (-q^2\alpha)/(2D^2r^6) \qquad (1.2)$$

where α is a constant, the polarizability. W_p thus varies with the inverse *sixth* power of distance. The interaction energy between a unit charge and a CH_2 ~ 5Å away would be ~0.002 kcal [13]. Induction forces are thus too small to function importantly in lipid–protein associations.

(c) *London–van der Waals' – dispersion forces.* These interactions operate between *all* groups of proteins and lipids. They are critically important in the interaction between *non-polar* residues because they are the *only* forces involved there. London–van der Waals' forces depend strongly on distance and steric factors. For identical groups with polarizability, α, interacting *in vacuo*, the attraction energy, W_{disp}, in the ideal case would be

$$W_{disp} = (3\Delta E\alpha^2)/(4r^6) \qquad (1.3)$$

where ΔE is an average electronic excitation energy.

For two CH_2 groups, the experimentally measured interaction is $-1340/r^6$ kcal/mole (r in Å). Thus, although W_{disp} varies as the inverse sixth power of distance, the attraction energy for groups 5 Å apart will be ~0.01 kcal/mole. When the attractive forces between many interacting pairs, (e.g. packed hydrocarbon residues) are summed, large total interaction energies are obtained. Indeed, the total interaction between hydrocarbon chains in packed crystalline lipid arrays, or between a long hydrocarbon chain and the apolar face of an amphipathic α-helix (p. 28), can reach 10–20 kcal/mole [13].

(d) *Hydrophobic interactions* (*entropic forces*). The fundamental driving force for any process is the unalterable tendency of matter to move to a state of lowest free energy. It is not certain that biomembranes, as complete entities, represent thermodynamically stable states, but the interactions between lipids and proteins must follow thermodynamic principles nevertheless [14].

The native association of membrane 'core' components involves many apolar associations. A principal impetus for these associations comes from the tendency of apolar groups to order water about them, causing an unfavorable loss of entropy. Therefore, other factors permitting, apolar groups will tend to cluster together, thereby excluding water, increasing the entropy of the system as a whole and decreasing its free energy. Accordingly, a mixture of monomeric membrane components should, under appropriate conditions, assemble spontaneously to form native membrane domains.

Recent membrane reconstitution studies fit this concept, but it is difficult to develop the hypothesis beyond the heuristic level. Thus, for the elementary case, for example, lipid + protein $\leftrightarrows$ membrane, one might envisage application of the equations:

$$\begin{aligned} \mu_{M_L} &= \mu_{W_L} + RT \ln [Q_L] \\ \mu_{M_P} &= \mu_{W_P} + RT \ln [Q_P] \end{aligned} \tag{1.4}$$

where μ_{M_L}, μ_{M_P}, μ_{W_L} and μ_{W_P} are the free energies for the membrane, M, and dissolved state, W, of lipid, L, and protein, P, respectively. $[Q_L]$ and $[Q_P]$ are appropriate measures of lipid and protein concentration. The free-energy differences should be negative to favor membrane formation.

The real situation is vastly more complex.

First, although there may be an overall decrease in free energy in membrane formation, the reactants may need to pass through an *activated* state.

Second, membranes usually contain more than one type of protein and more than one type of lipid and both proteins and lipids can occur in multiple states.

We reason, as before [15], that the formation and dissociation of biomembranes involves a complex set of interacting equilibria. These can be heuristically represented in the following way:

$$\begin{array}{c} M \underset{1}{\rightleftarrows} m \underset{2}{\leftrightarrows} \sum_{i=1}^{n} L_{x(i)} P_{y(i)}^{(i)} \underset{3}{\rightleftarrows} \left[\sum_{i=1}^{n} x_{(i)} L \right] + \left[\sum_{i=1}^{n} y_{(i)} P^{(i)} \right] \\ \swarrow\!\nearrow_4 \qquad\qquad \swarrow\!\nearrow \qquad \searrow\!\nwarrow \\ \sum_{i=1}^{r} L_{A(i)} \qquad \sum_{i=1}^{n} y_{(i)} P_{(\mathrm{sol})}^{(i)} \quad \sum_{k=1}^{t} P_{w(k)}^{(k)} \end{array} \tag{1.5}$$

Here M represents the intact membrane and m small fragments thereof;

$$\sum_{n=1}^{n} L_{x(i)} P_{y(i)}^{(i)}$$

represents a summation over the n liproprotein subunits in the small fragments;

$x_{(i)}$ and $y_{(i)}$ represent the number of molecules of lipid and protein, respectively, in the ith subunit. L does not refer to one particular lipid, but to all the classes of lipid in the membrane. (Some of these, for example cholesterol, are loosely held, while others are strongly bound.) Also the $y_{(i)}$ protein molecules within a subunit may be different molecules.

$$\sum_{i=1}^{r} L_{A(i)}$$

represents r different aggregated states of lipid with $A_{(i)}$ molecules per aggregate. Similarly,

$$\sum_{k=1}^{t} P_{w(k)}^{(k)}$$

represents t aggregates of protein containing $w_{(k)}$ molecules of protein per aggregate. Finally,

$$\sum_{i=1}^{n} y_{(i)} P_{(\text{sol})}^{(i)}$$

represents the n protein species in states permitting solubility in water. We suspect that possible *soluble* states for true membrane proteins would involve different quaternary, tertiary and even secondary structures than apply to their membrane state. Thus, if a given architecture provides a condition of minimum free energy in the membrane environment (with its abundance of apolar interactions) this structure will not yield minimum free energy in water.

Step 1 represents the disruption of large membrane sheets, for example, intact plasma membranes into small fragments, such as occurs with liquid shear. The dissociation of membrane fragments into their component lipoproteins (Step 2) does not proceed readily.

Step 3 – subunit dissociation into lipid and protein – cannot readily be separated from Step 2. We suspect that it involves unstable intermediates indicated by [ ]. This is to be anticipated from the known behavior of membrane lipids and apolar proteins in aqueous media. Thus cholesterol and phosphatides form micelles in water, and one would expect that released lipid would tend to aggregate into micelles (Step 4). Also, if the structure of the proteins in the membrane is such as to expose large hydrophobic regions, for example, murein–lipoprotein, the proteins will either aggregate through apolar associations or rearrange into structures that are thermodynamically stable in aqueous media, that is, with hydrophobic residues concentrated in the interior and polar groups at the surface of the molecules.

1.1.3.3. The secondary structure of membrane proteins

1.1.3.3.1. Introduction

Studies of the secondary structures of the proteins in membranes were initiated by Maddy and Malcolm [16, 17], on erythrocyte ghosts, and Wallach and Zahler

[18] on tumor cell plasma membranes by conjoint application of infrared (IR) spectroscopy and optical activity measurements. Both groups (and others since then) attempted to quantify the conformational content of various biomembranes, proceeding on the then common assumptions that:

(a) the spectral properties of diverse synthetic homopolypeptides in certain conformations closely match the spectral properties of protein peptide segments in the same conformation; and

(b) the right-handed α-helix, β-pleated sheet and 'random' conformations are the major biologically significant peptide arrays, and the only ones present in membrane proteins.

These assumptions, as well as the variable light scattering contributions of diverse membrane isolates, have generated considerable controversy, particularly concerning the interpretation of optical activity measurements [4]. However, the extensive and intensive experimentation in this area has revealed some important general patterns.

1.1.3.3.2. Infrared spectroscopy

Polypeptides and proteins. IR measurements in some regions of amide absorption are useful in the conformational analysis of polypeptides and proteins [4]. The Amide bands of polypeptides arise principally from in-plane C–O, C–N and C–H stretching, OCN bending and CNH bending, as well as out-of-plane C–N twisting, and C–O and N–H bending.

The Amide I absorption bands of polypeptides are sensitive to folding of the peptide backbone and to hydrogen bonding with other peptide linkages, that is, to conformation or 2° structure. Indeed, polypeptides in α-helical and/or unordered conformations exhibit IR spectra very different from those of β-structures, particularly in the Amide I region (80% C=O stretch, 10% C–N stretch, 10% N–H deformation). Thus, the Amide I band of α-helical polypeptides lies at 1652 cm^{-1}–1656 cm^{-1}, but occurs near 1630 cm^{-1} in the β-structured polymers. In the antiparallel β-structure, an additional band appears near 1690 cm^{-1}. Unfortunately, the Amide I band does not readily permit distinction between 'unordered' and α-helical peptide without measurement of dichroism. This is because of the proximity of the Amide I absorptions in the α-helical and unordered conformations, small spectral variations from one peptide to the other, and band broadening due to side chain absorption.

Studies of proteins whose detailed structures are known by X-ray crystallography have demonstrated that it is valid to use Amide I absorption in conformational analyses [4]. The presence in proteins of β-structures together with α-helical and/or unordered conformations can readily be distinguished. For example, lysozyme, which has only about 10% of its peptide linkages in the anti-parallel β-structure, exhibits clearly discernible bands at 1632 cm^{-1} and 1687 cm^{-1}, in addition to the principal Amide I peak at 1650 cm^{-1}.

Polypeptides at air–water interfaces. Many synthetic poly-L-amino acids when spread at air–water interfaces, form films with an area of 20.5 Å^2/residue, typical for α-helices [19, 20]. Bilayers form at high surface pressures. When uniquely oriented films are inspected by infrared dichroism, they reveal spectra characteristic of *right-handed α-helix*: specifically, the frequencies of the ester C=O band (1740 cm^{-1}), Amide I band (1658 cm^{-1}) and Amide II band (1552 cm^{-1}) fit the established values for *right*-handed α-helical polypeptides, not *left*-handed helical polymers. The parallel polarization of the C=O band confirms these conclusions since the left-handed form exhibits perpendicular dichroism at this frequency. The infrared data are fully supported by electron diffraction studies [19,20]. When poly-*β-benzyl* L-*aspartate* (mol. wt. 130,000 and 250,000) is spread at an air–water interface, from dichloroacetic acid:chloroform 1:99 it also produces a film composed of *right-handed* α-helices, *although* the polypeptide is in the form of *left-handed* α-helices in the spreading solvent – 1:99 dichloroacetic acid. Moreover, when the subphase is made 1% in iso-propanol, the interfacial film exhibits the IR characteristics of *left-handed* helices. *The interfacial condition can therefore produce a change of helical sense*!

The shift from left- to right-handed helix does not require a large amount of energy (e.g. as for unwinding of the backbone) since, if the H-bonds along short helical segments are opened, refolding of that segment in the opposite sense can occur by a simple rotation about the bond to the α-carbons. The formation of *right-handed* helix at the interphase can reasonably be attributed to an interference by the aqueous, polar substrate with the forces normally favoring the left-handed helix. The effect of isopropanol is attributed to altered entropic interactions between the polymer and the water subphase.

These experiments indicate that, depending on amino acid sequence and composition, entropic factors and microenvironment, some α-helical segments in an unknown protein might have a *left-handed* sense. This possibility is particularly pertinent to *membrane proteins*, where the membrane lipids would provide an apolar microenvironment for penetrating peptide segments.

Membranes. The IR spectra of cellular membranes have been correlated with those of synthetic polypeptides and proteins with known conformation to yield information concerning the secondary structure of membrane proteins and their possible conformational transitions.

The Amide I region of erythrocyte membranes lyophilized from 7 mM phosphate buffer, pH 7.4, shows maximal absorption at 1652 cm^{-1}, indicative of protein in the α-helical and/or unordered conformation; that no band is observed at 1630–1640 cm^{-1} implies that there is little β-structure. The Amide I spectra of erythrocyte ghosts that have been freshly suspended in D_2O closely resemble those of dried films.

IR studies on plasma membranes purified from Ehrlich ascites carcinoma, dried at neutral pH, also yield no evidence of β-structured peptide, but the spectra of membranes dried from acid solvents indicate some transition to the β-structure [18,21]. Myelin, in which most of the membrane protein probably lies at the surface of lipid bilayers, also demonstrates an Amide I band near 1655 cm^{-1} with no unusual absorption near 1630 cm^{-1} [22].

In contrast, according to IR criteria, isolated adipocyte plasma membranes [23], as well as purified thymocyte plasma membranes [24], contain an appreciable proportion of peptide in the antiparallel β-conformation.

IR spectroscopy indicates that the proteins of conventionally isolated rat-liver mitochondrial membranes contain a significant proportion of peptide in the antiparallel β-conformation [25]; the exact proportion depends upon metabolic state [26–28]. The spectra exhibit a peak around 1650 cm^{-1} with a pronounced shoulder near 1635–1640 cm^{-1} and a shoulder near 1690 cm^{-1}. A comparison of outer and inner mitochondrial membranes after elution of soluble 'matrical' mitochondrial proteins by osmotic lysis indicates that the β-structured peptide is associated with the 'inner' membranes.

1.1.3.3.3. Measurement of optical activity

Optically active substances exhibit different refractive indices for left- and right-circularly polarized light, producing a change in optical rotation with the wavelength of light. This phenomenon is referred to as optical rotatory dispersion (ORD). A related phenomenon, circular dichroism (CD), derives from the different absorption of left- and right-circularly polarized light as a function of wavelength.

The optical activity of peptide bonds arises from π^0–π^- electronic transitions near 190 nm and n–π^* transitions near 220 nm. The overall optical activity of polypeptides depends upon the spatial relationships of the individual peptide bonds, and α-helical, β-, and 'unordered' conformations yield distinct ORD and CD spectra. For this reason, optical activity measurements have served usefully in estimating the conformational proportions of *some* soluble polypeptides and proteins.

Polypeptides and proteins. Solutions of 'unordered' homopolypeptides generally exhibit intense negative CD at 198 nm and a large negative ORD extremum at 205 nm. However, this simple pattern probably does not apply directly to constrained 'unordered' regions in globular proteins. Thus, when unordered homopolypeptides are immobilized in films, the major negative CD band shifts to 200–205 nm. Data on partly unfolded proteins also document the influence of steric constraints on optical activity spectra. Polypeptides in the *right-handed* α-helical conformation display two large negative CD bands near 222.5 nm and 208 nm and a positive one near 192 nm. The *left-handed α-helical* conformation gives an identical spectrum of opposite sign. The ORD spectrum of right-handed α-helix is characterized by a large minimum at 233 nm, a crossover to positive rotation at 223 nm, a shoulder near 210 nm, and maximum at 198 nm.

The CD of β-structured poly-L-lysine shows a single negative band at 218 nm, and a positive extremum at 195 nm. However, diverse β-structured polypeptides yield widely varied optical activity spectra, whose extrema may overlap those of α-helical polypeptides. For this reason, ORD and CD may be insensitive to the presence of β-structure.

One initially hoped that the contribution of a α-helical, β-structured, and unordered segments of globular proteins would sum, as in model systems, and permit conformational analysis by comparison of the protein ORD and CD spectra with those of reference polypeptides. These anticipations have faded because [29]:

(a) In α-helices, optical activity increases with the number of consecutive, helically-arrayed residues below 20, and also diminishes when the helices are distorted. However, X-ray analyses show that there are usually fewer than 20 residues per helical segment in globular proteins and that imperfections are common.
(b) Optical activity is side chain dependent.
(c) 'Unordered' segments in proteins, are irregular but not 'unstructured', they vary among proteins and are non-uniformly distributed elements of diverse size and array. Such segments are thus, not comparable to the 'unordered' states of homopolypeptides.
(d) Many peptide linkages of globular proteins reside in apolar, micro-environments, which tend to reduce the rotational strengths of the optically active bands and shift these to higher wavelengths.
(e) Conformations other than the conventional right-handed α-helical, β-, and 'unordered' structures might exist in proteins. For example, while the left-handed α-helix is intrinsically less stable than the right-handed helix structure, extrinsic forces may favor the former in certain situations (e.g. [19,20]).

The difficulties encountered in the conformational analysis of soluble globular proteins are further compounded in biomembranes, due to the particulate character of the latter. In the case of small vesicles, this may cause little more than minor flattening and broadening of the CD bands; with large membrane fragments, more severe artifacts can appear. Also, diverse membranes vary in their morphological, chemical and physical properties and, generally, lack any one, predominant protein. Optical activity measurements, therefore, can at best yield only the *average* conformational proportions in membranes. However, these might mirror structural homologies among the proteins of a given membrane or diverse membranes.

Membranes. When compared with the spectra of polypeptide standards, the ORD spectra of nearly all membrane types studied so far show the following peculiarities: (a) shapes close to that obtained with pure, right-handed α-helix, (b) low intensity, (c) displacement of the entire spectrum to longer wavelengths than those observed with α-helix. With mitochondria, extreme deviations are observed; these can be attributed to high proportions of β-structured peptide.

The CD spectra of most membranes also exhibit an 'α-helical' shape, low amplitudes and displacement of the major minima to the red. Part of this may arise from light-scattering artifacts. However, as we have shown previously [29], most features of membrane optical activity do not arise from the particulate nature of the membrane. Indeed, there is as yet no explanation of a major oddity in membrane optical activity, namely, of the fact that *if* the membrane proteins were partly in a right-handed α-helix and partly unordered, the spectra should not have the 'α-helical' shape. Conversely, *if* the proteins were primarily helical, they should show greater amplitude. The membrane ORD and CD spectra appear *as if* most of the optical activity of the membrane protein were masked, with only right-handed α-helix being expressed.

The major features of membrane optical activity can be explained as follows:

(a) Soluble globular proteins, as well as membrane proteins, tend to give low 'helical' signals because of short helical segments, distortions from perfect helicity and apolar environments.

(b) In native proteins, peptide linkages not in α-helical or β-structured disposition, are arranged in multiple, irregular, but non-random arrays, each yielding a spectrum other than that of unordered poly-L-amino acids. The summed contribution of such segments is likely to be diffuse, allowing the helical shape to dominate.

Measurements of membrane optical activity are thus plagued by the same difficulties that generally hinder conformational analyses of proteins by CD and ORD; they may be additionally complicated by artifacts due to the particulate nature of the membranes. Despite obstacles to accurate quantification, optical activity measurements indicate that the proteins of most biomembranes contain more than 40% of their peptide linkages in a right-handed α-helical conformation. This proportion is greater than that found in most globular proteins, and may at that be an underestimate.

1.1.3.3.4. Metabolic and physiological modifications of membrane protein conformation

IR studies. IR analyses show that addition of ATP and Mg^{2+} (0.5–1.0 mM) to erythrocyte membranes induces a shift to the β-conformation and away from the α-helical or unordered structure [27]. This transition is enhanced by addition of Na^+ plus K^+, which stimulates ATP hydrolysis. The spectral changes are blocked by agents which inhibit ATP hydrolysis. The conformation transition is prominent within two minutes after ATP addition when the reaction is carried out at room temperature and stopped rapidly with liquid nitrogen before lyophilization for IR examination. The transition is also observed when the metabolic reaction and IR measurements are carried out in D_2O [27].

Induction of electron transport in liver *mitochondria* by addition of succinate (0.1 mM), in the presence of oxygen, shifts the protein conformation towards the antiparallel β-structure [30]. This is shown by the increase in IR absorption around 1635 cm^{-1} and 1685 cm^{-1}. This process can be prevented or reversed by inhibiting electron transport with CN^- or antimycin-A, or by inducing phosphorylation (addition of ADP to 250 μM). The shift towards the antiparallel β-structure becomes extreme when electron transport is uncoupled from oxidative phosphorylation by addition of 100 μM dinitrophenol [30].

In both erythrocyte ghosts and mitochondria, the metabolically induced spectral changes observed in the Amide I region are considerable, however, as pointed out in [28], they could arise without a large rearrangement of the peptide backbone: A small change in the orientation of the peptide groups by rotations about certain bonds in the backbone of an antiparallel loop could, under favorable circumstances, engender increased antiparallel-β *H-bonding*; this would be reflected in the IR spectra.

Summary. Conformational analyses indicate that many membranes contain proteins of more than average helicity. However, other membranes, inner mitochondrial membranes in particular, contain proteins with appreciable proportions of β-structure. Both optical activity measurements and infrared spectroscopy indicate that the secondary structures of membrane proteins may vary with metabolic and physiologic activity. Early data further indicate that the antiparallel β-structure may play an important role in energy-linked changes of membrane protein conformation.

1.1.3.4. Membrane protein structure and lipid–protein interactions in biomembranes

1.1.3.4.1. Side chain penetration

Early molecular models of membrane structure (1.2.1.1.) proposed that biomembranes consist of bilamellar arrays of amphipathic lipids with the membrane proteins bound to the polar surfaces of the lipid bilayers.

To account for the extensive apolar interactions known to exist between many membrane proteins and membrane lipids, it was proposed that the proteins at membrane surfaces might associate hydrophobically through apolar residues extending into apolar lipid regions. However, use of space-filling models shows that if a protein is apposed to the surface of a phosphatide bilayer, even a bulky apolar amino acid side chain, for example, that of *Trp*, can only extend into the *glycerol* region of the bilayer; it cannot reach the apolar acyl chain region [4]. Such a side chain penetration would also be energetically improbable.

Apolar lipid protein interactions *therefore require that part of all of a membrane protein, rather than its apolar side chains, penetrate into or through the apolar membrane core* [4]. We will now examine structural devices which might satisfy this requirement.

1.1.3.4.2. Low overall polarity and/or high overall hydrophobicity of membrane proteins

Apolar associations play an important role in the interactions of many proteins with their membranes. Various authors have therefore proposed that 'intrinsic' membrane proteins contain disproportionately few polar amino acids and that this low polarity allows for apolar lipid protein associations. For example, Capaldi and Vanderkooi [31] argue that, whereas 85% of 205 soluble proteins have 'polarities' of 47 ± 6%, 47% of 19 membrane proteins have polarities of less than 40% (minimum 30%). These authors define 'polarity' as the sum of the mole percentages of 'polar' amino acids residues. They consider Asp, Asn, Glu, Lys, His, Arg, Ser and Thr as 'polar' and other amino acids as 'apolar'. However, 'polarity' should be expressed in terms of the ratio (volume of polar residues: volume of non-polar residues) [32] and one must also consider hydrophobicity. Moreover, one cannot classify side chains in an all-or-none fashion because there are gradations of 'polarity' between, for example, Lys and Thr, and of hydrophobicity between Phe and Leu.

Tanford's experimentation on the distribution of free amino acids between water and polar solvents [33] and Bigelow's computations [34] allow for reasonable assessments of protein hydrophobicity: each amino acid residue is assigned an experimentally determined 'hydrophobicity' value, $H\phi$; these range between 0.45 kcal/residue for Thr and 3.00 kcal for Trp. Knowing the mean residue hydrophobicities and the amino acid composition of a protein, one can compute its average hydrophobicity. The average hydrophobicities of soluble proteins range from 0.440 to 2.020 kcal/residue with 50% of proteins having $H\phi$ values between 1.0 and 1.2 kcal/residue [34].

The hydrophobicity scale comprises a set of experimentally determined values, rather than an arbitrary all-or-none classification. Thus, Arg and Thr appear weakly hydrophobic, not simply 'polar' as suggested in [31]. Also, Lys, although polar, exhibits a high $H\phi$ value (1.5 kcal/residue) due to its aliphatic chain. The hydrophobicity criterion has been applied to membrane proteins that can be eluted by *ionic manipulations* and those which require *detergents* or *organic* solvents for extraction. Ionically-elutable proteins generally exhibit distinctly *lower hydrophobicities* than the more tightly bound species. For example, the erythrocyte protein 'spectrin', *S. Faecalis* ATPase, and bovine F_1 ATPase, all of which are eluted by ionic manipulations, exhibit $H\phi$ values of ≤ 1.10 kcal/residue, while rhodopsin and bovine myelin proteolipid, which cannot be extracted ionically, yield hydrophobicities of ≤ 1.20 kcal/residue, respectively (Table 1.1). Interestingly, murein lipo-protein, which is not easily eluted ionically, exhibits an unusually *low* hydrophobicity, 0.73 kcal/residue. Also, except for the last

Table 1.1

Amino acid composition, average hydrophobicities and polarity indices of some membrane proteins that are readily eluted by aqueous solvents and others that resist elution[1].

	Elutable			Nonelutable		
Amino acid	Human RBC spectrin	*S. Faecalis* ATPase	Bovine F_1 ATPase	Bovine retina rhodopsin	Bovine myelin proteolipid	Murein-lipoprotein
Lys	6.7	6.1	6.2	4.3	4.5	8.8
His	2.6	1.7	1.7	1.7	2.3	–
Arg	5.8	4.5	5.8	2.6	2.6	7.0
Asp	10.9	10.0	7.9	6.4	4.0	24.5
Glu	20.5	13.0	11.7	8.9	5.9	8.8
Thr	3.6	6.7	5.8	7.2	8.4	3.5
Ser	4.1	6.3	6.2	5.1	5.2	16.5
Pro	2.4	3.9	4.2	5.5	2.8	–
1/2 Cys	1.1	0.3	0.4	2.1	2.9	–
Met	1.7	2.3	2.1	3.4	1.3	3.5
Gly	4.9	8.7	9.2	6.8	10.4	–
Ala	9.2	8.4	10.4	8.5	11.9	15.8
Val	4.7	6.8	7.5	8.5	7.4	7.0
Ile	4.0	6.2	6.2	5.5	5.1	1.8
Leu	12.4	9.3	8.7	8.5	11.5	7.0
Tyr	2.0	3.3	2.9	4.7	4.4	1.8
Phe	3.0	3.1	2.9	8.1	8.2	–
Trp	–	–	0	2.1	1.8	–
NH_3	–	–	–	–	–	8.0
$H\phi_{Av}$[2]	0.96	1.10	1.09	1.23	1.21	0.73
p[3]	1.54	1.12	0.96	0.73	0.65	2.04

[1] Amino acid compilation from [4].
[2] Mean residue hydrophobicity.
[3] Polarity index.

protein, *both the 'elutable' and 'non-elutable' proteins show hydrophobicities in the 0.9–1.2 kcal/residue range typical for most water-soluble proteins.*

The $H\phi_{Av}$ value of a protein is an index of *non-polarity.* It has been suggested [32,34] that one can obtain a measure of *polarity* from the polarity index, p, defined as $p = V_e/V_i$, where V_e and V_i are external ('shell') and internal volumes, respectively (Fig. 1.1). Taking the residue volumes [34] and assuming that the 'polar' residues, that is, Arg, His, Lys, Asp, Glu, Asn, Tyr, Ser and Thr, lie in V_e and all others in V_i, one can calculate p from the amino acid composition*. When

* X-ray analyses of globular proteins show that, while charged and polar residues are excluded from the interior of the macromolecules, *not all apolar amino acid residues* lie buried in the interior. For example, in subtilisin, 50 % of the apolar groups are accessible to water. For this reason, the polarity index cannot be considered more than an approximation.

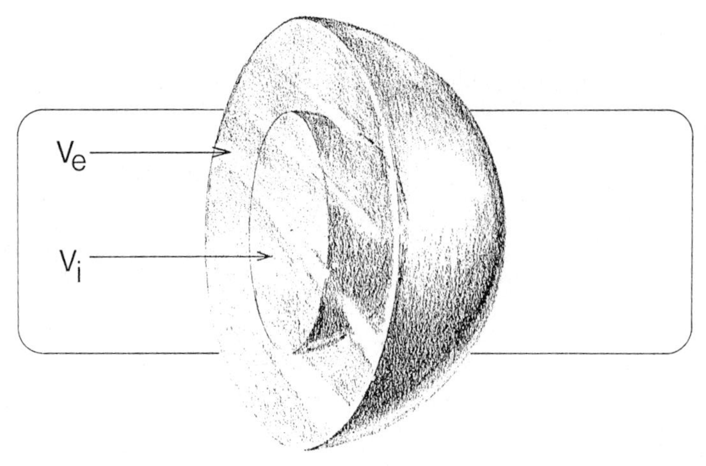

Fig. 1.1. Polar and apolar volumes of a protein.

the polarity-index criterion is applied to membrane proteins which are 'elutable' or non-elutable by aqueous solvents, one finds that all proteins except myelin proteolipid and murein-lipoprotein fall in the polarity range of many soluble globular proteins [34]. Interestingly, murein lipoprotein has an exceptionally high *p*-value.

These calculations indicate that one cannot unambiguously classify membrane proteins as highly hydrophobic or poorly polar. One must, therefore, search for features other than overall hydrophobicity or polarity to explain lipid–protein association in biomembranes.

1.1.3.4.3. Linear amphipathic sequences

One can synthesize linearly amphipathic polypeptides of high molecular weight by joining the N- and C-termini of a 'block' composed of polymerized, apolar L-amino acids, to two 'blocks' composed of copolymers of polar L- and D-amino acids [35], (Fig. 1.2).

This structural principle occurs in biomembranes and certain membrane-active proteins.

Melittin. A simple example of a linearly amphipathic sequence in a membrane active protein is found in melittin, a small, basic oligopeptide from bee venom

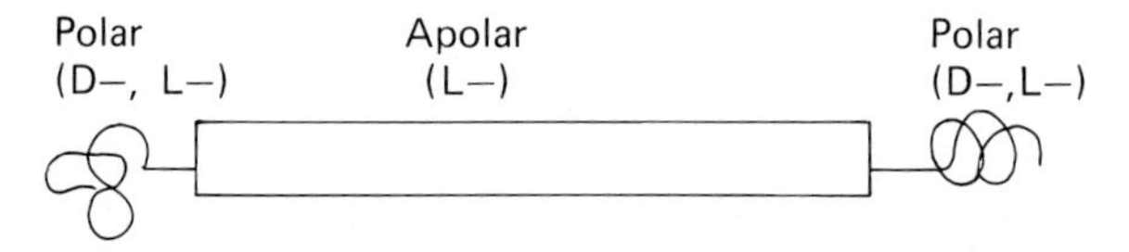

Fig. 1.2. An amphipathic 'block' copolymer.

(Fig. 1.3). This substance has a high overall hydrophobicity ($H\phi_{Av} = 1.25$ kcal/residue), and a low polarity index ($p = 0.68$). Proceeding from the N-terminus, the sequence contains 20 residues of which all but Gly^1, Gly^3, Gly^{12} and Ser^{18} exhibit appreciable hydrophobicity and only one, Lys^7, is charged. In contrast, the C-terminus consists of six polar residues, including four basic residues giving the peptide a net ⊕ charge. With the *Lys* at position 7 and a *Pro* at position 14, α-helical structuring is improbable.

```
 1              5                      10
 Gly—Ile—Gly—Ala—Val—Leu—Lys—Val—Leu—Thr— Thr—Gly—Leu—Pro—

   15                      20
   Ala—Leu—Ile—Ser—Trp—Ile     —Lys—Arg—Lys—Arg—Gln—Gln
```

Fig. 1.3. Amino acid sequence of melittin. Calculated hydrophobicity 1.25. The boxed-in portion indicates the apolar segment.

The interaction of melittin with phospholipids has been elucidated by Verma and Wallach [36,37] using infrared and laser-Raman spectroscopy as well as fatty acid 'spin probes'. The data show that the polypeptide can associate with both neutral and ⊖ charged phospholipids, but the interaction is much stronger with the latter. However, the major interation is a hydrophobic one between the fatty acid chains of the phospholipid and the irregularly-arrayed apolar peptide segment.

Membrane glycoproteins. Many membrane proteins are *glycoproteins*, that is, they bear covalently-linked sugar residues. These carbohydrate moieties exert an important influence on the physical and chemical properties of their peptide entities and of the membranes whereon they lie. They may be critically involved in cellular interactions (Chapter 8) and serve as receptors for lectins (Chapter 10), specific antibodies and probably other macromolecules. Very often similar or identical oligosaccharide units are found attached to both glycoproteins and glycolipids (Chapter 4) and frequently one observes different degrees of glycosylation in normal and neoplastic cells (Chapters 2 and 4). The general properties of membrane glycoproteins have been recently reviewed *in extenso* by Hughes [37a] and alterations of membrane glycoproteins in tumors are described in Chapter 2.

The major neutral sugars linked to membrane glycoproteins are D-galactose, D-mannose, D-glucose and D-fucose and the principal amino sugars are *N*-acetylglucosamine, *N*-acetylgalactosamine and *N*-acetylneuraminic acid. The number of sugar residues attached to amino acids ranges between 2 and 20–30, with an average number near seven. The carbohydrate chains are usually branched. The sugars link to the polypeptide chains through three different glycosidic bonds:

(a) *N*-β-glycosidic bonds to the amide of asparagine; (b) *O*-β-glycosidic bonds to the hydroxyl groups of serine, threonine, hydroxylysine and hydroxyproline residues; and (c) *S*-glycosidic bonds between cysteine and carbohydrate. The sugar residues are always linked to special amino acid sequences. The carbohydrate chains exhibit great variety in sugar composition and sugar sequence, and all evidence indicates that the sugars (and their associated peptide moieties) lie at the surface of membranes not in their cores. This architectural feature provides membrane glycoproteins with an unusual amphipathic character, exemplified by 'glycophorin'.

'Glycophorin' is the name given to the major glycopeptide of the human erythrocyte membrane, a glycoprotein thought to span the membrane thickness. Studies initiated by Winzler [38] show that this glycoprotein has a linearly amphipathic sequence. This is illustrated in Table 1.2 which gives the amino acid composition of the intact glycoprotein, of the soluble peptide segment released

Table 1.2

Amino acid composition of the major human erythrocyte membrane glycoprotein and its fragments[1,2]

Amino acid	Sialoglycoprotein		
	Intact	Soluble peptide	Insoluble peptide
Lys	3.5	4.3	2.0
His	3.8	4.9	3.8
Arg	4.1	3.3	3.9
Asp	6.0	7.6	2.4
Glu	10.0	4.7	7.0
Thr	13.8	23.8	6.2
Ser	13.6	23.8	6.0
Pro	6.5	4.0	4.5
1/2 Cys	0	0	0
Met	<1.0	0	2.1
Gly	6.8	3.5	10.7
Ala	6.8	6.1	7.9
Val	7.7	4.9	8.1
Ile	4.5	2.8	14 3
Leu	4.5	1.6	11.6
Tyr	3.6	1.5	2.4
Phe	3.5	0	5.1
Trp	–	–	–

[1] From Winzler [38].

[2] Values are in residues per 100 residues. This protein contains 64 % carbohydrate including most of the cell membranes' sialic acid in conjunction with galactose, mannose, fucose, *N*-acetylgalactosamine. and *N*-acetylglucosamine. These comprise oligoheterosaccharides bound to the frequent threonine and serine residues usually via the *N*-acetylgalactosamine.

by mild proteolysis of intact erythrocytes, and of the insoluble peptide fraction remaining after proteolysis. The protease-accessible portion (bearing the sugar moiety) contains a large proportion of polar amino acids, giving it an average Hϕ of only 0.50 kcal/residue, compared with 0.93 kcal/residue for the whole protein. The very high average hydrophobicity of the 'residue', 1.33 kcal/residue, is concordant with the concept that this hydrophobic portion 'anchors' the protein to the membrane [38].

As originally demonstrated by Winzler [38] the amino terminal segment, bearing the carbohydrate, lies extracellularly, while the apolar region lies within the hydrophobic membrane core, and the polar carboxyl terminal protrudes into the cytoplasmic domain. The tentative amino acid sequence for the apolar segment, shown in Fig. 1.4 shows an abundance of apolar residues (—). No ionizable residues occur for a stretch of 22 residues (numbers 12–34). However, two Glu residues (Glu^{9} and Glu^{11}) and two Arg residues (Arg^{35}, Arg^{36}) provide the apolar segment with two charged ends and an asymmetric charge distribution. If the 23 residue segment were in α-helical array, it could span about 35Å, that is, approximately half the width of the erythrocyte membrane. However, there is no conclusive proof that the apolar segment forms an α-helical array.

1 5 10 15
Val–Gln–Leu–Ala–His–His–Phe–Ser–Glu–Ile–Glu–Ile–Thr–Leu–Ile–
(Pro) (Pro) (Ala)

16 20 25 30
–Gly–Phe–Gly–Val–Met–Ala–Gly–Val–Ile–Gly–Thr–Ile–Leu–Leu–Ile

31 35 40 45
–Ser–Tyr–Gly–Ile–Arg–Arg–Leu–Ile–Lys–Lys–Ser–Pro–Ser–Asp–Val

46 50
–Lys–Pro–Leu–Pro–Ser–Pro–

Fig. 1.4. Tentative amino acid sequence for the apolar portion of the major glycoprotein of human erythrocyte membranes. ( ) indicate sequence uncertainties. (From Segrest et al. [39]).

It appears that the major glycoprotein of human erythrocyte membranes can participate in an aggregation–disaggregation equilibrium [40]. Monomer formation is favored at elevated temperatures ($> 37\,^{\circ}C$) and is inhibited by phosphate buffers.

Another membrane protein exhibiting linear amphipathic character is the cytochrome b_5 isolated from rabbit liver endoplasmic reticulum [41,42]. This enzyme has an overall molecular weight of 16,700 but consists of an enzymatically active portion, composed of 97 residues and a 'tail' composed of ~40 residues. The enzymatically active moiety has a very low hydrophobicity ($H\phi_{Av} = 0.89$ kcal/residue) whereas the tail is highly hydrophobic ($H\phi = 1.7$ kcal/residue), more so than the apolar segment of 'glycophorin'. It appears that the enzymatically active portion of the protein lies at the membrane surface and is anchored to the apolar portions of the membrane by the hydrophobic 'tail'. The endoplasmic reticulum of many hepatomas is depleted of cytochrome b_5 (Chapter 6).

1.1.3.4.4. Conformationally-induced segregation of polar and apolar amino acid residues

When the detailed architecture of myoglobin and hemoglobin became elucidated by X-ray crystallography (e.g. [3]) an important structural principle emerged which has since then been discovered in numerous other proteins. It was found that most of the helical segments of these proteins contain sequences where apolar residues occur regularly at every 3rd or 4th position. This gives the helices a *polar* face and an *opposite apolar* face; that is, the helices are *conformationally amphipathic.* As shown in Fig. 1.5, the conformational segregations of apolar and polar residues in hemoglobin helices is very pronounced. In the hemoglobins, as well as other soluble proteins containing such amphipathic helices, the tertiary and quaternary structures of the proteins are such as to engender association of the apolar surfaces and orientation of the polar domains toward water. These structural devices are conserved in all hemoglobins studied, despite extensive differences in amino acid composition. Thus the peptide chains of all vertebrate hemoglobins fold into virtually identical tertiary structures, although only 9 out of more than 140 residues are the same in all the globin chains analyzed so far. However, in all cases 30 residues are always nonpolar and in all cases the amphipathic helices are arrayed in apolar association between apolar helical faces. In all tetrameric hemoglobins the quarternary associations also generate a hydrophilic channel running through the center of the protein.

The associations of apolar surfaces with apolar surfaces in water soluble globular proteins are dictated by entropic factors: minimal free-energy and maximal entropy are obtained when apolar residues are sequestered away from water (in analogy with the micellarization of amphipathic lipids). In view of this fundamental principle and spectroscopic data pointing to a relatively high α-helicity in membrane proteins, Wallach and Zahler [18] proposed in 1966 that:

(a) There exist classes of membrane proteins with amphipathic helices penetrating the apolar cores of their membranes.
(b) The 'solvent' for these helices is not composed of water, but of apolar fatty acid chains of phosphatides.
(c) Minimum free energy and maximal entropy are obtained when the *polar faces of the helices associate with each other.*
(d) Association of the polar faces of *four* helices (with the structural qualities of the H-helices) with the helix axes more or less perpendicular to the membrane plane would generate *cylindrical proteins, penetrating the thickness of biological membranes, their apolar perimeters* interacting with the acyl chains of *phos-*

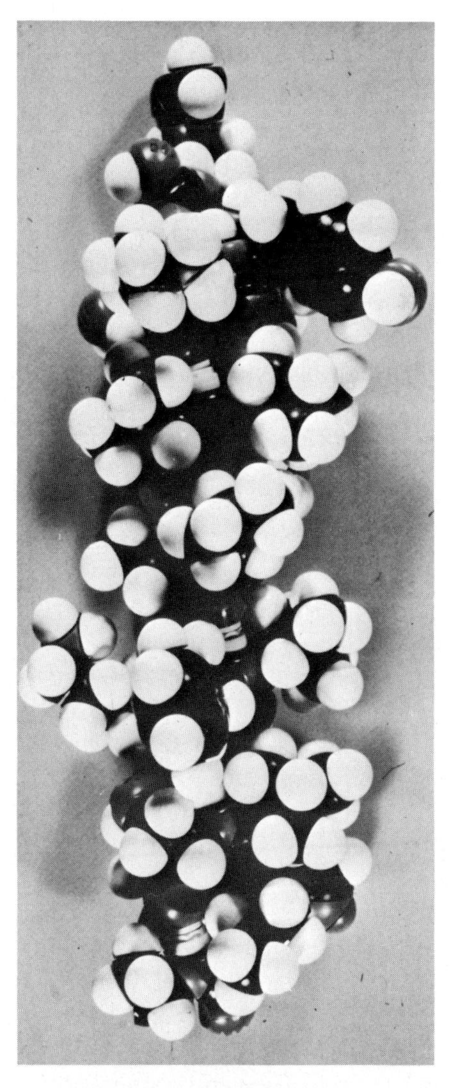

Apolar face

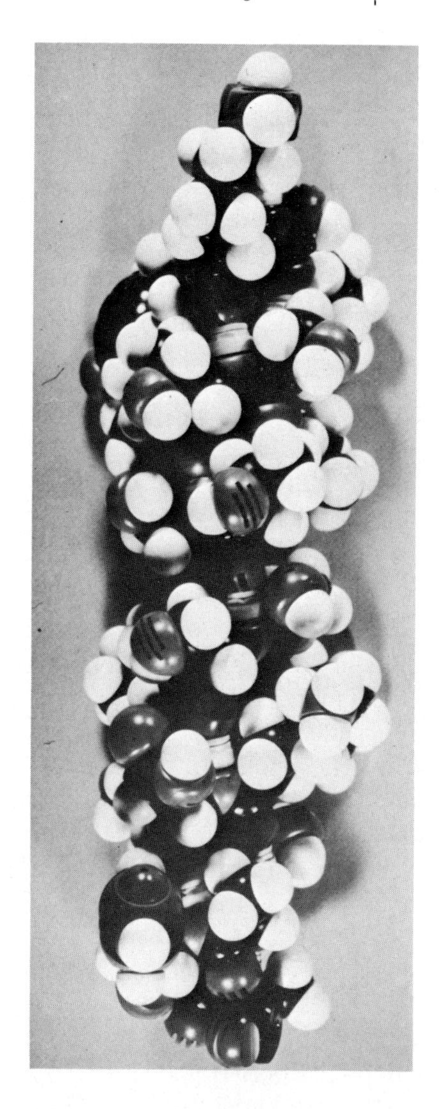

Polar face

Fig. 1.5. Conformationally-induced segregation of apolar and polar amino acid residues in the H-helices of hemoglobins.

pholipids and their *polar interiors* possibly forming small aqueous *channels penetrating through the membranes.*

(e) Spherical or ellipsoidal proteins with fully apolar perimeters, while possibly suited for residence in a membrane's apolar core, would not be properly structured for interaction also with the aqueous intra- and extracellular compartments. This would require (1) more or less cylindrical proteins, with (2) hydrophobic perimeters, (3) possibly polar channels running through the protein assembly (analogous to the 'pore' penetrating the hemoglobin tetramer) and (4) polar moieties projecting into aqueous spaces.

The feasibility of this concept was subsequently demonstrated by use of space filling models [12]. However, information indicating that the hypothesis might actually be correct for certain lipoproteins and biomembranes did not emerge until 1973–1974 when sufficient structural data had become available to develop meaningful structural models for high-density plasma lipoproteins and the murein lipoprotein in the outer membrane of *E. coli* (cf. [4]).

High density serum lipoproteins. Considerable information has accrued on the structure of the soluble high-density lipoproteins (HDL) of the blood plasma. This information is relevant to the topic of membrane structure because (a) similar structural principles may occur in HDL and biomembranes, (b) some evidence [6] suggests that HDL protein components occur also in erythrocyte membranes, (c) extensive information indicates that some of the lipid components of HDL can exchange with membrane lipids.

The HDL constitute a class of lipoproteins comprising the same protein subunits, but differing in lipid composition (hence size, density and molecular weight). Highly purified HDL from human plasma consists of predominantly (90%) two apoproteins, ApoA-I (ApoLp–Glu-I) and ApoA-II (ApoLp–Glu-II)/associated with phospholipids, cholesterol and cholesterol esters. The lipids typically comprise 35–38% of the lipoprotein. Unlike biomembranes, HDL contain cholesterol esters in equal proportions to cholesterol.

The amino acid sequence of ApoLp–Glu-II has been determined by Brewer et al. [43]. There are no extended sequences of apolar residues; rather long peptide segments bear apolar side chains that occur at every 3rd to 4th position along the peptide chain. The computed hydrophobicity of the 20-residue segments presented in Fig. 1.6 is 1.12 kcal/residue, very similar to that of the H-helix of hemoglobin (1.13 kcal/residue) but well below the value of 1.45 kcal/residue for complete ApoA-II.

HDL are highly α-helical (61%–70%) (e.g. [44–47]). The helix content of the apoproteins increases upon recombination with lipid [48]. Moreover ^{13}C nuclear magnetic resonance spectroscopy indicates that the affinity of the HDL apoproteins for phospholipids depends more on their fatty acid chains than their head groups. However, ApoA-I, which has a lesser phospholipid-binding capacity than ApoA-II, shows a preference for lecithin. ApoA-I, in contrast, shows a very high affinity for sphingomyelin [49].

Assmann and Brewer [50] point out that the ApoA-II sequence contains an apolar residue at every 3rd to 4th position and that, in the α-helical conformation, such a sequence will produce an apolar band running the length of the helix. A molecular model of the ApoA-II sequence in α-helical conformation (Fig. 1.6)

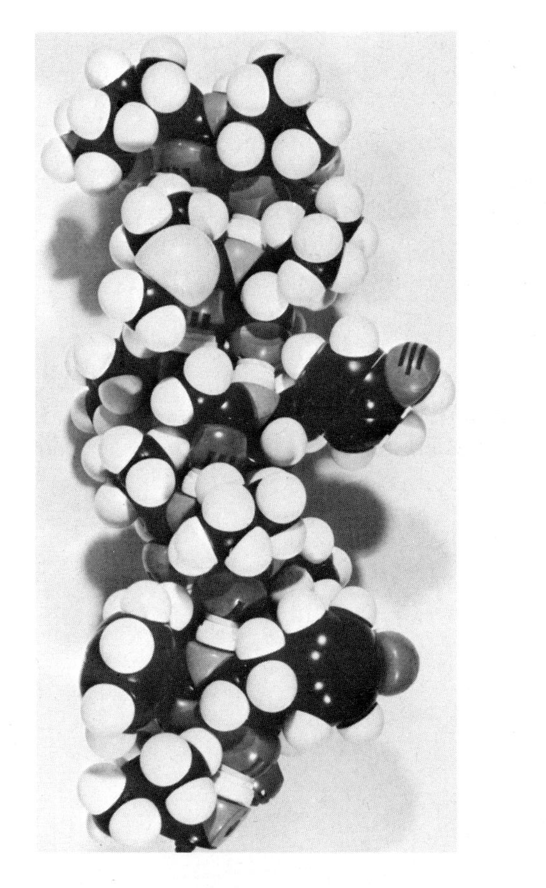

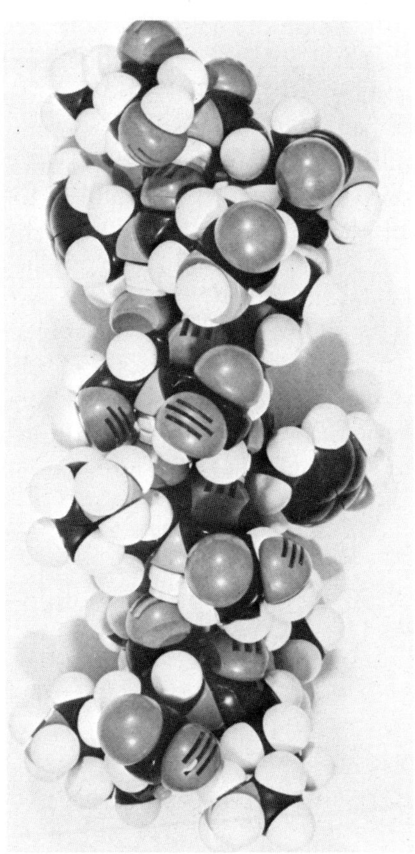

Apolar face

Polar face

Fig. 1.6. Conformational segregation of apolar and polar amino acid residues in apo-lipoprotein II of serum high density lipoprotein.

shows that this helix indeed has a polar and an apolar face, although the spatial segregation of polar groups is less marked than in the H-helix of hemoglobin (Fig. 1.5). Assman and Brewer [50] propose that the apolar face shown in Fig. 1.6 is the site of apolar protein–protein and protein–lipid interactions in HDL. NMR studies [49,51] are in accord with this notion. Moreover, the 20 residue segment shown in Fig. 1.6 has just the right length to accomodate a C_{18} fatty acid chain.

Although there is still no complete concensus as to the assembly of HDL, the structural details now known are clearly relevant to a discussion of membrane structure.

Murein lipoprotein. The principal lipoprotein in the outer membrane of *E. coli*, murein lipoprotein, constitutes the first documented example of conformational segregation of apolar and polar residues in an 'integral' membrane protein. The chemical structure of murein lipoprotein has been determined by Braun and associates [52–54].

The protein consists of 58 amino acid residues. The N-terminus consists of an unusual amino acid (glycerylcysteine [*S*-(propane-2′,3′-diol)-3-thio-2-amino-propanic acid]) which is linked to two fatty acids by two ester linkages and to another fatty acid by an amide bond. The protein contains no Pro, Trp, Phe, His or Gly (Table 1.1). The protein has a very low hydrophobicity ($H\phi_{Av} = 0.73$ kcal/residue) and a high polarity ($p = 2.04$).

As shown in Fig. 1.7, the N-terminus of murein lipoprotein consists of two consecutive oligopeptides (14 and 15 residues) with almost identical amino acid sequences (5 very conservative substitutions). Hydrophobic, uncharged residues occur at every 3rd or 4th position along the entire chain. The pattern is similar to that found in the C-terminal half of tropomyosin [55].

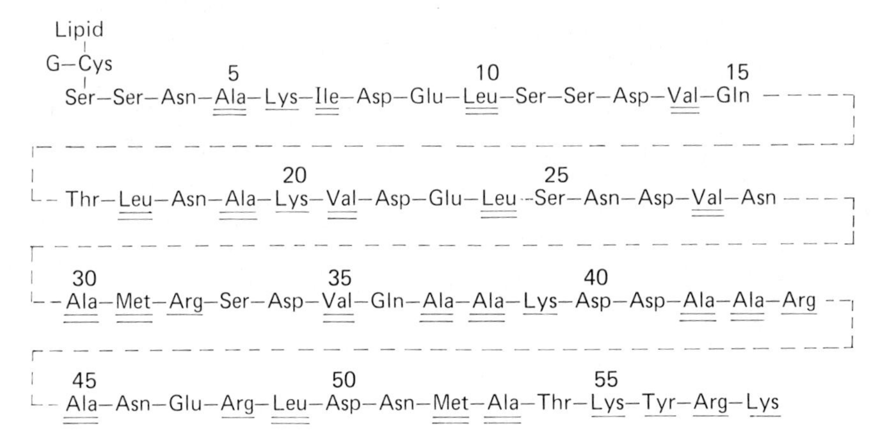

Fig. 1.7. Amino acid sequence of Murein lipoprotein. The homologous oligopeptides at the N-terminus have been aligned. (———) = Hydrophobic; (═══) = Hydrophobic and apolar. From [54].

Murein lipoprotein is more than 70% α-helical [56]. The sequence of repetitive apolar residues thus generates a non-polar band extending the length of the helix. This also is found in tropomyosin [55] where the apolar bands of two helices associate hydrophobically to form a coiled-coil structure with a hydrophilic perimeter.

Such a coiled-coil structure is not suited for hydrophobic interaction with membrane lipids. However, as pointed out by one of us [4,12,15,18] and by Inouye [57], an alternate arrangement, that is, polar association of several helices, would lead to an aggregate whose perimeter is formed by the apolar bands. Such an assembly would be well suited for hydrophobic association with membrane lipids.

Inouye [57] has presented a detailed model of such an array. He shows that six α-helical lipoprotein units can be arranged in a 'superhelix' with a 166° turn. This superhelix has an apolar perimeter and a central hydrophobic channel 12.5 Å in diameter. Glu^{9}, Asp^{13}, Glu^{23} and Asp^{27} face into the channel interior giving it an anionic character. The assembly is stabilized by a large number of ionic and other polar interactions between helices. The length of the assembly is 76 Å, corresponding to the thickness of the outer membrane and the thickness of a phospholipid bilayer. The surface area of the assembly perpendicular to the membrane plane is $\sim 1.4 \times 10^{8}$ Å^{2}.

Inouye's model [57] also accounts satisfactorily for the disposition of the fatty acid bound covalently at the N-terminus, and shows that the N-terminus must lie at the external surface of the membrane with the C-terminus linked covalently to the peptidoglycan layer at the inner surface of the membrane. Inouye [57] also demonstrates that, with greater twist in the superhelix, up to 12 subunits can be accommodated in a 75 Å membrane width. However, with 12 subunits the diameter of the central channel will become ~36 Å. In any event, the central channel would be large enough to account for the free movement of many small molecules through the outer membrane of *E. coli.*

1.1.4. Lateral diffusion of lipids

1.1.4.1. Lipid bilayers

Evidence obtained by NMR spectroscopy [58] and by use of spin-labelled lipid probes [59,60] shows that lipids can diffuse rapidly within the plane of an artificial lipid bilayer.

Above the gel → liquid-crystalline transition temperature of the lipid, lateral diffusion can be considered to occur through a hopping process. This might involve the formation of temporary vacancies in the bilayer or direct exchange between identical molecules. In either case, the hopping frequency, ν, can be related to the diffusion constant, D, by [60]

$$D = \theta \cdot \nu \cdot \lambda^2 \tag{1.6}$$

where λ is the length of one jump (~8 Å) and θ is a geometrical factor (~0.5 for hexagonal lattices).

NMR measurements give hopping frequencies in excess of 3×10^{3} sec^{-1}, that is, $D > 10^{-11}$ cm$^{2}\cdot$sec^{-1}, but ESR studies give D values of $\sim 10^{-8}$ cm$^{2}\cdot$ sec^{-1} for androstane spin-label in dipalmitoyl phosphatidylcholine [60] and $\sim 2 \times 10^{-8}$ cm$^{2}\cdot$sec^{-1} for dipalmitoyl tempophosphatidylcholine [59]. These values correspond to hopping frequencies of 3–6 $\times 10^{6}\cdot$sec^{-1} and translational rates of $\sim 10^{4}$ Å$\cdot$sec^{-1}. In the presence of cholesterol (1:1 molar ratio), D decreases by a factor of ~2.5 due to the mobility restrictions imposed by cholesterol (Chapters 3 and 5).

One should note that diffusion in lipid bilayers is not restricted to translation of entire molecules. Thus, in the fluid state, the lipid hydrocarbon chains will exist as various rotational isomers with a variety of 'kinks' [60]. These 'kinks' are mobile structural elements which can be propagated with a diffusion coefficient $\approx 10^{-5}$ cm$^2 \cdot$ sec^{-1}.

1.1.4.2. Biomembranes

Most attempts to measure rates of lateral diffusion in biomembranes have involved the use of nitroxide analogues of membrane lipids. This approach was introduced by Scandella et al. [61] who mixed a sonicated dispersion of spin-labelled phosphatidylcholine (fatty acid-labelled) with sarcoplasmic reticulum vesicles and followed the dilution of the label after fusion of the two vesicle types. The estimated diffusion coefficient contains several uncertainties, but its value, 6×10^{-8} cm$^2 \cdot$ sec^{-1}, is in the same range as that found for cholesterol-free, fluid lipid bilayers. Barnett and Grisham [62], using an essentially similar approach, obtained comparable data on plasma membrane isolates of kidney medulla. D values of $\sim 3 \times 10^{-8}$ cm$^2 \cdot$ sec^{-1} and $\sim 10^{-7}$ cm$^2 \cdot$ sec^{-1} have been reported for spin-labelled fatty acids incorporated respectively into *E. coli* membranes [63] and sarcoplasmic reticulum [64].

The diffusion constants obtained with spin labels fall within a rather restricted range, despite the differences between the membranes tested and the different properties of the labels employed. Nevertheless, the values reported must be viewed with some reserve [4,65].

First, there is evidence that spin label 'impurities' do not behave in exactly the fashion of the molecules they are supposed to represent. In fact, they tend to fluidize their immediate environment [4,65,66]. *Second*, Oldfield, et al. [67] show that in a phase-heterogeneous systems spin-labels preferentially probe fluid regions. *Third*, Keith et al. [65] demonstrate that the spectral analyses used to determine hopping frequencies by ESR measurements tend to overestimate these frequencies.

Moreover, direct experimental evidence shows that not all membranes exhibit rapid diffusion of lipids in the membrane plane. Thus, Morrison and Morowitz [68] pulse-labelled exponentially-growing *B. megaterium* with radioactive palmitic acid and observed a very non-uniform label distribution by radioautography. They demonstrated that the incorporated label was primarily in the phospholipid. The non-uniform distribution is clearly incompatible with rapid lateral diffusion over the whole of the membrane area. Rapid localized diffusion is not excluded.

1.1.5. Transmembrane diffusion of lipids

1.1.5.1. Lipid bilayers

Transverse diffusion, 'flip-flop', is a mechanism whereby lipid molecules cross from one half to the other of a lipid bilayer. In the first measurement of this process [69] a monolayer of stearic acid was transferred to a glass slide followed by a second monolayer of radioactive stearate to form a bilayer. This was frozen and fracturated parallel to the bilayer plane to allow radioactivity measurements on the two halves. A time sequence showed that exchange occurred with a half time of 50 min in the presence of calcium and 25 min in the absence of calcium.

Kornberg and McConnell [70] investigated transverse diffusion of phospholipids using a spin-label approach. They found that the exchange of spin-labelled phosphatidylcholine between the inside and outside of artificial phosphatidylcholine liposome membranes occurs with a long half time – about 6.5 h at 30 °C.

1.1.5.2. Biomembranes

In contrast to model membranes, the biomembranes examined exhibit rapid rates of transverse diffusion. Thus, McNamee and McConnell [71] find a half time of 4–7 min for the transmembrane diffusion at 15 °C of dioleoyl-phosphatidyl-tempoyl-choline (OPTC), a spin-label phosphatidylcholine analog, incorporated into excitable membrane vesicles derived from the electric organ of *Electrophorus electricus*. In contrast, the 'flip-flop' of OPTC in egg phosphatidylcholine vesicles has a half time in excess of 5 hr.

Furthermore, Huestis and McConnell [72] find a half time of 20–30 min for 'flip-flop' in red blood cells at 37 °. Also, the data of Grant and McConnell [73] suggest that the rate of 'flip-flop' in the membranes of *Acholeplasma laidlawii* is in the order of less than one minute at 0 °C. Finally, Smith and Green [74] find an average half time of 72 min for sterol 'flip-flop' in erythrocytes.

The available data thus indicate that the rate of transmembrane transfer of phosphatide molecules is rather rapid in biologic membranes compared with artificial membranes. An asymmetric distribution of phospholipids across biomembranes might therefore derive from an asymmetric distribution of membrane proteins.

1.1.6. Lateral diffusion of proteins in the membrane plane

1.1.6.1. Theory

Let us assume for the moment that a membrane constitutes a two-dimensional continuum whose protein and lipid components undergo random thermal motion. As shown in Section 1.1.4, the lateral diffusion of lipids in such systems can be described as a series of spatial exchanges occurring at a frequency of

10^7/sec, corresponding to a diffusion coefficient of $\sim 3 \times 10^{-8}$ cm$^2 \cdot$ sec^{-1} [63,75]. The lateral mobility of the proteins can be expressed in terms of lipid mobility, using the 'free volume' model proposed by Sackman et al. [69,75].

In this model the velocity of lateral protein diffusion depends on the rate at which lipid displacement at the protein perimeter creates 'free volume' (Fig. 1.8).

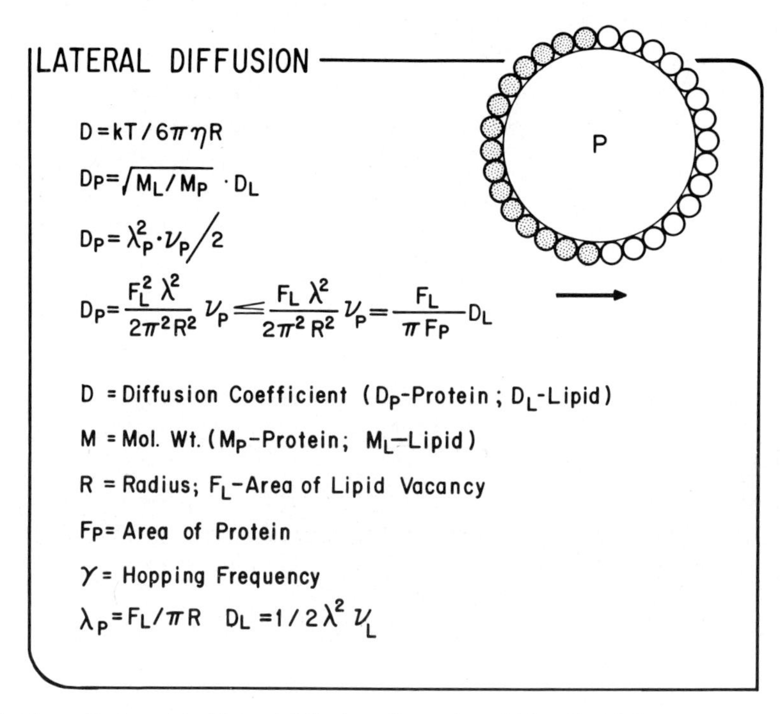

Fig. 1.8. Theoretical model of lateral diffusion of proteins within a lipid bilayer.

Accordingly, a cylindrical protein of mol.wt. $\sim 10^5$ ($R \sim 2 \times 10^{-7}$ cm) would diffuse $\sim 10^2$ times more slowly ($D_p \sim 3 \times 10^{-10}$ cm$^2 \cdot$ sec^{-1} at 25 °C) than the lipid molecule ($D \sim 3 \times 10^{-8}$ cm$^2 \cdot$ sec^{-1}). This corresponds to a mean translational distance of $\sim 3 \times 10^{-5}$ cm/sec. This value of D_p is consistent with a viscosity of $\sim$1–2 gm $\cdot$ cm$^{-1} \cdot$ sec^{-1}, which is reasonable for light biologic oils (e.g. olive oil) and fits the viscosity value obtained by Cone from rotational diffusion measurements on the rhodopsin of visual receptor membranes [76]. However, many membrane proteins have tightly associated membrane lipids and the radii of such lipoproteins may be greater than twice those of the naked protein. Since D_p varies as the inverse square of the effective radius, the diffusion coefficient of these proteins would be $\gtrsim 5 \times 10^{-11}$ cm$^2 \cdot$ sec^{-1}.

1.1.6.2. Measurement

Two basic approaches have been employed to test whether proteins can move parallel to the membrane plane. In the first, one starts with a non-random distribution of the proteins and follows the rate of randomization. In the second, one begins with a uniform distribution and follows clustering or aggregation induced by external stimuli.

Spectroscopic measurements. Poo and Cone [77] have measured the lateral mobility of *rhodopsin* in retinal rod outer segments. They first 'bleached' geometrically defined areas of rod outer segments by short pulses of light delivered in geometrically defined microscopic beams. They then followed the redistribution of unbleached rhodopsin molecules with time. They obtained half-times for redistribution of 35 ± 5 μsec for frog outer segments (8 μm in diameter) and 23 ± 5μsec for mudpuppy segments (12 μm in diameter). They estimate a lateral diffusion constant, D, from the equation

$$D = (0.69\text{b}\, L^2)/(\pi^2\, t_{\frac{1}{2}}) \tag{1.7}$$

where L is the diameter of the outer segment disc, $t_{\frac{1}{2}}$ is the half-time for redistribution, and b is a geometric constant. They compute $D = 3.5 \pm 1.5 \times 10^{-9}$ $\text{cm}^2/\text{sec}^{-1}$ and $3.9 \pm 1.2 \times 10^{-9}$ $\text{cm}^2/\text{sec}^{-1}$ for frogs and mudpuppies, respectively. The viscosity η ($\text{gm} \cdot \text{cm}^{-1} \cdot \text{sec}^{-1}$) for lateral diffusion is computed from the Einstein–Sutherland equation [78]

$$\eta = (kT)/(a\pi DR) \tag{1.8}$$

where k = the Boltzman constant, 1.4×10^{-16} erg $(^\circ\text{C})^{-1}$, T is in °K, R is the radius for spherical diffusion and $a = 6$ for diffusing molecules that are large compared to solvent molecules and $a = 4$ for smaller molecules. Since rhodopsin spans the membrane thickness, R, the radius for the lateral diffusion of rhodopsin, probably lies between 15–30 Å. This yields a η of ~1–4 poises, which is in reasonable agreement with values computed for the rotational diffusion of rhodopsin [76,79].

Peters et al. [80] have shown that the proteins in erythrocyte membranes are far less mobile than rhodopsin. These authors reacted erythrocyte ghosts covalently with fluorescein dithiocyanate, to yield diffuse fluorescence, uniform over the surfaces of the membranes. Using an elegant microspectrophotometer, the authors then photo-bleached one half of the stained membrane surface and monitored the distribution of fluorescence at varying intervals after 'bleaching'. They found *no* redistribution up to 20 min after bleaching. This indicates that erythrocyte membrane proteins cannot move appreciably parallel to the membrane plane.

The photometric data thus demonstrate two extreme cases: (a) very rapid lateral diffusion – in the case of rhodopsin in retinal rod outer segments, and (b) essentially no lateral diffusion – in the case of erythrocyte membranes.

Intermixing experiments. In 1967, Watkins and Grace [81] monitored the redistribution of antigenic components on the surfaces of Hela cells and Ehrlich ascites carcinoma cells after fusion of the cells. They found complete intermixing of the markers within an hour after fusion, using a mixed hemadsorption technique to distinguish between the antigens of the two cells.

The fusion approach was next employed in the classical experiment of Frye and Edidin [82]. These workers manufactured mouse antibodies against the $H2^{i}$ antigens of mouse cIID cells and rabbit antibody against human VA 2 cells. They then prepared goat anti-mouse immunoglobulin conjugated with fluorescein and goat anti-rabbit immunoglobulin conjugated with rhodamine. The mouse antigens on cell surfaces could thus be recognized by the green fluorescence of the complex between H2 antigen-mouse alloantibody and goat anti-mouse immunoglobulin. The human antigens could be similarly detected by the red fluorescence of anti-rabbit immunoglobulin bound to the complex between human antigens and anti-rabbit heteroantibody.

Frye and Edidin [82] then induced fusion of the mouse and human cells *in vitro*, using Sendai virus and used their two fluorescent markers to follow the redistribution of the mouse and human antigens on the surfaces of the hybrid cells as a function of time. Immediately after fusion, the species-specific surface antigens were segregated, each limited to one half of the hybrid. However, within 40 min the antigens had become uniformly intermixed. Redistribution was not affected by inhibitors of protein or ATP synthesis, but could be blocked by lowering the temperature from 20 °C to 15 °C. Frye and Edidin suggested that this redistribution could arise from lateral diffusion of the antigen-bearing sites *within* a fluid lipid environment. From the estimated migration rate of $\sim 5\ \mu m$ in ~ 40 min, Singer and Nicolson [5] have computed a diffusion coefficient of $5 \times 10^{-11}\ cm^2 \cdot sec^{-1}$. This value for D differs substantially from that predicted for the diffusion of proteins (of mol. wt. $\sim 10^5$) through a liquid crystalline milieu; that is, $D \sim 3 \times 10^{-10}\ cm^2 \cdot sec^{-1}$.

Edidin [83] has excluded (a) possible antigen redistribution via the cytoplasm and (b) progressive association and dissociation of antibody. However, available data do not exclude the possibility that the *antigen* molecules progressively associate and dissociate with the external surface of the plasma membrane, that is, that the 'lateral mobility' observed actually represents a 'hopping' process, *on, rather than within*, the membrane. Such a process cannot be easily dismissed since the antigens are not covalently linked to the membrane.

Edidin and Fambrough [84] have performed another type of intermixing experiment using cultured rat muscle fibres, which are typically 30 μm in diameter and as long as 1 mm. They applied focal spots of rhodamine-labelled anti-membrane antibody onto the surface of single muscle fibres and monitored the rate at which the fluorescence spread from the point of application. They used both intact antibody and univalent Fab fragments.

The rates of patch spreading with the Fab fragments are in the order of a few μm/min at room temperature and yield an apparent diffusion constant of 1–3 $\times 10^{-9}$ $cm^2 \cdot sec^{-1}$. With intact antibody the spreading rates are 100-fold slower. Cooling to 4 °C also reduces the spreading rate by a factor of $\sim$100 and no spreading was observed after fixation with 5% glutaraledehyde. To test the possibility that the spreading results from exchange of Fab fragments rather than receptor motion, newly labelled fibres were treated with an excess of unlabelled intact antibody. If exchange did occur, the fluorescent spot should fade as labelled Fab becomes replaced by unlabelled antibody. Since no fading was observed, the authors excluded exchange as a factor in spreading. However, it is difficult to understand why exchange should *not* occur because the Fab fragments and intact antibody should have identical binding sites and the binding affinity of the Fab fragments should not exceed that of antibody. Other unexplained aspects of these experiments are the lack of spreading of some spots and the nearly 10-fold lower spreading rate in the presence of unlabelled antibody [84].

'Capping' and 'Patch' formation. — Surface antigens and lectin receptors. Considerable attention has been directed to membrane transposition phenomena in *lymphoid* cells. Fluorescein-labelled and ferritin-labelled anti-immunoglobulins, alloantibodies or plant agglutinins have been used to monitor the distributions of their respective receptors.

Using fluorescein labelled anti-immunoglobulin, Taylor et al. [85] and Raff and de Petris [86] have shown a diffuse distribution of membrane *immunoglobulins* in resting lymphocytes. However, addition of anti-immunoglobulin induces a polar distribution ('cap formation') of the membrane immunoglobulins and their subsequent pinocytotic engulfment. Analogous redistribution of *surface antigens* has been demonstrated by immunofluorescence [87,88] and using immunoperoxidase [89]. These transpositions occur at 20–25 °C, but not at 4 °C [85,90]. Capping is also partially blocked by cytochalasin B, which is thought to interfere with contractile microfilaments involved in cell movement. When surface immunoglobulins are labelled with univalent Fab fragments, no capping occurs, and a uniform distribution of fluorescence is observed in contrast to the patchy distribution seen with divalent antibody.

Lectin receptors may also redistribute after binding to their specific ligands [90]. Inbar et al. [91] compared normal lymphoid cells from rat lymph nodes and mouse spleens with neoplastic lymphocytes from a Moloney virus-induced ascites mouse lymphoma. They measured concanavalin A binding by using ^{3}H-labelled lectin, (Chapter 10) and evaluated receptor mobility by using concanavalin A coupled to fluorescein. They also studied fluorescein-conjugated wheat germ agglutinin.

Their data show that, although lymphoma cells bind more concanavalin A per cell than normal lymphocytes do, each cell type binds the same amount of concanavalin A per mg cell protein. Using fluorescein-conjugated concanavalin A, they demonstrate cap formation in 30% of normal lymphocytes, but apparently none at all in lymphoma cells. They argue that the formation of caps shows that concanavalin A sites can move in the surface membrane and that the greater capping in normal lymphocytes indicates higher receptor mobility in untransformed cells. Fixation with glutaraldehyde or formaldehyde inhibits capping, clustering of concanavalin A receptors sites, and cell agglutination by concanavalin A, but does not change concanavalin A-*binding*. The authors suggest that that binding sites for concanavalin A are '*floating in a fluid membrane* in a random distribution' and that this distribution changes upon interaction with concanavalin A through lateral translation of the receptors,

If membrane transposition phenomena really involved extensive redistribution of *penetrating* membrane molecules, this should be seen by freeze-etch electron microscopy, which reveals the apolar cores of membranes. However, an early report to the contrary [92], there is no change *within* lymphocyte membranes during capping [90,94,95]. Accordingly, Karnovsky and Unanue [90,94], Unanue et al. [96] as well as Speth and Wunderlich [95], argue that redistribution phenomena represent movement of receptors *on* the *external lymphocyte surfaces, not within the membrane core*. This would occur if the 'mobile' receptors are bound at external membrane surfaces in thermodynamic association–dissociation equilibria normally favoring the membrane-bound state. Binding of lectin, etc. would perturb these equilibria and lead to migration of unbound receptors to the site of initial binding as in a typical nucleation phenomenon. Bivalency of the lectins, etc. would restrict the number of nucleation sites and might account for 'capping'. This hypothesis is fully consistent with the data in [90,94,96]. It cannot, however, be generalized any more than the 'fluid mosaic' hypothesis, since Edidin and Weiss [97] have recently demonstrated a highly restricted lateral mobility of $H2^d$ antigens on cultured mouse fibroblasts and of human species antigens on cultured human cells.

Freeze-fracture electron microscopy. The electron microscopic technique of freeze-fracturing (± freeze-etching) is a very powerful tool for the analysis of membrane structure. The method involves four basic steps: (a) Tissues, cells or membrane isolates, fixed or unfixed, but usually suspended in a cryoprotectant (typically glycerol), are frozen rapidly in a Freon that has been pre-cooled to $\sim -150\,^{\circ}C$; (b) The frozen specimen, maintained under conditions preventing surface contamination, that is, either under high vacuum or under liquid nitrogen ($-196\,^{\circ}C$), is fractured (cleaved) with a cold knife (chilled with liquid nitrogen), creating two fracture faces; (c) One or both of the fracture faces is coated by evaporating first platinum and then carbon onto them at a determined direction (−100 °C to −170 °C and high vacuum) causing the coating material to condense as a thin film(s), constituting a coherent replica(s) of the fracture face(s); (d) The specimens with the adherent replicas are removed from the vacuum and the organic material digested off the inert replicas. These are then examined by transmission electron microscopy.

Optionally, the sample can be 'etched' after fracturing. For this, the sample is 'heated' to −100 °C for 30–120 sec under high vacuum. Under these conditions, water sublimes from the fracture faces; the ice-containing faces thus recede, exposing large unfractured areas for replication. For etching, the sample must be fixed and suspended in distilled water prior to freezing; else the etching process merely produces layers of nonvolatile solutes.

The freeze-cleaving process *splits membranes in planes parallel to their surfaces, exposing the membrane interior.* A single fracture within a membrane thus generates two complementary fracture faces, which, in the case of plasma membranes, face the extra-cellular and intracellular spaces respectively. These fracture faces provide information about the organization of the central, apolar lamellae of biomembranes. The true *inner* and *outer* surfaces of membranes are not revealed by fracturing but can be visualized by etching. By combining freeze-fracturing with etching, one can examine the cores of membranes, represented by two fracture faces, and membrane surfaces, revealed by heat-etching.

The fracture faces of most biomembranes are populated with particles, the *intramembranous particles.* The images actually observed derive from (a) the deposition of replicating material on the particle, and (b) the shadow cast by the particle (Fig. 1.9). Without a shadow, one cannot be certain that a deposit represents a particle protruding from a fracture face. The shape of the shadow also provides information about particle architecture. Indeed, Weinstein et al. [98], have used this information to show that some of the particles in erythrocyte membranes are spherical and some are ellipsoids; none appear to be cylinders. The two fracture faces of biomembranes show dissimilar particle concentrations. In erythrocytes, for example, the faces of the juxtacytoplasmic membrane lamellae bear ~5 times as many particles as the faces of the outer layers.

There is substantial evidence that the intramembranous particles represent protein or protein complexes penetrating into and/or through the membranes:

(a) The particles are observed on the fracture faces of artificial lipid bilayers containing protein, but not those lacking protein (e.g. [99]). Thus the fracture faces of bilayers made with lipids of retinal rod outer segments lack particles, but the native membranes, as well as lipid bilayers with incorporated rhodopsin, show intramembranous particles [99]. It appears, therefore, that in rod outer segments, whose dominant protein is rhodopsin, the intramembranous particles represent rhodopsin or rhodopsin complexes.

Fig. 1.9. Fracture face of a freeze-cleaved plasma membrane. This electron micrograph, obtained from a human, transitional cell bladder carcinoma, shows membrane-associated particles varying in both size and shape. The direction of replication is from the bottom. Accumulation of carbon-platinum on the particle perimeters facing the source of replicating material appear black. Regions devoid of replicating material, that is, 'shadows', appear white. Magnification: 250,000 × (Courtesy of Dr. R. Weistein).

(b) Myelin, which contains little protein, most of it ionically associated with lipid head groups, lack intramembranous particles [100].
(c) When erythrocyte ghosts are treated with pan-proteases, one observes a convincing correlation between protein loss and the decrement in the number intramembranous particles.
(d) In the case of nuclear membranes, or the plasma membranes of some micro-organisms, one can experimentally redistribute intramembranous particles prior to freeze fracture. This particle redistribution can be blocked by

glutaraldehyde, a bifunctional reagent which combines with membrane proteins but not lipids.

(e) In the membranes of human erythrocytes, one observes a reasonable correlation between the distribution of intramembranous particles and that of 'glycophorin', which penetrates into (through) the membrane core. Surface labelling studies that employ ferritin-conjugated ligands for the glycoprotein receptors, in conjunction with freeze-fracturing, indicate that the receptors share the same distribution as the intramembranous particles [100]. Some data even suggest a one-to-one relationship between the erythrocyte intramembranous particles and the major ghost glycoprotein. However, the *penetrating* portion of glycophorin (mol. wt. $\sim 10^4$ [39]) is too small to appear as an intramembranous particle, and Guidotti [101] suggests that the particles represent multiprotein complexes, including the major glycoproteins.

(f) When one corrects for the contribution of the replication process to particle dimension [100], one arrives at a true diameter of ~5.0 nm for a typical intramembranous particle. This would correspond to a globular protein with an approximate molecular weight of $\sim 10^5$ daltons, a value representative of most erythrocyte membrane proteins [4]. However, observations on several membranes [102,103] including erythrocyte ghosts [104], indicate that some membrane proteins are encased by a shell of 2–3 molecular layers of tightly associated lipids. This would increase the true particle radius to 8–9 nm.

Intramembranous particles are generally described as uniform in shape, dimension (7.5–8.5 nm) and distribution. However, careful analyses show that particle distribution is often not statistical and that there can be large variations of particle size (4–13 nm in lymphoid cells). Even in erythrocytes the size variation is appreciable, although the majority of particles range between 7.5–8.5 nm. In human erythrocytes intramembranous particles occupy about 8% of the surface area. The particle number (10^6/cell) fits the amount of major glycoprotein/cell. Typical particle concentrations are $\sim 4 \times 10^{17} \cdot cm^{-3}$ (3×10^{11} particles/cm^2 membrane surface, 8×10^{-7} cm membrane thickness; [100]). In discussing particle size, it is important to note that only structures which have a diameter of >3.0 nm *and* which protrude sufficiently from the fracture face to cast a detectable shadow during replication (2 nm) can be visualized. To detect an intramembranous protein by freeze-fracture electron microscopy thus requires (a) that it be sufficiently large and (b) that it extend well into the membrane core. Nonparticulate areas *are thus not necessarily protein free* [100].

Particle aggregation. If the intramembranous particles represent proteins penetrating through the membrane and these penetrating proteins are free to diffuse laterally within the membrane plane, this mobility should be reflected in the distribution of the particles under various conditions.

If we assume that a membrane constitutes a two-dimensional dispersion of protein particles in a lipid continuum, the particle concentrations observed, together with the values for protein diffusion coefficients presented earlier, create interesting implications in terms of colloidal theory. Thus, a 'particle dispersion' as concentrated as a membrane, tends to be unstable because the high total free energy – due to the large interfacial area between the dispersed phase (particles) and continuous phase (lipid) – can be decreased by aggregation of the dispersed phase. The interfacial energy will be positive and aggregation will occur when forces of attraction between the particles exceed repulsion forces, that is, when the collisions are inelastic.

If n_0 is the particle concentration at time, t, $= 0$, and if $t_{\frac{1}{2}}$ is the time required to reduce the particle concentration to half of n_0, i.e. $n = n_0/2$ at $t = t_{\frac{1}{2}}$,

$$t_{\frac{1}{2}} = 1/(4\pi D_p R n_o) \tag{1.9}$$

where n_0 and n are particles/cm^3, D_p is the diffusion coefficient in $cm^2 \cdot sec^{-1}$, and R the collision radius in cm (approximately double the particle radius).

To estimate the theoretical $t_{\frac{1}{2}}$ for intramembranous particles we use the following numerical values:

$$D_p = 3 \times 10^{-10}\ cm^2 \cdot sec^{-1}$$

$$R = 2 \times 10^{-7}\ cm$$

$$n_0 = 4 \times 10^{17} \cdot cm^{-3}$$

Then

$$t_{\frac{1}{2}} \sim 3 \times 10^{-3}\ sec$$

that is, if the membranes were *unstabilized* dispersed systems (where every collision is *inelastic*), particle aggregation should occur very rapidly and one should not observe the typical fracture face appearance, where most particles are *unaggregated.*

A $t_{\frac{1}{2}}$ of 3×10^{-3} sec at 25 °C also indicates that unstabilized particles could readily aggregate during the freezing step of freeze-fracture microscopy. Thus, assuming an initial temperature of 30 °C, maximal cooling rate of 1000 °C/sec and particle immobilization when temperature reaches -10 °C, the 40 °C temperature drop would be achieved in $\sim 4 \times 10^{-2}$ sec, that is, $15 \times t_{\frac{1}{2}}$.

Two possibilities must be considered in attempts to explain the lack of aggregation of intramembranous particles.

(a) The particles are stabilized against aggregation by electrostatic and/or other repulsive forces. These can be estimated from:

$$t_{\frac{1}{2}} = (1/4\pi D R n_0) e^{W/RT} \tag{1.10}$$

where W is the net stabilization energy, R is the gas constant (1.986 cal $°K^{-1} \cdot mol^{-1}$) and T is the temperature in °K.

To increase $t_{\frac{1}{2}}$ from the calculated 3×10^{-3} sec to 10^5 sec (~1 day) would require an energy about 17 kcal/mol^{-1} at 37°C. This is a substantial value but can be realized through electrostatic repulsions in media of low dielectric constant such as the hypothicated lipid phase.

(b) The particles lack translational mobility and do not represent a disperse system in a fluid phase, as hypothisized.

Another point to be considered is the following: if intramembranous particles represent proteins that are dispersed in a lipid phase and undergo rapid elastic collisions, they should exhibit a random distribution. Many authors claim 'random distributions' but do not support this conclusion by statistical analyses. In fact, this point has only recently been tested experimentally by Weinstein [105] who finds a *non-random* particle distribution in the *plasma membrane* fracture faces of a number of animal cell types. Unfortunately, equivalent analyses do not yet exist for cytoplasmic membranes of animal cells or plasma membranes of microbial cells.

The information that *is* available is *not* consistent with the hypothesis that the *plasma membranes* of eukaryotic cells constitute two dimensional fluid lipid phases in which penetrating proteins can undergo free lateral diffusion. However, there is evidence that this state may exist to some extent in certain intracellular membranes as well as the membranes of micro-organisms, and that certain manipulations can induce lateral motion of intramembranous particles in plasma membranes of eukaryotic cells.

Effects of temperature. In the plasma membranes of *Acholeplasma laidlawii* [106,107] and *E. coli* [108] as well as in the alveolar and nuclear membranes of *Tetrahymena pyriformis* [109,110] and the nuclear membranes of diverse lymphocytes [111], chilling below a specific temperature induces the formation of large, particle-depleted areas on the fracture faces. This process can be reversed by heating the cells back to above the critical temperature. Glutaraldehyde-fixation prior to chilling prevents appearance of smooth areas and fixation after chilling blocks return to normal particle topology by reheating.

In the case of *Acholeplasma laidlawii* and *E. coli*, the emergence of smooth areas is accompanied by lateral segregation and massive aggregation of intramembranous particles. In *Tetrahymena* one also observes lateral segregation and aggregation, but, in addition, a decrease in particle frequency (i.e. some particles are displaced perpendicularly from the fracture faces). In lymphocyte nuclear membranes, one observes no particle aggregation – only a decrease in particle number and in mean particle diameter.

These phenomena can be reasonably attributed to reversible gel → liquid-crystalline transitions of membrane lipids. Chilling below the transition temperature crystallizes the lipids' hydrocarbon chains. Membrane proteins – that is, intramembranous particles – are excluded from the ordered lipid domains and are displaced laterally into still fluid regions producing particle aggregation. The extent of packing will depend on the magnitudes of the lattice forces in the ordered lipid regions relative to the repulsive forces between particles. The disappearance of particles from the fracture faces and reduction in particle size can be attributed to protein extrusion from ordered lipid in a direction perpendicular to the membrane plane, and may represent instances where repulsions between particles are too great to allow aggregation in the membrane plane.

Significantly, the nuclear pore complexes of nuclear membranes do not redistribute with chilling. This suggests that their location and distribution is stabilized by a framework within, or adjacent to, the membranes. Moreover, the temperature-induced nuclear membrane changes can be prevented by glutaraldehyde treatment. This reagent crosslinks proteins (span $<10^{-7}$ cm) but does not combine with lipids. Since the membrane particles lie $\lessapprox 10^{-6}$ cm apart, the stabilization induced by glutaraldehyde cannot be attributed to cross-linking *between* particles. It *can* be explained if the particles also lie near a *framework*, not visualized by freeze-fracture electron microscopy, and become linked thereto by glutaraldehyde.

The *plasma membranes* of lymphocytes, or *Tetrahymena* exhibit the same freeze fracture pattern at all temperatures, both above and below the transition temperatures of the nuclear membranes. Erythrocyte membranes also fail to show thermotropic changes of fracture face topology. Two (non-exclusive) explanations must be considered.

First, the plasma membrane lipids do not undergo a defined phase transition upon chilling. This may relate wholly or in part to the high cholesterol: phospholipid ratio in plasma membranes (~1.0, [112] versus ~0.10 in nuclear membranes [113]). *Second*, phase transitions do occur, but relocation of the particles is prevented by their firm association with an unvisualized framework; this would be analogous to the case of the glutaraldehyde fixed nuclear membranes.*,**

Effects of pH. A study by Pinto da Silva [114] suggests that electrostatic repulsions may play an important role in preventing aggregation of intramembranous particles of erythrocyte ghosts. These membranes exhibit a uniform (not *random*)

---

* Freeze-fracture electron microscopy of biomembranes reveals disorder–order transitions only when these cause particle displacement. The detection sensitivity for particle displacement (formation of smooth areas) depends on the minimum distance between particles. The smaller the mean inter-particle distance, the smaller the smooth area which can be detected. Taking a diameter of 80 nm for a recognizable smooth area, and assuming that it represents ordered lipid only, one can detect disorder–order transitions in clusters of $\sim 2 \times 10^4$ phosphatide molecules (taking a surface area projection of 5×10^{-1} nm per packed phospholipid).

** Raman-spectroscopic studies show that the acyl chains of thymocyte plasma membranes actually *do* undergo a phase transition near 20 °C, but that this transition is relatively non-cooperative [24].

particle distribution at pH 7.5 or above, but lowering pH to 5.5–5.0 produced large particle clusters with intervening particle-free zones. This particle aggregation was reversible and could be blocked by glutaraldehyde fixation. The observations appear to relate to the extraction *loss of certain membrane proteins* from the erythrocyte *ghosts* [115].

Effect of cryoprotectants. McIntyre et al. [116] find that when they expose lymphoid cells to 25% glycerol or dimethyl sulfoxide at 0 °C prior to fixation, they obtain particle aggregation in the plasma membranes. This effect is reversible, occurs within minutes, and can be blocked by prefixing with glutaraldehyde. Chilling alone does not induce aggregation. No extensive smooth patches appear and there appears to be no decrease in the number or size of particles as seen in cooled lymphocyte nuclei.

Proteolysis following melittin or lysolecithin. Simple trypsinization of sheep erythrocyte ghosts does not aggregate their intramembranous particles. However, trypsinization *after* treatment of the membranes with lysolecithin or the amphipathic polypeptide melittin, produces massive particle aggregation [117]. Neither of these membrane-active agents alone produces particle clumping. It appears that the intramembranous particles of sheep erythrocyte membranes cannot move laterally, unless the membrane lipids are fluidized *and* the membrane proteins mobilized by trypsinization. The trypsin effect could represent a reduction of electrostatic repulsions between proteins or a breakdown of a protein framework.

The effects of bifunctional reagents. In those membranes (e.g. [106–116]) where intramembranous particles can be caused to aggregate by chilling, and to redisperse by warming, particle coagulation can be blocked by glutaraldehyde fixation in the dispersed state, and redispersion by fixing in the aggregated state. The same applies to particle aggregation induced in erythrocyte ghosts by various mechanisms. Particle distribution can thus be spatially fixed by a short range cross-linking reagent.

As noted, the fluid membrane model requires that the particles undergo collisions at an extremely high frequency. In the presence of a rapidly-reacting cross-linking reagent, these should lead to massive particle aggregation; however, this is not observed. This fact implies that intramembranous particles are often restrained from undergoing the translational motion by as yet undefined stabilization mechanisms.

1.1.6.3. Significance of lateral translational motion

The translational motions discussed involve large distances in molecular terms (i.e. $>10^{-6}$ cm), and no information available proves that membrane proteins undergo such motion under physiologic conditions. Although lateral particle translation can be reversibly induced by chilling in the nuclear membranes of some eukaryotic cells and the surface membrane of some micro-organisms, collision theory and the effects of bifunctional agents suggest that lateral diffusion of proteins in these membranes may lack major physiological significance in these cases also. (This is not intended to exclude possible biologically significant lateral motion of non-visualized membrane proteins).

While there is some question, therefore, whether lateral translation actually occurs in biomembranes, Keith and Snipes [120] present evidence that such motion might have an important biological role. Using spin-label techniques they show that in cells with extensive internal membrane structures, the aqueous spaces exhibit a 'high local viscosity' and argue that diffusion tangential to the membrane plane on or/within membranes might be much more rapid than diffusion through the bulk cytoplasm. However, the maximal cytoplasmic microviscosity they report ($\sim$2 gm · cm^{-1} · sec^{-1}, 20 °C) lies in the low range of values reported for biomembranes [76]!

A number of other functions might be suggested. These include regulation of enzyme function by association–dissociation phenomena, lateral information transfer, the formation of induction gradients in development, spatial conservation of macromolecules. However, we lack specific information on any of these points.

1.1.7. Rotational motion of proteins

1.1.7.1. Theory and measurement

Certain membrane proteins can rotate rapidly about an axis normal to the membrane plane. In the case of proteins penetrating through a lipid layer, this rotational diffusion can be related to the hopping frequency, ν_L, of the lipid molecules using the model of Sackmann et al. [63,75]. In this model, a cylindrical protein of radius, R, rotates around an axis perpendicular to the membrane plane. The protein perimeter is not perfectly circular and bears at least one protrusion a few Å high (either due to amino acid side chains, attached lipid, or both). The protein can only rotate when a lipid vacancy occurs in the neighborhood of the protrusion. The rotational motion of the protein then proceeds in a series of angular steps, as diagrammed in Fig. 1.10.

A protein with mol. wt. $\sim$100,000, a radius, R, near 2.5 nm, embedded in a typical phospholipid bilayer with a mean distance between lipids $\sim 10^{-7}$ cm and a typical lipid hopping frequency, ν_L, $\sim 10^7$ sec^{-1} will exhibit rotational correlation times of 50–75 μsec. Lower values might occur with unusually high hopping frequencies, ν_L (e.g. highly unsaturated lipids).

The model gives virtually the same results if the protein bears more than one

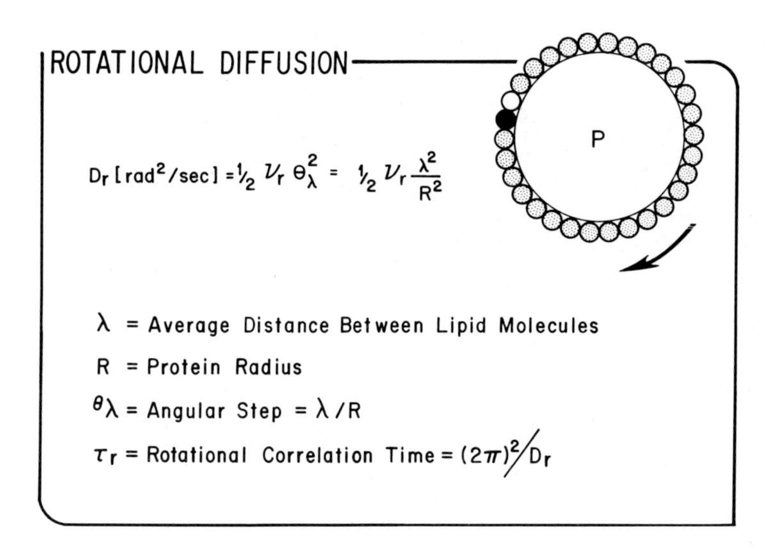

Fig. 1.10. Theoretical model for rotational diffusion of proteins within a lipid bilayer. After [75].

surface protrusion. This is because the relaxation time of free rotation of a globular protein with mol. wt $\sim 10^5$ daltons is $\sim 10^{-9}$ sec. Accordingly, with two protrusions, two lipid voids need to appear simultaneously for only $\sim 10^{-9}$ sec to allow a rotational step. This time is short compared to the lifetime of a lipid void, $\sim 10^{-7}$ sec.

Rotational diffusion of membrane proteins has been measured spectroscopically, using the photoselection process: linearly polarized light is preferentially absorbed by chromophores whose transition moments for absorption are parallel to, or nearly parallel to the electric vector of the light. Thus, if a sample is irradiated with linearly polarized light, and the polarization plane is rotated, absorption will be greater at some polarization angles than at others *if* there is a population of chromophores in some preferential orientations. If there is no preferential orientation, there is no preferential absorption.

In steady state methods, polarization, p, is defined by

$$p = \frac{A_{\parallel} - A_{\perp}}{A_{\parallel} + A_{\perp}} \tag{1.11}$$

where $A_{\parallel}$ and $A_{\perp}$ are the optical absorptions for parallel and perpendicular polarization of the exciting light. For fluorescent molecules where absorption leads to emission of fluorescence, that is, $A \propto I$, the fluorescence intensity,

$$p = \frac{I_{\parallel} - I_{\perp}}{I_{\parallel} + I_{\perp}} \tag{1.12}$$

and r, the molecular anistropy, is

$$r = \frac{I_{\parallel} - I_{\perp}}{I_{\parallel} + 2I_{\perp}} \tag{1.13}$$

where $I_{\parallel}$ and $I_{\perp}$ are the fluorescence emission intensities observed through a polarizer oriented parallel and perpendicular to the polarization plane of the incident light and $(I_{\parallel} + 2I_{\perp})$ = total fluorescence. For a rotating fluorescent sphere

$$r_0/r = (1/p_0 - 1/3)/(1 + 3\tau/\rho_0) \tag{1.14}$$

where r_0 and p_0 are the values of r and p when the fluorphores do not change orientation between excitation and emission, τ is the lifetime of the excited state and ρ_0 is the rotational relaxation time (average time of rotation through angle $\cos^{-1} \cdot 1/e$). Since ρ_0 for a rigid sphere of radius, R, is

$$\rho_0 = \frac{4\pi\eta R^3}{kT} \tag{1.15}$$

where η is the viscosity, k = Boltzmann's constant and T is in °K. Therefore, knowing τ one can estimate ρ_0 from Perrin plots, that is, $(1/p - 1/3)$ vs. T/η.

More recently pulse methods have been employed. If we define $r(t)$ the absorption anistropy at a time, t, after an exciting pulse by

$$r(t) = \frac{A_{\parallel}(t) - A_{\perp}(t)}{A_{\parallel}(t) + 2A_{\parallel}(t)} \tag{1.16}$$

then, for a rigid spherical molecule, the decay of dichroism is given by

$$r(t) = r_0 \exp. (-t/\rho_0) \tag{1.17}$$

where r_0 is the dichroism at $t = 0$.

The first evidence for rapid rotational diffusion in a membrane protein was published in 1959 by Hagins and Jennings [121]. These authors [121] measured the photo-dichroism [122] of the rhodopsin in retinal rod outer segment (ROS) membranes and demonstrated thereby that the long axis of the rhodopsin chromphore lies parallel to the membrane plane. However, they detected no dichroism when they viewed the ROS in a direction perpendicular to the membrane discs, indicating that the molecules lacked a preferred orientation in this direction. Hagins and Jennings [121] could also not induce dichroism by flash bleaching, and concluded that the rhodopsin rotates two rapidly to allow detection of dichroism within the 100 ms time resolution of their technique.

Brown [79] has recently confirmed that the rhodopsin in ROS lacks persistent photo-dichroism, but demonstrated that such can be induced by glutaraldehyde, This suggests that rhodopsin normally undergoes rapid rotational Brownian motion, and that this is blocked when glutaraldehyde couples the rhodopsin molecules to each other or to some other membrane component.

Coincidentally. Cone [76] using 5 nsec pulses of polarized neon laser light (540 nm), demonstrated that rhodopsin absorption can be *transiently* polarized. His measurements yield a rotational relaxation time of 20 μsec at 20 °C, suggesting a membrane viscosity of $\sim$2 gm $\cdot$ cm^{-1} $\cdot$ sec^{-1}. The rotation decreases sharply with temperature and, in agreement with [79], is abolished by glutaraldehyde fixation.

Similar measurements by Junge [123] demonstrate rapid rotational Brownian motion also for the cytochrome a_3–CO complex in inner mitochondrial membranes (rotational relaxation time $>$100 μsec).

In contrast to these examples, chlorophyll a_1 of chloroplasts (probably a protein complex) exhibits a photo-induced dichroism [124], which persists for at least 100 msec. This fact suggests either that chlorophyll a_1 is relatively immobile or that it rotates only around a single axis normal to the plane of the membrane and the porphyrin ring.

The purple pigment protein of *Halobacterium halobium* membranes also exhibits very slow rotational diffusion [125]. Although pulse irradiation reveals transient photodichroism in these membranes, the lower limit for the rotational relaxation rate lies near 20 msec (1000 times that observed for rhodopsin). Apparently, the purple protein is strongly immobilized in its membrane [125] as is also suggested by X-ray diffraction experiments [126].

Axonal membranes also appear to contain highly immobilized domains. This has been demonstrated by Tasaki et al. [127], who measured the fluorescence polarization of 2-*p*-toluidinylnaphthalene-6-sulfonate (TNS) bound to perfused squid giant axons. The intensity of TNS fluorescence changes measurably during nerve excitation. TNS binds hydrophobically to both proteins and lipids. If the membrane proteins and/or lipids that bind TNS and alter its fluorescence during excitation were highly mobile, one would expect no permanent polarization of TNS fluorescence. However, Tasaki et al. [127] find that the probes that change fluorescence during axon conduction lie with their absorption and emission dipoles oriented parallel to the long axis of the axon.

In summary, one cannot make generalizations about the rotational diffusion of membrane proteins: some membrane proteins are highly mobile, while others appear highly restrained. In addition, interpretation of measurable rotational relaxation times is complex. *First*, irregular molecules do not have a single relaxation time. *Second*, the structure of the membrane in the vicinity of a fluorescent or absorbing molecule can introduce unrecognized elements of anisotropy. *Third*, the molecules observed spectroscopically represent only a small part of the protein to which they are attached. Thus when rotation is observed, this may include rotation of the chromophore about an attachment site and/or rotation of a protein subunit.

1.1.7.2. Significance of rotational motion

The observation that certain membrane proteins possess free rotational mobility and others do not clearly has considerable implication for membrane structure. But what might be the functional significance of rotational diffusion? An obvious possibility is membrane transport. Clearly, transport proteins must at some times have access to the external and internal surfaces of membranes. Indeed, Cone [76] has suggested that proteins tumbling with a rotational relaxation time similar to that of rhodopsin could transport up to 10^4 molecules sec^{-1} by rotational diffusion.

However, rhodopsin spins about an axis *perpendicular* to the membrane plane and we know of no experiments showing specifically that membrane proteins can rotate around axes *parallel* to the membrane plane, as might be required for transport. Indeed, it is generally assumed that the limiting element for such rotation may be the energy required to transfer the hydrophilic groups of a hypothetical penetrating protein through the apolar membrane core. It is clear that a protein such as the 'glycophorin' which contains a substantial carbohydrate moiety located on the outside of the cell surface, would have a negligibly slow rate of rotation across the membrane. However, it is not reasonable to extend this argument to all, or most membrane proteins [5] because it depends on the assumption all membrane proteins are invariant amphipathic structures which present substantial polar regions to the aqueous phase. However, we have already pointed out that the architecture of a protein in an aqueous milieux is not likely to be identical to that of the same macromolecule in an apolar medium. Similarly, a membrane protein with polar groups clustered at the membrane–water interface might restructure during rotation, burying the previously exposed polar groups. Indeed, the structure of a transport protein might be such that all but a small part of its perimeter is hydrophobic. Such a protein could be immersed in the membrane and have only that portion interacting with the aqueous phase that is necessary for substrate binding. In addition, substrate binding might also induce a structural change which could render the protein even more hydrophobic and hence more free to rotate.

This issue can be put into perspective by comparing the polar and apolar volumes of various proteins with those of phosphatides, using the polarity-index criterion. As noted (p. 23) the ratio (polar volume: apolar volume) of rhodopsin is 0.65. We have evaluated this ratio for phosphatidylcholine, using the approach of Rothman and Engelman [128]; taking the glycerylphosphorylcholine-ester segment to represent the polar moiety and the hydrocarbon chains as the apolar portion, we find a ratio of 0.6. But phosphatidylcholine in *biomembranes* shows a 'flip-flop' rate with $t_{\frac{1}{2}}$ of ~ 1 min, even though this phospholipid is a very asymmetric molecule. We therefore conclude that the energy barrier to transverse rotation of a protein such as rhodopsin need not be significantly greater than that involved in phospholipid 'flip-flop'.

Bretscher [129] has argued that proteins which rotate across the membrane should be equally accessible to chemical reagents or proteolytic enzymes on both sides of the membrane. However, most of the major erythrocyte proteins distinguishable electrophoretically are inaccessible in intact cells, but accessible in ghosts, suggesting that they are located only on the cytoplasmic side of the membrane.

The distribution of proteins across the membrane thickness has been recently reviewed (e.g. [4]), and there is no doubt that many membranes exhibit a transverse asymmetry. This is apparent from X-ray diffraction studies (e.g. on myelin), from membrane inversion studies, from the differential susceptibilities of external and internal surfaces to proteases, phospholipases and covalent labelling procedures, from the asymmetric images obtained by freeze-etch electron microscopy, from studies on the activities of membrane bound enzymes, etc. However, some major issues remain unresolved. For example: Do the observed asymmetries reflect the reactivities of the various membrane proteins at the two membrane faces? Do they indicate differences of location, or both? Is the protein disposition static?

But even if Bretscher's interpretation is correct, available data only apply to the *major proteins of the erythrocyte membrane.* But, membrane transport proteins are very minor components and are not detected by the methods used to evaluate protein disposition.

Rotational diffusion need not be a *requirement* for transport. As first pointed out by Wallach and Zahler [18], transport could proceed via specific channels running within membrane proteins from one surface to the other. A theoretical exposition of ion transport by such a mechanism has been presented by Onsager [130] and several relevant model systems involving cyclic polypeptides have been described [131].

1.1.8. Translation perpendicular to the membrane plane

The concomitant appearance of smooth faces and decrease in particle concentration on the fracture faces of nuclear membranes, after cooling intact lymphocytes (and *Tetrahymena*) below a critical temperature, suggest that gel → liquid–crystalline transitions can extrude particles (partially or completely) from the fracture faces by an asymmetric translation, normal to the membrane plane.

While lateral translation of membrane proteins has been much discussed, the possible significance of motion normal to the membrane plane has not been evaluated. However, simple geometric considerations show that the motion of a spherical or ellipsoidal particle normal to the membrane plane can produce a change in the apparent size of the particle on a fracture face.

Thus, a spherical particle with a diameter 5×10^{-7} cm will cast a recognizable shadow 1.5×10^{-7} cm long if it protrudes 5×10^{-8} cm (5 Å) from the fracture face. The maximum diameter (uncoated) of the protruding portion will be 3×10^{-7} cm. If the particle moves out of the fracture plane by a few Å, it will no longer be recognized. In contrast, protrusion into the fracture face by an additional 5×10^{-8} cm (10 Å total) will yield a diameter of 4×10^{-7} cm. For a larger spherical particle, equivalently located, a 5×10^{-8} cm translation would produce a still larger change of apparent diameter ($3.5–5.0 \times 10^{-7}$ cm for a 7×10^{-7} cm diameter particle). An ellipsoid would show a smaller dependence of measured diameter on vertical displacement. Accordingly, the wide variation of particle size usually observed on fracture faces may not merely represent the presence of proteins of different size, but could also reflect the *trans-membrane* distribution of membrane proteins. A freeze-fracture image might then represent a 'snapshot' of transmembrane protein distribution at a given instant.

A simple extension of this reasoning shows that for ellipsoidal particles, with their short axes parallel to the membrane plane, *rotation about this axis* would lead to alterations of apparent particle size. In other words, variation of particle size on a fracture face might be a reflection of *rotational motion about an axis parallel to the membrane plane.*

1.1.9. Transmembrane alterations of membrane protein organization

Nicolson and Painter [132] have shown that perturbations of the internal surface of erythrocyte ghosts can be transmitted to the external surface, presumably via the major glycoprotein which spans the membrane thickness. These experimenters produced antibody to the 'spectrin' components of erythrocyte membranes, proteins which lie at the cytoplasmic surface. When erythrocytes are lysed osmotically in the presence of 'antispectrin', the antibody becomes sequestered within the ghosts, binds to the spectrin and cross-links this. This results in aggregation of the major glycoprotein within the membrane plane, manifested by a clustering of its carbohydrate moieties on the external surface; this redistribution can be demonstrated electronmicroscopically by use of ferritin-labelled ligands for the carbohydrate receptors.

Ji and Nicolson [133] have also shown the reverse, that is, perturbation of components located on the external surfaces of erythrocyte ghosts can be transmitted to the internal surface. In these experiments erythrocyte ghosts were reacted with castor bean lectin (Chapter 10). This agglutinin combines with the carbohydrate moiety of a membrane glycoprotein, probably not 'glycophorin'. As a result, several erythrocyte membrane proteins defined by SDS-polyacrylamide electrophoresis become more susceptible to cross-linking by the bifunctional reagent, dimethyl-malonimidate. The 'spectrin' components of the ghosts become particularly susceptible. Since castor bean lectin does not penetrate to the ghost interior and spectrin does not extend to the membrane exterior, the data can be taken as another example of a trans-membrane linkage.

1.2. Membrane models

The membrane literature contains a profusion of membrane models. Most of these address membrane structure rather than function. It is doubtful that any single model can apply to all membranes, to all parts of a given membrane or to all states of a membrane system.

1.2.1. Structural models

1.2.1.1. The paucimolecular model

The 'paucimolecular' membrane model of Danielli and Davson [134] proposes the following: Biomembranes consist of bilayers of phosphatides (and associated lipids). The polar lipid headgroups lie at the membrane surfaces and their hydrocarbon chains constitude an internal apolar lamella. Membrane proteins lie at the polar surfaces and interact with the lipids primarily through ionic and/or hydrogen bonds. As pointed out elsewhere [4], the experimentation which originally led to this model is consistent with many other hypotheses. Moreover, most current work on model bilayers relies on inference and correlation. Nevertheless, it appears highly probably that phospholipid bilayers exist in some form in many biomembranes. Also, membrane phosphatides ($\pm$ other membrane lipids) can spontaneously form such structures, and recent X-ray diffraction data convincingly demonstrate the existance of bilayer structures in multilamellar biomembranes.

The great weakness of the Danielli–Davson hypothesis (and related models) is its failure to describe the disposition of membrane proteins and the interactions of membrane lipids with membrane proteins.

1.2.1.2. Mosaic models

1.2.1.2.1. The hypothesis of Parpart and Ballentine

Although Danielli and Davson presented the first detailed membrane model in 1935, a lipid–protein mosaic structure had been proposed by Nathanson [135] already in 1904. However, the first detailed mosaic hypothesis was presented by Parpart and Ballentine in 1952 [136]. The authors developed this model in order to account for the state of proteins in the erythrocyte membrane. They proposed that the membrane consists of *protein continuum.* The protein continuum extends through the membrane thickness but is interrupted by cylindrical channels 50 Å in diameter with hydrated walls. The channels are filled with cylindrical lipid plugs. The head groups of the amphipathic lipids orient toward the hydrated walls of the channels and their hydrocarbon chains form an apolar core. The authors propose strong polar associations between phosphatide headgroups and protein and weak apolar associations of cholesterol and some other lipids with the apolar residues of the strongly bound phosphatides.

Present information lends little support for the details of the Parpart–Ballentine model. However, there is increasing evidence that significant proportions of membrane protein penetrate into or through the membrane core. Parpart and Ballentine [136] do not envisage apolar lipid–protein associations, but these play a major role in modern mosaic hypotheses.

1.2.1.2.2. Lipid–globular protein mosaic models (*LGPM*)

Modern mosaic models have been proposed by Wallach and associates [4,12, 15,18] and Singer and associates [137,138]. Both groups conclude that some proteins penetrate into or through the membrane core. Both suggest that the penetrating portions of these proteins bear apolar residues which form hydrophobic associations with hydrocarbon chains of membrane lipids. Wallach and associates [4,12,15,18] have described the LPGM as follows:

Membranes contain domains, assembled as tangentially mobile patchworks, penetrated to varying depths by both protein and lipid. Contact between protein subunits and between protein and lipid occurs in both polar and apolar regions of the membrane. Although in a bilayer array, the lipid adjacent to the protein has a composition and organization determined by the surrounding protein, the biologic 'ordering' influence of the protein falling off with distance.

Membrane peptide lies on both of the membrane surfaces and also within the apolar core of the membrane. Surface-located peptide may be coiled irregularly, but other conformations may exist. Penetrating protein segments may be predominantly helical rods, each with opposite hydrophobic and hydrophilic faces, packed to form subunit assemblies, whose perimeters bear apolar aminoacid residues. Helicity is not an absolute requirement; unordered or β-structured peptide segments with suitable amino acid sequence and properly arrayed could also provide the apolar perimeters required of penetrating protein segments. High protein hydrophobicity is not a requirement and lipid–protein interactions are not exclusively hydrophobic.

The apolar amino acid residues, which form the perimeters of the subunits, are relatively specific binding sites for the hydrocarbon residues of tightly retained membrane lipids. Polar lipid head groups can also participate in polar associations with surface-located protein side chains. The association of tightly bound lipids with membrane proteins is analogous to the apolar hemeprotein interactions in hemoglobin. Lipids, lying more distant from the protein, for example, cholesterol, are bound less tightly and less specifically. The axes of the protein subunits lie normal or nearly normal to the membrane surfaces. The distribution of *polar* amino acids in the protein is conjectured to be such that their side chains lie at the membrane surfaces or cluster round the central axis of each subunit, or both, possibly forming apolar channels penetrating through the membrane. The permeability of the hypothicated channels would be highly sensitive to the conformational state of the protein.

Wallach et al. [139] have recently extended the original Wallach–Zahler hypothesis in a synthesis of their new spectroscopic data on erythrocyte- and thymocyte-plasma membranes. Their evidence indicates that much of the membrane cholesterol lies in *clusters* (cf. also Chapter 5), that much of membrane phospholipid is *protein-bound* (cf. also Chapter 3) and that appreciable membrane phospholipid lies in microdomains, where the lipid can undergo *cooperative* transitions of state. The assembly of the lipoprotein domains is constrained and stabilized by a protein framework within the membrane.

The listed mosaic hypotheses, like all other membrane models, must be viewed with some reserve. However, three predictions of the model of Wallach and associates [4,12,15,18,139] have been born out: (a) proteins in several membranes are surrounded by shells or lipid in apolar association with the proteins (e.g. 102–104), and, in at least one membrane [104,139], these lipid layers can be perturbed by subtle protein modifications; (b) amphipathic helices have been detected in soluble and membrane lipoproteins; (c) there is evidence that association of amphipathic helices can create polar trans-membrane channels; (d) cooperative lipid phase transitions can occur in erythrocyte and thymocyte plasma membranes.

1.2.1.2.3. The fluid lipid–globular protein mosaic model (FLGPM)

This hypothesis, first proposed by Singer and Nicolson [5], is closely related to the LGPM. It suggests that the matrix of membranes is not stabilized by protein–protein interactions (whether within the membrane or through a cytoskeleton). Instead, membrane lipid is proposed to form the membrane continuum. Since membrane phosphatides tend to have fluid hydrocarbon chains at physiologic temperatures, the continuum would be fluid. Thus, membrane lipids and proteins could undergo free translational diffusion in the membrane plane. Except for short range interactions, the protein distribution parallel to the membrane plane would be random.

The 'fluid mosaic' hypothesis attempts to integrate the behavior of pure lipid systems (described above and in Chapters 3 and 5) with the membrane transposition phenomena describe earlier in this chapter. Indubitably, some membranes contain fluid lipid domains and some membrane proteins exhibit rapid rotational motion. On the other hand, the plasma-membranes of animal cells contain high proportions of cholesterol. This does not engender high fluidity. Also, as we have already pointed out, some membrane proteins show remarkable little mobility (e.g. [125]). Furthermore, recent electron-microscopic evidence [109–111] strongly suggests that the lateral mobility of membrane proteins can be severely restricted. The 'fluid mosaic' hypothesis may be valid for nuclear membranes and also the membranes of some microorganisms. However, it cannot be freely generalized and many phenomena ascribed to 'membrane fluidity' most likely constitute movements of molecules *on* rather than *within* membranes. Finally, the model contains a major weakness in that it *does not allow for stable surface patterns* such as are probably required for specific cell recognition (Chapter 8); that is, the FLGPM model does not allow spatial encoding of information.

1.2.1.2.4. Cooperativity models

Most membrane models derive from structural features and static properties. In contrast, concepts of Changeux and associates [140–141], further extended by Wyman [142] focus on membrane functions and the interaction of functional components. Their dynamic model is built on the behavior of regulatory enzymes. These molecules and membranes share certain crucial properties:

(a) Both associate with specific, regulatory, structure- and function-determining 'ligands' which alter the properties of their receptors without involving the active sites in catalytic or covalent reactions. Both respond cooperatively (in state and function) to critical levels of regulator substances. Therefore, both exhibit transduction and amplifying characteristics.
(b) Regulator enzymes, and membranes differ in their symmetry properties. Enzyme activation always requires the stereospecific reaction between enzyme and activator. Specific recognition in membranes may also involve such processes (Chapter 8). However, it could also involve a specific, topologic subunit array within the membrane plane (Chapter 8). Biological responsiveness may thus depend upon structural asymmetries within the membrane.
(c) Membranes constitute plastic, condensed, non-covalently bonded arrays of diverse proteins, plus their associated lipids, separating two aqueous phases by an apolar one. They differ from other macromolecular assemblies in that they comprise virtually *unlimited, two-dimensional* multimolecular aggregates. Moreover, whereas soluble proteins are ordinarily encompassed by a symmetrical solvent environment, the *membrane* forms an unlimited barrier separating two physiochemically and metabolically dissimilar phases. Biomembranes are asymmetrical because their environments are asymmetrical.

In the Changeux–Wyman model each membrane subunit can exist in at least two interchangeable structural states. Transitions between these two states depend cooperatively upon the states of near neighbors. The subunits could be distributed irregularly over the membrane surface as *small, localized, oligomeric* clusters or form part of a *lattice* with homologous associations between neighbors. Only the second case can explain both the *graded* and *all-or-none responses* observed in membranes.

The Changeux–Wyman hypothesis focusses on membrane proteins and largely ignores membrane lipids. It has enormous heuristic value, but is deceptively general and virtually impossible to test as such. On the other hand, while cooperative state transitions have not been demonstrated for membrane *proteins* they have been documented for membrane lipids ([60]; Chapters 3–5) including plasma membrane phosphatides [139]. This is illustrated in Fig. 1.11, showing a highly cooperative thermotropic lipid state transition in thymocyte plasma membranes [143].

The fact that a highly cooperative change in lipid state can occur in a cholesterol-containing membrane, where much of the phospholipid is protein bound, must be considered highly significant. This is *not* because the state change occurs in response to a temperature change, *but* because it is so cooperative. As detailed in Chapters 3 and 8, a system capable of a cooperative thermotropic lipid-phase transition can also undergo a cooperative state transition in response to altered *ligand binding*, for example, Ca^{2+} (cf. Chapter 6). Moreover, the data in [139] indicate that a cooperative *lipid*-phase transition in biomembranes can induce a concomitant alteration in *protein* state.

Viewed from this angle the cooperative lattice hypotheses become more accessible. This is of major importance because these hypotheses are the only ones presented to date that provide a unifying explanation for the diversity of membrane anomalies observed in neoplasia. We will return to this matter in Chapters 8 and 11.

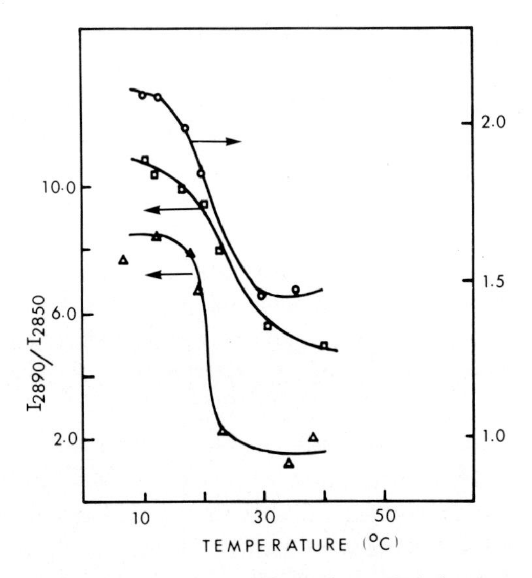

Fig. 1.11. Raman-spectroscopic evidence for a cooperative acyl chain state transition in thymocyte plasma membranes. The figure shows plots of ratio of integrals between 2860–2900 cm^{-1} (I_{2890}) and 2830–2860 cm^{-1} (I_{2850}) vs. temperature. In the case of plasma membranes I_{2890} represents the integral between 2875–2895 cm^{-1}. The points marked in the curves are an average value of four different recordings at a particular temperature. ○ = plasma membrane; □ = egg lecithin + cholesterol; △ = egg lecithin. From [143] through the courtesy of Elsevier Publishing Company.

1.3. Modified membrane organization in neoplasia

1.3.1. Micromorphology

There exists a vast body of descriptive information concerning possible anomalies of membrane micromorphology in tumors. However, Hagnenau's review [144] of the earlier literature (dealing primarily with thin section electron microscopy) leads to the conclusion that there is no single morphologic abnormality characterizing tumor cells; however, the *plasma membrane* commonly loses the differentiated character of its parental cell type, *mitochondria* become irregular in size and morphology and *endoplasmic reticulum* fragments. However, no change in micromorphology is consistently associated with malignancy.

The introduction of scanning electron microscopy [145–147] has also not proven very informative. All tumor cells examined exhibit 'extraordinary' surface activity which greatly exceeds that displayed by normal cells of an equivalent type grown under identical conditions. This activity is expressed in the numbers and lengths of microvilli, in the numbers of surface blebs, in the frequency of ruffles and other types of lamellipodia. Except for this generalization, each type of cell appears distinctive in its form and surface topology.

A number of recent publications deal with micromorphologic analyses to certain specific membrane alterations in neoplasia, that is, cell coupling and lectin reactivity. We will report on this information in Chapters 9 and 10 and will here address the limited literature concerned with overall anomalies in the membrane organization of neoplastic cells.

McNutt et al. [148] report a comparison of three cloned cell lines, untransformed Balb/c 3T3 mouse fibroblasts, an SV-40 transformant and a revertant, using thin-section electron microscopy in combination with a surface replication method. They find that microfilaments, 7 nm in diameter, called α-filaments, abound in the normal and revertant clones where they are most prominent in the small anterior pseudopodia and microvilli. The α-filaments are much less prominent in the anterior expansions of the transformed cells.

Since the α-filaments bind heavy meromyosin and thus bear a resemblance to F-actin [149], McNutt et al. [148] suggest that they may function in cellular motion.

It is possible that α-filaments actually link to intramembranous particles and one should therefore view alterations observed by McNutt et al. [148] in the light of the experiments of Nicolson and Painter [132] and Ji and Nicolson [133] on erythrocyte ghosts. As was mentioned before, these membranes bear protein filaments – spectrin – not dissimilar to the α-filaments on their internal surfaces. Moreover, perturbation of these proteins modifies the structure of the external membrane surface and vice versa. The altered numbers and distributions of α-filaments described by McNutt et al. [148] may thus relate to an altered membrane organization.

1.3.1.1. Distribution of intramembranous particles

Four recent studies by freeze-fracture electron microscopy [150–153] suggest an important difference between the general plasma membrane micromorphology of Balb/c 3T3 mouse fibroblasts (clone A31), and spontaneously as well as virally transformed variants thereof. The authors report that noncontacted 3T3 cells contain 'randomly' distributed intramembranous particles and that development of cell-to-cell contacts during the logarithmic phase of growth produces aggregation of these particles. The authors' data indicate a direct correlation between the degree of cell contact and the percentage of cells showing intramembranous particle aggregation (Table 1.1). However, transformed cells show no evidence of intramembranous particle aggregation, even at cell densities where cell-to-cell contact is extensive.

TABLE 1.1

Aggregation of intramembranous particles in the plasma membranes of normal and neoplastic mouse fibroblasts at various culture densities[1]

Cell line[2]	Proportion of fracture faces bearing particle aggregates[3]
Balb/c 3T3 (A31) (10^3 cells/cm^2)	0.25
Balb/c 3T3 (A31) (10^5 cells/cm^2)	0.75
Balb/c SVA31	0.05
Balb/c PyA31	0.05
Balb/c MSVA31	0.10
Balb/c MCA31	0.10

[1] From [150, 151].
[2] A31 = clone A31; SV = SV40 transformed; Py = polyoma transformed; MSV = transformed by murine sarcoma virus; MC = methylcholanthrene transformed.
[3] Particle density was established by evaluation of two or more fracture faces per cell (1 μm^2 area per face) on at least 40 individual cells in each category. Particle aggregation is defined as the presence of at least 10 clustered intramembranous particles.

The aggregation of intramembranous particles in transformed cells is a function of culture density and does not appear related to inhibition of locomotor activity [153]. Moreover, dibutyryl-cAMP does not restore the freeze-fracture plasma membrane morphology of diverse transformed cells to that seen in untransformed cells [154].

The possibility that the divergent clustering in normal and neoplastic cells is artefactually produced by the use of cryoprotectants [116], has been excluded by Furcht and Scott [154]. These workers show that the normal cells show aggregates whether quenched for freeze-fracture using cryoprotectants or not. If anything, the presence of glycerol tends to increase aggregate size in 3T3 cells.

In any event, transformed cells showed no particle aggregation, regardless of preparative procedures.

The authors initially proposed [150] that cell-to-cell contact of non-transformed 3T3 cells might produce a change in the distribution of intrinsic membrane proteins associated with intramembranous particles and that these changes may influence control of cell proliferation. In their second publication [151] they argue that the neoplastic cells' plasma membranes have 'fluid membrane lipids', whereas the normal cells' membrane lipids are 'ordered'. We will return to this suggestion in Section 1.3.2, but here point out that the published data provide no statistical assurance that the different clustering patterns described are actually representative of the non-transformed *versus* transformed state.

We have pointed out (Eq. 1.9, *et sequitur*) that, if intramembranous particles represent proteins dispersed in a fluid lipid matrix, they should be randomly distributed, provided that the repulsive interactions between the particles are sufficient to prevent aggregation. While it is generally stated that intramembranous particles normally exhibit a 'uniform' or 'random' distribution, statistical evaluations of particle distributions have not been carried out prior to the analyses of Weinstein [105]. This author applied the analytical procedures perfected for cluster analysis in astronomy to high resolution freeze-fracture electron micrographs, to determine whether the scattering of intramembranous particles was such as to fit a statistical distribution or not.

His data show unequivocally that, in normal cells, what is generally described as 'random' or 'uniform' distribution of intramembranous particles is, in fact,

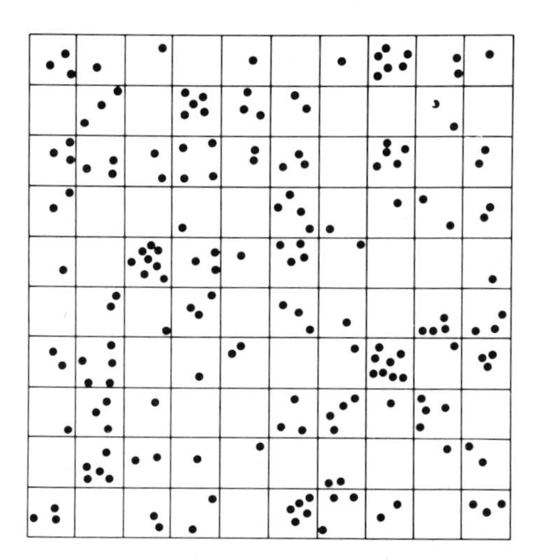

Fig. 1.12. Typical distribution of intramembranous particles on a fracture face. The distribution depicted is what is generally described as 'random' or 'uniform'. However, statistical evaluation yields a coefficient of dispersion equal to 1.74, indicating a non-random distribution. From Weinstein [105]; to be published).

not a random distribution; that is, the particles tend to cluster more than would be expected from a dispersion of particles undergoing random, elastic collisions (Fig. 1.11).

Weinstein [105] has applied his analytical procedures to a comparison of the distribution of intramembranous particles in the plasma membranes of normal human bladder epithelial cells with the distributions in the membranes of invasive bladder cancers (Table 1.2). Significantly, the normal cells' membranes show a *non-statistical distribution*, whereas those of the *invasive tumor cells exhibit* a statistical distribution of intramembranous particles. One can therefore suggest that the process of invasiveness is, in this case, associated with an altered restraint on the distribution of intramembranous particles. A possible relation of this finding to the depletion of α-filaments found in some neoplastically converted fibroblasts [148] remains to be tested.

TABLE 1.2

Altered intramembranous particle distribution during neoplastic transformation of urinary bladder epithelium[1]

Tissue	Dispersion coefficient for intramembranous particles[2]	Percentage of test fields fitting a Poisson (random) distribution ($p < 0.005$)[3]
Normal Bladder	1.450	79 %
Invasive Carcinoma	1.235	98 %

[1] From Weinstein [105].
[2] The coefficient of dispersion is defined as variance/mean. As determined by computer Monte Carlo simulation, values in this system below 0.71 indicate ordering into a lattice; 0.71–1.36 indicates 'statistical randomness' and values above 1.36 indicate particle aggregation.
[3] Chi-square test.

1.3.2. Studies on membrane 'fluidity'

1.3.2.1. Spin-label studies

Barnett et al. [151] have attempted to correlate the different distribution of intramembranous particles in the plasma membranes of normal and neoplastically converted fibroblasts with estimates of membrane 'fluidity', based on spin-label measurements. These authors exposed the various cell types to a nitroxide derivative of heptadecanoic acid and then recorded the ESR spectra after washing the cells. They observed that the molecular motions of the lipid probe were somewhat less 'ordered' in the neoplastic cells than in the normal fibroblasts, and attribute this to a greater 'fluidity' of the lipids in the neoplastic cells' plasma membranes.

Unfortunately, the authors' experiments contain serious flaws which prevent realistic interpretation of the small differences observed between normal and transformed cells and make any conclusions about plasma membrane 'fluidity' unjustified: *First*, the authors do not measure label uptake. This is essential in comparing multicompartment systems. *Second*, nitroxide lipid analogues tend to perturb and 'fluidize' their micro environment [4,65,66]. *Third*, these probes tend to seek out 'fluid' environments [67]. *Fourth*, nitroxide spin-labels are inactivated (reduced) to different degrees by different cells [155]. *Fifth*, and most important, nitroxide-labelled fatty acids bind to all kinds of biomembranes, liposomes, oil droplets, as well as diverse proteins, for example, albumin [104] and melittin [36]. Some membranes give 'fluid' signals, others do not. Fatty acid spin probes in oil droplets give 'fluid' signals, but the same probes bound to hydrophobic domains in proteins indicate 'immobilization'.

The data presented thus give no information as to the location(s) of the probes and there is no justification to assume that they are restricted to plasma membranes. A slightly greater amount of adsorbed albumin (from medium) in the *normal* cells would give the appearance of greater 'fluidity' in the *transformed* cells. An increased oil droplet content (common in many neoplastic cells) would also give the appearance of greater membrane 'fluidity'. Differential uptake of the spin probes by the two cell categories together with different reduction of the free radicals (e.g. secondary to enhanced aerobic glycolysis in neoplastic cells) further confuse the issue.

We conclude therefore that the spin-label technique, despite its great potential, has not yet provided any reliable information about membrane alterations in neoplasia.

1.3.2.2. Fluorescence studies

1.3.2.2.1. Lipid probes

Inbar et al. [156] report on the use of the aromatic fluorophore 1,6-diphenyl-1,3,5-hexatriene (DPH) as a probe of the 'fluidity' of the plasma membranes of normal and neoplastic lymphocytes. They exposed intact normal rat or mouse lymph-node cells, normal human peripheral blood lymphocytes, leukemic human blood lymphocytes (chronic lymphocytic leukemia) and cells of a Moloney virus-induced mouse T-cell lymphoma, to aqueous dispersions of DPH and thereafter measured steady state polarization, p (Section 1.1.7.1), eq. 1.12, and made estimates of membrane microviscosity, η, by reference to Perrin plots (Section 1.1.7.1) for unbound DPH. Their data show that the polarization of DPH taken up by leukemic mouse and human lymphocytes was 84% and 92%, respectively, of that found in normal rodent or human lymphoid cells. The viscosities in the DPH microenvironment would therefore be 43% and 25% lower in neoplastic rodent and human lymphoid cells than in the normal cells at 4 °C. The trend at 37 °C is very similar and the authors conclude that the plasma membrane lipids of the neoplastic cells are more 'fluid' than those of normal lymphoid cells.

The results of Inbar et al. [156] appear quantitatively more impressive than reported comparisons of normal and neoplastic cells, using spin-labelling. However, the experimentation again contains basic defects which preclude secure interpretation. *First*, there are no binding measurements; these are needed for comparisons between cells. *Second*, fluorescent probes such as DPH, like many spin-labels, tend to disorder (fluidize) their microenvironment and tend to localize in pre-existing fluid domains. *Third*, the cell types used are not comparable, particularly rodent lymph node cells and a T-cell population. *Fourth*, and most crucial, the experiments provide no clue as to the location(s) of the DPH. The fact that DPH fluorescence intensifies after addition of the cells does not mean penetration 'into the surface membrane', but partition into various sites within the cell. The observed differences in p could reflect a greater proportion of neutral fat in the neoplastic cells, the known greater nuclear surface membrane mass in these cells or many other variables.

There is no justification for the conclusion that the measurements reflect an altered plasma membrane mobility in neoplastic lymphocytes, or that this has anything to do with cholesterol (cf. Chapter 5).

1.3.2.2.2. Fluorescent lectins

Shinitsky et al. [157] and Inbar et al. [158] have attempted to probe the 'mobility' of plasma membrane proteins by use of fluorescein-labelled concanavalin A (F-ConA), wheat germ agglutinin (F-WGA) and soybean agglutinin (F-SBA).

These experiments were designed to compare the decay rates of fluorescence polarization of free and cell-bound fluorescein-lectin conjugates. Since the relaxation rate of fluorescein ($\tau \sim 4$ μsec) is vastly less than that of the proteins, these workers constructed plots of $1/r$ vs. $\eta_0\ \tau/\eta$ (cf. Eqs. 1.14 and 1.15) where η_0 is the viscosity of aqueous buffer and η that of a buffer solution containing sucrose (up to 70%). At high values of $\eta_0\ \tau/\eta$, $1/r$ reflects predominantly the mobility of the protein rather than that of the fluorophore, thereby allowing evaluation of ρ_0, the relaxation rate of the coupled proteins, as well as the anisotropies, r_0, of the coupled proteins. The ρ_0 and r_0 values were then compared with the corresponding values, ρ and r, or the F-lectins absorbed by diverse cells. The effects of medium viscosity on the fluorescence polarization of cell-bound lectins was not evaluated because of the destructive effects of high sucrose concentrations. However, the experiments could have been carried out using polysucrose (Ficoll), which is commonly used to purify viable lymphocytes.

The results obtained with various F-lectins are given in Table 1.3. In all cases tested, glutaraldehyde fixation increased ρ to >200 μsec and, when done, trypsinization of the cells produced a ρ between 63–68.

The authors claim that the ρ-values reflect those of the membrane lectin

TABLE 1.3

Fluorescence polarization of fluorescein-conjugated lectins in solution and cell-bound[1]

	Unbound lectin[2]			Lectin bound to: Rat lymphnode cells[3]			Mouse lymphoma[4]			Hamster fibroblasts[5]			SV40 transformed hamster fibroblasts[6]		
Lectin	τ (nsec)	r	ρ_0 (nsec)	r	ρ (nsec)	ρ_0/ρ_0	r	ρ (nsec)	ρ_0/ρ_0	r	ρ (nsec)	ρ_0/ρ	r	ρ (nsec)	ρ_0/ρ
F-ConA	3.9	0.174	58	0.179	70	0.83	0.184	160	0.36	0.190	120	0.48	0.180	73	0.79
F-WGA	4.1	0.124	22	0.160	59	0.37	—	—	—	—	—	—	—	—	—
F-SBA	3.5	0.095	53	0.150	123	0.43	—	—	—	—	—	—	—	—	—

[1] From [157,158].
[2] In buffer solution.
[3] Normal; CR/RAR strain.
[4] Moloney virus-induced; A-strain.
[5] Tertiary culture; Syrian golden hamster.
[6] Not derived from 5.

receptors and that differences in ρ indicate alterations in receptor mobility. Importantly, they argue that the receptor mobility is greater in neoplastic fibroblasts than in normal cells and lower in neoplastic lymphocytes than in normal lymphoid cells. They interpret the data to indicate major changes of membrane 'fluidity' with neoplastic conversion! This conclusion is not justified for the following reasons.

(a) The relaxation times lie almost two orders of magnitude below that of the 20 μsec for rhodopsin [76] which is a small protein with a proven high rotational mobility. Also, as pointed out already, many membrane proteins show a lower mobility than rhodopsin, and the ρ-values lie well below the the theoretical limits computed by Sackman et al. [75].

(b) The ρ-values for concanavalin A 'bound' to normal lymphocytes and SV40 transformed fibroblasts do not differ significantly from the solution value (in contrast to F-WGA and F-SBA).

(c) There is no information on the effect of fluorescein coupling on the subunit association of the lectins; a change could modify binding.

(d) The cells are not comparable. Rat lymph node cells are not appropriate controls for mouse T-cells. Cultured fibroblasts should be compared with neoplastic variants derived therefrom.

(e) The measurements were carried out at 24 °C where concanavalin A is rapidly interiorized into cells.

(f) There were no binding studies. These are essential for comparisons since one cannot, *a priori*, assume a single type of binding site for each cell.

(g) The ρ-values of bound lectins are lower for lymphoma cells than for normal cells, suggesting greater viscosity in the former. However, Inbar et al. [156] argue the opposite on the basis of the measurements of DPH polarization (see above), i.e. lower viscosity in lymphoma cells.

(h) At the concanavalin A levels used (20 μg/ml) concanavalin A receptors of transformed fibroblasts aggregate. This should restrict diffusional rotation. But the authors argue for *greater* mobility in F-concanavalin A treated transformed fibroblasts than in normal cells.

It appears that the concanavalin A experiments considered here do not give any insight into '*membrane mobility*'. Indeed, there might be trivial reasons for the observed differences, for example:

(a) Under the conditions employed, part of the lectin is no longer membrane bound, but on the way to intracellular digestion; this process can vary from one cell to the next.

(b) Membrane-bound lectins rotate independently about sites of attachment to receptors, but this rotation varies from cell to cell depending on receptor distribution.

(c) Receptors are bound to membranes with varying affinities.

Whatever the interpretation of the data, the authors' conclusion that the observed relaxation rates reflect *membrane* 'fluidity' appears premature and certainly cannot be generalized.

1.3.2.3. Electron diffraction studies

We have already commented on the use of electron diffraction in the conformational analysis of polypeptides spread at air–water interfaces [19,20]. In a promising new development, Hui et al. [159] have examined purified 'bile front' isolates from normal rat liver and hepatoma H35 by electron diffraction, using a special 'hydration stage' for transmission electron microscopes. Their data indicate that the membranes from both cell types can undergo a sharp lipid phase transition below 20 °C. In freshly prepared membranes from both cell types $T_t \sim$ 10–11 °C ± 1 °C [160]. With time at 0 °C or freeze-thawing, T_t shifts to ~ 17 °C ± 1 °C. No consistent alteration in freeze fracture pattern accompanied the thermotropic changes in electron diffraction.

The change of electron diffraction observed is attributed to a segregation below T_t of pure phospholipid domains from phospholipid–cholesterol domains [160]. Since both membrane types have 1:1 molar cholesterol: phospholipid ratios, the data also imply 'cholesterol clustering'.

The results of Hui and associates [159,160] relate to those of Wallach and associates [139,143,161] who found very sharp lipid transitions in isolated erythrocyte ghosts and lymphoid cell plasma membranes by Raman and resonance-Raman spectroscopy, despite the facts that *neither* membrane exhibits any thermotropic protein-lipid segregation, detectable by conventional freeze-fracture electron microscopy and that both membrane categories exhibit cholesterol:phospholipid molar ratios of 1:1. The explanation for these findings, supported by paramagnetic studies [139,161], is that plasma membrane cholesterol is sequestered into clusters.

Although the original report of Hui et al. [159] suggests a different thermotropism for normal vs. neoplastic membranes, this does not appear a consistent phenomenon. Indeed, one might not expect such a difference in plasma membranes since these have high cholesterol: phospholipid ratios in both normal and neoplastic cells. Mitochondrial membranes, which do become cholesterol-enriched in hepatocarcinogenesis (Chapter 5) may yield more distinctive results.

1.4. Coda

Intensive efforts in recent years have greatly advanced many aspects of membrane technology and have thus produced significant insights into the principles that govern the structure and organization of biomembranes. The stage appears set for the serious application to the problem of malignancy.

References

[1] Van Deenen, L.L.M., *Prog. Chem. Fats Other Lipids*, 1965, *8*, pt. 1, 1965.
[2] Davenport, J.B. in *Biochemistry and Methodology of Lipids*, Wiley, New York, 1971.
[3] Perutz, M.F., Kendrew, J.D. and Watson, H.C., *J. Mol. Biol.*, *13*, 669 (1965).
[4] Wallach, D.F.H. and Winzler, R., *Evolving Strategies and Tactics in Biomembrane Research*, Springer Verlag, New York, 1974.
[5] Singer, S.J. and Nicolson, G., *Science*, *175*, 720 (1972).
[6] Langdon, R.G., *Biochim. Biophys. Acta*, *342*, 213 (1974).
[7] Wallach, D.F.H., *J. Theoret. Biol*, *39*, 321 (1972).
[8] Dunham, P.B. and Hoffman, J.F., *Proc. Natl. Acad. Sci. U.S.*, *66*, 936 (1970).
[9] Dryer, W.J., Papermaster, D.S. and Kühn, H., *Ann. N.Y. Acad. Sci.*, *195*, 61 (1972).
[10] Knüfermann, H., Schmidt-Ullrich, R., Ferber, E., Fischer, H. and Wallach, D.F.H., in *Erythrocytes, Thrombocytes, Leukocytes*, E. Gerlach, K. Moser, E. Deutsch and E. Wilmanns, eds., G. Thieme, Stuttgart, 1973, p. 12.
[11] Schmidt-Ullrich, R., Ferber, E., Knüfermann, H., Fischer, H. and Wallach, D.F.H., *Biochim. Biophys. Acta*, *332*, 175 (1974).
[12] Wallach, D.F.H. and Gordon, A.S. in *Regulatory Functions of Biological Membranes*, J. Jarnefelt, ed., Elsevier, Amsterdam, 1968, p. 87.
[13] Salem, L., *Can. J. Biochem. Physiol.*, *40*, 1287 (1962).
[14] Tanford, C. in *The Hydrophobic Effect*, John Wiley, New York, 1973.
[15] Wallach, D.F.H. and Gordon, A.S., *Fed. Proc.*, *27*, 1963 (1968).
[16] Maddy, A.H. and Malcolm, B.R., *Science*, *150*, 1616 (1965).
[17] Maddy, A.H. and Malcolm, B.R., *Science*, *153*, 213 (1966).
[18] Wallach, D.F.H. and Zahler, H.P., *Proc. Natl. Acad. Sci. U.S.*, *56*, 1552 (1966).
[19] Malcolm, B.R., *Nature*, *227*, 1358 (1970).
[20] Malcolm, B.R., *Biopolymers*, *9*, 911 (1970).
[21] Wallach, D.F.H. and Zahler, P.H., *Biochim. Biophys. Acta*, *150*, 186 (1968).
[22] Jenkinson, T.J., Kamat, V.B. and Chapman, D., *Biochim. Biophys. Acta*, *183*, 427 (1969).
[23] Avruch, J. and Wallach, D.F.H., *Biochim. Biophys. Acta*, *233,344*, 180 (1971).
[24] Verma, S.P., Wallach, D.F.H. and Schmidt-Ullrich, R., *Biochim. Biophys. Acta*, *394*, 633 (1975).
[25] Wallach, D.F.H., Graham, J.M. and Fernbach, B.R., *Arch. Biochem. Biophys.*, *131*, 322 (1969).
[26] Wallach, D.F.H., Gordon, A., Graham, J. and Fernbach, B. in *Physical Principles of Biological Membranes*, F.M. Snell, J.J. Wolken, G.I. Iverson and J. Lam, eds., Gordon and Breach, London, 1970, p. 345.
[27] Graham, J.M. and Wallach, D.F.H., *Biochim. Biophys. Acta*, *241*, 180 (1971).
[28] Wallach, D.F.H. and Graham, J.M. in *International Symposium on the Biochemistry and Biophysics of Mitochondrial Membranes*, F. Azzone and N. Siliprandi, eds., Academic Press, New York, 1972, p. 231.
[29] Wallach, D.F.H., Low, A. and Bertland, A.V., *Proc. Natl. Acad. Sci. U.S.*, *70*, 3235 (1973).
[30] Graham, J.M. and Wallach, D.F.H., *Biochim. Biophys. Acta*, *193*, 225 (1969).
[31] Capaldi, R.A.C. and Vanderkooi, G.V., *Proc. Natl. Acad. Sci. U.S.*, *69*, 930 (1972).
[32] Fisher, H., *Proc. Natl. Acad. Sci. U.S.*, *51*, 1285 (1964).
[33] Tanford, C., *J. Am. Chem. Soc.*, *84*, 4240 (1962).
[34] Bigelow, C.C., *J. Theoret. Biol.*, *16*, 187 (1967).
[35] Auer, H.E. and Doty, P., *Biochemistry*, *5*, 1708 (1966).
[36] Verma, S.P. and Wallach, D.F.H., *Biochim. Biophys. Acta*, *345*, 129 (1974).
[37] Verma, S.P. and Wallach, D.F.H., *Biochim. Biophys. Acta*, to be published.
[37a] Hughes, R.C., *Prog. Biophys. Mol. Biol.*, *26*, 191 (1973).
[38] Winzler, R., *Int. Rev. Cytol.*, *29*, 77 (1970).
[39] Segrest, J.P., Jackson, R.L., Andrews, E.P. and Marchesi, V.T., *Biochem. Biophys. Res. Commun.*, *44*, 390 (1971).
[40] Tuech, J.K. and Morrison, M., *Biochem. Biophys. Res. Commun.*, *59*, 352 (1974).

[41] Strittmatter, P., Rogers, M.J. and Spatz, L., *J. Biol. Chem.*, *247*, 7188 (1972).
[42] Strittmatter, P., *Proc. Natl. Acad. Sci. U.S.*, *68*, 1042 (1971).
[43] Brewer, H.B., Jr., Lux, S.E., Rowan, R. and John, K.M., *Proc. Natl. Acad. Sci. U.S.*, *69*, 1304 (1972).
[44] Shore, B. and Shore, V., *Biochemistry*, *8*, 4510 (1969).
[45] Scanu, A.M., *Biochim. Biophys. Acta*, *265*, 471 (1972).
[46] Jackson, R.L. and Gotto, A.M., *Biochim. Biophys. Acta*, *285*, 36 (1972).
[47] Lux, S.E., Hirz, R., Shrager, R.I. and Gotto, A.M., *J. Biol. Chem.*, *247*, 2598 (1972).
[48] Assman, G. and Brewer, H.B., Jr., *Proc. Natl. Acad. Sci. U.S.*, *71*, 989 (1974).
[49] Stoffel, W., Ziesenberg, O., Tunggal, B. and Schreiber, E., *Proc. Natl. Acad. Sci. U.S.*, *71*, 3696 (1974).
[50] Assman, G. and Brewer, H.B., Jr., *Proc. Natl. Acad. Sci. U.S.*, *71*, 1534 (1974).
[51] Assman, G. and Brewer, H.B., Jr., *Proc. Natl. Acad. Sci. U.S.*, *71*, 3701 (1974).
[52] Braun, V. and Bosch, V., *Proc. Natl. Acad. Sci. U.S.*, *69*, 970 (1972).
[53] Braun, V. and Bosch, V., *Eur. J. Biochem.*, *28*, 51 (1972).
[54] Hantke, K. and Braun, V., *Eur. J. Biochem.*, *34*, 284 (1973).
[55] Sodek, J., Smillie, L.B. and Jurasek, L., *Cold Spring Harbor Symp. Quant. Biol.*, *37*, 299 (1972).
[56] Braun, V., *J. Inf. Dis. Suppl.*, *128*, 9 (1973).
[57] Inouye, M., *Proc. Natl. Acad. Sci. U.S.*, *71*, 2396 (1974).
[58] Kornberg, R.D. and McConnell, H.M., *Proc. Natl. Acad. Sci. U.S.*, *68*, 2564 (1971).
[59] Devaux, P. and McConnell, H.M., *J. Am. Chem. Soc.*, *94*, 4475 (1972).
[60] Träuble, H. and Sackmann, E., *J. Am. Chem. Soc.*, *94*, 4499 (1972).
[61] Scandella, C.J., Devaux, P. and McConnell, H.M., *Proc. Natl. Acad. Sci. U.S.*, *69*, 2056 (1972).
[62] Barnett, R.E. and Grisham, C.M., *Biochem. Biophys. Res. Commun.*, *48*, 1362 (1972).
[63] Sackmann, E., Träuble, H., Galla, H.J. and Overath, P., *Biochemistry*, *12*, 5360 (1973).
[64] Stier, A., Sackmann, E., *Biochim. Biophys. Acta*, *311*, 400 (1973).
[65] Keith, A.D., Sharnoff, M. and Cone, G.E., *Biochim. Biophys. Acta*, *300*, 379 (1972).
[66] Bieri, V., Wallach, D.F.H. and Lin, P.S., *Proc. Natl. Acad. Sci. U.S.*, *71*, 4797 (1974).
[67] Oldfield, E., Keough, K.M. and Chapman, D., *FEBS Letters*, *20*, 344 (1972).
[68] Morrison, D.L. and Morowitz, H.F., *J. Mol. Biol.*, *49*, 441 (1970).
[69] Deamer, D.W. and Branton, A., *Science*, 158, 655 (1967).
[70] Kornberg, R.D. and McConnell, H.M., *Biochemistry*, *10*, 1111 (1971).
[71] McNamee, M.G. and McConnell, H.M., *Biochemistry*, *12*, 2951 (1973).
[72] Huestis and McConnell (quoted by MacNamee and McConnell, 1973).
[73] Grant, C.W.M. and McConnell, H.M., *Proc. Natl. Acad. Sci. U.S.*, *70*, 1238 (1973).
[74] Smith, R.J.M. and Green, C., *FEBS Letters*, *42*, 108 (1974).
[75] Sackmann, E., Träuble, H., Galla, H.-J. and Overath, P., *Biochem.*, *12*, 5360 (1973).
[76] Cone, R.A., *Nature*, *New Biol.*, *236*, 39 (1972).
[77] Poo, M.M. and Cone, R.A., *Nature*, *247*, 438 (1974).
[78] Sutherland, G.B., *Phil. Mag.*, *9*, 781 (1905).
[79] Brown, P.K., *Nature*, *New. Biol.*, *236*, 35 (1972).
[80] Peters, R., Peters, J., Tews, K.H. and Bahr, W., *Biochim. Biophys. Acta*, *367*, 282 (1974).
[81] Watkins, J.F. and Grace, D.M., *J. Cell Sci.*, *2*, 193 (1967).
[82] Frye, L.D.F. and Edidin, M.J.E., *J. Cell Sci.*, *7*, 319 (1970).
[83] Edidin, M.J.E., in *Membrane Research*, C.F. Fox, ed., Academic Press, New York, 1972, p. 15.
[84] Edidin, M.J.E. and Fambrough, D., *J. Cell Biol.*, *57*, 27 (1973).
[85] Taylor, R.B., Duffus, W.P.H., Raff, M.D. and DePetris, S., *Nature*, *233*, 225 (1971).
[86] Raff, M.C. and DePetris, *Fed. Proc.*, *32*, 48 (1973).
[87] Kourilsky, F.M., Silvestre, D., Neauport-Sautes, C., Loosfelt, Y. and Dausset, J., *Eur. J. Immunol.*, *12*, 249 (1972).
[88] Binns, R.M., Symons, D.A. and White, D.J., *J. Physiol.*, *226*, 33 (1972).
[89] DePetris, S. and Raff, M.C., *Nature*, *241*, 257 (1973).
[90] Karnovsky, M.J. and Unanue, E.R., *Fed. Proc.*, *32*, 55 (1973).
[91] Inbar, M., Ben-Bassat, H. and Sachs, L., *Int. J. Cancer*, *12*, 93 (1973).

[92] Loor, F., *Eur. J. Immunol.*, *3*, 112 (1973).
[93] Unanue, E.R., Karnovsky, M.J. and Engers, H.D., *J. Exd. Med.*, *137*, 675 (1973).
[94] Karnovsky, M.J., Unanue, E.R. and Leventhal, M., *J. Exp. Med.*, *136*, 907 (1972).
[95] Speth, V. and Wunderlich, F., personal communication.
[96] Unanue, E.R., Ault, K.A. and Karnovsky, M.J., *J. Exp. Med.*, *139*, 295 (1974).
[97] Edidin, M. and Weiss, A., in *Control of Proliferation in Animal Cells*, B. Clarkson and R. Baserga, eds., Cold Spring Harbor Laboratory, 1974, p. 213.
[98] Weinstein, R., Gurrick, R.R. and Witchett, C.E., *J. Cell Biol.*, *55*, 276a (1972).
[99] Hong, K. and Hubbell, W.L., *Proc. Natl. Acad. Sci. U.S.*, *69*, 2617 (1972).
[100] McNutt, N.S. and Weinstein, R., *Prog. Biophys. Mol. Biol.*, *26*, 45 (1973).
[101] Guidotti, G., *Annu. Rev. Biochem.*, *41*, 731 (1972).
[102] Träuble, H. and Overath, P., *Biochim. Biophys. Acta*, *307*, 491 (1973).
[103] Jost, P.C., Griffith, O.H., Capaldi, R.A. and Vanderkooi, G., *Proc. Natl. Acad. Sci. U.S.*, *70*, 480 (1973).
[104] Wallach, D.F.H., Verma, S.P., Weidekamm, E. and Bieri, V., *Biochim. Biophys. Acta*, *356*, 68 (1974).
[105] Weinstein, R., *J. Cell Biol.*, *63*, 367a (1974).
[106] Verkleij, A.J., Ververgaert, P.H.J., van Deenen, L.L.M. and Elbers, E., *Biochim. Biophys. Acta*, *288*, 326 (1972).
[107] James, R. and Branton, D., *Biochim. Biophys. Acta*, *323*, 378 (1973).
[108] Pinto da Silva, P. Douglas, S.D. and Branton, D., *Nature, New Biol.*, *232*, 194 (1971).
[109] Speth, V. and Wunderlich, F., *Biochim. Biophys. Acta*, *291*, 621 (1973).
[110] Wunderlich, F., Batz, W., Speth, V. and Wallach, D.F.H., *J. Cell Biol.*, *61*, 633 (1974).
[111] Wunderlich, F., Batz, W., Speth, V. and Wallach, D.F.H., *Biochim. Biophys. Acta*, *373*, 34 (1974)
[112] Ferber, E., Resch, K., Wallach, D.F.H. and Imm, W., *Biochim. Biophys. Acta*, *266*, 1194 (1972).
[113] Kleinig, H., *J. Cell Biol.*, *46*, 396 (1970).
[114] Pinto da Silva, P., *J. Cell Biol.*, *53*, 777 (1972).
[115] Elgsaeter, A. and Branton, D.A., *J. Cell Biol.*, *63*, 1018 (1974).
[116] McIntyre, J.A., Gilula, N.B. and Karnovsky, M.J., *J. Cell Biol.*, *60*, 192 (1974).
[117] Bhakdi, S., Speth, V., Knüfermann, H., Wallach, D.F.H. and Fischer, H., *Biochim. Biophys. Acta*, *356*, 300 (1974).
[118] Kumins, C.A. and Kwei, T.K. in *Diffusion in Polymers*, J. Crank and G.S. Park, eds., Academic Press, London, 1968.
[119] Weinstein, R. and McNutt, N.S., *Seminars in Hematol.*, *7*, 259 (1970).
[120] Keith, A.D. and Snipes, W., *Science*, *183*, 666 (1974).
[121] Hagins, W.A. and Jennings, W.H., *Dis. Farady Soc.*, *27*, 180 (1959).
[122] Cherry, R. in *Biological Membranes*, D. Chapman and D.F.H. Wallach, Academic Press, London, Vol. 3, 1975, in press.
[123] Junge, W., *FEBS Letters*, *25*, 109 (1972).
[124] Junge, W. and Eckhoff, A., *FEBS Letters*, *36*, 207 (1973).
[125] Razi-Naqvi, K., Gonzales-Rodriguez, J., Cherry, R.J. and Chapman, D., *Nature, New Biol.*, *245*, 249 (1973).
[126] Blaurock, A.E. and Stoeckenius, W., *Nature, New Biol.*, *233*, 152 (1971).
[127] Tasaki, I., Watanabe, A. and Hallett, M., *J. Memb. Biol.*, *8*, 109 (1973).
[128] Rothman, J.E. and Engelman, D.M., *iature, New Biol.*, *237*, 42 (1972).
[129] Bretscher, M., *Science*, *181*, 622 (1973).
[130] Onsager, L. in *Physical Principles in Biological Membranes*, F. Snell, J. Wolken, G. Inverson and J. Lam, eds., Gordon and Breach, New York, 1970, p. 137.
[131] Lev, A., Malev, V. and Osipov, V. in *Membranes*, G. Eisenman, ed., Dekker, New York, in press.
[132] Nicolson, G.L. and Painter, R.G., *J. Cell Biol.*, *59*, 395 (1973).
[133] Ji, T.H. and Nicolson, G.L., *Proc. Natl. Acad. Sci. U.S.*, *71*, 2212 (1974).
[134] Danielli, J.F. and Davson, H., *J. Cell. Comp. Physiol.*, *5*, 495 (1935).
[135] Nathenson, A., *Jahrb. wiss Botan.*, *39*, 607 (1904).

[136] Parpart, A.K. and Ballentine, R. in *Trends in Physiology and Biochemistry*, E.S.G. Barron, ed., Academic Press, New York, 1952, p. 135.
[137] Lenard, J. and Singer, S.J., *Proc. Natl. Acad. Sci. U.S.*, *56*, 1828 (1966).
[138] Singer, S.J. in *Structure and Function of Biological Membranes*, L.I. Rothfield, ed., Academic Press, New York, 1971, p. 146.
[139] Wallach, D.F.H., Bieri, V., Verma, S.P., and Schmidt-Ullrich, R., *Ann. N. Y. Acad. Sci.*, 1975, in press.
[140] Changeux, J.P., Blumenthal, R., Kasai, M. and Podleski, T. in *Molecular Properties of Drug Receptors*, R. Porter and M. O'Connor, eds., Churchill, London, 1970, p. 197.
[141] Changeux, J.P. and Thiery, J. in *Regulatory Functions of Biological Membranes*, J. Jarnefelt, ed., Elsevier, Amsterdam, 1968, p. 115.
[142] Wyman, J. in *Symmetry and Function of Biological Systems at the Macromolecular Level*, A. Engström and B. Strandberg, eds., Wiley Interscience, New York, 1969, p. 267.
[143] Verma, S.P., Wallach, D.F.H. and Schmidt-Ullrich, R., *Biochim. Biophys. Acta*, *394*, 633 (1975).
[144] Hagnenau, F. in *The Biological Basis of Medécine*, E.E. Bittar and N. Bittar, eds., Academic Press, New York, 1969, p. 433.
[145] Porter, K.R., Todaro, G.J. and Fonte, V., *J. Cell Biol.*, *59*, 633 (1973).
[146] Vesely, P. and Boyde, A., *Proceedings of the Workshop on Scanning Electron Microscopy in Pathology*. IIT Research Institute, Chicago, I11, 1973, p. 689.
[147] Porter, K.R. and Fonte, V.G., *Proceedings of the Workshop of Scanning Electron Microscopy in Pathology*. IIT Research Institute, Chicago, Ill., 1973, p. 683.
[148] McNutt, N.S., Culp, L.A. and Black, P.H., *J. Cell Biol.*, *56*, 412 (1973).
[149] Pollard, T.D. and Korn, E.D., *J. Cell Biol.*, *48*, 216 (1971).
[150] Scott, R.E., Furcht, L.T. and Kersey, J.H., *Proc. Natl. Acad. Sci. U.S.*, *70*, 3631 (1973).
[151] Barnett, R.E., Furcht, L.T. and Scott, R.E., *Proc. Natl. Acad. Sci. U.S.*, *71*, 1992 (1974).
[152] Furcht, L.T. and Scott, R.E., *Exp. Cell. Res.*, *81*, 311 (1974).
[153] Scott, R.E. and Furcht, L.E., *Proc. Soc. Exp. Biol. Med.*, in press, 1975.
[154] Furcht, L.T. and Scott, H.E., *J. Cell Biol.*, *63*, 106a (1974).
[155] Kaplan, J., Canonico, P.G. and Caspary, W.J., *Proc. Natl. Acad. Sci. U.S.*, *70*, 66 (1973).
[156] Inbar, M., Schinitsky, M. and Sachs, L., *FEBS Letters*, *38*, 268 (1974).
[157] Shinitsky, M., Inbar, M. and Sachs, L., *FEBS Letters*, *34*, 247 (1973).
[158] Inbar, M., Schinitsky, M. and Sachs, L., *J. Mol. Biol.*, *81*, 245 (1973).
[159] Hui, S.W., Parsons, F. and Schulman, N., *Biophys. J.*, *15*, 107a (1975).
[160] Hui, S.W., personal communication.
[161] Bieri, V. and Wallach, D.F.H., *Biochim. Biophys. Acta*, *406*, 415 (1975).

2

Membrane proteins

2.1. General introduction

Membrane proteins remain among the more intractable macromolecules. However, one cannot achieve a full understanding of the membrane anomalies of neoplasia without detailed knowledge about the membrane protein-complement, composition, metabolism and exchange of normal *versus* neoplastic cells. We do not now possess this knowledge, but have probably reached the stage where we can obtain it. The primary aim of this chapter, accordingly, is to provide a base for future excursions.

As pointed out in Chapter 1, there is no clear definition as to what constitutes a 'membrane' protein and information as to the number of proteins in a given membrane category, or the diverse categories of membrane proteins, remains controversial and incomplete.

Nevertheless, as reviewed in [1–7], much attention has focussed on membrane protein *metabolism*, particularly the synthesis of membrane proteins by liver endoplasmic reticulum and transfer of these proteins to plasma membranes, and the synthesis and metabolism of membrane-bound immunoglobulin in lymphocytes. Moreover, intensive efforts are now directed toward comparisons of the membrane protein composition and metabolism of normal and neoplastic cells.

Because most cells contain multiple membrane systems and these systems tend to differ in composition, analysis of the composition and metabolism of the protein components of membranes from normal and neoplastic cells demands the application of rigorous, quantitative membrane fractionation procedures and the use of reliable membrane markers. Unfortunately, very few studies comparing membrane protein composition of normal and neoplastic cells meet these requirements and major conclusions thus appear unwarranted at this time. The same is true for metabolic investigations. Moreover, one cannot easily transfer information obtained on, for example, lymphocytes to hepatocytes, or data obtained by *in vivo* experimentation to the *in vitro* situation.

Because the study of membrane proteins depends so critically on methodology, we must devote the first section of this chapter to the major techniques used in experiments on the composition and metabolism of membrane proteins. We will not go into great detail, but we will point out important assumptions, weaknesses and complications, referring the reader to more comprehensive treatises where necessary.

Following our overview of methodology we will turn to a treatment of important general principles of membrane proteins dynamics. We will deal with comparative studies on diverse normal and neoplastic cells where possible. In this we cannot provide a very uniform picture because very different kinds of information are obtained depending on the cell system used. This is partly due to tissue differences and partly from the disparate strategies applied to diverse systems. For example, studies on fibroblasts are carried out *in vitro*, focus on comparisons of normal cells with virally-induced transformants, emphasize plasma membranes, but employ very divergent, often inconsistent techniques. In contrast, most studies on hepatocytes are carried out *in vivo*, use chemically-induced hepatomas (deviating from normal to varying degrees) and tend to concentrate on fractions enriched in endoplasmic reticulum or 'bile-fronts'.

We will deal rather little with membrane enzymes in this chapter, but will treat this general topic in Chapter 6; we will comment on certain specific membrane-associated enzymes also in other chapters [3, 4, 5, 7, 8, 10].

2.2. Analytical methods

2.2.1. Membrane fractionation

Several recent, comprehensive reviews [8–10] have dealt with the complex problems involved in membrane isolation and purification. These are particularly relevant to comparison of membranes isolated from normal and neoplastic cells for the following reasons: Tumors very commonly show anomalies of membrane structure and differentiation, for example surface membrane specializations disappear, mitochondria vary abnormally in size, endoplasmic reticulum generally dedifferentiates. The membranous organelles of tumors may thus behave differently in a given fractionation scheme than do the particles from normal cells. It is essential, therefore, to utilize fractionation criteria which are insensitive to such variables and to keep careful balance sheets. Unfortunately, these cautions are often neglected.

The greatest difficulties are encountered in *plasma membrane* fractionations, for reasons detailed in [8, 10–12]:

First, most cells do not have a uniform surface topology. This is true even for 'free' cells such as leukocytes, but is most marked in solid tissues, where different areas of the cell surface become highly specialized.

Second, neoplastic transformation, both *in vivo* and *in vitro* nearly always involves some change in surface topology. This circumstance is particularly problematical when cells are propagated *in vivo*, since the manner of plasma membrane fractionation then depends critically on cell interactions. At the same time one is not free to assume that the problem is obviated by working *in vitro*, since explants of normal and neoplastic cells do not yield a uniform surface topology. Moreover, surface topology can vary markedly with culture conditions. For example, murine C1300 neuroblastoma cells in *suspension* culture are round and depleted of surface protrusions [13], whereas the same cells, *in monolayer* culture show a profusion of surface processes.

Another difficulty arises from changes in membrane composition, for example, altered levels of mitochondrial diphosphatidylglycerol (Chapter 3), cholesterol (Chapter 5) or changed complement of glycolipid (Chapter 4). Such chemical modifications tend to alter the physical properties of the membranes involved, producing differing stabilities and varying sedimentation characteristics during differential and/or isopycnic ultracentrifugation.

The impact of these complications is well illustrated by experiments carried out on the *liver-hepatoma* system [14–20]. Most authors equate 'liver plasma membrane' with the 'bile-front' moiety of hepatocytes. This can be isolated in the form of rather large fragments using a method originally developed by Neville in 1960 [14,21–23], in which liver is homogenized in hypotonic (1 mM) $NaHCO_3$, pH 7.5.

Even with normal hepatocytes this approach involves a number of hazards [8, 10–12]. *First*, the method neglects the vast sinusoidal hepatocyte surface ('blood fronts'); *second*, the method induces nuclear rupture and adventitious adsorption of nuclear material onto non-nuclear membranes; *third*, the method causes disruption of at least 10% of mitochondria and 15% of lysosomes [24]; *fourth*, the isolated membranes tend to fragment further; *fifth*, the method produces poor yields; *sixth*, the low ionic strength employed tends to elute some membrane-associated proteins.

Most important in the present context is the fact that the method depends upon the various connections between apposed hepatocytes which together lead to the formation of bile canaliculi and 'bile-fronts'. These intercellular associations are disturbed to a greater or lesser extent *in all hepatomas*. Thus, application of a procedure designed to isolate a surface specialization which is defective or lacking in a tumor will not yield the same product as is obtained with a normal cell. Simple modifications, such as the use of 2 mM $CaCl_2$ for the disruption of hepatoma 484 [18] do not alter the basic dilemma.

The different behavior of the membranes in bile-front isolates from normal liver and hepatomas is illustrated by the deviant distribution of marker enzymes in sucrose gradients. Thus, the protein and ATPase activities of rat hepatoma 'bile-fronts' equilibrate at higher buoyant densities than is the case for the components of normal liver 'bile-fronts' [25], and studies on a series of mouse hepatomas [26] show a rather inconsistent distribution of plasma membrane alkaline phosphatase.

As noted, the fractionation difficulties are not restricted to plasma membrane isolates. Thus, hepatoma mitochondria exhibit different sedimentation characteristics than normal mitochondria ([26]; cf. also Chapter 6). The same appears true for lysosomes and peroxisomes [27]. Smooth and rough 'microsomal membranes' from hepatomas also fractionate differently from their normal equivalents. For example, Moyer and associates [20, 28] report a reduction of smooth microsomal membranes in hepatoma cells and host liver cells compared with normal liver. As is common they equate 'smooth microsomal membranes' with smooth endoplasmic reticulum, although 'smooth microsomes' are known to be a mixture of membrane vesicles arising from endoplasmic reticulum, plasma membrane, Golgi apparatus and other sources [8, 10–12]. Moyer and associates [20, 28] use the same fractionation procedure for the membranes of normal and neoplastic cells and define both according to their phospholipid/protein ratios and cholesterol/protein ratios. However, as pointed out in Chapter 3 and Chapter 4, carcinogenesis is commonly associated with abnormal phospholipid/protein ratios and/or cholesterol/phospholipid ratios.

For these reasons, such experimentation cannot be easily evaluated. 'Rough microsomal membranes' *do* arise predominantly from endoplasmic reticulum, with some contribution from the outer nuclear envelope. However, neoplastic cells can differ from normal in terms of membrane-ribosome affinity. This, in turn, influences fractionation patterns. The variable association of ribosomes with 'rough microsomal membranes' introduces additional complications, particularly into experiments on membrane protein composition and turnover. These obstacles can be avoided by applications of methods such as developed by Siekevitz and associates [29, 30], involving washes with 0.15 M KCl or dilute EDTA with or without mild detergent treatment.

Ambiguities arising from membrane fractionation procedures can be equally serious in the case of fibroblasts. One of the most common methods used for such cells is that of Warren and associates [31]. Here suspended cells are fixed with Zn^{2+}, acetic acid or fluorescein mercuric acetate and thereafter sheared mechanically in hyposmotic media. The nuclei and adherent, small cytoplasmic organelles are extruded and the plasma membranes separated as 'ghosts'. Among the major disadvantages of the 'fixation' approach are the following [8–12]:

(a) The fixation process tends to inactivate functional membrane entities. This is particularly severe when glutaraldehyde is used to stabilize the membranes [32] but can be partly reversed, in the case of Zn^{2+} fixation, by addition of

EDTA to the membrane isolates [33–35]. Not all membrane enzymes can be reactivated, however.

(b) The 'fixation' process makes estimates of yield and purity difficult and forces undue reliance on non-quantitative, microscopic criteria.

(c) The 'fixation' step causes association with plasma membranes of various entities which are not so associated prior to fixation. Ribosomes constitute a particularly troublesome example, and their presence seriously complicates membrane protein analyses.

(d) The method precludes analyses of cytoplasmic membranes.

A relatively satisfactory, if laborious, general method for cell fractionation and membrane isolation is that originally developed by Wallach and Kamat [36] and further elaborated as reviewed in [8, 10]. Cells are disrupted by nitrogen cavitation and membrane fragments isolated by differential centrifugation and isopycnic ultracentrifugation in polymer density gradients. The method allows isolation of both specialized and unspecialized regions of plasma membrane, as well as concurrent separation of nuclei, mitochondria, lysosomes and endoplasmic reticulum. Functional inactivation is slight, and one can readily establish yields and balance sheets.

This approach has been successfully extended (by zonal techniques) to the large scale isolation of membranous organelles from cultured normal and polyoma-transformed BHK21 fibroblasts by Graham et al. [37]. Not unexpectedly, the membranes from normal and neoplastic cells exhibit a different fractionation behavior.

Also, when normal and SV40-transformed hamster lymphocytes are disrupted by nitrogen cavitation and fractionated in polymer density gradients, the plasma membranes from the tumor cells equilibrate at a higher density ($\rho \sim 1.10$) than those from normal cells [38, 39]. This appears to be due to a different charge density at the inner surfaces of the membrane vesicles isolated from the two cell types [8, 10]. The endoplasmic reticulum markers of the tumor cells also sediment at lower centrifugal fields than those from normal lymphoid cells.

In conclusion:

(a) One cannot *a priori* use identical fractionation schemes for the isolation of a given membrane class from normal cells and from neoplastic cells.

(b) Intrinsic, functional membrane markers are not necessarily reliable for comparisons between normal and neoplastic cells. As pointed out in Chapters 2–6, the composition and enzyme complement of diverse membranes can vary dramatically during differentiation and between the normal and neoplastic state. Extrinsic labels [8, 10], for example, use of lactoperoxidase radioiodination, specific antibodies, lectins or hormones would be more reliable, but one cannot *a priori* assume identical reactivities in normal and neoplastic cells. Clearly multiple criteria are essential and complete balance sheets must be maintained.

(c) Due to the 'membrane dedifferentiation' common in neoplasia fractionation, methods relying on membrane specializations cannot be considered reliable until proven so.

2.2.2. Membrane protein solubilization

There exists a large armamentarium of techniques for the solubilization of membrane proteins, but no one method can be considered universally satisfactory. Moreover, no method, or set of methods, known to give certain results with one tissue type (e.g., normal cells), can be assumed to give identical results when applied to another tissue type (e.g., neoplastic cells).

The principles underlying the diverse solubilization methods, have been treated by one of us [8] together with some technical detail. We will therefore provide only a terse overview. This is given in Table 2.1.

2.2.3. Separation and characterization of membrane proteins

2.2.3.1. Introduction

Soluble proteins can be fractionated according to their charges, isoelectric points, sizes, shapes and specific surface sites. The methods and principles used in these separations have also been applied to membrane proteins, albeit with only limited success until recently [8]. Accordingly, only few membrane proteins have been isolated in sufficient purity to allow the determination of their primary structure. Some of these proteins (murein lipoprotein, cytochrome b_5 from liver endoplasmic reticulum, the major glycoprotein of human erythrocyte membranes and high density apo-lipoprotein) have been discussed in Chapter 1, in terms of membrane organization. Another well-characterized membrane protein (or membrane-associated protein) is the highly helical, high-molecular weight protein that is easily eluted from liver bile front membranes [22]. However, no comparable success has been achieved with tumor membrane proteins.

The field of membrane protein fractionation and characterization has been recently reviewed in detail [8, 78] and we will here address primarily those topics which are particularly germane to comparisons of normal and neoplastic cells. We will also discuss several important and relevant methodological advances which have been achieved in the past year.

2.2.3.2. Separation of membrane proteins

Virtually all protein separation methods have been designed for the fractionation of proteins soluble in aqueous solvents. The application of these methods to membrane proteins, therefore, requires the use of solubilizing agents (e.g. urea, detergents) throughout the separation procedure. Disregard for this requirement,

TABLE 2.1

An overview of methods for membrane protein solubilization

(1) Solubilizing agent	(2) Principle	(3) Solubilization efficiency; criteria	(4) Conditions; selectivity	(5) Protein separation; protein function	(6) Applications; remarks; [Ref.]
Ionic manipulations	Modification of ionic interactions				
Acetic acid		30–40 %	1 vol membrane pellet + 6 vol 0.26 M acetic acid		HuE BoE ShE [40]
			20 min, 20 °C; 2 × reextraction	Separation by charge	
			No lipids, no glycoproteins	Multiple bands upon PAGE in presence of 26 % acetic acid	
				N.D.	
K_2CO_3		70 %	Extraction of membrane pellet in 0.05 M K_2CO_3 at 25 °C for 30 min	25 bands in PAGE	Liver PM (bile-front) [22]
		5×10^5 g · min		N.D.	
Low ionic strength		10–20 %	Dialysis against d H_2O, 4 °C 1–4 days	EM: Erythrocyte 'torus' and hollow cylinder proteins	HuE [41, 42]
		$\sim 6 \times 10^5$ g · min		One homogeneous band in PAGE	
EDTA + TMAB		90 %	Dialysis for 4–5 days against 0.005–0.1M EDTA and/or	No separation	HuE [43, 44]

TABLE 2.1 (*continued*)

(1) Solubilizing agent	(2) Principle	(3) Solubilization efficiency; criteria	(4) Conditions; selectivity	(5) Protein separation; protein function	(6) Applications; remarks; [Ref.]
EDTA + TMAB		3×10^6 g · min	0.1M TMAB	N.D.	
ATP + EDTA		1) 20 %	Low ionic strength, slightly alkaline pH:	ATPase +	1) and 2) HuE [45]
		< 1.5×10^6	1) 0.3 mM ATP, pH 7.5, 5 mM mercaptoethanol 4 °C, 24-48 h	No separation without additional solubilizing agents, e.g., acetic acid or 8 M urea	
		2) 30 %			
		N.D.	2) 0.5 mM EDTA, pH 7.5, 4 °C, 48 h	'Spectrin' ATPase	
High salt		1) N.D.	1) 0.6–1.2 M NaCl	N.D.	1) HuE [46]
		1.65×10^5 g · min	2) 0.15 M NaCl	Acetylcholine esterase and SDS-PAGE band 6	2) HuE [48]
Chaotropic agents	Interference with entropically driven, structural interactions. H-Bond rupture				
Organic		Up to 60 %	1) GHCl, 6 M mercaptoethanol 1 %, pH 8.5.	No separation	HuE [48] HuE, ShE, BoE [40] HuE [49]
		1.2×10^7 g · min Insufficient separation of protein and lipids	2) 2 % CTAB, 5.5 M GHCl. 1 % mercaptoethanol	N.D.	

Inorganic $SCN^- > C10_4^- > NO_3 >$ $I^- > Br^- > Cl^- >$ $SO_4{}^{2-} > CH_3COO^- > F^-$ (decreasing effectivity)		28 % (HuE) 31 % Liver Microsomes 9–12 × 10^6 g · min	Chaotropic agent + 50 mM Tris–HCl, pH 8	No separation Major lipoprotein aggregates No glycoproteins N.D.	HuE [50] Bo liver microsomes
Neutral organic solvents	Reduction of dilectric constant of medium; increase in coulombic interactions				
N-Butanol		90 % proteins aqueous phase. Aggregates (~300,000 Mol. wt.)	De-ionized ghost membranes 0–4 °C	Nucleoside triphosphatase activity. *O*-phosphate monoesterphosphorylase Glycoproteins preferentially	BoE [51] HuE [52]
Chloroform–methanol		N.D. 6 × 10^6 g · min	0.01 M Tris–EDTA HCl, pH 7.4	Glycoprotein No separation N.D.	Hu, Bo, Pig, Ho, ShE [53] HuE [54]
Ethanol		~10 % Aggregates > 200,000 D	1) Extraction with hot acetone and 100 % hot ethanol 2) Equilibration with H_2O	Glycoproteins MN-Specificity + N.D.	HuE [55] HuE [56] HoE ShE BoE

TABLE 2.1 (*continued*)

(1) Solubilizing agent	(2) Principle	(3) Solubilization efficiency; criteria	(4) Conditions; selectivity	(5) Protein separation; protein function	(6) Applications; remarks; [Ref.]
2-Chlorethanol		100 %		No separation	HuE, ShE [57]
		After dialysis against urea (conc.), GHCl and KI, non-sedimentable	9:1 2-chlorethanol; H_2O, pH 2.0–7.0	— 1) Species specific phospholipid-binding retained 2) Promotion of α-helix 3) Carboxylic acid ester formation	HuE [58]
n-Pentanol		80–85 % protein 80 % lipids	De-ionized ghost membranes	No separation	Mouse liver [59]
		→ aqueous phase lipoprotein aggreg.	Phase-separation	N.D.	
Phenol			1) 1:1 phenol: H_2O, 65 °C	No separation	1) HuE [60]
			2) Phenol: acetic acid H_2O 1:1:1 (w/v/v) Biogel P100 → separation protein–lipid	Glycoprotein N.D.	2) HuE [49]
		Aggregation			
Aqueous pyridine		40 %	33 % Aqueous pyridine		[61] HuE
		4×10^6 g · min	16 h dialysis against H_2O	No Separation	
Organic acids	Unfolding of peptide chains				

Acetic and formic acids		N.D. N.D.	Conc. acetic or formic acid	1) Drastic conformation changes (β-structure) 2) Formation of *O*-formates	1) HuE [62] 2) HuE [63]
Trifluoroacetic acid (TFA)		> 80 %	98 % TFA N.D. Separation protein–lipid	No separation on Biogel P150 or Biogel P300 in presence of 98 % TFA N.D.	[64]
Nonionic surfactants	Amphiphilic compounds form complexes with protein and/or lipids				
Triton	X-100 [9–10] X-114 [7–8] X-45 [5]	up to 90 %	Low ionic strength buffer, pH 8.0–9.0 containing 1 % detergent	Separation by charge Inhibition or stimulation of enzymes. Immunologic antigenicity retained	
Tween 40	(n = 20; palmitate)	10^7 g · min		Glycoprotein; bands 3, 4, 2, 6	HuE [65] Rabbit Thymocyte plasma membranes [66]
Tween 80	(n = 20; oleate)				
Tergitol					
Lubrol					

TABLE 2.1 (*continued*)

(1) Solubilizing agent	(2) Principle	(3) Solubilization efficiency; criteria	(4) Conditions; selectivity	(5) Protein separation; protein function	(6) Applications; remarks; [Ref.]
Anionic detergents					
Desoxycholate (DOC)		85 % (at phospholipid/DOC molar ratio 1/1) 2.4×10^6 g · min	Varying phospholipid/DOC ratio, 30', 0 °C	ATPase inactivation reversible after DOC removal	HuE [67]
Lithium diiodosalicylate (LIS)		N.D. 4×10^6	0.3 M LIS, 0.05 M Tris, pH 7.5. 20 °C, 15 min	Glycoproteins A, B, MN Blood Group activity	HuE [68]
Sodium dodecylsulfate (SDS)		Complete solubilisation 5×10^7 g · min	Membrane pellet SDS added to final concentration of at least 1 % (± reducing agent, s.a. mercaptoethanol or dithiothreitol). Incubation at 100 °C, 5 min. ~ 20 % lipids stay attached protein. Protein bind SDS in varying amounts	Activation of some proteases at conc. 1 % Complete denaturation; in general no immunological reactivity High resolution in SDS–PAGE. Mol. sieving Artefacts due to isopeptide formation	All types of Membranes (e.g. [47])
Sodium *N*-dodecyl-sarcosinate					Bacterial Membranes [69] Mouse Liver [59]

Cationic detergents (Cetyltrimethyl-ammonium bromide (CTAB)		Complete solubilisation	1 % CTAB	N.D. N.D.	HuE [70] Bo Retina [71]
Chemical modification	E.g., reduction of –S–S– linkages or blockage of complexing–SH groups. Inversion of intrinsic protein charges. Electrostatic repulsions				
PCMB, PCMBs		66 %	1–5 mM pCMBS	SDS-components 1, 2, 2.1, 4.1, 5, 6, No glycoprotein N.D.	HuE [72] HuE [73] HuE [74]
Succinylation, maleylation		25 % 1.6×10^7 g · min <10 % 7.2×10^7 g · min	Membrane protein (ghosts) in 0.03 M Na-borate, pH 8.8, 0.3 M maleic or succinic anhydride 1 % phospholipid contamination	No ATPase or Rh activity. Separation into 7 components in absence of detergents ABO antigenicity + PHA binding activity	HuE [75] HuE [76] HuE [77]

Footnote
HuE – Human erythrocytes
ShE – Sheep erythrocytes
BoE – Bovine erythrocytes
HoE – Horse erythrocytes
N.D. – not done

PM – plasma membrane
SDS – sodium dodecyl sulfate
CTAB – cetyl tetramethyl ammonium bromide
GHCl guanidine – HCl
PAGE – polyacrylamide electrophoresis

or use of a different solubilizer for membrane disruption and protein fractionation, can produce artifact.

2.2.3.2.1. Separation according to molecular weight

Gel-permeation chromatography. As reviewed in [79], gel permeation chromatography of membrane proteins solubilized in detergents or chaotropic agents can sometimes separate these proteins into broad classes; however, the technique does not yield high resolution.

Sodium dodecyl sulfate polyacrylamide gel electrophoresis (SDS-PAGE). At this writing SDS-PAGE is the most effective method for membrane protein separation [8]. In the presence of excess SDS (i.e. above 0.23%, the critical micelle concentration), most (but not all) proteins dissociate, unfold and bind SDS to form negatively-charged complexes of a size proportional to the protein molecular weight. When placed in an electric field, all of the complexes migrate toward the anode, but if the electrophoresis is carried out using cross-linked polyacrylamide of *suitable* porosity, the electrophoretic mobility generally varies exponentially with molecular weight, i.e.

$$\text{mol. wt.} = k \cdot 10^{-bx} \tag{2.1}$$

where k is a constant, x is the migration distance and b is the slope of a calibration curve obtained by use of proteins with known molecular weight. In general, SDS-PAGE can separate proteins ranging in molecular weight from $\sim 2.5 \times 10^3$ to 2.5×10^5 daltons. SDS-PAGE is usually performed in the presence of SH-reducing agents so as to disrupt intra- or inter-chain S–S linkages and to prevent adventitious formation of disulfide bonds. Until recently, SDS-PAGE was essentially an analytical tool, but high-resolution fractionations can now be achieved on a preparative scale [80].

Certain cautions must be exercised when applying SDS-PAGE to complex, unknown situations, for example, comparisons of the membrane proteins in normal and neoplastic cells. *First*, not all polymeric soluble proteins dissociate into monomers in the presence of SDS [8]. The same is true for membrane proteins, such as the 'spectrin' of erythrocytes [81] and a concanavalin A binding protein of thymocytes [66]. Indeed, the work of Maddy and Dunn [80] indicates that SDS can actually foster aggregation of proteins under certain conditions. *Second*, different proteins can migrate very closely, provided their SDS-complexes have similar shapes and sizes. That this is the case for erythrocyte membrane components has been shown by N-terminal analyses [82,83], by isoelectric focussing prior to SDS-PAGE [84] and by immunological methods [85]. *Third*, some proteins, glycoproteins in particular, migrate anomalously [45,86]. In glycoproteins this depends upon size and character of the sugar moieties. Since glycosylation processes can be abnormal in neoplastic cells, SDS-PAGE analyses of the proteins in the membranes of normal and neoplastic cells can produce ambiguous results.

2.2.3.2.2. Separation according to charge

Zonal Electrophoresis. Some membrane proteins can be separated by zonal electrophoresis in polyacrylamide using area plus fairly concentrated acetic or propionic acids as solubilizers [87–89]. Nonionic detergents have also been employed [90]. The organic acids act as solubilizers, confer a positive charge on the proteins and block conversion of urea to cyanate (which reacts with protein amino groups).

Isoelectric focussing. In this potent analytical technique [91], proteins are separated according to their isoelectric points in pH gradients stabilized by inert 'ampholytes'. This method has now been successfully modified for membrane analyses.

As first applied to membrane proteins by Merz et al. [92], using urea as principal solubilizing agent, the method did not yield satisfactory resolution. Much better results were obtained by Bhakdi et al. [84] in their separation of the erythrocyte membrane proteins, extracted by dilute EDTA, on polyacrylamide laden with 8 M urea. However, neither study controlled possible modification of proteins by cyanate formation at the acid end of the pH gradient [91]. However, this obstacle has been overcome by Bhakdi et al. [93] using Triton X-100, both to solubilize erythrocyte membranes and to maintain protein solubility during focussing. Excellent separations particularly of glycoproteins, are achieved in this fashion, even on a preparative scale [93].

Isoelectric focussing has also been applied successfully to the separation of membrane proteins from plasma membranes of thymocytes and SV40-transformed lymphocytes [94]. The proteins are extracted, using 1 % Triton X-100 ethanol (0.001 M/0.001 M) and focussed in Triton/8 M urea in 5 % polyacrylamide containing 2 % ampholyte (pH range 3.5–10). Equilibrium focussing conditions were established in these experiments.

Two dimensional systems. Much additional resolution, as well as structural information, can be obtained by combining different separation principles in a two-dimensional approach.

(a) *SDS-PAGE ± reducing agents:* Here the membranes are separated in the first dimension by SDS-PAGE without reducing agent, using a conventional cylindrical gel. This gel is then applied to one edge of a square polyacrylamide slab, laden with SDS + reducing agent, and electrophoresis repeated in the second dimension, that is, at 90°. This approach defines proteins containing S–S linkages and has been successfully used to characterize complement components bound to erythrocyte membranes during complement-induced cytolysis [95,96].

(b) *PAGE-SDS-PAGE:* In this approach, proteins are separated by charge in the first dimension (e.g., using urea-acetic acid) and according to molecular weight (SDS-PAGE) in the second dimension.

(c) *Isoelectric focussing-SDS-PAGE:* This powerful two-dimensional method has been developed recently by Bhakdi et al. [84,93], using first urea and now Triton X-100 to dissolve membrane proteins and to keep them solubilized during focussing. Focussing is carried out in the first dimension and SDS-PAGE in the second dimension. It has also proven possible to focus certain proteins isolated by preparative SDS-PAGE [93].

(d) *Crossed-immunoelectrophoresis (CIE):* CIE is another very recently developed method which allows separation and analysis of proteins according to their antigenic properties (as well as their complement of specific receptors, e.g., for lectins). The method thus seems particularly suited for the analysis of membrane antigens, lectin receptors, hormone receptors, etc.

CIE is best carried out using non-ionic detergents (e.g., Triton X-100, Lubrol, Berol) as solubilizers; these can, in some cases, solubilize as much as 90% of membrane protein (Section 2.2.2.; Table 2.1), do not generally interfere with the antigenic properties of membrane proteins, permit immunoprecipitation reactions, and conserve many enzymatic functions. CIE has already been successfully applied to the high-resolution analysis of erythrocyte membrane proteins [65,96] and to the characterization of the concanavalin A receptors of thymocyte plasma membranes [66]. In CIE, membrane proteins solubilized in detergent are electrophoresed in the first dimension (according to charge), using agarose as support. The separated proteins are then electrophoresed at right angles *into* an agarose containing suitable antibody. Electrophoresis in the second dimension is carried out under identical conditions as used in the first separation, *except* that the pH is adjusted to the isoelectric point of γ-globulins, which thus remain immobile. Accordingly, a set of characteristic precipitation arcs appear, which can be recorded at high sensitivity.

CIE has major advantages over conventional immunoelectrophoresis, such as applied to liver membranes by Raftell et al. [97]. *First*, the height of the precipitation bands gives an accurate measure of the amount of antigen present. *Second*, the method readily detects immunologic cross reactivity between proteins. *Third*, electrophoresis in the second dimension can be carried out using polyvalent ligands other than immunoglobulins. (Cf. also Chapter 6, ref. 10.)

2.2.3.2.3. *Affinity chromatography*

Certain membrane proteins bear receptor sites for specific ligands (e.g., for lectins or immunoglobulins). When solubilized in a fashion that does not destroy the receptor sites, these proteins can be adsorbed onto columns, covalently charged with the respective ligand, and specifically eluted by using an appropriate receptor equivalent (e.g., a hapten). This method is now widely utilized in membrane studies, and has recently permitted the selective purification of the concanavalin A receptor in thymocyte plasma membranes [66]. The method can be applied to both analytical and preparative separations.

2.2.4. Methods for the analysis of membrane protein turnover

Turnover studies are intended to measure biologic lifetimes. They involve measurement of the relative rates of incorporation and release of radioactively labelled membrane precursors by functionally intact cells or tissue. For this, the specific activity of purified membranes or membrane components are determined at appropriate intervals after labelling. For protein labelling, essential ^{3}H- or ^{14}C-labelled amino acids are used most commonly. To ensure adequate label uptake *in vivo*, animals are fasted for about one day before the label pulse; for satisfactory *in vitro* incorporation, the cells are precultivated in media depleted in the component(s) to be used for labelling. (The metabolism of lipids and carbohydrates, the two other major membrane constituents are analyzed, using appropriate radioactive lipid precursors [e.g., acetate, glycerol, or choline] or labelled monosaccharides, respectively.)

These methods can give much information about the extent of membrane protein and lipid turnover, as well as about the kinetics and mechanisms involved. The flow of radioactivity after a label pulse yields information about the sites of synthesis of membrane components, their transport within cells, sites of glycosylation, etc.

2.2.4.1. Standard pulse-labelling methods

The easiest way to label newly synthesized proteins *in vivo* and *in vitro* is to equilibrate functionally intact cells with essential, radioactively-labelled amino acids. Because leucine is a prevalent residue in most proteins and most proteins contain similar proportions of this amino acid, uniformly labelled L-[^{3}H]- or L-[^{14}C]leucine is the most commonly employed label. However, use of a single amino acid for turnover studies on different membrane proteins may generate error, since not all membrane proteins have precisely identical amino acid proportions. This source of error does not exist in turnover measurements of a single component under different conditions.

Another source of error in pulse-labelling experiments derives from label-reutilization; this gives artifactually high protein lifetimes. Reutilization can be avoided (and/or its importance assessed) by several techniques. For example, one can employ arginine ^{14}C-labelled in the guanidino moiety (L-[^{14}C]guanidino-arginine), as precursor [98,99]. In tissues with high arginase activities, the radiolabel is reutilized to a negligible degree, because of hydrolysis of guanidine to urea, which cannot be reutilized.

Another approach is to label with [^{14}C]carbonate. This method [100, 101] avoids reutilization, since ^{14}C-incorporation is restricted to amino acids derived from Krebs cycle intermediates.

As still another means for reducing label reutilization, the cells or tissues can be incubated with inhibitors of protein synthesis after the label pulse. However, these chemicals will also inhibit synthesis of proteins active in the biodegradation of membrane proteins and their use will thus interfere with turnover studies.

To determine the kinetics of biosynthesis and biodegradation by methods involving single isotopes, the membrane or membrane proteins have to be isolated for each time point. This is time consuming and requires amounts of membrane material which are too large for most systems. Thus, Gurd and Evans [102] could not obtain sufficient sensitivity to determine whether various SDS-PAGE components of liver bile front membranes exhibit distinctive turnover rates. Schmidt-Ullrich et al. [103], in their study of thymocyte membrane metabolism, overcame this problem by use of a mixture of labelled amino acids; this was acceptable because reutilization did not constitute a problem in their study.

2.2.4.2. Double labelling

The double labelling approach was designed by Arias et al. [99] to measure the different degradation rates of various membrane proteins. It can be used for *in vivo* studies or for *in vitro* experiments where the cells can be maintained in culture for at least 10 days. Amino acids labelled with different isotopes (^{3}H or ^{14}C) are supplied to the cells sequentially. The first isotope, ^{14}C, is supplied as a pulse and, after a defined time, the system is pulsed with the second isotope. A few hours later the membranes under study are isolated and their ^{14}C/^{3}H ratio determined.

This method does not yield *absolute* lifetimes, but can provide a precise measure of relative turnover under certain conditions: *First*, if turnover is defined in terms of biosynthesis and biodegradation, all proteins studied must be in the process of isotopic dilution. This requirement is critical in cases where proteins are transported from one subcellular domain to another prior to degradation. With liver microsomes, this factor is not important when the membranes are isolated within 6–8 hr after the second pulse [99, 104]. *Second*, the isotopes must not be transferred to non-protein membrane components. *Third*, protein labelling must follow exponential decay kinetics; these have been established for liver microsomes, their isolated protein component and many other membrane proteins. *Fourth*, the synthetic rates of the various protein constituents must be the same at the time the isotopes are administrated. This may be important, since the synthesis of various proteins can follow a circadian rhythm [105].

2.2.4.3. Carbohydrate labelling

The principles outlined for measurement of protein turnover also apply to metabolic studies on the carbohydrate moieties of glycoproteins. The precursors used most commonly are [^{14}C]galactosamine, [^{14}C]glucosamine, a precursor for neuraminic acid and [^{14}C]fucose (fucose is a component of many blood group substances; Chapter 4). For the study of carbohydrate addition to immunoglobulins, other labelled sugars (e.g. galactose and mannose) have also been employed [6]. The conjoint use of carbohydrate label-pulses with subsequent subcellular fractionation and/or radioautography has served to elucidate the glycosylation processes of surface membrane proteins; in many cell types this occurs predominantly in the Golgi apparatus, but some peptide-carbohydrate linkages may also be formed in the rough endoplasmic reticulum.

2.2.4.4. Labelling of surface membrane proteins

2.2.4.4.1. Enzyme-catalyzed radioiodination

This method, developed by Phillips and Morrison [106] and Marchalonis et al. [107] radioiodinates protein tyrosyl residues (usually with ^{125}I) in an H_2O_2 – requiring reaction catalyzed by lactoperoxidase. Because molecules as large as the peroxidase do not rapidly or freely penetrate the plasma membranes of viable cells, reactive tyrosines on the external surfaces of the membrane are labelled preferentially. The peroxidase method has been used by several groups [5,108] to study the turnover of membrane proteins and immunoglobulins in lymphocytes. In these studies, the radioiodinated cells are washed repeatedly, placed in culture and the release of labelled proteins measured. The release of immunoglobulins or other well-defined surface components (e.g., H2 antigens, lectin receptors) can be quantified by immunoprecipitation, using specific antibodies [109,110]. Surface-radioiodination has also been combined with preliminary metabolic labelling with [^{3}H]- or [^{14}C]leucine in studies of fibroblast membrane metabolism [111].

2.2.4.4.2. Labelling of galactosyl residues on glycoproteins

Gahmberg and Hakomori [112] have developed an elegant method for the tritiation of certain galactosyl residues exposed at the surfaces of intact cells. The cells are first treated with galactose oxidase and the modified galactosyl residues subsequently reduced with tritiated borohydride. Glycolipid galactosyl residues react in addition to galactose residues on glycoproteins.

2.2.4.4.3. Covalent labelling with small molecules

A large variety of radioactively-labelled or fluorescent charged, low-molecular weight reagents have been tested as possible covalent markers for the external surfaces of animal cells, erythrocytes in particular. Several of these reagents are discussed in Chapter 8 in terms of their effects on surface charge. The use of these agents in membrane structural analyses has been critically reviewed in [113] and it appears that they do not constitute suitable labels for metabolic studies: *First*, the labels penetrate into the cell interior to a greater or lesser degree. *Second*, the reagents cause severe structural damage to the plasma membranes of most cells.

2.2.4.5. Critical aspects of turnover measurements

Until recently, protein lifetime determinations have been carried out rather empirically. However, Poole [114] has developed a more rigorous statistical, approach which we will briefly describe.

Using leucine labelling as an example, Poole [114] shows that the radioactive amino acid leaves the 'free' amino acid pool very rapidly during the first day and rather slowly during the following nine days. The kinetics involved can be described by the formula

$$F(t) = \frac{16}{0.25 + t} + \frac{1.02}{(0.055 + t)^2} \tag{2.2}$$

where $F(t)$ = the specific activity of 'free' pool and t = the time after injection of precursor. Poole [114] follows the decay of radioactive leucine in rat liver catalase for 1 to 10 days after the tracer injection. When the decay curve is plotted linearly, a straight line results, suggesting a half-life ($t_{\frac{1}{2}}$) of 3.5 days. However, if all precursor is incorporated before starting the measurements, decay should follow the equation

$$\frac{dP(t)}{dt} = KP(t) \tag{2.3}$$

where $P(t)$ = specific activity and K = rate of turnover, or in the case of significant reutilization, the equation

$$\frac{dP(t)}{d(t)} K[F(t) - P(t)] \tag{2.4}$$

where $P(t)$ = specific activity of precursor in the protein and $F(t)$ = the quantity of the pool of free precursor.

Poole [114] used a digital computer to test the data and obtained a value of 0.471 day, corresponding to a true $t_{\frac{1}{2}}$ for liver catalase of 1.47 days rather than 3.5 days.

Label reutilization significantly complicates interpretation of $t_{\frac{1}{2}}$ data. Thus, computations, in which various K values are inserted into equations 2.3 and 2.4, show that the initial increase of specific radioactivity of a given protein does not occur *immediately* upon tracer pulsing, but can be variably delayed, depending on the $t_{\frac{1}{2}}$ of the protein. For example, proteins with $t_{\frac{1}{2}}$ values of one and five days show maximal labelling after 18 hrs and sixty hours respectively.

Moreover, the computations indicate that proteins with true $t_{\frac{1}{2}}$ values <2 days, will yield apparent $t_{\frac{1}{2}}$ values of 3–4 days when the specific activity is measured only one day after pulse labelling. Furthermore, proteins with $t_{\frac{1}{2}}$ values >1 day yield rather undiscriminating decay curves.

The complications due to small differences in decay rates a long time after isotope administration apply also to the double labelling approach. Indeed, the timing of the second pulse is critical and, contrary to the suggestion of Arias et al. [99], equation 2.4 indicates that the second isotope should be supplied as late as possible, that is, briefly before cell disruption. The optimum time for administration of the first isotope should allow about six half-lives before cell disruption.

Poole's data [114] further indicate that many published data about intracellular transport of newly synthesized proteins labelled *in vivo*, cannot be interpreted unambiguously. Moreover, due to its extensive reutilization, a 'pulse' of leucine cannot really 'pulse label' most membrane (or other) proteins.

As a general principle, one can best determine the half-lives of proteins that turn over rapidly, either by pulse labelling with leucine and measuring decay during the first day, or by use of the double-label method. For proteins having half-lives longer than one day, the situation becomes very complex. The specific activity decay tends to be pseudo-exponential over a period of ten days and the decay-constants depend on the half-life of each protein. Accordingly, the differences in $t_{\frac{1}{2}}$ values can often not be established within statistically reliable limits.

The double-labelling method is complicated by the possibility that many proteins analysed are not absolutely homogenous and may therefore be contaminated by components with substantially different turnovers. The only reliable way to establish meaningful $t_{\frac{1}{2}}$ values for proteins with very slow turnovers is the use of L-[^{14}C]guanidino-arginine or [^{14}C]carbonate as precursors.

Even though surface labelling by lactoperoxidase-catalyzed iodination of reactive tyrosines is rather easy and sensitive, it is not without its own sources of error: *First*, the procedure must be performed on cells *washed free of 'soluble' proteins*, which would otherwise react in preference to proteins that are tightly membrane-associated (cf. Section 2.2.5). This necessity introduces an arbitrary element into what is defined as a 'surface' protein. *Second*, the technique, as employed, does not distinguish between proteins which are transiently associated with the cell surface and those which are 'intrinsic' plasma membrane components. Labelling of the former category could yield anomalously low $t_{\frac{1}{2}}$ values, such as reported for some lymphocyte 'surface' components (Section 2.5). *Third*, the lactoperoxidase technique requires washing of the treated cells (serum is often used in this step). and some labelled surface components can be eluted by washing. The contention of Gates et al. [115] that one can identify 'surface peptides' of complex cells unambiguously without subcellular fractionation thus appears premature.

Comparisons between normal and neoplastic cells may engender additional ambiguities, since two different cell categories need not have identical affinities for reactive 'membrane-associated' proteins (cf. Section 2.2.5).

The complications discussed here may also occur in studies using galactose oxidase since glycoproteins (Section 2.2.5) and glycolipids (Chapter 4) can adsorb to plasma membranes and/or exchange between such membranes and other compartments.

2.2.5. Interchange of proteins between membranes and their soluble environment

2.2.5.1. Introduction

Membrane proteins associate with membrane lipids and with each other through non-covalent interactions. Because of this fact, one of us [8] has suggested the possibility that some membrane proteins might exchange to some degree with their soluble intra- and extra-cellular environments. As detailed in Chapter 1, such postulated transitions from the 'membrane state' to a 'soluble state' might face formidable energetic barriers, would probably require structural rearrangements and might therefore require metabolic energy.

There is now increasing evidence, for many cells, that membrane components can interchange with identical entities in the intra- and extracellular space. Such exchange can seriously complicate evaluation of turnover data, as well as comparisons between membranes from normal and neoplastic cells (cf. Section 2.4.3). We will, therefore, treat the phenomenon in some detail.

2.2.5.2. Erythrocytes

Several workers have presented evidence concordant with the existence of protein exchange between *erythrocyte* membranes and their environment. Erythrocytes are particularly suited for such studies, since their mature forms are not capable of protein synthesis. However, if membrane protein exchange or interchange occurs with these cells, one must anticipate similar processes in other cells, both normal and neoplastic.

At least some of the membrane proteins of adult erythrocytes are synthesized by reticulocyte ribosomes [116,117]. Thus, Lodish and Deslau [117], using [^{35}S]methionine as precursor find, that both intact reticulocytes and mRNA-containing lysates thereof synthesize identical non-hemoglobin proteins. Two of these, that is, components B_2 and E, with mol. wt. of 53,000 $\pm$ 2000 and 33,000 $\pm$ 2000, respectively, are associated exclusively with the membrane fraction from intact cells [116]. Component B_2 is synthesized by intact reticulocytes; in contrast, the corresponding component B_1, synthesized by reticulocyte lysates, has a mol. wt. of some 2000–4000 (20–40 amino acids) greater than B_2, suggesting that a small peptide segment might be split off before the remainder is incorporated into the membrane.

Further evidence indicating that the [^{35}S]methionine labelled component B_1 is the precursor of B_2 was obtained by trypsin digestion of both components; analysis of the resultant peptides by paper ionophoresis at pH 3.5, showed very similar peptide profiles for components E, B_1 and B_2.

Lodish and Deslau [117] report a time-dependent increase in the synthesis of component B_2, but Deutsch and Blumenfeld [118], who used [^{14}C]isoleucine as precursor found only very minor label incorporation into spectrin. These authors' data also indicate that reticulocytes can synthesize not only two, but a whole series of membrane proteins (mol. wt. 20,000–200,000). Ribonuclease treatment of membranes did not alter the pattern of membrane protein synthesis, indicating that peptides bound to ribosomes do not significantly contribute to the membrane protein pattern observed. Significantly, the [^{14}C]isoleucine specific activity increased at the same rate in hemoglobin as in the membrane proteins, suggesting that both categories of protein are synthesized on the same polysomes and processed in a similar fashion [117].

Once inserted into their membrane environment, erythrocyte membrane proteins probably do not turn over metabolically. This has been investigated by Morrison et al. [119], who studied rat erythrocytes *in vivo*. They tagged red cells with ^{51}Cr and labelled their surface membrane proteins by lactoperoxidase-catalyzed iodination. The 5 membrane components (mol. wt. 90,000–120,000) labelled by ^{125}I seemed to decay at the same rate *in vivo*, indicating that these components at least do not exhibit differential turnover rates (such as found for many plasma membrane components in nucleated cells). The decay curves for ^{51}Cr and ^{125}I were similar, indicating that the decay of radioactivity in blood samples taken after 1, 19 and 404 hr is due to *cell* turnover and not to turnover of membrane components.

On the basis of the very slow *in vivo* loss of ^{125}I from intact rat erythrocytes after *in vitro*, lactoperoxidase-catalyzed iodination one might suspect that erythrocyte membrane proteins stay with their cell for the duration of its life. However, a recent report by Yarrison and Choules [120] indicates exchangeability of the major glycoprotein of human erythrocytes; this protein is one of the principal components labelled when intact cells are radioiodinated by the peroxidase technique.

Yarrison and Choules [120] isolated the protein by the method of Marchesi and Andrews [68] and radioiodinated it with ^{125}I, using the chloramine T-method [121]. They then exposed intact, washed human erythrocytes to the labelled glycoprotein and observed an 80-fold greater uptake than of human serum albumin, similarly iodinated. Binding followed typical adsorption kinetics; it was first order during the first hour and reached saturation in 3–4 hr. A 175-fold excess of serum albumin did not change adsorption of the glycoprotein. The data do not allow the conclusion that the glycoprotein is inserted normally into the erythrocyte membrane and it will be important to determine whether it exhibits as stable a membrane association as the ^{125}I-labelled proteins of lactoperoxidase-iodinated erythrocytes [119].

Recent evidence also indicates that certain plasma proteins can associate very intimately with erythrocyte membranes. Thus, experiments using radioiodinated albumin [122] and immunologic techniques [65] show that erythrocyte membranes produced by the standard method of Dodge et al. [123] bind traces of serum albumin, which cannot be eluted by other than drastic measures.

Even more significantly, Langdon [83] has reported immunological cross reactivities between membrane proteins from human erythrocytes and certain serum lipoproteins. Langdon [83], moreover, shows identical N-terminal amino acid residues in several erythrocyte membrane SDS-PAGE bands and in certain apo-lipoproteins. (As earlier reported by Knüfermann et al. [82], Langdon finds multiple N-terminal residues in all of the SDS-PAGE components, attesting to their heterogeneity.)

Langdon [83] finds that about half of the erythrocyte membrane proteins react immunologically with antisera raised against 'pure' plasma lipoproteins. He concludes that the apo-lipoproteins represent a quantitatively significant proportion of the membrane. However, the data of Bjerrum et al. [124], showing no immunologic cross reaction between Triton X-100-extracted human erythrocyte ghosts and human β-lipoproteins by crossed immune electrophoresis are not in accord with Langdon's results.

2.2.5.3. Cultured fibroblasts

Kapeller et al. [125] have presented evidence that cultured chick embryo fibroblasts continuously shed glycoproteins bearing receptors for wheat germ agglutinin. This shedding seemed to be largely responsible for the rapid turnover (1–8 hr; glucosamine labelling) of surface glycoproteins. Kidney fibroblasts from C3H mice also shed surface material containing H2 antigens.

An important example of exchange has been documented by Yamada and Weston [126]. These authors show that a major, 220,000 dalton glycoprotein synthesized by fibroblasts, particularly in high density cultures and associated with their external cell surfaces, very readily exchanges into the culture medium and on to other cells. These observations are particularly significant, since many *neoplastic* fibroblasts show low levels of this protein (cf. Section 2.5.3).

2.2.5.4. Mouse mammary ascites adenocarcinoma TA3

Cooper et al. [127] have recently shown that a sialo-glycoprotein, bearing N-type blood group determinants, is shed into the ascites fluid and serum of animals bearing the Ha subline of TA3. This protein, as purified by perchloric acid extraction (0.6 M) and gel permeation chromatography, has a mol. wt. ~ 600,000. A large fragment of the protein (~460,000 daltons), bearing N-blood group receptors, can also be released by mild proteolysis of intact cells.

2.2.5.5. Lymphoid cells
2.2.5.5.1. Immunoglobulins (Ig)
A huge literature exists dealing with the metabolism, turnover and release ('shedding') of Ig and Ig-chains attached to the surface of diverse lymphoid cells. Much of this literature has been recently reviewed by Marchalonis and Cone [5].

Surface labelling studies. Several workers have attempted to evaluate the dynamics of lymphocyte plasma membranes in experiments relying exclusively on lactoperoxidase-catalyzed radioiodination on intact cells.

According to Cone et al. [128], when mouse bone marrow-derived (B) cells, thymus-derived (T) lymphocytes, mouse SIAT 4 lymphoma or mouse MOPC 460 myeloma cells are iodinated with ^{125}I using the lactoperoxidase procedure and then maintained in culture, one observes a continuous release of label into the medium. This 'shedding' process requires active cell metabolism and is blocked by inhibitors of energy metabolism (e.g. azide; iodoacetamide), low temperature (4 °C) and, to a lesser extent, by inhibitors of protein synthesis (e.g. puromycin).

Cone et al. [128] have measured the loss of ^{125}I-labelled surface proteins from thymic and splenic lymphocytes kept *in vitro* for 1–3 hr after lactoperoxidase-mediated iodination of whole cells. The same group [129] also used specific immunoprecipitation with rabbit anti-mouse Ig G (RAM IgG) to evaluate Ig release after iodination. For this, they extracted cells using 9 M urea/1.5 M acetic acid/0.2 % mercaptoethanol, at various intervals after iodination, and fractionated the extracts by PAGE, using the same medium. They found rapid release of ^{125}I-labelled material from cells into the medium, without loss of viability. A large proportion of the released surface-labelled material was Ig, as documented by specific precipitation of the released material with RAM IgG [129].

Cone et al. [128] and Marchalonis et al. [129] found a quasi-linear release of total, labelled cell-surface Ig, over a period of at least 2 hr, but B-cells from athymic mice yield the fastest release, and T-cells shed Ig at one fifth this rate. Normal splenic and thymic cells fell in between. Double labelling methods showed that the released immunoglobulins were in fact synthesized by the cells, and that the shedding phenomenon represents rapid membrane turnover. The rate of

Ig-release observed fits the observations of others [6,108] and is 100-fold less than the rate of Ig-secretion by plasma cells.

'Shedding' of the cell surface material by lymphocytes is not limited to Ig. Thus Cone et al. [128] have measured the overall *in vitro* release of ^{125}I-label from the surfaces of mouse thymocytes and splenic lymphocytes after lacto-peroxidase-catalyzed radioiodination. They find that more than 50% of the surface label is 'shed' into the medium within *one hour* after iodination in both of these cell categories, a mouse lymphoma (SIAT 4) and a mouse myeloma (MOPC 460). The data of Cone et al. [128] indicate that surface material is released three times as fast from thymic cells as from splenic cells, and the rate in splenic cells is reduced to one third when RAM Ig is bound to the cell surface.

Whatever the difference between the thymic and splenic cells, it cannot be explained by different proportions of membrane-bound immunoglobulin, since the distribution of ^{125}I upon PAGE is very different in extracts from the two cell types. It appears more probable that the iodination procedure does not label membrane proteins of splenic and thymic cells, either representatively or equivalently.

Using surface radioiodination, Vitetta et al. [130], have shown that the predominant Ig molecule on the surface of mouse splenic lymphocytes is monomeric 8S IgM. Under tissue culture conditions, this molecule is rapidly released *attached to plasma membrane fragments* [131] containing non-covalently bound lipids and other plasma membrane proteins. This shedding process takes place in the presence of inhibitors of protein synthesis and cytochalasin B (concentration 10^{-4}–10^{-6} M). On the other hand, the release requires metabolic energy and can be inhibited by low temperature. The half-lives determined were 8 hr for non-Ig plasma membrane proteins, 6–8 hr for membrane associated Ig, much longer than indicated by the data of Cone et al. [128], even though the same labelling method was used. A lag phase in release of 1–3 hr is explained by exchange with an intracellular pool.

It is difficult to reach firm conclusions from experiments relying solely on lactoperoxidase iodination for reasons given in Section 2.2.4.4. Moreover, in the case of lymphoid cells, evaluation is further complicated by the rapid internalization of macromolecules at temperatures above 25°C (cf. Chapter 10). This complication is documented on rabbit thymocytes by Schmidt-Ullrich et al. [38], who find that, when the cells are radioiodinated by the lactoperoxidase procedure as used by Marchalonis et al. [129] and Cone et al. [128], about 30% of the covalently-bound ^{125}I ends up in the cytoplasmic proteins within less than 30 min after iodination at 30°C.

The turnover rates suggested by the radioiodination experiments of Cone et al. [128] and Marchalonis and Cone [129] are much much greater than the value, of 8 hr for macrophage plasma membranes [32] and hepatocyte bile fronts [105, 140] or 24 hr for fibroblast membranes [155], all obtained by metabolic labelling with radioactive amino acids. They are also very different from the mean half-life of thymocyte plasma membrane proteins, 24–28 hr, computed for resting cells by Schmidt-Ullrich et al. [103], again using metabolic labelling with radioactive amino acids. The most reasonable explanation for the discrepancy is that radioiodination by the lactoperoxidase method tends to preferentially label *loosely-bound* surface proteins [8]. That this is indeed so in the case of fibroblasts has already been pointed out (Section 2.2.5.3).

Studies involving metabolic-plus surface-labelling. Vitetta and Uhr [131,132] have compared the release of intrinsically-labelled-([^{3}H]tyrosine)-Ig with that of extrinsically-labelled [^{125}I]Ig from normal mouse-spleen lymphocytes. Their data suggest that two distinct mechanisms are involved. Thus, whereas the shed radioiodinated material is 8S [^{125}I]IgM and is attached to membrane fragments, the released [^{3}H]tyrosine Ig consists of IgG, 19S IgM and 8S IgM, all not membrane attached.

Vitetta and Uhr [133], Vitetta et al. [134] and Uhr and Vitetta [135] have extended these studies, using axenic mice which synthesize only IgM, of which 8S IgM becomes surface bound and 19S IgM is secreted. Their data indicate that, after synthesis on the rough endoplasmic reticulum, IgM is transferred to the Golgi apparatus. Thereupon, the processing of the two categories of IgM differs, as reflected by the longer half-life of 'shed' 8S IgM, 5–6 hr, compared with secreted 19S IgM, 2–3 hr. Part of the IgM bound to the Golgi apparatus remains attached to the inner surfaces of the Golgi membranes, while the rest is released into Golgi vesicles and secreted into the medium by fusion of these vesicles with plasma membranes. This intracellular processing is reflected by the 1–2 hr lag between a [^{3}H]tyrosine pulse and appearance of the IgM at the cell surface. The authors' data also show that labelled Ig, released from the surface within 2 hr after a 90 min [^{3}H]tyrosine pulse, comes primarily from plasma cells which make up 1–5% of a splenic lymphocyte population. IgM released later than 2 hr after the end of the tyrosine pulse represents membrane-associated Ig.

No IgG or α-chains are detected (in contrast to cells from nonaxenic mice). Most of the intracellular [^{3}H]tyrosine-labelled Ig is 8S IgM, but 10–25% is 19S and 5–10% of the ^{3}H is in free chains. However, 90% of the labelled Ig recovered from the culture medium is 19S IgM. This finding is in accord with the results of Melchers and Anderson [6], showing that polymerization of IgM occurs intracellularly.

Finally, using lactoperoxidase-catalyzed radioiodination, Vitetta and Uhr [131] demonstrate that Ig accounts for about 20% of the ^{125}I-labelled protein on the surfaces of splenic lymphocytes from axenic mice. Most of this Ig is 8S IgM, with only a small proportion of 19S IgM, or free chains. Only 7.8 % of [^{3}H]tyrosine-coprecipitated material is in μ-, γ- and α-chains [134]. Significantly, only 8% of the ^{125}I-labelled material coprecipitated by anti-Ig can be identified as Ig. This demonstrates that Ig is shed, together with other membrane proteins. One of these proteins is a 30,000–40,000 dalton component, synthesized and secreted after antigenic stimulation and bound to the antigen–antibody complex.

Because inhibitors of protein synthesis do not influence the release of preformed surface proteins and because the release kinetics are linear rather than exponential, it is unlikely that enzymes participate critically in the 'shedding' process. On the contrary, the data suggest that membrane-associated Ig (and some proteins) exist in an equilibrium or steady state distribution between the 'membrane phase' and the extra- and intra-cellular fluid phases (cf. Chapter 1).

This concept has been analysed in detail by Lerner et al. [136] in their studies on the differentiation of Ig-bearing B-type lymphocytes into Ig-secreting plasma cells. In these experiments, Lerner et al. [136] used a clone of continuously-replicating human lymphocytes. The differentiation of this clone is arrested at a stage between the G_0 lymphocyte and a Ig-secreting plasma cell. The cells both *secrete* Ig and bear Ig on their surfaces.

Using [^{3}H]leucine to measure Ig production and puromycin to block protein synthesis, Lerner et al. [136] find that the half-lives of κ-chains and Fc fragments in logarithmically growing cells are about 45 min. The initial half-disappearance time for cytoplasmic κ-chains and Fc fragments is 2 and 1.5 hr, respectively. Stationary cells arrested in the G_1-phase of the cell cycle have the same half-life for membrane Ig, but are depleted in cytoplasmic Ig when compared to logarithmically growing cells; the latter exhibit about 10-fold more cytoplasmic Ig than membrane Ig, even 6 hr after puromycin treatment. In contrast, actinomycin D prolongs the persistence of cell surface Ig but depletes the cytoplasmic pool by 90%. The data further indicate that at least 90% of the Ig synthesized by logarithmically-growing cells is secreted and therefore does not saturate membrane Ig-binding sites. Finally, the results show that synthesis of Ig and its attachment to the plasma membrane are regulated independently.

The immunological analysis by Hüttenroth et al. [137] of the surface expression and secretion of Ig during the cell cycle of 10 different human lymphoid cell lines, supports the results of Lerner et al. [136]. These workers show that synthesis of secreted Ig occurs predominantly in late G_1 and early S, whereas membrane bound Ig remains nearly constant throughout the cell cycle (except for small fluctuations in bound κ- and μ-chains). In cells synchronized by double-thymidine block, the surface density of κ-chains drops slightly before mitosis, but there are no changes in surface κ-chains or μ-chains during G_1 or early S.

2.2.5.5.2. HL-A Antigens

An abundance of evidence indicates that HL-A antigens, which are surface components of human lymphocytes, are shed into the serum and even the seminal plasma. HL-A antigens are also readily released into the culture medium by a variety of human lymphoid lines. Pellegrino et al. [138] have purified these antigens by extracting membranous debris in exhausted culture fluid with 3 M KCl, followed by preparative PAGE.

More recently, Oh et al. [139], were able to isolate and purify large amounts of HL-A from normal serum. SDS-PAGE analyses of the serum HL-A material showed four components with molecular weights of 24,000, 36,000, 50,000 and 90,000. Unlike the HL-A material extracted from culture media, the serum components were stable, even at pH 2.0 and could be renatured after treatment with 4 M urea or 0.1 % SDS.

Billings and Terasaki [140] have combined QAE Sephadex ion exchange chromatography with affinity chromatography on immobilized concanavalin A and preparative PAGE to isolate a homogeneous glycoprotein (10–12 % sugar) with a molecular weight of ~130,000. This accounted for ~17 % of the HL-A in the serum.

Miyakawa et al. [141] have demonstrated that there is a diversity of HL-A positive proteins in plasma. Using gel permeation chromatography and specific assays for $HL\text{-}A_{common}$, $HL\text{-}A_1$ and $HL\text{-}A_2$, they separated and identified three classes of HL-A-positive proteins with molecular weights of $2–8 \times 10^5$, 48,000 and 10,000. The $2–8 \times 10^5$ dalton molecules were identified as distinct membrane domains, carrying the intact HL-A antigen, and possibly also other membrane entities. The 48,000 dalton component appeared to be identical to HL-A fragments released from cultured lymphocytes by papain digestion. Indeed, the 48,000 dalton fragment can be released also from the $2–8 \times 10^5$ dalton component by papain treatment. The 10,000 dalton fragments do not carry $HL\text{-}A_1$ or $HL\text{-}A_2$ receptors and only a portion of the $HL\text{-}A_{common}$ activity of the larger fragments. It thus appears that the $2–8 \times 10^5$ dalton entity is a shed components, and that the 48,000 and 10,000 dalton fractions are fragments thereof, possibly arising from proteolytic degradation.

Surface antigens are also shed from mouse thymocytes [142]. When these cells are radioiodinated by the lactoperoxidase procedure, both H2 and Thy-1 antigens become labelled and Thy-l (mol. wt. ~30,000) is released in substantial amounts during 6 hr incubation *in vitro*. As in the case of Ig, material labelled by surface iodination is shed at a greater rate than antigen labelled metabolically.

2.2.5.5.3. The effects of various specific macromolecular ligands on the release of lymphocyte surface membrane components

Interaction of antigens with surface immunoglobulins. The surface immunoglobulins of bone-marrow derived lymphocytes and, to a lesser extent, thymus-derived lymphoid cells can bind specific antigens, leading to their activation and differentiation.

Rolley and Marchalonis [143] have investigated the interaction of DNP-hemoglobin (9, 12 or 20 moles DNP/mole protein, DNP_9 Hb, DNP_{12} Hb and DNP_{20} Hb) or hemocyanin (Hc), both labelled either with ^{125}I or ^{121}I, with the surface-Ig of splenic lymphocytes or thymocytes obtained from unsensitized normal mice and B-cells obtained from normal mice and congenitally athymic *nu/nu* mice. They measured the release of Ig and antigen-Ig complexes into the medium using two immunoprecipitation methods. In their first approach, they interacted uniodinated lymphocytes with $[^{125}I]DNP_9Hb$, added unlabelled, purified mouse IgG to the material released by the cells, and precipitated with RAM IgG. In their second approach, they labelled cell surface proteins via lactoperoxidase-catalyzed ^{125}I-iodination, and precipitated material released into the medium after addition of carrier $DNP_{20}Hb$ or Hc, using rabbit antisera specific for DNP-Hb or Hc. In some experiments they also treated cells with trypsin (2–10 μg/ml; 20 min, 37°C) and stopped the reaction with trypsin inhibitor.

They found that ~25% of $[^{125}I]DNP_9Hb$ bound to the cells was released within 1–2 hr cultivation at 37°C, but only very small amounts at 4°C. Trypsinization liberated 84% of the $[^{125}I]DNP_9Hb$ remaining after 37°C. Moreover, radioautography after trypsinization showed that only 8% of the cells still bore labelled antigen. Comparison of results obtained with labelled surface proteins and with labelled antigen showed that most of the naturally released antigen is released as such. However, the data also indicate that a significant proportion of antigen is released as an Ig-complex.

Rolley and Marchalonis [143] further document that incubation of radioiodinated splenic lymphocytes with Hc, led to the release of not only Hc–Ig but also of Ig specific for the DNP–hapten. The presence of this specificity, in material released into the medium, was documented by adsorption on to particulate DNP–gelatin complexes, and by immunoprecipitation with DNP_{20} Hb.

These data support the result of Cone et al. [128] and indicate that the presence of antigen augments the turnover of surface Ig below its normal $t_{\frac{1}{2}}$ of 1–3 hr [129, 136,144]. Results of others [136,144] also indicate that the normal $t_{\frac{1}{2}}$ values of the Ig on the surfaces of lymphocytes from diverse species lie between 1 and 3 hr.

Release of surface Ig after addition of anti-Ig. Wilson et al. [145] have employed radioiodinated anti-Ig, -κ and -μ to monitor the behavior of the Ig on mouse spleen lymphocytes. They found that polyvalent anti-Ig, anti-κ and anti-μ desorbed rapidly and in two phases. The initial rapid elution (30 min; 30% of adsorbed label) appears due to antigen-antibody dissociation. The subsequent phase ($t_{\frac{1}{2}}$ = 1–3 hr) does not involve antigen–antibody dissociation and represents the shedding of Ig, which is continously regenerated metabolically (inhibition by both NaN_3 and antimycin A). Release of surface-Ig was found slower in thymic than splenic cells. According to molecular sieving on Sepharose 6B, 60–70% of the ^{125}I-labelled antibody elutes in the form of free molecules and the rest as anti-Ig–Ig complexes. On the basis of mass-law considerations [128,136] one would expect a more rapid release of Ig–anti-Ig complexes than of free Ig. Since the opposite occurs, one most conclude that the Ig–anti-Ig complexes have a greater membrane affinity than Ig per se.

Release of lectin receptors after addition of lectin. The relative reactivities of normal and tumor cells with diverse lectins now constitute an important part of tumor biology. We deal with most of this topic in Chapter 10; and here address only the matter of metabolic turnover of membrane proteins bearing lectin receptors.

We have recently investigated the turnover of plasma membrane proteins in quiescent and concanavalin A-stimulated rabbit thymocytes [103]. Using a mixture of ^{14}C-amino acids as labels and SDS PAGE of highly purified plasma membrane fractions as separative procedure, we were able to determine the $t_{\frac{1}{2}}$ values for each major SDS-PAGE protein and glycoprotein. Most of the membrane proteins showed separate turnover rates. However, 24 hr after stimulation (when RNA synthesis had increased 30 times), the half-lives of some higher molecular weight glycoprotein components were changed from more than 24 hr down to 24–18 hr. One component, a glycoprotein with a molecular weight of 55,000, stood out; it yielded a half-life of 10 hr after stimulation, instead of 24 hr. Recent results [146] have confirmed our previous suggestions that the increase in turnover might be induced by interaction between concanavalin A and the membrane receptor. By affinity chromatography on immobilized concanavalin A, we were able to show that the 55,000 dalton glycoprotein is the monomeric concanavalin A receptor molecule. This component appears to form multimers which cannot be totally dissociated by sodium dodecyl sulfate, suggesting that some of the higher molecular weight glycoproteins with an increase protein turnover are multimers of the monomeric 55,000 dalton component. The multimers must have a different structure within the membrane, since they turn over at a different rate than the monomer. Even though there was no concavalin A bound to the isolated membrane, the cultivation medium of stimulated thymocytes contained concanavalin A, as well as 'shed' membrane protein. The larger amount of monomeric receptor in the supernatant of stimulated cells in comparison to the unstimulated control suggests that the increased receptor turnover is correlated with an increased release. Since our experiments were performed under dissociating conditions, we could not distinguish whether receptors and ligands were released as complexes or as individual components.

However, this question has been answered by Jones [147] working on mouse thymocytes. After radioiodination of surface proteins by the lactoperoxidase technique, about 90% of the labelled surface material is released within 1 hr. This includes receptors for concanavalin A and phytohemagglutinin. Both mitogens increased the release of surface receptors when equilibrated with the living radioiodinated cells. Moreover, Jones' [147] experiments show that the shed material includes free lectins, free receptor and receptor–lectin complexes.

As expected, the $t_{\frac{1}{2}}$ values for ^{125}I-labelled material are well below those obtained by metabolic labelling.

The interaction of insulin with insulin receptors on lymphoid cells. Insulin binds with high affinity to the plasma membranes of numerous cells [148, 149] and very high association constants have been reported (e.g. 2×10^{10} M^{-1} for adipocytes; [149]). The plasma membranes of lymphoid cells also bear insulin receptors and insulin modifies a number of lymphocyte membranes functions (Chapters 7 and 10).

DeMeyts et al. [150] have investigated the interaction between low concentrations of [^{125}I]insulin ($\sim 2 \times 10^{-9}$ M) and cultured human lymphocytes (IM9). Their data suggest: (a) Heterogeneity of binding sites (cf. also [151]); this might also reflect clonal variation. (b) A decrease of apparent affinity for insulin with increasing saturation (i.e. negative cooperativity). The authors suggest that increasing site–site interaction with increasing saturation, produces lower affinity in other sites.

Gavin et al. [152] have recently presented interesting information about the altered membrane dynamics following insulin binding. They incubated cultured human lymphocytes (1M-9; in exponential growth phase) at 37 °C with 10^{-8} M [^{125}I]insulin. Insulin binding reached a steady state within 30 min (cf. also [148]), but the bound, iodinated hormone was readily eluted from the membrane by addition of excess cold insulin.

In a separate set of experiments, [^{125}I]insulin binding was measured after preincubation of the cells with 10^{-8} M unlabelled insulin. Using pre-incubation times up to 2 hr followed by cell washing and incubation with [^{125}I]insulin at 15 °C for 90 min, Gavin et al. [152] found no difference of insulin binding in comparison to non-preincubated cells. However, when preincubation was extended to 5 hr, insulin binding was reduced by 30 %, and after 16 hr pre-incubation by 55 %.

This decrease in insulin binding was temperature dependent. Thus, when cells were preincubated at 23 °C and 37 °C with or without 10^{-8} M insulin, washed and then incubated with [^{125}I]insulin at 15 °C for 90 min, the binding decrement was much smaller at 23 °C than at 37 °C. Moreover, binding returned to normal after additional incubation at 37 °C for 16 hr. Subsequent data by Gavin et al. [153] indicate that the insulin receptor is shed into the medium as a soluble, active protein.

Morphological correlates of the interaction between surface receptors and their ligands. In Chapter 10 we will address the micromorphological studies on the differing interactions of normal and neoplastic cells with diverse lectins. Here we will deal only with several recent publications which are directly pertinent to the 'shedding' problem.

Specifically, Karnovsky and associates [154] and Unanue and associates [155–157] have evaluated the interaction of several spleen cell surface components (Ig, concanavalin A receptors and H2 antigens), using electron microscopy and radioautography to detect ferritin- or radioactively-labelled ligands. They have also employed the hemocyanin technic to localize concanavalin A (Chapter 10).

The various ligands generally induce a movement of receptors on the cell surface combined with, or followed by endocytosis [156]. Ig and concanavalin A receptors move and 'cap' quickly, but H-2 antigens redistribute very slowly. The redistribution is temperature dependent, increases with higher concentration of ligand, and depends on the number of different ligands reacted at one time. Multivalent antigens, for example, DNP_{12}-guinea pig albumin, sheds, redistributes and internalizes within 30 min, leaving the cell surface free of receptors for 4 hr [157]. After this elimination, the Ig-antigen complexes and the Ig progressively reappears.

Antigenic modulation. The term 'antigenic modulation' refers to the loss of *phenotypic* but not genotypic expression of certain surface antigenic components after exposure of the cells to specific antibody in the absence of complement, c′ [158]. In the presence of c′, cytolysis would occur. This phenomenon has been well studied for the case of the TL antigens on mouse lymphoid cells [159], but the unique antigens of human Burkitt lymphoma cells and human melanoma cells are also depressed upon exposure of the cells to specific antibody [160]; certain myelomas also exhibit antigenic modulation. Mouse H2 antigens do not modulate after simple antibody exposure, but can do so when anti-mouse Ig is added subsequent to treatment with anti H2 [158].

The thymus leukemia antigen (TL) system. The best studied model of antigenic modulation involves the antigens of the TL system (TL 1, 2 and 3) of mice determined genetically by locus *Tla* adjacent to H2 in linkage groups IX (158–162). The TL antigens *normally* occur only on thymocytes. Some mouse strains are TL+ and some are TL−. The latter can produce specific antibody (anti-TL). Many lymphocytic leukemias, including ones originating in TL− animals are TL+. TL+ leukemias passaged in susceptible TL− animals (pre-immunized) grow and lose their TL+ surface specificity. When returned to TL+ hosts, the cells again become TL+. Similarly, when TL+ cells are incubated *in vitro* with *anti*-TL, they become TL−. This is well demonstrated in the case of ascites leukemia RADA1 [161], where modulation occurs more rapidly than in normal thymocytes. The cells completely lose expression of TL 1, 2 and 3 on their surfaces within 10 min after exposure to anti-TL (in a wide range of concentrations) at 37°C. This phenomenon is accompanied by development of resistance to complement-mediated cytolysis by anti-TL. Once the TL− phenotype is established, it remains stable for at least 4 hr. Thereafter, the cells again become TL+.

Modulation does *not* occur at 0 °C and is abolished by actinomycin D (5 μg/ml; 1 hr) and iodoacetamide, but not by puromycin and cycloheximide or radiation. However, studies with radioactive precursors show no change in overall DNA, RNA or protein synthesis during modulation. This need not be surprising, since the TL system involves only a small proportion of the cell mass.

Lamm et al. [163] have dissociated the cytotoxic activity of anti-TL from its modulating capacity by papain treatment of anti-Tl γ-globulin. Thus, univalent anti-TL fragments (Fab) retain the modulating activity of the intact S7 γ-globulin, although they lack the complement-fixing capability of Fc-fragment necessary for cytolysis.

Yu and Cohen [164] have developed methods for the isolation of TL antigen and for the metabolic analysis of the modulation phenomenon. First, Yu and Cohen showed that TL- and $H2^a$-antigens could be released in solubilized form (passing through a 0.22 μ filter) from intact cells by extracting these with 0.5% Nonidet NP40 (a non-ionic detergent). They precipitated the TL– and antigen-bearing fragments specifically using either mouse anti-TL and RAMIg or anti-$H2^a$ and RAMIg. They quantified the antigens by prelabelling the cells either with a 5 h-pulse of [^{3}H]fucose or via lactoperoxidase-catalyzed radio-iodination. SDS-PAGE of the precipitates obtained with anti-Tl yielded two major ^{125}I-labelled peaks, of which the more rapidly migrating one (~40,000–45,000 daltons) could be identified as TL-containing membrane protein. SDS-PAGE of the anti-$H2^a$ precipitates gave only one peak ([^{3}H]fucose) with a molecular weight of also ~40,000–45,000.

Yu and Cohen [165] then used these techniques to elucidate events occurring during modulation. For this they analyzed both cell extracts and culture media of ASL-1 lymphocytes, exposed either to normal mouse serum (controls), anti-$H2^a$ (non-modulating) or anti-TL (modulating). Their results show that (a) material precipitable by anti-TL disappears more rapidly from modulating, prelabelled cells ([^{3}H]fucose or ^{125}I) than from prelabelled controls. Indeed, 10 hr after initiation of modulation, the TL+ SDS-PAGE peak had decreased by 80%, and after 20 hr it was no longer distinguishable. There was no alteration in the non-TL-containing peak precipitated by anti-TL, and incubation with anti $H2^a$ caused no change in the level of $H2^a$ positive material precipitable by anti-$H2^a$. However, the amount of TL material released into the medium was the *same* under modulating and non-modulating conditions. Moreover, ASL-1 cells incubated with modulating serum incorporated as much [^{3}H]fucose into their membranes as cells treated with non-modulating sera.

The control point for modulation is thus *not* at the level of biosynthesis or shedding. One possibility is enhanced degradation, either *in situ* or subsequent to endocytosis. Another possibility to be considered is a *reorganization of the membrane.* Thus, Stackpole et al. [166] found that, after modulation of TL antigens on the surfaces of normal mouse thymocytes or RADA1 leukemia cells,

TL antigens can still be detected in appreciable concentrations on the cells surfaces by use of specific immunofluorescence and immuno-electron microscopy (sandwich techniques). Bivalent antibody specific to TL tended to be displaced into patches or caps during modulation, while *monovalent*, *modulating* fragment showed little aggregation. (Some pinocytosis was also observed.)

Modulation of membrane Ig. Hüttenroth et al. [167] have demonstrated modulation of surface Ig on lymphoid cells of four established human cell lines (MW, HG, Tm, JK) derived from normal individuals. Fifty to seventy percent of the cells of these lines bear κ- and μ-chains on their surfaces. Using a rosetting technique to monitor the presence of these molecules, Hüttenroth et al. [167] have shown that anti-κ serum caused the disappearance (or sequestration) of κ- and μ-chains from the cell surfaces within 2 hr of *in vitro* incubation at 37°C. This process did not occur at 0°C, and was inhibited by NaN_3, but *not* by actinomycin D, cycloheximide or puromycin. Anti-μ caused the deletion of μ-chains only. The 'cleared' surfaces were restored to their normal complement of κ- and/or μ-chains within 12 hr after removal of modulating antibody.

These data parallel results obtained with the TL system in many ways, including the observation that antiserum direct against one specificity can also modulate other specificities [158, 161]. However, there are important differences, that is, TL modulation occurs with monovalent antibody and is blocked by actinomycin D [161, 163], whereas this is not the case for κ- and μ-chain modulation.

Clearly, the process of antigenic modulation might allow tumor cells to avoid immunological rejection. However, the phenomenon has a more general significance, as is apparent from the work of Hüttenroth et al. [167]. It represents a control mechanism for membrane structure which may be representative of many processes involving surface membranes, and which must be considered in all comparisons of normal- *vs.* malignant-cell surface structure. Its relevance to cells other than lymphocytes has already been established by Aoki et al. [160].

2.3. *Membrane proteins of normal and neoplastic lymphocytes*

2.3.1. *Normal cells*

The proteins of lymphocyte membranes, particularly of plasma membranes, have been much studied of late, particularly in terms of turnover [5, 6]. Less information is available concerning composition and function, and only very few studies address possible alterations of membrane protein composition, related to the neoplastic state of malignant lymphocytes.

Plasma membrane isolates from lymphoid cells are generally prepared after disrupting the cells by low shear forces, for example, Dounce homogenization [168, 169] passage of cell suspension through metal screens or nitrogen decompression [38]. SDS-PAGE analyses of such membrane isolates vary somewhat in detail, but the following general patterns appear consistent: (a) All show proteins of molecular weights ranging between 10,000 and 200,000. (b) The plasma membrane isolates contain a much greater proportion of high molecular weight proteins than endoplasmic reticulum [38, 168]. (c) Glycoproteins are more concentrated in plasma membranes than endoplasmic reticulum and can be iodinated by the lactoperoxidase procedure [38]. (d) Identically prepared plasma membrane isolates of a given lymphocyte category from different species show similar complements of non-glycosylated proteins; glycoproteins, however, vary considerably between species.

Membrane-bound immunoglobulins can be detected by comparing SDS-PAGE patterns obtained without and with disulfide reducing agents, which split the intact, large Ig molecules into their smaller heavy and light chains.

In carrying out such analyses, it is important to ascertain the composition of the cell population to be analyzed, since even trace contamination of a preparation with plasma cells can lead to significant immunoglobulin levels.

Using the SDS-PAGE criterion, membrane-associated immunoglobulin has been detected in plasma membranes from an IgG_2 producing mouse myeloma, pure B-type lymphocytes from *nu-nu* mice and the GD 248 SV40-induced hamster lymphoma [170]. These data indicate that Ig is not necessarily eluted from plasma membranes during isolation. However, Schmidt-Ullrich et al. [38] and others [168] could detect no membrane-Ig in plasma membranes isolated from rabbit thymocytes.

Ladoulis et al. [169] claim that significant amounts of radioiodinated Ig or Ig fragments can be extracted by Triton X-100 from plasma membrane isolates prepared from both rat thymocytes and splenic lymphocytes which had been labelled by the lactoperoxidase procedure. These authors precipitated the ^{125}I-labelled, Triton-extracted membrane proteins with rabbit anti-rat immunoglobulin and analyzed the precipitates by SDS-PAGE. The iodinated components comprised a 117,000 dalton entity, apparently 1/2 8S IgM subunits and a 27,000 dalton glycoprotein, possibly kappa chains. Unfortunately, the studies of Ladoulis et al. [169] are not unequivocal in terms of B-cell or plasma cell contamination of their thymocyte preparation.

2.3.2. *Neoplastic lymphocytes*

Neoplastic lymphocytes, particularly myeloma cells and Burkitt lymphoma cells have been rather extensively studied in terms of the immunoglobulin components associated with their cell surfaces. There also exists considerable, if scattered, information about transport mechanisms (e.g., [171, 172]), electrophoretic mobility (Chapter 8) and lectin receptors (Chapter 10). However, we lack substantial information about the general compositional and metabolic features of neoplastic lymphoid cells. This situation is, in part, due to the different fractionation characteristics of normal *vs.* transformed lymphocytes [170].

However, a recent study by Kessel and Bosmann [173] is relevant here. These workers describe line, 1210/A, of the L1210 mouse lymphocytic leukemia. These cells adhere to glass surfaces, unlike another variant, L1210/AS, which grows in suspension culture. L1210/A is less malignant than L1210. Both L1210/A and L1210/AS release more sialic acid than L1210, upon sialidase treatment. When the three cell types were subjected to mild papain-treatment, similar amounts of glycopepetide were released from L1210 and L1210/AS, but L1210/A yielded only about one third as much material (in terms of carbohydrate). Moreover, the electrophoretic mobility of (Chapter 8) of L1210A cells was less than that of L1210 lymphocytes (before and after sialidase) and the incorporation of fucose and leucine into glycoproteins was also lower in L1210/A than in L1210. The decrease in surface glycoprotein was attributed to a depletion of a glycosyltransferase involved in the addition of sialate to a glycoprotein acceptor. L1210/A is also depleted in a variety of glycosidases (cf. Chapter 7). Unfortunately the authors performed only perfunctory protein analyses on their plasma membrane isolates.

Evidence that L1210 cells bear a sialyltransferase on their external cell surfaces has been presented by Bernacki [174]. After sialidase treatment of the lymphocytes, the cells incorporated six times as much *N*-[^{14}C]acetylneuraminic acid into cell surface components (primarily glycolipids) as untreated cells.

2.4. *Membrane proteins and membrane protein metabolism in normal and neoplastic hepatocytes*

2.4.1. *Introduction*

Liver membrane systems are extremely complex and their functions are highly diversified. We cannot, accordingly, categorize liver membrane proteins in the simple way possible with erythrocytes, or even lymphocytes. However, a large variety of liver membrane proteins have been very well characterized. This is most obviously true for mitochondrial enzymes, but includes entities, such as Neville's bile front protein ([22]; p. 70), the protein components of bile front nexuses (Chapter 9), endoplasmic reticulum cytochrome b_5 (Chapter 6) and many others.

No qualitative differences between normal and neoplastic cells have been detected in terms of well-defined membrane components; and more general comparisons, for example, of bile-front isolates [20] have been quite unrevealing, if only because of the different fractionation properties of normal and neoplastic hepatocytes.

On the other hand, much of present knowledge of membrane protein metabolism comes from studies on normal liver, and some interesting information has accrued concerning membrane protein metabolism in neoplastic hepatocytes. We will treat these aspects of protein metabolism in the following section.

2.4.2. Normal hepatocytes

2.4.2.1. Turnover

Most data on the general mechanisms involved in biosynthesis of membrane proteins, as well as their incorporation into and disappearance from the membrane, have been obtained with hepatocytes. Different, highly purified membrane fractions, particularly rough and smooth endoplasmic reticulum, as well as bile front plasma membranes have been used. In addition, several studies analyse the turnover of specific membrane proteins under various metabolic conditions. Rough endoplasmic reticulum membranes usually contain adherent polysomes and must be freed of ribosomal RNA by desoxycholate treatment [175], to avoid contamination by non-membranous proteins.

Omura et al. [29] have determined the specific radioactivity of rat liver microsomal membranes, as well as several enzymes isolated therefrom, 23 to 335 hr after intraperitoneal injection of D,L-[1-^{14}C]leucine. To evaluate the differential turnover of diverse membrane proteins, cytochrome b_5 and NADPH-cytochrome *c* reductase were isolated from KCl-washed, trypsinized microsomes. The authors found very similar half-lives for total membranes (108 hr), smooth membranes (113 hr), and rough (111 hr) membranes. In comparison, washed trypsinized microsomes yielded a half-life of only about 80 hr. The same was found for NADPH-cytochrome *c* reductase, but cytochrome b_5, similarly isolated, had significantly longer half-life, that is, 117–123 hr. The $t_{\frac{1}{2}}$ values for the membrane lipids were much shorter, that is, 44 hr for the fatty acid moieties ([2-^{14}C]sodium acetate) and only 27 hr for the polar head groups ([1,3-^{14}C]glycerol). The data of Omura et al. [29] clearly indicate that the proteins and lipids of hepatocyte microsomal membranes turn over at very different rates and that different membrane proteins also turn over separately. Additional evidence in support of the last point comes from studies on catalase [176], tryptophan pyrrolase [177] and arginase [178].

The effects of phenobarbital and other drugs provide additional proof that diverse liver membrane proteins turn over at different rates. Thus Kuriyama et al. [30] report that 5–7 hr after rats are given a single phenobarbital injection, there is a 4-fold increase in specific NADPH-cytochrome *c* reductase activity. This increase is due to newly synthesized enzyme with the same functional and immunologic properties as the enzyme derived from untreated animals. In contrast, the activity of cytochrome *P*450 increases only slightly, and cytochrome b_5 not at all within that time period. Cytochrome b_5 activity does increase with continuous phenobarbital administration, but its $t_{\frac{1}{2}}$ is always less than that of NADPH-cytochrome *c* reductase. The mean half-life of the microsomal membrane proteins, NADPH-reductase and cytochrome b_5, are 70 hr, 70 hr and 100 hr, respectively, in control animals, compared to ~70 hr, 30 hr, and ~80 hr, respectively, after phenobarbital treatment (these values represent rough approximations, calculated from the data given). Kuriyama et al. [30] provide convincing evidence that the increases in enzyme activity arise from augmented synthesis and decreased degradation in the case of the reductase, and decreased degradation in the case of cytochrome b_5. Their results cannot be explained by the increased amino acid reutilization in phenobarbital treated animals [30], because the half-life of total microsomal protein and the two purified proteins is only 20–30% shorter when non-reutilized L-[*guanidino*-^{14}C]arginine is used as precursor. Moreover, phenobarbital-feeding does not change the *overall* degradation-rate of microsomal membrane proteins. This is shown by application of the double-labelling method of Arias et al. [99] and by use of L-[*guanidino*-^{14}C]arginine as precursor.

A selective increase of precursor incorporation into NADPH-cytochrome *c* reductase, in comparison to other microsomal membrane proteins, occurs also in the livers of phenobarbital treated mice [181]. However, the drug does not noticeably affect protein degradation. Use of [*guanidino*-^{14}C]arginine or the double label approach gave identical results.

Arias et al. [99] document how label reutilization can influence lifetime measurements. In contrast to the $t_{\frac{1}{2}}$ values of Omura et al. [29] for rough and smooth microsomal membranes (111 hr and 113 hr respectively, p. 102), Arias et al. [99] obtain a value of 48–50 hr, using the double label method, 79–108 hr by pulsing with uniformly labelled leucine and 125–139 hr using L-[^{14}C]arginine. These results are not surprising, since 50% of the hepatic amino acid pool derives from protein, catabolism [182].

Less information is available concerning protein turnover in membranes other than those of the microsomal fraction. However, Arias et al. [99], using the double label technique obtain a very short half-life, $t_{\frac{1}{2}}$ = 43 *hr*, for liver plasma membrane. Swick et al. [101], using ^{14}C-labelled carbonate as precursor (to avoid reutilization), report a mean $t_{\frac{1}{2}}$ value of 3–4 days for liver mitochondrial proteins, and data on nuclear membranes [193] indicate that the proteins of these membranes are continuously degraded and replenished (mean $t_{\frac{1}{2}}$ = 85 hr using [^{3}H]leucine as radioactive precursor).

In summary, the proteins of all hepatocyte membranes studied appear to exist in a dynamic metabolic state and are turned over more rapidly than their cells. Moreover, the evidence suggests that various individual membrane proteins turn over at different rates and that the mean half-lives of membrane proteins are longer than those of membrane lipids. Membranes thus do not turn over *en bloc*.

We will now treat some important features of membrane proteins synthesis and degradation, as well as the mechanisms participating in the insertion of proteins into membranes.

2.4.2.2. Biosynthesis

Membrane proteins are manufactured on polysomes. However, it is not yet clear whether they are synthesized exclusively by membrane-bound polysomes (i.e. on rough endoplasmic reticulum) and are transferred into a membrane-milieu concurrently (or immediately upon completion), or whether the membrane proteins pass into a soluble pool and thereupon equilibrate with the membrane pool. There is experimental evidence suggesting that both mechanisms may exist, but as pointed out in Hendler's recent review [7], one cannot now decide whether membrane proteins are *preferentially* synthesized on membrane-bound ribosomes. All that can be concluded is that proteins *exported* from the cell are predominantly synthesized on membrane-bound polyribosomes, that membrane bound ribosomes also produce cellular proteins, and that free and bound ribosomes synthesize different categories of protein.

2.4.2.3. Biodegradation of membrane proteins

The presence and concentration of various proteins in their membranes is regulated not only by their rates of synthesis, but also by their rates of degradation. (Little is known about the role of exchange.) The topic of membrane protein catabolism has been clarified through a stimulating series of experiments by Schimke and associates [180, 184, 185]. These studies indicate that, in general, membrane proteins of high molecular weight are degraded more rapidly than smaller molecules. These conclusions rest on experiments where proteins of membranes, isolated at varying intervals after labelling, were solubilized by sonication in 0.1 % SDS and fractionated using SDS-PAGE or gel permeation chromatography. Moreover, all degradation kinetics of membrane-bound proteins and isolated membrane components are first-order, leading to the conclusion that proteolytic enzymes are responsible for membrane protein degradation.

The matter of proteolytic degradation of membrane proteins has been recently reviewed by Siekevitz [3]. No definite mechanisms have been established, but degradation kinetics suggest that proteins, once synthesized and incorporated into a membrane, are degraded randomly. It is possible that membrane proteins undergo reversible folding and unfolding and that degradation occurs more readily in the unfolded state. Another mechanism to be considered is that membrane proteins exchange into the juxta-membranous milieu, where they are degraded by soluble proteases. Dice and Schimke [185] have shown that this process does occur in the case of ribosomal proteins and probably membrane proteins.

According to Dice and Schimke [185] ribosomal proteins, like membrane proteins, do not turn over *en bloc*, but are synthesized and replaced as individual, independent entities. In these studies polysomes isolated from rough microsomal membranes were split into 40 S, 60 S and 80 S subunits by combined application of RNAase, EDTA and ionic manipulation. About 30% of the ribosomal protein, mainly entities of high molecular weight were solubilized by 0.4 M KCl or 0.4 M LiCl ('split proteins'). The more tightly-bound proteins are assumed to be 'core proteins' and, in analogy to membrane proteins, show greater susceptibility to proteolytic degradation, the higher the molecular weight. However, there is no correlation between the turnover rates of ribosomal or microsomal membrane proteins and their relative susceptibilities to tryptic digestion [184].

The greater susceptibility of *larger* proteins to degradation may derive from their larger surface/volume ratio and the greater probability of such molecules to bear an unfolded, protease-accessible region at a given time. If proteases are involved in the degradation of membrane proteins, the source(s) or place of action of these enzymes have not been established. Lysosomal cathepsins are obvious candidates, as are the cytoplasmic proteases active at neutral pH [186]. However, membrane-bound proteases have also been invoked since the description of such putative enzymes in erythrocyte membranes by Morrison and Neurath [187].

2.4.2.4. Turnover of the proteins and glycoproteins of bile-front plasma membranes

As is the case for most membrane isolates, SDS-PAGE of bile-front preparations reveals a large number of protein components ranging in molecular weight between 15,000 and 300,000. However, these SDS-PAGE patterns are not very informative *per se.*

Ray et al. [188] first examined the biosynthesis of bile-front membrane proteins by following the radioactivity in smooth microsomal membranes and bile-front isolates after a pulse of [^{3}H]leucine. Labelling in the microsomes peaked after 20 min and declined thereafter, whereas the specific activity of the bile-fronts increased for up to 4 hr, even after inhibition of protein synthesis by cycloheximide. The data suggest a transfer of protein from the endoplasmic reticulum to the plasma membrane. The results of Barancik and Lieberman [189] support this interpretation. These authors followed the increase in the specific activity of various bile-front SDS-PAGE components for 2 hr after injection of [^{3}H]leucine into rats. Protein synthesis was blocked by injection of cycloheximide 15 min after the pulse. The authors claim that the larger SDS-PAGE components (mol. wt. $>$50,000) appear more slowly in the plasma membrane than proteins of lower molecular weight, but their counts increased only 25% between 30 min and 120 min, leaving large room for error. However all available data indicate that appreciable time is necessary for the transfer of new protein to the plasma membrane.

Gurd and Evans [102] have applied the double labelling technique [99], using

[^{3}H]- and [^{14}C]leucine to determine the relative turnover rates of their hepatocyte plasma membrane fractions and microsomes. They found that the proteins of their plasma membranes are degraded more slowly than those of the microsomes. These results contradict those of Arias et al. [99] who, using L-[*guanidino*-^{14}C]arginine as precursor, found half-lives of 48-50 hr for rough and smooth endoplasmic reticulum and a shorter $t_{\frac{1}{2}}$ for liver bile-fronts. The discrepancy might be due to the different membrane isolation methods used. Gurd and Evans [102] employ the method originally described by Neville [21], and homogenize the liver tissue in 1 mM $NaHCO_3$, pH 7.6. However they isolate plasma membranes and membranes of the endoplasmic reticulum by zonal centrifugation of the whole homogenate in a 6–54% (w/v) sucrose gradient. Gurd and Evans [102] find that the *glycosylated* SDS-PAGE components remaining after EDTA extraction of their membranes are more rapidly degraded than the bulk of the membrane proteins. In accord with other experimentation, they observe the highest breakdown rates in the high molecular weight components. Since the glycosylated proteins are of high molecular weight, their susceptibility to degradation may thus be a function of size rather than of glycosylation. Gurd and Evans [102] could not detect any differential turnover among diverse SDS-PAGE components, but this could be because their approach lacks the sensitivity of the double-labelling method.

2.4.2.5. 'Flow' of membrane proteins between subcellular organelles

While there is considerable information on the biosynthesis and biodegradation of hepatocellular membrane proteins, rather little is known about the mechanisms involved in the transport or 'flow' of membrane proteins from their sites of synthesis to their ultimate locations.

Dallner et al. [190, 191], studying the livers of young adult rats and using NADH-cytochrome *c* reductase and glucose-6-phosphatase as markers, show that 6 hr after a [^{14}C]leucine pulse the specific activities of the two enzymes are identical in their rough and smooth microsomal membranes. This indicates a 'flow' of newly manufactured protein from the rough endoplasmic reticulum to the smooth membranes (which include plasma membrane fragments). This 'flow' might occur either by a 'directed' translocation of protein components in a membrane continuum, or by exchange via a soluble pool.

More recent studies by Kawasaki and Yamashima [192] indicate a 'flow' of glycoproteins from rough endoplasmic reticulum via smooth endoplasmic reticulum and Golgi apparatus to the plasma membrane. These authors used D-[1–^{14}C]glucosamine to pulse-label carbohydrate residues and followed the appearance of the label in various subcellular organelles at varying times using the cell fractionation methods of Neville [88] for isolation of plasma membranes and of Ernster et al. [175] for microsomal membranes. Their studies suggest that the glycoproteins are completed in the Golgi apparatus (which contains most of cells' glycosyltransferases) and are incorporated into the plasma membrane thereafter. The mean half-life obtained ($t_{\frac{1}{2}}$ = 37 hr) is close to that obtained with L-[*guanidino*-^{14}C]arginine.

Studies more systematic than those of Ray et al. [188] and Barancik and Lieberman [189] have been carried out by Francke et al. [193]. These authors followed the time course of radiolabelling in rough microsomes, nuclear membranes, Golgi apparatus and plasma membranes for up to 65 hr after pulse labelling of liver cells *in vivo* with L-[*guanidino*-^{14}C]arginine. The time course of specific activity in the subcellular fractions identifies the rough endoplasmic reticulum as the locus of protein synthesis. From there the label appears in the Golgi apparatus and finally in the plasma membrane, whose continuous increase in specific activity during 12 hr parallels the decrease in the Golgi apparatus.

The increase of specific activity in the plasma membrane exhibits two phases. The first (10–20 min) is as rapid as the appearance of newly synthesized albumin in serum [194]. The second coincides with the disappearance of activity from the Golgi apparatus and endoplasmic reticulum membranes, and suggests a continuous flow of membrane proteins to plasma membrane. The transfer of material from the Golgi apparatus to plasma membrane may involve mechanisms similar to those operating in secreting cells [195].

The rather simple protein flow concept of Francke et al. [193] has certain weaknesses. Thus, the authors report a similar overall turnover of proteins in plasma membrane and endoplasmic reticulum, while Arias et al. [99] report a more rapid turnover in plasma membrane. Moreover, Francke et al. [193] do not evaluate the significance of differential turnover of individual membrane proteins.

Bennett et al. [196] have evaluated the attachment of *carbohydrate* residues and the extension of the carbohydrate chain in about 50 different cell types, including hepatocytes. These authors used [^{3}H]fucose as label and followed its incorporation by radioautography. Their data demonstrate a 'flow' of glycoproteins from the smooth endoplasmic reticulum, via Golgi apparatus to the plasma membrane. Fucose is attached to the proteins in the Golgi apparatus, suggesting that some glycoproteins are completed in this organelle. Essentially, no fucosyltransferase activity could be detected in the plasma membrane; other glycosyltransferases seem to be transferred to the plasma membrane together with other Golgi material [197,198,199].

Similar mechanisms of glycoprotein 'flow' are postulated for both secretory and non-secretory cells, since both show similar kinetics. Up to 10 min after injection of the radioactive sugar the label is localized exclusively in the Golgi apparatus. Between 20 and 25 min it appears in the surface membrane and, in the case of secretory cells, is released for up to 4 hr. In non-secreting cells, the specific activity in the plasma membrane increases up to 4 hr and decreases thereafter.

2.4.3. Membrane protein metabolism in neoplastic hepatocytes

Neoplastic hepatocytes can differ considerably from normal liver cells in their complement of membrane-bound enzymes. We will deal with this topic in Chapter 6 and will here focus on protein synthesis and degradation.

Few experiments have been designed to specifically compare the protein complement of microsomal membranes from hepatomes with membranes isolated from liver. However Chiarugi [200] has used a PAGE technique for just such a study on hepatoma 5123C. Normal liver proteins were labelled using [^{3}H]-dimethyl sulfate and hepatomas by [^{14}C]dimethyl sulfate. The microsomal membrane proteins from the two sources were then compared by 'split-gel' electrophoresis in polyacrylamide containing 8 M urea at pH 4.5, and the ^{3}H/^{14}C ratios determined. Significantly, the labelling patterns of liver microsomal membranes showed two high-molecular weight components, which were lacking in the hepatoma membranes. However, these results have not been followed up, or confirmed, for example, by subfractionation of microsomes, SDS-PAGE and comparisons with other hepatomas.

As already pointed out (Section 2.2.1), one of the difficulties complicating experimentation on membrane protein metabolism in neoplastic hepatocytes is the deviant fractionation behavior of these cells. This problem is evident in the studies of Moyer and Pitot [20] comparing normal rat livers with Morris hepatomas 7800 (rapidly-growing; undifferentiated) and 9618 A (slow-growing; differentiated). The authors used the same fractionation procedure for all three cell types and utilized a discontinuous sucrose gradient for the ultracentrifugal separation of the microsomal pellet into three fractions, namely: S_1 ('very smooth'), consisting of Golgi apparatus and secretory vesicles, according to electron microscopic criteria, S_2, consisting of smooth endoplasmic reticulum primarily and R, comprising rough endoplasmic reticulum. The authors did not consider the fate of small plasma membrane fragments which, in sucrose gradients, behave very much like smooth endoplasmic reticulum vesicles.

Moyer and Pitot [20] report that in normal liver and hepatoma 9618A, 50% of the membrane protein goes into S_1 and 25% each into S_2 and R. In contrast, fractions S_1 and S_2 from hepatoma 7800 contain similar amounts of protein and together account for 85% of the protein applied to the gradient. Unfortunately, the data do not allow one to determine whether these findings reflect membrane 'dedifferentiation' (Chapter 3), depletion of R in the hepatomas and host livers, or an alteration in the fractionation behavior of the various cell types *prior* to the sucrose gradient step (cf. Chapter 6).

After pulse-labelling with [^{14}C]leucine, a biphasic decay curve is obtained with total microsomes, regardless of cell type. The second phase probably derives from the substantial reutilization of isotope after 36 hr. The $t_{\frac{1}{2}}$ values of the subfractions from hepatoma 7800 are 30% shorter than those of host liver, whereas the half-lives of the fractions from the differentiated hepatoma 9618 A, were actually some 25% longer than those of host liver.

SDS-PAGE revealed no qualitative differences between equivalent fractions isolated from the various cell types. In all cases the SDS-PAGE component ranged between molecular weights of 45,000 and 200,000, with most of the material in the 45,000–100,000 dalton range. The various membranes show only minor differences in the qualitative distribution of these SDS components. Moyer and Pitot [20] have measured the decay rates of various SDS-PAGE components in the S_1, S_2 and R fractions derived from host livers and the two hepatomas (Tables 2.2, 2.3). Different half-lives are found for a given molecular weight class in the three fractions. Also, each protein appears to turn over at its own rate, but the hepatoma fractions exhibit lower rates of decay than those of host liver (Tables 2.2, 2.3). In contrast to data on normal liver [20], and on host liver, turnover in tumor membranes appears unrelated to molecular weight (Tables 2.2, 2.3).

Unfortunately, the data of Moyer and Pitot [20] cannot be fully evaluated because of the variable $t_{\frac{1}{2}}$ values of host livers (Tables 2.2, 2.3) and because reutilization can occur even in the initial phase of decay. However, the results fit those of Narayan and Chandrasekhara [201], who report that the *in vivo* $t_{\frac{1}{2}}$ of the bile-front and microsomal proteins of 2-*N*-fluorenylacetamide-induced hepatomas is twice that of the same fractions from normal liver.

Hepatomas constitute a more or less rapidly-dividing cell population, but normal hepatocytes replicate very infrequently. It is therefore necessary to evaluate the effect of cell cycle on membrane protein metabolism. This might be achieved by experimentation on regenerating liver and/or by evaluating membrane changes in hepatoma as a function of cell cycle or stage of tumor growth. This has been done by Olivotto and Paoletti [202], who have examined the dependence of microsomal- and nuclear-membrane protein metabolism on growth state in the rat Yoshida ascites hepatoma *in vivo*. They monitored protein syn-

TABLE 2.2

Protein turnover in smooth microsomal membranes from Morris hepatoma 7800 and host liver[1]

Tissue	$t_{\frac{1}{2}}$ (hr) of SDS-PAGE component[2]		
	115,000 daltons	75,000 daltons	60,000 daltons
Host liver	10.8	14.5	19.5
Hepatoma 7800	24.0	13.5	25.5

[1] From Moyer and Pitot [20].
[2] Initial phase of decay.

TABLE 2.3

Protein turnover in smooth and rough microsomal membranes from morris hepatoma 9618A and host liver[1]

Tissue fraction	$t_{\frac{1}{2}}$ (hr) of SDS-PAGE component					
	2×10^5 D	1.35×10^5 D	1.15×10^5D	7.5×10^4D	6×10^4D	4.3×10^4D
Host liver						
S_1	14.1	28.8	16.6	27.7	21.1	34.8
S_2	15.0	30.3	28.6	35.2	25.6	26.0
R	11.7	18.5	29.7	21.1	29.1	26.6
Hepatoma 9618A						
S_1	22.3	25.4	59.2	56.3	42.1	44.1
S_2	46.8	46.0	24.8	19.4	34.3	74.5
R	82.1	44.4	39.4	28.4	35.0	18.9

[1] From Moyer and Pitot [20].

thesis by [^{3}H]lysine incorporation for 80 min while the degradation was assessed by following the decrease of specific activity [^{14}C]lysine labelled proteins. Protein synthesis proceeded at a constant rate throughout the early phase of logarithmic growth and the resting phase, even though it was reduced by 30% in the resting phase (in which the tumor mass remains constant). Since the protein mass in the resting phase is constant, protein degradation must proceed at a rate that just balances synthesis. Protein turnover occurred at similar rates in nuclei and microsomes, both significantly higher than the rates in mitochondria. Further experimentation is clearly indicated before one can reach an understanding of the lesser turnover rates of membrane proteins in hepatomas.

At this time one cannot distinguish whether the turnover changes reported by Moyer and Pitot [20] reflect an increased rate of synthesis compared with degradation, perhaps due to altered 'template stability' (Chapter 6), decreased degradation *vs.* synthesis or both. It is also unclear to what extent the changes relate to the neoplastic state.

2.5. *Membrane proteins in normal and neoplastic fibroblasts*

2.5.1. *Introduction*

Although mechanisms of protein synthesis and degradation have been best characterized in hepatocytes, cultured fibroblasts have proven more suitable for experimentation on the effects of growth and neoplastic transformation on membrane proteins and their metabolism. However, experimentation has focussed on *plasma membrane* proteins, and there is rather little information on intracellular membranes. Moreover, there are a number of inconsistencies in the literature. These stem in part from differences between the cell systems employed. Equally important, however, are the divergent methods used by various experimenters and the inadequacies of these methods, particularly when applied to cells differing in membrane properties (Section 2.2.1).

2.5.2. Turnover of plasma membrane proteins during the cell cycle

Studies on the turnover of surface membrane components in replicating and resting fibroblasts were initiated by Warren and Glick [203]. L-Cell fibroblasts were labelled *in vitro* using [^{14}C]leucine, -valine, -glucose, and -glucosamine. The cells were then transferred to isotope-free media and the disappearance of radioactivity followed in plasma membrane isolates prepared by the method of Warren et al. [31]. All labels yielded half-lives of 8–9 days for growing cells, but in resting cells the radioactivity disappeared more rapidly ($t_{\frac{1}{2}} \sim 1$ day). Because of reutilization, the true $t_{\frac{1}{2}}$ values are probably shorter. A recent, comprehensive study by Roberts and Yuan [111], using a variety of labelling procedures to avoid reutilization artifacts also indicates that plasma membrane proteins do not turn over rapidly in quickly-dividing cultured fibroblasts.

However, the type of plasma membrane isolate analyzed in these studies is known to be contaminated with intracellular membranes, as well as ribosomes, preventing unambiguous evaluation of the data. The slightly greater turnover rate of [^{14}C]glucosamine, calculated from the decline in the radioactivity of membrane neuraminic acid could be due to a lesser degree of reutilization (as established for rat liver bile-fronts [192], or to a faster turnover of membrane glycoproteins. The latter possibility is supported by the release of considerable amounts of protein-bound [^{14}C]neuraminate into the culture medium.

More recently, Gerner et al. [204] synchronized cultured KB cells, using the double thymidine block [205,206] and followed the incorporation of [^{14}C]leucine, -glucosamine and -choline into surface membrane isolates [31], subcellular organelles and soluble cytoplasmic proteins. All three precursors reveal a striking increase of cell surface labelling after cell division. This effect is most marked with [^{14}C]glucosamine and probably reflects the increase in cell surface material after cell division.

The rates of incorporation for choline and leucine are very similar (in contrast to results obtained with hepatocytes). No equivalent changes in isotope incorporation into soluble cytoplasmic proteins are observed.

The variation of plasma membrane metabolism with cell cycle has been more rigorously studied by Graham et al. [207] employing cultured mouse mastocytoma cells and virus transformed fibroblasts (NIL-2HSV), both synchronized by a thymidine pulse. The fractionation approach employed by these experimenters constitutes an extension [208,209] of the methods developed by Wallach and associates [12] and the authors employed exceptional care to define their membranes analytically and to avoid artifact. The cells double their volume between the G_1 and G_2 phase [210]. Simultaneously, the cell surface increases by 60%; a further extension of surface area (to double) occurs during mitosis.

The authors show that between the G_1 and G_2 phase of mitosis, the amount of plasma membrane protein per mg cellular *protein* stays constant. There are no qualitative changes in membrane protein composition during the cell cycle, according to SDS-PAGE, and the phospholipid/protein ratio stays constant.

In contrast, plasma membrane-*carbohydrate* composition varies with cell cycle. Thus, the carbohydrate/protein ratio increases during G_1, then decreases and returns to interphase levels in G_2. The most prominent carbohydrate change during G_1 is in the proportion of fucose (early in G_1), although mannose, galactose and glucose also increase slightly. Glucosamine is added late in G_1 and sialic acid in the late S-G_2 phase, as already noted by Mayhew [211] in RPMI (No. 41) cells and by Glick et al. [212] in KB cells. It appears that cells in mitosis generally contain more sialic acid than cells in interphase.

Graham et al. [207] find that the activity of plasma membrane 5′-nucleotidase increases in correlation with plasma membrane formation. In contrast, adenylate cyclase activity increases up to mitosis and then declines and ATPase activity builds up after completion of the new membrane.

Finally, Graham et al. [208] observe that the concentration of H2 histocompatibility antigens is minimal during the S-phase and reaches a maximum in early interphase.

To further characterize changes in the carbohydrate moieties of plasma membrane proteins during the cell cycle, Glick and Buck [213], split off peptides from cultivated BHK_{21}/C_{13} and transformed C_{13}/B_4 fibroblasts by trypsinization. They analysed cells arrested in metaphase by vinblastine-sulfate, vinblastine-treated cells *not* in metaphase, and cells arrested with thymidine (G_1-S boundary). They fractionated the trypsinates on Sephadex G-50 and showed that cells in the metaphase have 43% less sialic acid than vinblastine-treated cells *not* in metaphase. On the other hand, metaphase cells contained three times more L-fucose than vinblastine-treated cells *not* in metaphase. The carbohydrate differences did not correlate with the activities of various glycosyltransferases in cells (determined in Triton homogenates during various stages of the cell cycle).

Although this point has not been raised, it is conceivable that the different levels of membrane sialic acid found at different stages of the cell cycle may reflect altered susceptibilities of membrane glycoproteins and glycolipids to glycosidases at different stages of the cell cycle (or altered glycosidase activity or release, Chapter 7). This suggestion is in accord with the reported variation of membrane fragility as a function of cell cycle [207]. It also fits the results of Gahmberg and Hakomori [214] on hamster NIL fibroblasts, showing greater accessibility of neutral glycolipids to galactose oxidase during the G_1 phase in comparison to the

S phase (Chapter 5). Since the quantity of these glycolipids during the cell cycle is nearly constant, the reported labelling differences have to be explained by alterations in membrane structure.

2.5.3. Complement of plasma membrane proteins in normal and neoplastic fibroblasts

Three principal strategies are currently employed to elucidate changes in the plasma membrane protein-composition of cultured fibroblast subsequent to neoplastic transformation by oncogenic DNA or RNA viruses:

(a) Selective radiolabelling of surface proteins and/or glycoproteins followed by SDS-PAGE.
(b) Isolation and purification of plasma membranes, followed by separation and quantification of membrane proteins and glycoproteins, using SDS-PAGE and/or chromatography.
(c) Selective release of surface peptide and/or glycopeptide by proteases, for example, trypsin, followed by chromatographic separation of the peptides.

These approaches are often used in combination with metabolic labelling.

The maturation of tumor viruses includes the synthesis of nucleocapsid protein(s) and, in the case of RNA viruses, virus-envelope protein. We must therefore comment on some properties of the oncogenic viruses used in the studies to be described.

Cells transformed by the DNA viruses, SV40 and Py, do not produce virus. In contrast, cells transformed by oncogenic RNA viruses may or may not produce virus, depending on cell origin, virus strain, and, in some cases presence or absence of 'helper virus' (e.g., RAV). In the case of *chick* fibroblasts infection, with Rous sarcoma virus including the Schmidt-Ruppin strain (SR-RSV), the Prague strain (P-RSV), the Bryan high titre strain (BH-RSV) plus helper virus, and temperature-sensitive mutants thereof, (ts), at permissive temperatures, *transformation* is *accompanied by virus production.* The same is true for avian sarcoma virus B77.

In cells of *rodent* origin, the situation is complicated. B77 infection leads to transformation *and* production in hamster, rat and mouse cells. P-RSV yields transformation plus production in hamster cells, but not in rat or mouse cells. However, transformation by RSV, B-RSV and SR-RSV does not lead to virus production. This is also true for (ts) mutants at permissive temperatures.

Sheinin and Onodura [215] have used SDS-PAGE to analyze the proteins and glycoproteins of mouse 3T3 fibroblasts and an SV40-transformant thereof, both prelabelled with [^{3}H]glucosamine and [^{14}C]leucine. Membrane isolates [216] of both normal and neoplastic cells showed SDS-PAGE components ranging between 15,000 and 200,000 in mol. wt. The authors detected only minor differences in overall protein composition, but the plasma membranes isolated from

TABLE 2.4

Altered membrane protein composition in fibroblasts neoplastically transformed by various oncogenic viruses

Cell type	Transformant	Type of membrane isolate	Labelling	Membrane protein effect	Reference
NIL2E HF NIL 1 HF NIL 8 HF	 HSV Py	Whole cells solubilized in 2 % SDS + phenylmethylsulfonyl-fluoride.	[^{125}I]lactoperoxidase surface-labelling	Disappearance of a 200,000 GP	[217]
BALB/C 3T3 MF Swiss 3T6 MF	SV40 Py MSV	High shear 'membrane' from 1.55 and 2.30 M sucrose barrier.	[^{125}I]lactoperoxidase surface-labelling.	Disappearance of a 250,000 GP	[218]
a–d CEF a, b RKF	a) P-RSV b) ts 339 of B77 virus c) SR-RSV-A d) SR-RSV-ts68	Low shear; Na-iodoacetate/Tris (0.015/0.01), pH 8.0, 2 % DMSO dextran-polyethylene glycol.	[^{125}I]lactoperoxidase surface-labelling; [^{14}C]valine; [^{14}C]glucosamine.	Decrease: 250,000 daltons GP 39,000 daltons P Increase: 95,000 daltons P 73,000 daltons P Increase of band d at non-permissive temp.: decrease 39,000 daltons component upon transfer to permissive temperature	[220]
CEF	B-RSV (RAV-1) SR-RSV SR-RSV-ts 68 SR-RSV-ts SR-RSV-ts 25	Intact cells solubilized in 2 % SDS, 2 % mercaptoethanol, 1 mM phenylmethylsulfonyl fluoride.	[^{35}S]methionine	Decrease 210,000 daltons GP 145,000 daltons GP GP (SF antigen) Increase 80,000 daltons P 90,000 daltons P	[219]
CEF	RSV-ts 68 changes at permissive temp 36 °C	a) Plasma membrane vesicles b) Cell ghosts, Zn^{2+} fixation.	[^{35}S]methionine; [^{125}I]lacto-peroxidase surface-labelling.	Reduction of 1) >200,000 dalton GP (in cell ghosts) 2) 45,000 mol. wt. P (SDS-PAGE).	[222]

CEF	B-RSV (RAV)	Low shear Na-iodoacetate (0.015 M). Crude membrane fraction from sucrose gradient	[^{3}H]fucose; ^{3}H-/^{14}C-labelled amino acids	Disapperance of a 142,000 dalton GP (SDS-PAGE)	[224]
3T3 MF	SV40	Low shear, hypertonic Tris; rough and smooth membrane fraction, 1.1–1.18 density sucrose barrier. Excreted and EDTA extracted proteins; not pelleted at 1.2×10^5 g · min	[^{3}H]glucosamine/ ^{14}C-labelled amino acids [^{14}C]glucosamine/ ^{3}H-labelled amino acids	Non-excretion of a 150,000 dalton GP (SDS-PAGE)	[225]

HF = Hamster fibroblast; MF = mouse fibroblast; CEF = chick embryo fibroblasts; RFK = rat kidney fibroblast.
Py = Polyoma virus; SV40 = simian virus 40.
RSV = Rous sarcoma virus; P-RSV = Prague strain of RSV; SR-RSV = Schmidt Rupping strain of RSV; RSV (RAV-1) = Bryan-high titre strain of RSV associated with helper virus (RAV-1) tumorigenic in rodents; B77 = avian sarcoma virus.
P = Protein.
GP = Glycoprotein.

transformed cells showed 8.7%–27.7% greater labelling of the SDS-PAGE components by *glucosamine* than by leucine. Sheinin and Ondura [215] suggest that transformation with SV40 virus and with Py virus alter the glycoprotein composition of plasma membranes, that the changes produced by different variants of SV40 and Py are the same, but the SV40 produces different plasma membrane protein modifications than Py.

Hynes [217] has compared normal hamster NIL fibroblasts (clones 1, 8 and 2E) with neoplastic clones derived from the parental cells by transformation with polyoma virus (Py) or hamster sarcoma virus (HSV). He used the lactoperoxidase procedure to radioiodinate proteins exposed at the external plasma membrane surfaces, thereafter dissolved the cells in SDS, separated the cellular proteins by SDS-PAGE and localized the iodinated components in the gels by radioautography.

Radioautographs of the electropherograms showed similar SDS-PAGE banding between 20,000 and 200,000 daltons for both normal and transformed cells except for one *major difference:* significant reduction in the transformed cells of a ~200,000 dalton, glycosylated component present in parental cells (Table 2.4). Instead of the strong band in this molecular weight range, SDS-PAGE of transformed cells shows three faint bands of somewhat greater electrophoretic mobility. It is assumed that lactoperoxidase does not penetrate into either normal cells or their transformants (or does not penetrate differentially), that extracellular proteins bind identically to the two cell types and that the depleted protein component therefore represents a plasma membrane component. Moreover, since mild trypsinization eliminates the protein from normal cells, and neoplastic cells generally exhibit an unusual 'secretion' of proteases (Chapter 7), it is suggested that the action of such proteases accounts for the low levels of the 200,000 dalton glycoprotein in the neoplastic cells.

Very similar results have been reported by Hogg [218] in a comparison of BALB/C 3T3 and Swiss 3T6 mouse fibroblasts with 3T3 cells transformed by SV40, polyoma or mouse sarcoma virus. The cells were ^{125}I-labelled by use of the lactoperoxidase method, disrupted by high-shear homogenization, a crude membrane fraction isolated at the interface of a 1.66–2.0 M sucrose gradient, and this material analyzed by SDS-PAGE, using both radioautography and conventional staining. The important difference between the normal and transformed cells was the deletion in the latter of a 250,000 dalton component that is strongly labelled in normal cells. This component co-migrates with a band heavily stained by Coomassie blue in normal cells and is replaced by three faint bands in transformed cells. It is not established whether the 250,000 dalton component is glycosylated, but it is assigned to the plasma membrane by virtue of its iodination with peroxidase. However, unlike the 200,000 dalton protein of NIL cells, the 250,000 dalton protein of normal mouse cells cannot be released by mild trypsinization of intact cells.

A series of independent, but technically similar experiments on oncogenic RNA viruses extend the results of Hynes [217] and Hogg [218]. Thus, Vaheri and Ruoslati [219] report loss of a 210,000 dalton glycoprotein, of a 145,000 dalton glycoprotein and of a surface antigen in *whole cell extracts* of chick embryo fibroblasts infected by Rous sarcoma virus (and temperature sensitive variants thereof – at permissive temperatures). Cells infected by a non-transforming virus did not differ from normal cells.

Moreover, Stone et al. [220] used a variation of the membrane isolation of Warren et al. [31] to collect plasma membrane fractions from normal chick or rat fibroblasts infected by several strains of Rous sarcoma virus, all metabolically labelled with [^{14}C]glucosamine and [^{14}C]valine as well as surface-labelled by the lactoperoxidase method. All transformed cells were depleted of a large, iodinated glycoprotein (200,000–250,000 daltons by SDS-PAGE). When temperature-sensitive virus mutants were used, the glycoprotein loss occurred only when the infected cells were grown at temperatures permissive for transformation. A non-glycosylated component of mol. wt. 39,000 was also found lacking in the membranes from transformed cells.

Stone et al. [220] also report a large increase in the amount of polypeptide with mol. wt. ~ 73,000 in transformed cells and an increased rate of [^{14}C]valine incorporation into protein of this mol. wt. A smaller increase was observed for polypeptide of mol. wt. ~95,000, but neither protein is labelled by [^{14}C]glucosamine. Neither of these components were iodinated by lactoperoxidase, whether intact cells or isolated membranes were used and neither protein appears to be part of the viral envelopes. Finally, the experiments of Stone et al. [220] suggest that the protein alterations observed reflect altered rates of biosynthesis (rather than biodegradation).

Equivalent experiments on chick embryo fibroblasts, infected with a temperature-sensitive mutant of Rous sarcoma virus are reported by Wickus and Robbins [221] and Wickus et al. [222]. The cells were metabolically labelled with [^{35}S]methionine, radioiodinated using lactoperoxidase and plasma membrane vesicles isolated from 'Warren isolates' by the method of Perdue et al. [34]. SDS-PAGE of cells grown at a temperature permissive for transformation (and production) showed depletion of a 200,000 dalton component, which is iodinated in cells grown at non-permissive temperatures. A 45,000 dalton component, present under non-transforming conditions but not iodinated, is also depleted in cells grown at permissive temperatures.

Significantly, Wickus and associates [221, 222] show that the decrease in the 45,000 dalton component occurs in both Zn^{2+}-fixed ghosts and purified membrane vesicles derived therefrom, whereas the changes in the *high molecular* weight component appear to be restricted to ghosts. They suggest that this protein is 'bound loosely' to the external surface of the cell, and that its depletion in transformed cells may derive from an increased protease release by these cells (Chapter 7).

The conclusions of Wickus et al. [222] are supported by a recent report [125] describing the isolation of a high-molecular weight glycoprotein from cultured chick fibroblasts. Fibroblasts derived from chick embryos or chick heart were grown in media containing mixed [^{14}C]- and [^{3}H]-labelled amino acids, and washed cells were radioiodinated by the lactoperoxidase procedure. The cells were then extracted once with a buffered salt solution and once more with the same medium containing 0.2 M urea. The extraction procedure did not impair viability (cell proliferation). After low-speed centrifugation (3.75×10^5 g · min), the extracts were lyophilized, analyzed by SDS-PAGE and the SDS-PAGE electropherograms compared with those of whole-cells, both intact and trypsinized.

SDS-PAGE of intact cells revealed eight bands iodinated by the lactoperoxidase reaction, of which one with apparent mol. wt. ~220,000 daltons was the most prominent. This band was identified as a glycoprotein by the periodate-Schiff reaction and could be eliminated by gentle trypsin treatment, but *not* by collagenase. Appreciable amounts of this protein could be extracted with serum-free medium, and extraction was markedly improved by use of 0.2 M urea. The extracted protein remained largely unsedimented by centrifugation at $\sim 10^7$ g · min, but was not retarded on Sephadex G-200. The amount of extractable protein *increased* with the density of cells in culture. Significantly, appreciable amounts of labelled glycoprotein could be transferred *in vitro* to unlabelled chick fibroblasts.

The data indicate that the trypsin-sensitive ~200,000 dalton glycoprotein is loosely associated with the outer membrane surface, that it can exchange into the medium and/or to other cells, and that its level is depleted in low-density cultures.

The low levels of this type of protein in transformed cells may thus derive from one or more of the following: low culture density, cell protease secretion and decreased membrane affinity.

At variance with the rather consistent pattern reported up to now is the report of Poduslo et al. [223]. These workers radioiodinated mouse L-cell fibroblasts, BHK21 cells and Rous sarcoma transformant thereof (C_{13}/B_4), isolated 'plasma membranes' [31] and analyzed these by SDS-PAGE. They report *identical* ^{125}I-labelling for the membranes from BHK21 cells and their transformants ([223], Fig. 2), although iodination was restricted to high molecular weight components (mol. wt. ~ 230,000 dalton) corresponding to *galactoprotein A* or the 200,000 dalton and 250,000 dalton components reported by others [112, 217–221].

Bussell and Robinson [224] have compared the glycoprotein composition of chick embryo fibroblasts (CEF) following infection by transforming viruses (RSV) and non-transforming viruses (Rous-associated virus, RAV and Newcastle Disease virus, NDV). The cells were labelled with [^{3}H]fucose or ^{3}H- (^{14}C)-labelled amino acids. The cells were homogenized, large particles removed and a crude membrane fraction isolated by sucrose gradient centrifugation. SDS-PAGE analyses revealed a glycoprotein, MPl, of mol. wt. ~142,000 daltons in membranes from normal cells. This component is depleted in membranes from transformed cells but *also* in membranes from cells infected by RAV and NDV.

One cannot evaluate these data unambiguously, since the membrane fractions are very poorly defined and, since it is established that, during the maturation of even non-oncogenic RNA viruses, host membrane protein synthesis is reduced in favor of virus protein production and membrane insertion [224].

Sakiyama and Burge [225] have used a different approach. They pre-labelled mouse 3T3 fibroblasts and their SV40 transformants differentially using [^{14}C]- and [^{3}H]glucosamine and examined the glycoproteins *secreted* into the medium by the two cell types as well as the protein composition of membranes from cells which are disrupted in hypertonic Tris/NaCl/EDTA (0.5 M/0.01 M/0.01 M), pH 7.4. A fraction of rough and smooth membranes is collected from a sucrose density gradient between the density of 1.10 and 1.18. Proteins excreted or extracted from intact cells by PBS/EDTA (0.01 M) are defined as components that do not pellet at 1.2×10^5 g · min. According to their SDS-PAGE analyses, normal cells excrete a 150,000 dalton glycoprotein that is not released by the transformants. Moreover, a 60,000 dalton glycoprotein present in the membranes of normal cells is lacking from the transformed variants. However, glycopeptides prepared by pronase-digestion of EDTA membrane extracts from normal and neoplastic cells showed *no* differences in sialic acid content. *In toto*, the data of Sakiyama and Burge [225] suggest that transformed cells are depleted in the *protein* moiety of certain glycoproteins, not merely the carbohydrate entity, as often suggested.

However, other data suggest that glycosylation of membrane proteins can be incomplete in transformed cells. Thus, Chiarugi and Urbano [226] have shown that the glycosylation of membrane proteins isolated from normal BHK 21 fibroblasts is less than that of Py transformed cells according to the ratio of [^{14}C]glucosamine incorporation *vs.* [^{3}H]-labelled amino acid incorporation. Equivalent results have been obtained by Minnikin and Allen [227] in their comparison of BHK 21 cells and a Py transformant. They found that the proportion of [^{3}H]glucosamine labelling in transformed cells is only 30–40% that of normal cells; the label is incorporated predominantly into glucosamine, galactosamine and sialic acid.

The experiments described above (Table 2.4) relate to the elegant experiments of Gahmberg and Hakomori [112,228] and Gahmberg et al. [229]. These workers used galactose oxidase to oxidize galactose residues of glycoproteins and glycolipids exposed at external surfaces, and then tritiate these residues by reduction with [^{3}H]borohydride. They find that normal NIL fibroblasts bear a surface galactoprotein, *galactoprotein A* (SDS-PAGE mol. wt. ~ 200,000 daltons) which is either lacking or inaccessible to galactose oxidase in Py transformants. They also show that the exposure of *galactoprotein A* varies with culture conditions; it is not labelled in BHK 21 hamster fibroblasts growing at low density, but becomes prominent as the cultures reach confluence. The labelling of another protein, *galactoprotein B*, (SDS-PAGE mol. wt. 140,000) is independent of culture conditions and transformation. That the presence or exposure of *galactoprotein A* on transformed BHK cells depends upon expression of the transformed state, rather than the presence of the viral genome, has been documented by use of a temperature sensitive polyoma mutant (Py ts 3) (Table 2.5). BHK cells infected with Py ts 3 and grown at the permissive temperature (32 °C) show no labelling of *galactoprotein A*, whereas cells grown at non-permissive temperature show normal labelling.

Interestingly, galactose-oxidase labelling of *galactoprotein A* is supressed in normal cells after these are first exposed to high levels of castor bean lectin or concanavalin A, but low concentrations of these lectins enhance the labelling of the galactoprotein, and the labelling of *galactoprotein B* is relatively insensitive to lectin concentration [228]. This is in contrast to Py transformants, where labelling of *galactoprotein B* is enhanced at low lectin concentrations and suppressed at high lectin levels. These results suggest that the lectins react primarily with *galactoprotein A* in normal cells and with *galactoprotein B* in transformants. The authors suggest that *low* lectin concentrations induce *clustering* of the galactoproteins, resulting in an *increased* reactivity with galactose oxidase, although one might also expect clustering to sterically interfere with galactose oxidase activity. In any event, their data suggest *a different plasma membrane organization* in normal *versus* transformed fibroblasts.

A substantial body of information thus indicates that the glycosylation of

TABLE 2.5

Galactose-oxidase labelling of galactoprotein A in BHK 21 cells and cells transformed by wild-type polyoma virus (Py) and a temperature-sensitive mutant thereof (tsPy)[1]

Cells	Temperature[2] (°C)	Galactoprotein labelling[3]		
		Sparse culture	Subconfluent	Confluent
BHK	32	—	—	+ +
	39	—	+	+ +
PyBHK	32	ND	—	—
	39	ND	—	—
tsPyBHK	32	ND	—	—
	39	ND	+	+ +

1 From Gahmberg et al [229].
2 32 °C = permissive for tsPy.
3 ND = not done; — = absent; + = obviously present; + + = strong.

membrane glycoproteins can be defective in neoplastic fibroblasts as is common for *glycolipids* (Chapter 4). Meezan et al. [230], employing a different approach, have arrived at the same conclusion. These workers examined material released by trypsinization of normal and SV40 transformed cells, differentially pre-labelled using [^{3}H]- and [^{14}C]glucosamine. They digested the trypsinates with pronase and separated the peptide fragments by molecular sieving on Sephadex G-150 in the presence of SDS. They co-chromatographed hydrolysates obtained from surface-, endoplasmic reticulum- and nuclear membranes, and analyzed for various sugars relying on alterations of ^{3}H/^{14}C ratios to detect differences between normal and neoplastic cells. They obtained four major peaks of differing carbohydrate contents, but could discern no drastic difference between normal and neoplastic cells; however, they found a decrease of label incorporation into galactosamine and sialic acid in the case of transformed cells. They suggest that such differences might arise from altered glycosyl transferase patterns.

Buck et al. [231,232] have applied the approach of Meezan et al. [230] to intact cells. They released surface glycopeptides from various normal fibroblasts, as well as their Py, MSV and RSV transformants by trypsinization and fractionated the pronase-digested trypsinates by molecular sieving on Sephadex G-50. They report that, regardless of species, the membrane digests from transformed cells always eluted ahead of the material from normal cells, that is, the carbohydrate fragments from transformed cells are larger than those from normal cells.

Warren et al. [233] have extended these studies using a temperature-sensitive mutant (Ts) of the Schmidt-Ruppin sarcoma virus. Chick embryo fibroblasts infected with Ts and grown at 35°C (permissive temperature, exhibit all characteristics of transformation, while cells grown at non-permissive temperature (41 °C) behave as normal, uninfected cells. Digests obtained from cells grown at the permissive temperature are enriched in high molecular weight carbohydrate moieties.

Further information is presented by Warren et al. [233], who incorporated the double label approach to compare glycopeptide fragments that can be released from normal and neoplastic fibroblasts by trypsinization. They compared normal BHK 21 (clone 13) cells with RSV and Py transformants thereof. The normal and transformed cells were cultivated in the presence of [^{14}C]fucose and [^{3}H]fucose respectively, both during the log-phase of growth and the stationary phase. Trypsinates were prepared from the two cell categories, digested with pronase and co-chromatographed on Sephadex G-50. The label ratios indicated presence of larger carbohydrate moieties on the surfaces of the transformed cells. An increase of sialic acid appeared to be involved, and this was attributed to a 2.5–11-fold higher sialyl transferase in Triton X-100 extracts from neoplastic cells.

2.5.4. Plasma membrane metabolism in neoplastic fibroblasts

Before enquiring into the possible significance of membrane protein anomalies in neoplastic fibroblasts we must examine some metabolic correlates.

In this, one must discriminate between events induced by DNA viruses and those due to the RNA viruses. Cells transformed by RNA viruses continuously produce complete, enveloped virions, that is, they synthesize virus-membrane proteins, which in many cases are inserted into the host-cell plasma membrane during virus maturation. In the process, the synthesis of normal membrane proteins is modified and the membrane properties are altered [234].

A recent study by David [235] on the incorporation of viral polypeptides into the plasma membranes of HeLa (S3) cells injected with non-transforming vesicular stomatitis virus (VSV) illustrates the complexity of the problem. David [235] followed the kinetics of protein incorporation using the double-label approach ([^{3}H]- and [^{14}C]leucine) and also evaluated glycoprotein synthesis by conjoint use of [^{3}H]fucose and [^{14}C]leucine. David used plasma membrane isolates obtained after cell fixation by fluorescein mercuric acetate and analyzed their protein content and metabolism by SDS-PAGE. However, he excluded ambiguities due to the presence of adsorbed ribosomes by use of puromycin; this stops protein synthesis within seconds and releases nascent polypeptide chains from membrane-bound polysomes [235].

The cells were pre-labelled with [^{14}C]leucine for 2 hr, starting 2 hr after infection. They were then exposed to [^{3}H]leucine pulses 0.5–5 min in duration. The data show that synthesis of host cell proteins ceases within 2 hr after infection, whereas generation of virus protein proceeds after 2 hr. This includes protein N, a nucleocapsid component, protein G, a glycoprotein of the viral envelope and protein M, a low-molecular weight envelope protein, that is, these three proteins are labelled by the [^{3}H]leucine pulses. Incorporation of the G and M proteins into plasma membrane is rapid and can be detected after 30 sec; within 2 min the specific activity of G and M is the same in the membranes as in the cytoplasm. In contrast, component N, remains concentrated in the cytoplasm for 20–40 min and only then appears in the membranes. The incorporation of G and M into the membrane isolates was not blocked by puromycin, indicating that (a) their appearance in the membrane isolates is not due to ribosomal contamination, and (b) and intracellular pool of the completed proteins develops despite interference with normal peptide chain completion.

The effects of herpes simplex virus (HSV) infection on the plasma membrane protein composition and metabolism of Hep-2 cells (human epidermoid carcinoma-2) are equally pertinent (even though Hep cells are not fibroblasts). Heine et al. [236] infected Hep-2 cells with HSV (subtype 1) and analyzed the protein composition and metabolism of highly purified plasma membrane vesicles isolated by the procedure of Wallach and Kamat [36]. They measured amino acid incorporation using [^{14}C]leucine, [^{14}C]isoleucine and [^{14}C]valine, and glycoprotein synthesis using [^{3}H]fucose. In separate experiments, the labels were added either before or after viral injection and the kinetics of membrane protein synthesis measured by use of radioautographs of membrane SDS-PAG electropherograms. The electropherograms were also analyzed by conventional staining procedures.

The data indicate that the synthesis of normal membrane proteins ceases shortly after virus infection and that a series of normal membrane proteins are depleted at the expense of twelve new SDS-PAGE components (25,000–126,000 daltons). These new components appear within 18 hr of infection and are all also detectable in purified virus envelopes, albeit in different proportions than found in the membrane isolates. Three of the 12 components are intensely labelled by [^{14}C]glucosamine and can be designated as glycoproteins. Six SDS-PAGE bands also exhibit labelling due to [^{14}C]glucosamine.

A more recent paper by Heine and Roizman [237] provides strong evidence that virus membranes are assembled in contiguity with normal membrane domains. The authors used the affinity density perturbation principle [238] in these studies: The specific gravity of membranes bearing virus components is increased over that of normal membranes by reacting them with antibodies specific for the virus components. This allows separation of the two types of membranes in density gradients. By labelling with ^{3}H-amino acids before infection and ^{14}C-amino acids after infection, host-directed protein synthesis can be distinguished from virus-directed biosynthesis.

By use of SDS-PAGE analyses of the membrane isolates, Heine and Roizman [237] show that the membrane domains bearing antigenic virus components also contain host proteins.

Turning to DNA viruses, a promising approach to elucidate changes in protein synthesis induced by these agents has been presented by Lodish et al. [239]. These experimenters propagated [^{3}H]methyl methionine-labelled SV40 (strain SV-S) in African green monkey kidney fibroblasts (VERO) and purified the virus. They further isolated a poly (A)-containing mRNA from the infected VERO cells using affinity chromatography on poly(U)-Sepharose. They then evaluated the template function of this mRNA using cell extracts from Chinese hamster ovary (CHO) cells or reticulocytes; the CHO extracts were preincubated to allow decay of endogenous mRNA. [^{35}S]methionine was used to measure protein synthesis, and the synthesized proteins were identified by autoradiography of SDS-PAG electropherograms and of trypsin treated peptides separated by paper ionophoresis and/or chromatography on hydroxylapatite (HAC).

Lodish et al. [239] found that mRNA purified from SV40 infected VERO cells causes CHO extracts to synthesize a ~45,000 dalton SDS-PAGE component corresponding to the major SV40 capsid protein; this correspondence was documented biochemically and immunologically. Reticulocyte extracts also synthesized the capsid protein in the presence of SV40 mRNA, although at a lesser rate than CHO extracts.

An interesting finding is the fact that upon HAC chromatography the ~45,000 dalton protein manufactured by cell lysates elutes 3 fractions later than the capsid protein isolated from purified viruses. The reason may be that, as in the case of erythrocyte membrane proteins, a precursor larger than the final protein is synthesized first and a small fragment split off before its incorporation. However, the SV40 capsid protein is not known to act as membrane protein at any stage. Moreover, the critical question appears to be how mRNA from SV40 *transformed* cells influences host membrane protein synthesis.

Available data do not provide a clearcut metabolic basis for the membrane protein differences reported to occur subsequent to viral transformation. We here indicate some possible general mechanisms.

As far as glycoproteins are concerned, one should consider deviant glycosyl transferase activities in neoplastic cells (cf. also Chapters 4 and 8). This issue has been evaluated in some detail by Bosmann et al. [240] who compared the glycosyltransferase complement of mouse 3T3 fibroblasts with that of transformants induced by Py, SV40 and Py + SV40. Triton X-100 extracts (0.1 %) of transformed cells showed increased activities of polypeptide: *N*-acetylgalactosaminyltransferase, glycoprotein: galactosyltransferase and glycoprotein : fucosyltransferase. The increase in polypeptide: *N*-acetylgalactosaminyltransferase activity is of particular interest, since this enzyme is responsible for the attachment of the first sugar, that is, NAcGal to Ser or Thr residues of polypeptide chains [198].

Whatever the role of glycosyl transferases, there is no evidence that protein synthesis in fibroblasts varies appreciably with cell density [241,242]. However, Poste [243] argues that the 'cell coat' of trypsinized, virus transformed fibroblasts grows more rapidly and to greater thickness than is the case for normal cells. This worker measured the thickness of the cell coat by ellipsometry. In the presence of actinomycin D, cell coat synthesis of normal and chemically transformed cells ceases within 15 min, whereas, actinomycin D-treated BHK and NIL fibroblasts transformed by polyoma and Rous sarcoma viruses synthesize cell coat material for up to 135 min. A bigger pool of protein precursor in virus-transformed cells appears improbable, since cycloheximide does not block coat thickening. It is possible that the mRNA template involved is more stable [244] in the virus-transformed cells.

A final topic to be considered is the fact that membrane protein phenotype can vary with differentiation and that oncogenic transformation might lead to expression or deletion of normal membrane proteins as a distal effect and that membrane phenotype can vary with environmental conditions, even in neoplastic cells. This last phenomenon is well illustrated by the behavior of neuroblastoma cells.

N_2aE cells derived from neuroblastoma 1300 grow as differentiated neuroblasts when cultivated in monolayers, but as non-differentiated cells when maintained in suspension. Truding et al. [13] have isolated Zn^{2+}-fixed ghosts from cells grown in both ways, surface-labelled by the lactoperoxidase technique and metabolically labelled using [^{3}H]- or [^{14}C]glucosamine, [^{3}H]fucose and [^{3}H]- or [^{14}C]leucine. SDS-PAGE revealed similar patterns for membranes from differentiated and undifferentiated cells using chemical staining or the distribution of leucine radioactivity as indices; the molecular weight range of the membrane proteins was 30,000–120,000. However, with glucosamine as precursor, a 105,000 dalton component appears in the fractions from differentiated cells; this glycoprotein is lacking in membranes from undifferentiated cells. The 105,000 dalton protein can exist in a fucosylated – and non-fucosylated form. The latter can be extracted from membranes by 0.15 M NaCl, 0.02 M Na-phosphate, pH 7.4. The 105,000 dalton protein appears thus to be 'membrane-associated' and to require completion of the carbohydrate moiety for this association. The 105,000 dalton protein is apparently not exposed at the external surface of the cell, since it is not iodinated by the lactoperoxidase procedure. However, differentiated cells contain a 78,000 dalton surface component which can be iodinated in intact cells.

2.6. *Coda*

The past decade of research has provided us with powerful strategies and techniques for meaningful experimentation on membrane protein composition, dynamics and metabolism. Even though the need for further sophistication cannot be denied, we have assuredly reached the stage where we can conduct discriminating explorations into possible membrane protein anomalies in neoplasia.

We have learned how to fractionate diverse cells in rational ways and now know how to purify diverse membrane categories. Moreover, we have developed an effective armamentarium of techniques for membrane protein fractionation. Finally, we have developed potent strategies for the selective identification of proteins in *plasma membranes* (Fig. 2.1), the membranes which *seem* most immediately implicated in *malignant* neoplasia.

In addition, we have acquired an extensive knowledge of the metabolism and dynamics of plasma membrane proteins (Fig. 2.2). A crucial development in this area has been the discovery that certain proteins dissociate rather readily from plasma membranes, either as molecular entities or as parts of membrane fragments.

Turning to possible *general* membrane protein anomalies in neoplastic cells (rather than specific enzyme changes; Chapters 3–8 and 10), we lack cohesive information. However the data now available indicate that membrane proteins are synthesized by the identical processes in normal and neoplastic cells. Moreover no *universal tumor-specific* membrane proteins have been identified.

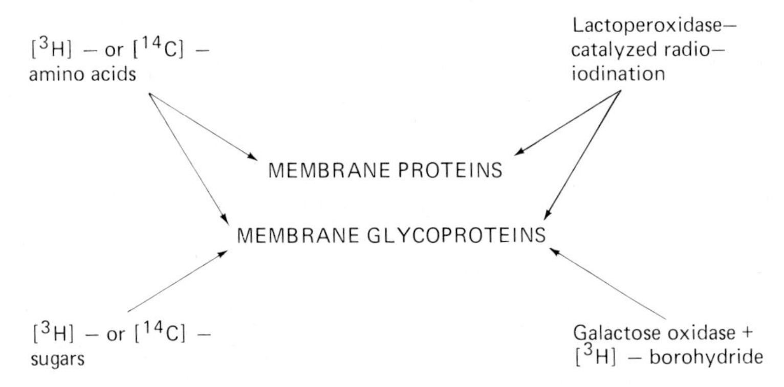

Fig. 2.1. Strategies for the labelling of membrane proteins and glycoproteins.

One of the few consistent findings is the depletion of certain neoplastic cells of certain high-molecular weight membrane (or membrane-associated) proteins. This phenomenon is well documented for various virally transformed fibroblasts cultured *in vitro*, which are deficient in ~200,000 dalton glycoproteins loosely associated with the external surfaces of non-transformed cells. The lack of certain high-molecular weight proteins in the microsomal membranes of neoplastic hepatocytes may represent the same phenomenon. However it is not clear at present how the observed depletion of large membrane-associated proteins comes about. Rapid proteolytic degradation appears a likely possibility, but increased 'shedding' (decreased membrane affinity) must also be considered since, at least in fibroblasts, these proteins dissociate rather freely. Both enhanced degradation and increased 'shedding' could account for the apparently enhanced protein turnover in the membranes of some neoplastic cells.

Extensive data indicate an anomalous degree and/or manner of glycosylation of membrane glycoproteins in neoplastic cells, but no consistent pattern has emerged. However, recent data indicate that virally transformed fibroblasts

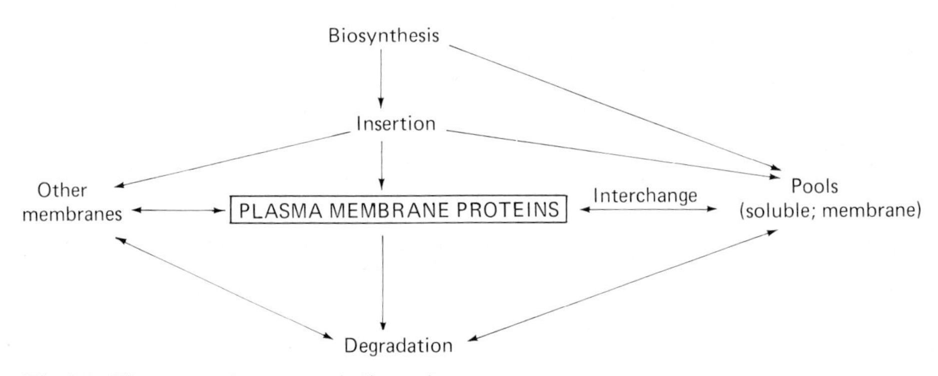

Fig. 2.2. Plasma membrane protein dynamics.

either lack a specific plasma membrane glycoprotein, *galactoprotein A*, or that this protein is somehow 'buried' in the neoplastic cells. Additional studies on plasma membrane galactoproteins strongly suggest that the plasma membrane *organization* of transformed fibroblasts differs from that of normal cells.

In summary, while one cannot reach a general conclusion about general membrane protein anomalies in neoplastic cells, available data indicate a diversity of abnormalities and one can anticipate extensive elucidation of the field in the near future.

References

[1] Rechcigl, M., Jr., in *Enzyme Synthesis and Degradation in Mammalian Systems*, M. Rechcigl, Jr. ed., University Park Press, Baltimore, 1971, p. 237.
[2] Schimke, R.T. and Dehlinger, P.J. in *Membrane Research*, C.F. Fox, ed., Academic Press, New York, 1972, p. 115.
[3] Siekevitz, P., *J. Theoret. Biol.*, *37*, 321 (1972).
[4] Warren, L. in *Membrane and Viruses in Immunopathology*, S.B. Day and R.A. Good, eds., Academic Press, New York, 1972, p. 89.
[5] Marchalonis, J.J. and Cone, R.E., *Transplant. Rev.*, *14*, 3 (1973).
[6] Melchers, F. and Andersson, J., *Transplant. Rev.*, *14*, 76 (1973).
[7] Hendler, R.W. in *Biomembranes*, L.A. Manson, ed., Plenum Press, New York, 1974, Vol. 5, p. 147.
[8] Wallach, D.F.H. and Winzler, R. in *Evolving Strategies and Tactics in Biomembrane Research*, Springer-Verlag, New York, 1974, Chapters 1 and 2.
[9] DePierre, J.W. and Karnovsky, M.L., *J. Cell Biol.*, *56*, 275-303 (1973).
[10] Wallach, D.F.H. and Lin, P.S., *Biochim. Biophys. Acta*, *300*, 211 (1973).
[11] Wallach, D.F.H. in *Specificity of Cell Surfaces*, B. Davis and L. Warren, eds., Englewood Cliffs, New York, Prentice-Hall, 1967, p. 129.
[12] Steck, T.L. and Wallach, D.F.H. in *Methods in Cancer Research*, V.H. Busch, ed., Academic Press, New York, 1970, Vol. 5, p. 93.
[13] Truding, R., Shelanski, M.L., Daniels, M.P. and Morell, P., *J. Biol. Chem.*, *249*, 3973 (1974).
[14] Emmelot, P., Bos, C.J., Benedetti, E.L. and Rümke, P., *Biochim. Biophys. Acta*, *90*, 126 (1964).
[15] Lieberman, I., Lansing, I.A. and Lynch, W.E., *J. Biol. Chem.*, *242*, 736 (1967).
[16] Emmelot, P. and Bos, C.J., *Int. J. Cancer*, *4*, 705 (1969).
[17] Emmelot, P. and Bos, C.J., *Int. J. Cancer*, *4*, 723 (1969).
[18] Emmelot, P., Feltkamp, C.A., Vaz Dias, H., *Biochim. Biophys. Acta*, *211*, 43 (1970).
[19] Bergelson, L.D., *Prog. Chem. Fats Other Lipids*, *13*, 1 (1972).
[20] Moyer, G. and Pitot, H.C., *Cancer Res.*, *33*, 1316 (1973).
[21] Neville, D.M., *J. Biophys. Biochem. Cytol.*, *8*, 413 (1960).
[22] Neville, D.M., *Biochim. Biophys. Acta*, *133*, 168 (1967).
[23] Emmelot, P. and Bos, C.J., *Biochim. Biophys. Acta*, *58*, 374 (1962).
[24] de Duve, C., *J. Cell Biol.*, *50*, 20D (1967).
[25] Emmelot, P. and Bos, C.J., *Biochim. Biophys. Acta*, *121*, 375 (1966).
[26] Wattiaux, R. and Wattiaux-de Coninck, S., *Eur. J. Cancer*, *4*, 201 (1968).
[27] Wattiaux, R. and Wattiaux-de Coninck, S., *Eur. J. Cancer*, *4*, 193 (1968).
[28] Moyer, G.H., Murray, R.K., Khairallah, L.H., Suss, R. and Pitot, H.C., *Lab. Invest.*, *23*, 108 (1970).
[29] Omura, T., Siekevitz, P. and Palade, G.E., *J. Biol. Chem.*, *242*, 2389 (1967).
[30] Kuriyama, Y., Omura, T., Siekevitz, P. and Palade, G.E., *J. Biol. Chem.*, *244*, 2017 (1969).
[31] Warren, L., Glick, M.C. and Nass, M.K., *J. Cell. Physiol.*, *68*, 269 (1966).
[32] Nachman, R.L., Ferris, B. and Hirsch, J.G., *J. Exp. Med.*, *133*, 807 (1971).

[33] Boone, C.W., Ford, L.E., Bond, H.E., Stuart, D.C. and Lorenz, D., *J. Cell Biol.*, *41*, 378 (1969).
[34] Perdue, J.F. and Sneider, J., *Biochim. Biophys. Acta*, *196*, 125 (1970).
[35] Perdue, J.F., Warner, D., Miller, K., *Biochim. Biophys. Acta*, *298*, 817 (1973).
[36] Wallach, D.F.H. and Kamat, V.B., *Methods in Enzymol.*, *8*, 164 (1966).
[37] Graham, J.M., Higgins, J.A. and Green, C., *Biochim. Biophys. Acta*, *150*, 303 (1968).
[38] Schmidt-Ullrich, R., Ferber, E., Knüfermann, H., Fischer, H. and Wallach, D.F.H., *Biochim. Biophys. Acta*, *332*, 175 (1974).
[39] Schmidt-Ullrich, R., Wallach, D.F.H. and Davis, F.D.G., *Biochim. Biophys. Acta*, in press.
[40] Maddy, A.H. and Kelly, P.G., *Biochim. Biophys. Acta*, *241*, 290 (1971).
[41] Harris, J.R., *Biochim. Biophys. Acta*, *188*, 31 (1969).
[42] Harris, J.R., *J. Mol. Biol.*, *46*, 329 (1969).
[43] Reynolds, J.A. and Trayer, H.J., *J. Biol. Chem.*, *246*, 7337 (1971).
[44] Reynolds, J.A., *Ann. N. Y. Acad. Sci.*, *195*, 75 (1972).
[45] Marchesi, V.T. and Steers, E., Jr., *Science*, *159*, 203 (1968).
[46] Mitchell, C.D. and Hanahan, D.J., *Biochemistry*, *5*, 51 (1966).
[47] Fairbanks, G., Steck, T.L. and Wallach, D.F.H., *Biochemistry*, *10*, 2606 (1971).
[48] Gwynne, J.T. and Tanford, C., *J. Biol. Chem.*, *245*, 3269 (1970).
[49] Triplett, R.B., Summers, J., Ellis, D.E., and Carraway, V.L., *Biochim. Biophys. Acta*, *266*, 484 (1972).
[50] Hatefi, Y. and Hanstein, W.G., *Proc. Natl. Acad. Sci. U.S.*, *62*, 1129 (1969).
[51] Maddy, A.H., *Biochim. Biophys. Acta*, *117*, 193 (1966).
[52] Rega, A.F., Weed, R.I., Reed, C.F., Berg, G.G. and Rothstein, A., *Biochim. Biophys. Acta*, *147*, 297 (1967).
[53] Hamaguchi, H. and Cleve, H., *Biochem. Biophys. Res. Comm.*, *47*, 459 (1972).
[54] Hamaguchi, H. and Cleve, H., *Biochim. Biophys. Acta*, *278*, 271 (1972).
[55] Fletcher, M.A. and Woolfolk, B.J., *J. Immunol.*, *107*, 842 (1971).
[56] Fletcher, M.A. and Woolfolk, B.J., *Biochim. Biophys. Acta*, *278*, 163 (1972).
[57] Kramer, R., Schlatter, C. and Zahler, P.H., *Biochim. Biophys. Acta*, *282*, 146 (1972).
[58] Zahler, P.H. and Wallach, D.F.H., *Biochim. Biophys. Acta*, *135*, 371 (1967).
[59] Goodenough, D.A. and Stoeckenius, W., *J. Cell. Biol.*, *54*, 646 (1972).
[60] Winzler, R.J. in *Red Cell Membrane : Structure and Function*, G.A. Jamieson and T.Z. Greenwalt, eds., Lippincott, Philadelphia, 1969, p. 157.
[61] Blumenfeld, O.O., Gallop, P.M., Howe, C. and Lee, L.T., *Biochim. Biophys. Acta*, *211*, 109 (1970).
[62] Wallach, D.F.H. and Zahler, P.H., *Biochim. Biophys. Acta*, *150*, 186 (1968).
[63] Wallach, D.F.H. and Zahler, P.H., unpublished.
[64] Drey, W.J., Papermaster, D.S. and Kühn, H., *Ann. N. Y. Acad. Sci.*, *195*, 61 (1974).
[65] Bjerrum, O.J. and Lundahl, P., *Biochim. Biophys. Acta*, *342*, 69 (1974).
[66] Schmidt-Ullrich, R., Wallach, D.F.H. and Ferber, E., *Ztschr. Immunitätsforsch.*, *147*, 337 (1974).
[67] Philippot, J., *Biochim. Biophys. Acta*, *225*, 201 (1971).
[68] Marchesi, V.T. and Andrews, E.P., *Science*, *174*, 1247 (1971).
[69] Tremblay, G.Y., Daniels, M.J. and Schaechter, M., *J. Mol. Biol.*, *40*, 65 (1969).
[70] Fairbanks, G. and Avruch, J., *J. Supramol. Structure*, *1*, 66 (1972).
[71] Heller, J., *Biochemistry*, *7*, 2906 (1968).
[72] Carter, J.R., Jr., *Biochemistry*, *12*, 171 (1973).
[73] Steck, T.L. and Yu, J., *J. Supramol. Structure*, *1*, 220 (1973).
[74] Cantrell, A.C., *Biochim. Biophys. Acta*, *311*, 381 (1973).
[75] Rosenberg, S.A. and Guidotti, G., *J. Biol. Chem.*, *243*, 1985 (1968).
[76] Moldow, C.F., Zucker-Franklin, D., Gordon, A., Hospelhorn, V. and Silber, R., *Biochim. Biophys. Acta*, *255*, 133 (1972).
[77] Lundahl, P. and Hjerten, S. in *Protides in the Biol. Fluids*, *21. Coll.*, H. Peeters, ed., Pergamon Press, Oxford, 1974, p. 27.
[78] Steck, T.L., *J. Cell Biol.*, *62*, 1 (1974).
[79] Weber, K., Pringle, J.R. and Osborn, M. in *Methods in Enzymol.*, vol. XXVI, C, C.H.W. Hirs and S.N. Timasheff, eds., Academic Press, New York, 1972, pp. 3-27.

[80] Knüfermann, H., Bhakdi, S. and Wallach, D.F.H., *Biochim. Biophys. Acta*, *389*, 464 (1975).
[81] Maddy, A.H. and Dunn, M.J. in *Protides of Biol. Fluids, 21 Coll.*, H. Peeters, ed., Pergamon Press, Oxford, 1974, p. 21.
[82] Knüfermann, H., Bhakdi, S., Schmidt-Ullrich, R., Wallach, D.F.H., *Biochim. Biophys. Acta*, *330*, 356 (1973).
[83] Langdon, R.G., *Biochim. Biophys. Acta*, *342*, 213 (1974).
[84] Bhakdi, S., Knüfermann, H. and Wallach, D.F.H., *Biochim. Biophys. Acta*, *345*, 448 (1974).
[85] Bjerrum, O.J., Bhakdi, S., Knüfermann, H. and Bøg-Hansen, T.C., *Biochim. Biophys. Acta*, *373*, 44 (1974).
[86] Segrest, J.P., Jackson, R.L., Andrews, E.P. and Marchesi, V.T., *Biochem. Biophys. Res. Commun.*, *44*, 390 (1971).
[87] Takayama, K., Maclennan, D.H., Tzagaloff, A. and Stoner, C.F., *Arch. Biochem. Biophys.*, *114*, 223 (1966).
[88] Neville, D.M., Jr., *Biochim. Biophys. Acta*, *154*, 540 (1968).
[89] Tuppy, H., Swetly, P. and Wolff, I., *Eur. J. Biochem.*, *5*, 339 (1968).
[90] Panet, R. and Selinger, Z., *Eur. J. Biochem.*, *14*, 440 (1970).
[91] Righetti, P.G. and Drysdale, J.W., *J. Chromat.*, *98*, 271 (1974).
[92] Merz, D.C., Good, R.A. and Litman, G.W., *Biochem. Biophys. Res. Commun.*, *49*, 84 (1972).
[93] Bhakdi, S., Knüfermann, H. and Wallach, D.F.H., *Biochim. Biophys. Acta*, *394*, 550 (1975).
[94] Schmidt-Ullrich, R., Wallach, D.F.H. and Davis, F.D.G., *Biochim. Biophys. Acta*, in press.
[95] Bhakdi, S., Knüfermann, H., Schmidt-Ullrich, R., Fischer, H. and Wallach, D.F.H., *Biochim. Biophys. Acta*, *363*, 39 (1974).
[96] Bhakdi, S., Knüfermann, H., Wallach, D.F.H. and Fischer, H., *Biochim. Biophys. Acta*, *373*, 295 (1974).
[97] Raftell, M., Blomberg, F. and Perlmann, P., *Cancer Res.*, *34*, 2300 (1974).
[98] Swick, R.W. and Handa, D.T., *J. Biol. Chem.*, *218*, 577 (1956).
[99] Arias, I.M., Doyle, D. and Schimke, R.T., *J. Biol. Chem.*, *244*, 3303 (1969).
[100] McFarlane, A.S., *Biochim. J.*, *89*, 277 (1963).
[101] Swick, R.W., Rexroth, A.K. and Stange, J.L., *J. Biol. Chem.*, *243*, 3581 (1968).
[102] Gurd, J.W. and Evans, W.H., *Eur. J. Biochem.*, *36*, 273 (1973).
[103] Schmidt-Ullrich, R., Wallach, D.F.H. and Ferber, E., *Biochim. Biophys Acta*, *356*, 288 (1974).
[104] Higashi, T. and Peters, T., Jr., *J. Biol. Chem.*, *238*, 3952 (1963).
[105] Wurtman, R.J. and Axelrod, J., *Proc. Natl. Acad. Sci. U.S.*, *57*, 1594 (1967).
[106] Phillips, D.R. and Morrison, M., *Biochemistry*, *10*, 1766 (1971).
[107] Marchalonis, J.J., Cone, R.E. and Santer, V., *Biochem. J.*, *124*, 921 (1971).
[108] Vitetta, E.S. and Uhr, J.W., *Transplant. Rev.*, *14*, 50 (1973).
[109] Atwell, J.L., Cone, R.E. and Marchalonis, J.J., *Nature, New Biol.*, *241*, 251 (1973).
[110] Jones, G., *J. Immunol.*, *110*, 1526 (1973).
[111] Roberts, R.M., Yuan, B.O.-Ch., *Biochemistry*, *13*, 4846 (1974).
[112] Gahmberg, C.G. and Hakomori, S.-I., *Proc. Natl. Acad. Sci. U.S.*, *70*, 3329 (1973).
[113] Wallach, D.F.H., *Biochim. Biophys. Acta*, *265*, 61 (1972).
[114] Poole, B., *J. Biol. Chem.*, *246*, 6587 (1971).
[115] Gates, R.E., McClain, M. and Morrison, M., *Exptl. Cell Res.*, *83*, 344 (1974).
[116] Lodish, H.F., *Proc. Natl. Acad. Sci. U.S.*, *70*, 1526 (1973).
[117] Lodish, H.F. and Deslau, O., *J. Biol. Chem.*, *248*, 3420 (1973).
[118] Deutsch, J. and Blumenfeld, O.O., *Biochem. Biophys. Res. Comm.*, *58*, 454 (1974).
[119] Morrison, M., Michaels, A.W., Phillips, D.R. and Choi, S.-I., *Nature*, *248*, 763 (1974).
[120] Yarrison, G. and Choules, G.L., *Biochem. Biophys. Res. Comm.*, *52*, 57 (1973).
[121] Hunter, W. in *Handbook of Exptl. Immunology*, B.M. Weir, ed., Davis, Philadelphia, 1967, p. 608.
[122] Weidekamm, E., Wallach, D.F.H. and Fischer, H., *Biochim. Biophys. Acta*, *241*, 770 (1971).
[123] Dodge, J.T., Mitchell, C. and Hanahan, D.J., *Arch. Biochem. Biophys.*, *110*, 119 (1963).
[124] Bjerrum, O.J., Bhakdi, S., Bøg-Hansen, T.C., Knüfermann, H. and Wallach, D.F.H., *Biochim. Biophys. Acta*, in press.
[125] Kapeller, M., Gal-Oz, R., Grover, N.B. and Doljanski, F., *Exp. Cell Res.*, *79*, 152 (1973).
[126] Yamada, K.M. and Weston, J.A., *Proc. Natl. Acad. Sci. U.S.*, *71*, 3492 (1974).

[127] Cooper, A.G., Codington, J.F. and Brown, M.C., *Proc. Natl. Acad. Sci. U.S.*, *71*, 1224 (1974).
[128] Cone, R.E., Marchalonis, J.J. and Rolley, R.T., *J. Exp. Med.*, *134*, 1373 (1971).
[129] Marchalonis, J.J., Cone, R.E. and Atwell, J.L., *J. Exp. Med.*, *135*, 956 (1972).
[130] Vitetta, E.S., Baur, S. and Uhr, J.W., *J. Exp. Med.*, *134*, 242 (1971).
[131] Vitetta, E.S. and Uhr, J.W., *J. Exp. Med.*, *136*, 676 (1972).
[132] Vitetta, E.S. and Uhr, J.W., *J. Immunol.*, *108*, 577 (1972).
[133] Vitetta, E.S. and Uhr, J.W., *J. Exp. Med.*, *139*, 1559 (1974).
[134] Vitetta, E.S., Grundke-Iqbal, I., Holmes, K.V. and Uhr, J.W., *J. Exp. Med.*, *139*, 862 (1974).
[135] Uhr, J.W. and Vitetta, E.S., *J. Exp. Med.*, *139*, 1013 (1974).
[136] Lerner, R.A., McConahey, P.J., Jansen, I. and Dixon, F.J., *J. Exp. Med.*, *135*, 136 (1972).
[137] Hütteroth, T.H., Litwin, S.D. and Cleve, H., *Cell. Immunol.*, *5*, 446 (1972).
[138] Pellegrino, M.A., Pellegrino, A., Ferrone, S., Kahan, B.D. and Reisfeld, R.A., *J. Immunol.*, *111*, 783 (1973).
[139] Oh, S.K., Pellegrino, M.A., Ferrone, S. and Reisfeld, R.A., *Fed. Proc. Abst.*, *2875*, 719 (1974).
[140] Billing, R.J. and Terasaki, P.I., *J. Immunol.*, *112*, 1124 (1974).
[141] Miyakawa, Y., Tanigaki, N., Kreiter, V.P., Moore, G.E. and Pressmann, D., *Transplantation*, *15*, 312 (1973).
[142] Vitetta, E.S., Uhr, J.W. and Boyse, E.A., *J. Immunol.*, *4*, 276 (1974).
[143] Rolley, R.T. and Marchalonis, J.J., *Transplantation*, *14*, 734 (1972).
[144] Dwyer, J.M., Warner, N.L. and Mackay, I.R., *J. Immunol.*, *108*, 1439 (1972).
[145] Wilson, J.D., Nossal, G.J.V. and Lewis, H., *Eur. J. Immunol.*, *2*, 225 (1972).
[146] Schmidt-Ullrich, R., Wallach, D.F.H. and Hendricks, J., *Biochim. Biophys. Acta*, *282*, 295 (1975).
[147] Jones, G., *J. Immunol.*, *110*, 1526 (1973).
[148] Freychet, P., Roth, J. and Neville, D.M., Jr., *Proc. Natl. Acad. Sci. U.S.*, *68*, 1833 (1971).
[149] Cuatrecasas, P., *Proc. Natl. Acad. Sci. U.S.*, *68*, 1264 (1971).
[150] DeMeyts, P., Roth, J., Neville, D.M., Jr. and Gavin, J.R., III, *Biochim. Biophys. Res. Comm.*, *55*, 154, 1973.
[151] Kahn, C.R., Freychet, P. and Roth, J., *J. Biol. Chem.*, *249*, 2249 (1974).
[152] Gavin, J.R., III, Roth, J. Neville, D.M., Jr., DeMeyts, P. and Buell, D.N., *Proc. Natl. Acad. Sci. U.S.*, *71*, 84 (1974).
[153] Gavin, J.R., III, Buell, D.N. and Roth, J., *Science*, *178*, 168 (1972).
[154] Karnosvky, M.J., Unanue, E.R. and Leventhal, M., *J. Exp. Med.*, *136*, 907 (1972).
[155] Unanue, E.R. and Karnovsky, M.J., *Transplant. Rev.*, *14*, 184 (1973).
[156] Unanue, E.R., Perkins, W.D. and Karnovsky, M.J., *J. Immunol.*, *108*, 569 (1972).
[157] Aolt, K.A. and Unanue, E.R., *J. Exp. Med.*, *139*, 1110 (1974).
[158] Takahashi, T., *Transplantation Proc.*, *3*, 1217 (1971).
[159] Boyse, E.A., Stockert, E. and Old, L.J., *Proc. Natl. Acad. Sci. U.S.*, *58*, 954 (1967).
[160] Auki, T., Guring, B., Beth, E. and Old, L.J. in *Recent Advances in Human Tumor Virology*, W. Nakahara et al., eds., University Park Press, Baltimore, 1971, p. 425.
[161] Old, L.J., Stockert, E., Boyse, E.A. and Kim, J.H., *J. Exp. Med.*, *127*, 523 (1968).
[162] Smith, R.T., Klein, G., Klein, E. and Clifford, P. in *Advances in Transplantation*, J. Dausset and G. Mathe, eds., Munksguard, Copenhagen, 1968, p. 483.
[163] Lamm, M.E., Boyse, E.A., Old, L.J., Lisowska-Bernstein, B., and Stockert, E., *J. Immunol.*, *101*, 99 (1968).
[164] Yu, A. and Cohen, E.P., *J. Immunology*, *112*, 1285 (1974).
[165] Yu, A. and Cohen, E.P., *J. Immunol.*, *112*, 1296 (1974).
[166] Stackpole, C.W., Jacobson, J.B. and Lardis, M.P., *J. Exp. Med.*, *140*, 939 (1974).
[167] Hütteroth, T.H., Cleve, H. and Litwin, S.D., *J. Immunol.*, *110*, 1325 (1973).
[168] Demus, H., *Biochim. Biophys. Acta*, *291*, 93 (1973).
[169] Ladoulis, C.T., Misra, D.N., Estes, L.W. and Gill, T.J., *III*, *Biochim. Biophys. Acta*, *356*, 27 (1974).
[170] Schmidt-Ullrich, R. and Wallach, D.F.H., to be published.
[171] Bosmann, H.B., *Nature*, *233*, 566 (1971).
[172] Kessell, D., *Cancer Res.*, *31*, 1883 (1971).
[173] Kessel, D. and Bosmann, H.B., *Cancer Res.*, *34*, 603 (1974).

[174] Bernacki, R.J.B., *J. Cell Physiol.*, *83*, 457 (1974).
[175] Ernster, L., Siekevitz, P. and Palade, G.E., *J. Cell Biol.*, *15*, 541 (1962).
[176] Price, V.E., Sterling, W.R., Tarantola, V.A., Hartley, R.W., Jr. and Rechcigl, M., Jr., *J. Biol. Chem.*, *237*, 3468 (1962).
[177] Feigelson, P., Dahman, T., Margolis, F., *Arch. Biochem.*, *85*, 478 (1959).
[178] Schimke, R.T., *J. Biol. Chem.*, *239*, 3808 (1964).
[179] Gan, J.C. and Jeffay, H., *Biochim. Biophys. Acta*, *148*, 448 (1967).
[180] Schimke, R.T., Ganschow, R., Doyle, D. and Arias, I.M., *Fed. Proc.*, *27*, 1223 (1968).
[181] Jick, H. and Shuster, L., *J. Biol. Chem.*, *241*, 5366 (1966).
[182] Loftfield, R.B. and Harris, A.J., *J. Biol. Chem.*, *219*, 151 (1956).
[183] Widnell, G.G. and Siekevitz, P., *J. Cell. Biol.*, *35*, 142A (1967).
[184] Dehlinger, P.J. and Schimke, R.T., *J. Biol. Chem.*, *246*, 2574 (1971).
[185] Dice, J.F. and Schimke, R.T., *J. Biol. Chem.*, *247*, 98 (1972).
[186] Fruton, J.S. in *The Enzymes*, P.D. Boyer, H. Lardy, and K. Myrback, eds., Academic Press, New York, 1960, Vol. 4, p. 233.
[187] Morrison, W.L. and Neurath, H., *J. Biol. Chem.*, *200*, 39 (1953).
[188] Ray, T.K., Lieberman, I. and Lansing, A.I., *Biochem. Biophys. Res. Comm.*, *31*, 54 (1968).
[189] Barancik, L.C. and Lieberman, I., *Biochem. Biophys. Res. Comm.*, *44*, 1084 (1971).
[190] Dallner, G., Siekevitz, P. and Palade, G.E., *J. Cell. Biol.*, *30*, 73 (1966).
[191] Dallner, G., Siekevitz, P. and Palade, G.E., *J. Cell. Biol.*, *30*, 97 (1966).
[192] Kawasaki, T. and Yamashima, I., *Biochim. Biophys. Acta*, *225*, 234 (1971).
[193] Francke, W.W., Merré, D.J., Deumling, B., Cheetham, R.D., Kartenbeck, J., Jarasch, E.-D. and Zentgraf, H.-W., *Z. Naturforsch.*, *266*, 1031 (1971).
[194] Peters, T., Jr., *J. Biol. Chem.*, *237*, 1186 (1962).
[195] Morré, D.J., Mollenhauver, H.H. and Bracker, C.E. in *Results and Problems in Cell Differentiation*, W. Reinert and H. Ursprung, eds., Springer, Berlin, 1970, Vol. 2, p. 82.
[196] Bennett, G., Leblond, C.P. and Haddad, A., *J. Cell. Biol.*, *50*, 258 (1974).
[197] Bosmann, H.B., *Biochem. Biophys. Res. Commun.*, *48*, 523 (1972).
[198] Hughes, R.C., *Prog. Biophys. Mol. Biol.*, *26*, 189 (1973).
[199] Roth, S., McGuire, E.J. and Roseman, S., *J. Cell. Biol.*, *51*, 536 (1971).
[200] Chiarugi, V.P., *Cancer Res.*, *32*, 2707 (1972).
[201] Narayan, K.A. and Chandrasekhara, N., *Biochem. J.*, *124*, 953 (1971).
[202] Olivotto, M. and Paoletti, F., *J. Cell. Biol.*, *62*, 585 (1974).
[203] Warren, L. and Glick, M.C., *J. Cell. Biol.*, *37*, 729 (1968).
[204] Gerner, E.W., Glick, M.C. and Warren, L., *J. Cell. Physiol.*, *75*, 275 (1970).
[205] Bello, L.J., *Biochim. Biophys. Acta*, *157*, 8 (1968).
[206] Bootsma, D., Budke, L. and Vos, O., *Exptl. Cell Res.*, *33*, 301 (1964).
[207] Graham, J.M., Summer, M.C.B., Curtis, D.H. and Pasternak, C.A., *Nature*, *246*, 291 (1973).
[208] Graham, J.M., *Biochem. J.*, *130*, 1113 (1972).
[209] Graham, J.M., in *Methodological Developments in Biochemistry*, E. Reid, ed., Longman, London, 1973, Vol. 3, p. 205.
[210] Warmsley, A.M.H. and Pasternak, C.A., *Biochem. J.*, *119*, 493 (1970).
[211] Mayhew, E., *J. Cell. Physiol.*, *69*, 305 (1967).
[212] Glick, M.C., Gerner, E.W. and Warren, L., *J. Cell Physiol.*, *77*, 1 (1971).
[213] Glick, M.C. and Buck, C.A., *Biochemistry*, *12*, 85 (1973).
[214] Gahmberg, C.G. and Hakomori, S., *Biochem. Biophys. Res. Comm.*, *59*, 283 (1974).
[215] Sheinin, R. and Onodera, K., *Biochim. Biophys. Acta*, *274*, 49 (1972).
[216] Brunette, D.M. and Till, J.E., *J. Memb. Biol.*, *5*, 215 (1971).
[217] Hynes, R.O., *Proc. Natl. Acad. Sci. U.S.*, *70*, 3170 (1973).
[218] Hogg, N.M., *Proc. Natl. Acad. Sci. U.S.*, *71*, 489 (1974).
[219] Vaheri, A. and Ruoslahti, E., *Int. J. Cancer*, *13*, 579 (1974).
[220] Stone, K.R., Smith, R.E. and Joklik, W.K., *Virology*, *58*, 86 (1974).
[221] Wickus, G.G. and Robbins, P.W., *Nature, New Biol.*, *245*, 65 (1973).
[222] Wickus, G.G., Branton, P. and Robbins, P. in *Control of Proliferation in Animal Cells*, B. Clarkson and R. Baserga, eds., Cold Spring Harbor Laboratory, 1974, vol. 1, 541.
[223] Poduslo, J.F., Greenberg, C.S. and Glick, M.C., *Biochemistry*, *11*, 2616 (1972).

[224] Bussell, R.H. and Robinson, W.S., *J. Virol.*, *12*, 320 (1973).
[225] Sakiyama, H. and Burge, B.W., *Biochemistry*, *11*, 1366 (1972).
[226] Chiarugi, V.P. and Urbano, P., *J. Gen. Virol.*, *14*, 133 (1972).
[227] Minnikin, S.M. and Allen, A., *Biochem. J.*, *134*, 1123 (1973).
[228] Gahmberg, C.G. and Hakomori, S.-I., *J. Biol. Chem.*, *250*, 2438, 2447 (1975).
[229] Gahmberg, C.G., Kiehn, D. and Hakomori, S.-I., *Nature*, *248*, 413 (1974).
[230] Meezan, E., Wu, H.C., Black, P.H. and Robbins, P.W., *Biochemistry*, *8*, 2518 (1969).
[231] Buck, C.A., Glick, M.C. and Warren, L., *Biochemistry*, *9*, 4567 (1970).
[232] Buck, C.A., Glick, M.C. and Warren, L., *Science*, *172*, 169 (1971).
[233] Warren, L., Fuhrer, J.P. and Buck, C.A., *Proc. Natl. Acad. Sci. U.S.*, *69*, 1838 (1972).
[234] Klenck, H.-D. in *Biological Membranes*, D. Chapman and D.F.H. Wallach, eds., Academic Press, New York, 1973, Vol. 2, p. 145.
[235] David, A.E., *J. Mol. Biol.*, *76*, 135 (1973).
[236] Heine, J.W., Spear, P.G. and Roizman, B., *J. Virol.*, *9*, 431 (1972).
[237] Heine, J.W. and Roizman, B., *J. Virol.*, *11*, 810 (1973).
[238] Wallach, D.F.H., Kranz, B., Ferber, E. and Fischer, H., *FEBS Letters*, *21*, 29 (1972).
[239] Lodish, H.F., Weinberg, R. and Ozer, H.L., *J. Virol.*, *13*, 580 (1974).
[240] Bosmann, H.B., Hagopian, A. and Eylar. E.H., *J. Cell. Physiol.*, *72*, 81 (1968).
[241] Baenziger, N.L., Jacobi, C.H. and Thach, R.E., *J. Biol. Chem.*, *249*, 3483 (1974).
[242] Aviv, H., Boime, I. and Leder, P., *Proc. Natl. Acad. Sci. U.S.*, *68*, 2303 (1971).
[243] Poste, G., *Exptl. Cell Res.*, *77*, 264 (1973).
[244] Pitot, H.C., Shires, T.K., Moyer, G. and Garett, C.T. in *The Molecular Biology* of *Cancer*, H. Busch, ed., Academic Press, New York, 1974, p. 523.

3

Phospholipids

3.1. Introduction

Together with protein, phospholipids constitute the major components of biomembranes. Some of the major membrane phosphatides are listed in Figs. 3.1 and 3.2. A great deal is known about phospholipids, their biosynthesis, biodegradation, as well as their physical and chemical properties. Indeed, many membrane models (Chapter 1) rest heavily on the properties of phospholipids and most model membranes consist of phospholipids primarily. Finally, many of the functional features of biomembranes, of both normal and neoplastic cells, are commonly interpreted in terms of phospholipid behavior. Despite this emphasis, there is amazingly little experimentation comparing normal with neoplastic cells in terms of phospholipid composition, phospholipid metabolism, phospholipid exchange, molecular motions of phospholipids, etc. We will later deal with what information is available, but will first treat some important general properties of phospholipids and of phospholipid metabolism.

The formulas of the important membrane phospholipids are given in Figs. 3.1 and 3.2. The structural characteristics of some of the major fatty acid chains in phosphatides are presented in Figs. 3.3 and 3.4.

The paraffin chains of membrane lipids may be unbranched or branched. Single branching may be of the *iso-paraffin* type, i.e.

$$\begin{array}{c} \quad CH_3 \\ \quad | \\ CH_3\text{–}CH\text{–}CH_2\text{–} \end{array}$$

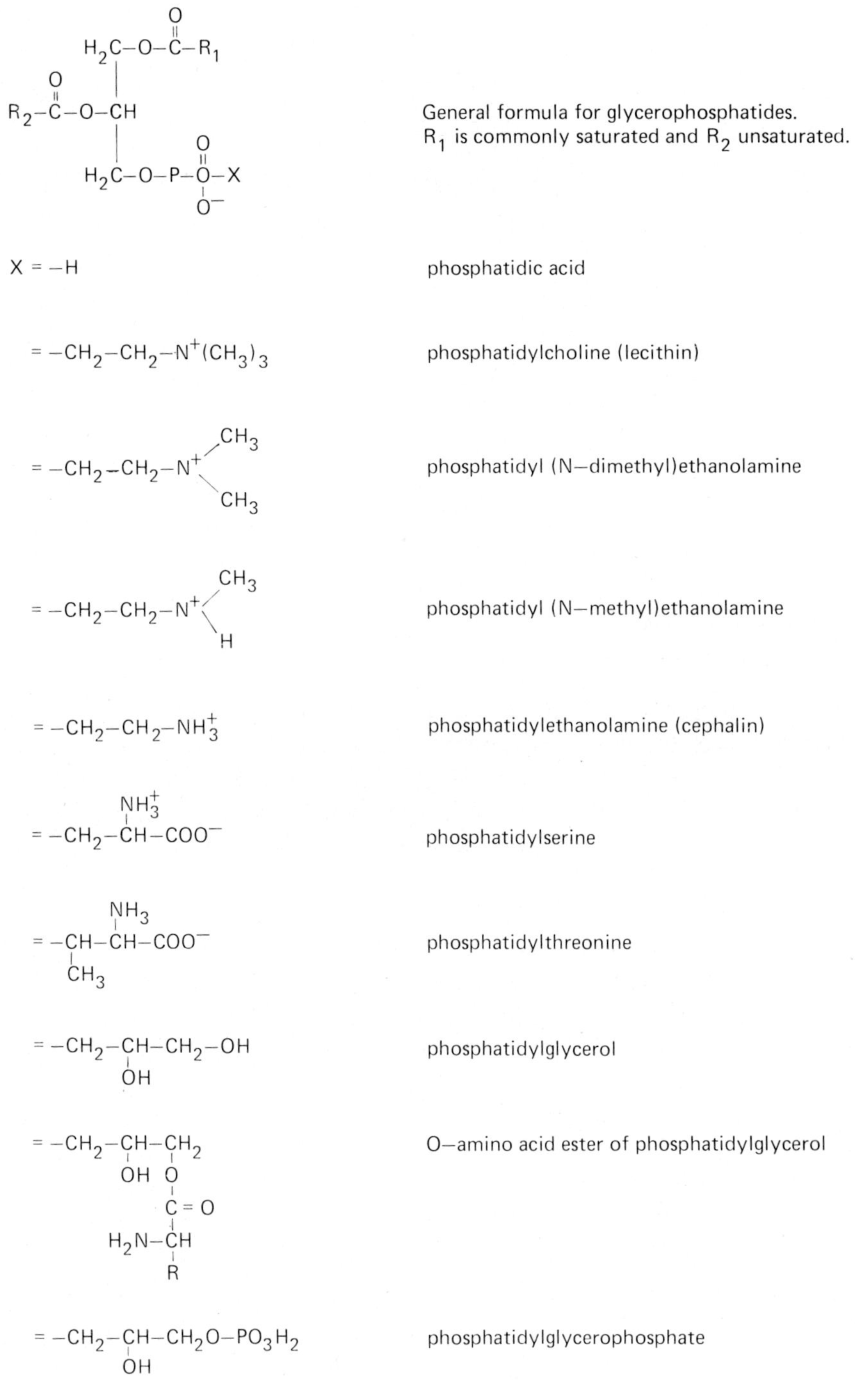

$$\begin{array}{l} \quad\quad\quad\quad\quad\quad O \\ \quad\quad\quad\quad\quad\quad \| \\ \quad\quad\quad H_2C{-}O{-}C{-}R_1 \\ \quad O \quad\quad | \\ \quad \| \quad\quad\, | \\ R_2{-}C{-}O{-}CH \\ \quad\quad\quad\quad | \quad\quad\quad O \\ \quad\quad\quad\quad | \quad\quad\quad \| \\ \quad\quad\quad H_2C{-}O{-}P{-}O{-}X \\ \quad\quad\quad\quad\quad\quad\quad\quad | \\ \quad\quad\quad\quad\quad\quad\quad\quad O^- \end{array}$$

General formula for glycerophosphatides. R_1 is commonly saturated and R_2 unsaturated.

X	
$= -H$	phosphatidic acid
$= -CH_2-CH_2-N^+(CH_3)_3$	phosphatidylcholine (lecithin)
$= -CH_2-CH_2-N^+(CH_3)(CH_3)$	phosphatidyl (N–dimethyl)ethanolamine
$= -CH_2-CH_2-N^+(CH_3)(H)$	phosphatidyl (N–methyl)ethanolamine
$= -CH_2-CH_2-NH_3^+$	phosphatidylethanolamine (cephalin)
$= -CH_2-CH(NH_3^+)-COO^-$	phosphatidylserine
$= -CH(CH_3)-CH(NH_3)-COO^-$	phosphatidylthreonine
$= -CH_2-CH(OH)-CH_2-OH$	phosphatidylglycerol
$= -CH_2-CH(OH)-CH_2-O-C(=O)-CH(NH_2)-R$	O–amino acid ester of phosphatidylglycerol
$= -CH_2-CH(OH)-CH_2O-PO_3H_2$	phosphatidylglycerophosphate

Fig. 3.1. Glycerophosphatides.

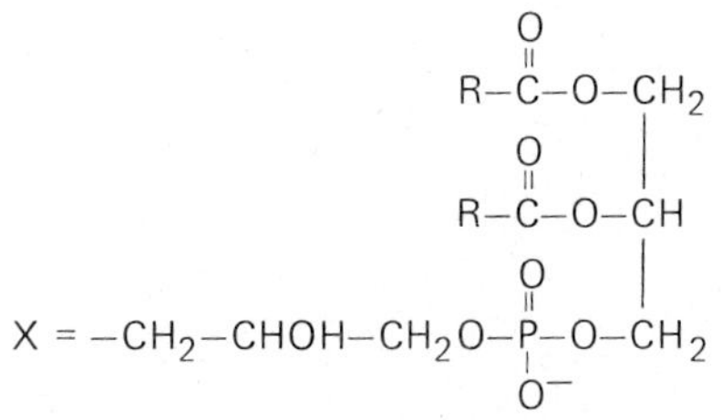

$$X = -CH_2-CHOH-CH_2O-\overset{O}{\overset{\|}{P}}(O^-)-O-CH_2-CH(O-CO-R)-CH_2-O-CO-R$$

diphosphatidylglycerol

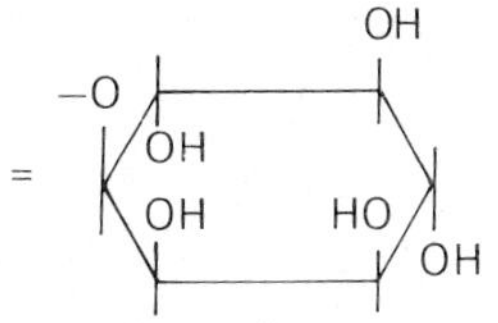

= –O (inositol ring: OH, OH, OH, HO, OH, OH)

Phosphatidyl(myo)inositol

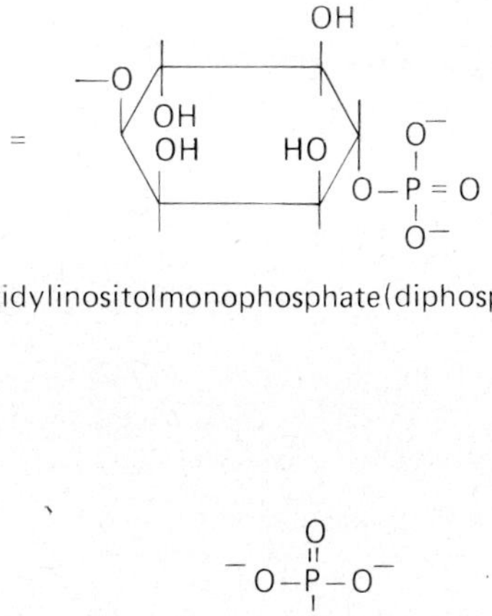

= –O (inositol ring: OH, OH, OH, HO, $O-P(O^-)(=O)-O^-$)

Phosphatidylinositolmonophosphate(diphosphoinositide)

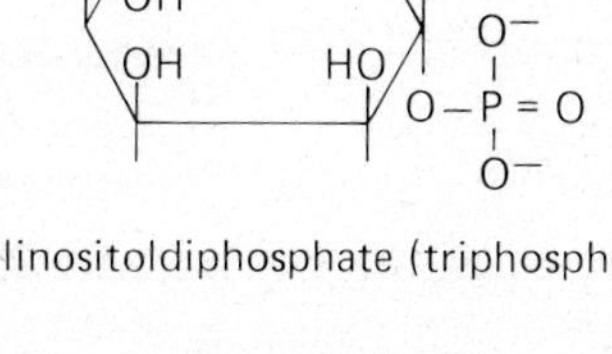

–O (inositol ring: $O-P(=O)(O^-)-O^-$, OH, OH, HO, $O-P(O^-)(=O)-O^-$)

Phosphatidylinositoldiphosphate (triphosphoinositide)

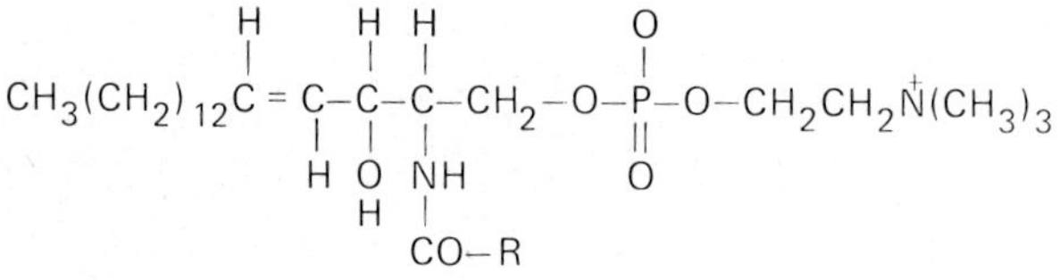

$$CH_3(CH_2)_{12}\overset{H}{C}=\underset{H}{C}-\underset{OH}{\overset{H}{C}}-\underset{\underset{CO-R}{NH}}{\overset{H}{C}}-CH_2-O-\underset{O}{\overset{O}{P}}-O-CH_2CH_2\overset{+}{N}(CH_3)_3$$

Sphingomyelin

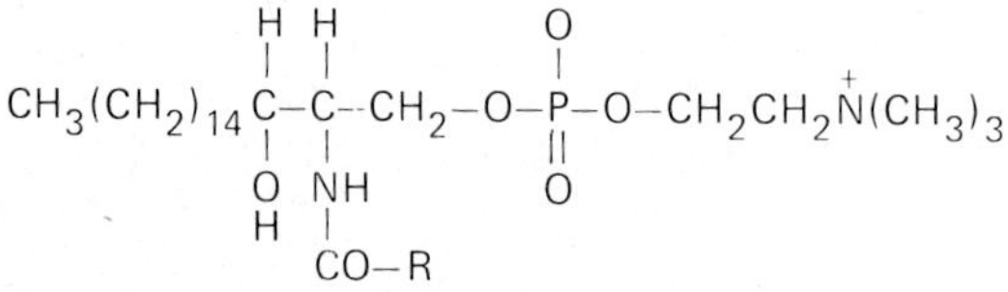

$$CH_3(CH_2)_{14}\underset{OH}{\overset{H}{C}}-\underset{\underset{CO-R}{NH}}{\overset{H}{C}}-CH_2-O-\underset{O}{\overset{O}{P}}-O-CH_2CH_2\overset{+}{N}(CH_3)_3$$

Dihydrosphingomyelin

Fig. 3.2. Sphingosyl phospholipids.

No. of carbons	Structural formula	Systemic name	Common name
4	$CH_3\overset{H}{C}=\underset{H}{C}-COOH$	*trans*–2–butenoic	Crotonic
6	$CH_3CH_2CH=CHCH_2COOH$	3–hexenoic	Dihydrosorbic
10	$CH_3(CH_2)_5\overset{H}{C}=\overset{H}{C}CH_2COOH$	*cis*–3–decenoic	
11	$CH_2=CH(CH_2)_8COOH$	10–undecenoic	Undecylenic
12	$CH_3(CH_2)_5\overset{H}{C}=\overset{H}{C}(CH_2)_3COOH$	*cis*–5–dodecenoic	Denticetic
14	$CH_3(CH_3)_3\overset{H}{C}=\overset{H}{C}(CH_2)_7COOH$	*cis*–9–tetradecenoic	Myristoleic
16	$CH_3(CH_2)_5\overset{H}{C}=\overset{H}{C}(CH_2)_7COOH$	*cis*–9–hexadecenoic	Palmitoleic
18	$CH_3(CH_2)_7\overset{H}{C}=\overset{H}{C}(CH_2)_7COOH$	*cis*–9–octadecenoic	Oleic
18	$CH_3(CH_2)_7\overset{H}{C}=\underset{H}{C}(CH_2)_7COOH$	*trans*–9–octadecenoic	Elaidic
18	$CH_3(CH_2)_5\overset{H}{C}=\overset{H}{C}(CH_2)_9COOH$	*cis*–11–octadecenoic	*cis* –Vaccenic
22	$CH_3(CH_2)_7\overset{H}{C}=\overset{H}{C}(CH_2)_{11}COOH$	*cis*–13–docosenoic	Erucic

Fig. 3.3. Some mono-unsaturated fatty acids. General structure = $CH_3(CH_2)_n\ CH=CH(CH_2)_m\ CO$.

No. of carbons	Structural formula	Systematic name	Common name
	Dienoic acids		
6	$CH_3CH=CHCH-CH-COOH$	2, 4–hexadienoic acid	Sorbic acid
18	$CH_3(CH_2)_4[\overset{H}{C}=\overset{H}{C}-CH_2]_2-(CH_2)_6COOH$	*cis, cis*–9, 12–octadienoic acid	Linoleic acid
18	$CH_3(CH_2)_7[\overset{H}{C}=\overset{H}{C}-CH_2]_2-(CH_2)_3COOH$	*cis, cis*–6, 9–octadienoic acid	
20	$CH_3(CH_2)_4[\overset{H}{C}=\overset{H}{C}-CH_2]_2-(CH_2)_8COOH$	*cis, cis*–11, 14–eicosadienoic	
	Trienoic acids		
18	$CH_3CH_2[\overset{H}{C}=\overset{H}{C}CH_2]_3-(CH_2)_6COOH$	all *cis*–9, 12, 15–octadecatrienoic	α–Linolenic acid
18	$CH_3(CH_2)_4[CH=CHCH_2]_3-(CH_2)_3COOH$	all *cis*–6, 9, 12–octadecatrienoic acid	γ–Linolenic acid
18	$CH_3(CH_2)_3[\overset{H}{C}=\underset{H}{C}]_2-\overset{H}{C}=\overset{H}{C}-(CH_2)_7COOH$	*cis*–9–*trans*–11–*trans*–13–octadecatrienoic acid	α–Eleostearic acid
20	$CH_3(CH_2)_4[\overset{H}{C}=\overset{H}{C}CH_2]_3-(CH_2)_5COOH$	all *cis*–8, 11, 14–eicosatrienoic acid	
20	$CH_3CH_2[\overset{H}{C}=\overset{H}{C}-CH_2]_3-(CH_2)_8COOH$	all *cis*–11, 14, 17–eicosatrienoic acid	
	Tetra-, pentra- and hexaenoic acids		
20	$CH_3(CH_2)_4[\overset{H}{C}=\overset{H}{C}-CH_2]_4-(CH_2)_2COOH$	all *cis*–5, 8, 11, 14–eicosaletraenoic acid	Arachidonic acid
20	$CH_3CH_2[\overset{H}{C}=\overset{H}{C}-CH_2]_5-(CH_2)_2COOH$	all *cis*–5, 8, 11, 14, 17–eisosapentaenoic acid	
22	$CH_3CH_2[\overset{H}{C}=\overset{H}{C}-CH_2]_5-(CH_2)_4COOH$	all *cis*–7, 10, 13, 16, 19–docosapentaenoic acid	
22	$CH_3(CH_2)_4[\overset{H}{C}=\overset{H}{C}-CH_2]_5-CH_2COOH$	all *cis*–4, 7, 10, 13, 16–docosapentaenoic acid	

Fig. 3.4. Some poly-unsaturated fatty acids. General structure = $CH_3(CH_2)_n$ $CH=CHCH_2-CH=CH\ldots CO_2H$, usually methylene-interrupted.

or of the *anteiso paraffin* variety, i.e.

$$CH_3-CH_2-\underset{}{\overset{\displaystyle CH_3}{\overset{|}{C}}}H-CH_2-$$

Multiple branching commonly occurs in *isoprenoid* units, i.e.

$$[CH_2-\overset{\displaystyle CH_3}{\overset{|}{C}}H-CH_2-CH_2]-$$

The paraffin acid moieties, whether branched or unbranched, may be saturated, singly unsaturated or multiply unsaturated. In unsaturated acyl chains the double bonds can be in the *cis* or *trans* configurations:

cis- [structural diagram: H–C=C–H, cis configuration]

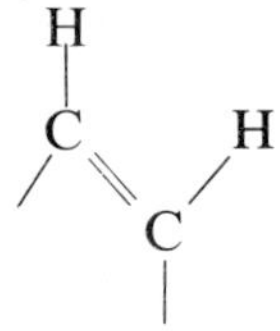

trans- [structural diagram: H–C=C–H, trans configuration]

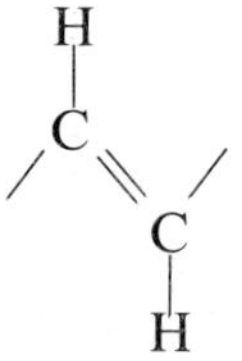

The most important branched, unsaturated paraffin chains constitute *isoprenoid* polyenes with the following unsaturated units:

$$-[CH_2-\overset{\displaystyle CH_3}{\overset{|}{C}}=CH-CH_2]-$$

These occur infrequently in phospholipids but comprise an essential feature of the carotenoids.

3.2. Physicochemical properties

All important structural lipids of biomembranes (phospholipids, as well as cholesterol and glycolipids) are amphipathic substances, that is, the molecules have polar and apolar portions. In aqueous media, the polar 'head groups' (for example, the hydroxyl group of cholesterol, the phosphorylcholine residue of lecithin or the sugar residues of glycolipids) tend to associate with water molecules, whereas the apolar entities (e.g. the sterol nucleus of cholesterol and the hydrocarbon chains of phospholipids and glycolipids) tend to aggregate with each other in some kind of micellar or bilamellar array, away from the water molecules. The phospholipid levels at which molecular aggregation occurs (critical micelle concentration; CMC) are exceedingly low; for long chain phosphatides, such as prevail in biomembranes, CMC values of $\sim 10^{-9}$–10^{-10} M are typical. The most important arrays comprise bimolecular leaflets consisting of two monolayers, each with its polar groups oriented towards a lipid-water interphase and its hydrocarbon residues oriented towards those of the other monolayer.

The exact arrangements of membrane lipids depend very much on environmental conditions and on such properties of the lipids as the length of the hydrocarbon chains and their unsaturation, as well as the size and charge of the polar head group. This matter has been very extensively studied by X-ray diffraction [1].

This method has not only established the common arrangement of amphipathic lipids into lamellar arrays, consisting of alternate sheets of lipid bilayer and water, but has also distinguished between a 'gel phase', in which the hydrocarbon chains are rigid and packed in an essentially crystalline array, and a 'liquid-crystalline phase', in which the hydrocarbon chains are disordered as in a liquid. Appreciation of this last fact has had an important impact on thinking about biological membranes and forms a cornerstone of the 'fluid-mosaic' concepts [2], which suggests that biological membranes consist of a fluid lipid continuum interspersed with membrane proteins.

The gel (ordered; crystalline) → liquid-crystalline transitions of lipid water systems has been studied in detail, not only by X-ray diffraction but also by nuclear magnetic resonance (NMR), thermal techniques and vibrational spectroscopy [3]. It is now established that the gel → liquid-crystalline transition of a pure lipid occurs over a very narrow temperature range and that the temperature of this phase transition depends markedly on the length and degree of unsaturation of the lipid hydrocarbon chains. The transition temperature also depends upon the nature of the polar head group of the lipid and, in the case of lipids containing ionizable residues in the head group, upon pH, ionic strength and ionic composition. This last point has been well illustrated in a recent study by Träuble and Eibl [4], who show that in the case of bilayers of phosphatidic acid, a change in the bulk pH from 7 to 9, which increases the charge per polar group from one to two, lowers the gel → liquid-crystalline transition temperature by about 20°C. Inversely, very small pH changes can introduce a phase transition at a constant temperature. Such effects are not limited to phosphatidic acid; they are also found in the case of phosphatidylserine, where protonation of the carboxyl residue increases the transition temperature, and in phosphatidylethanolamine where protonation of the amino group does the reverse. Moreover, the greater the ionic strength of mono-monovalent ions, the more fluid the system at a given temperature and pH. In contrast, divalent cations such as Mg^{2+} and Ca^{2+} tend to increase transition temperature under otherwise identical conditions due to charge neutralization. Divalent cations thus induce a liquid-crystalline → gel transition under given conditions of temperature and pH.

Another factor of major importance is the presence or absence of cholesterol. Increasing proportions of cholesterol incorporated into a phospholipid bilayer progressively broaden the thermal transition as measured by differential scanning calorimetry. At a cholesterol/phospholipid molar ratio of 1, no transition can be detected by thermal techniques, but Raman and nuclear magnetic resonance-spectroscopy show that in this situation the phospholipid hydrocarbon chains are in a state of 'intermediate fluidity' at all temperatures, and that cholesterol reduces the *cooperativity* of the acyl chain transitions [3]. This means that cholesterol tends to 'fluidize' the hydrocarbon chains of phospholipids (and glycolipids) below the transition temperature, but, by interfering with the normal, free mobility of the chains occurring above the transition temperature, tends to 'defluidize' these chains above the transition temperature. Cholesterol can thus be considered a 'fluidity-buffer' and one can expect that cholesterol-rich membranes, such as plasma membranes, would be insensitive to changes in phospholipid hydrocarbon chain fluidity induced by changes of pH or local divalent cation concentration. In contrast, cholesterol-poor membranes, such as mitochondria, would be more subject to such changes. (pH might become critical when metabolic production of non-volatile acids is high, for example, during glycolysis; the possible consequences of the high anaerobic glycolysis common in tumors should be viewed in this light. The role of calcium would depend critically upon calcium pumping by the mitochondria and plasma membrane).

Finally, as pointed out by Chapman and associates [3], cholesterol can play an important role, by permitting the intermixing of diverse lipids into a fluid phase, at a temperature where some of these lipids (for example, sphingomyelin) might otherwise be in a gel state. This role is of obvious importance in the membranes of animal cells which contain many, diverse phospholipids.

In simple lipid systems a switch from the gel state to the liquid-crystalline phase represents an 'order–disorder' transition. However, 'order' of the hydrocarbon chains of membrane lipids refers to a physical-chemical state and does not consider the possible *biological* ordering effects of membrane proteins. Thus, a membrane protein penetrating the apolar core of the membrane and interacting with membrane lipid hydrocarbon chains at its apolar perimeter will physically disorder interacting hydrocarbon chains. The extent of this effect will depend on the apolar surface topology of the protein, which in turn depends upon amino its acid distribution and its tertiary and quaternary structur. In any case, it actually constitutes a 'biological ordering' of the membrane lipid hydrocarbons.

In view of the long-standing suspicion that biological membranes may contain substantial proportions of lipid in a bilayer array, information gained about the physical-chemical properties of membrane lipids in model-lipid water systems has been extended to biological membranes. Thus, Steim et al. [5] and Blazyk and Steim [6] have demonstrated thermotropic phase transition in several biological membranes. These transitions were broad, if only due to the heterogenity of the lipids in these membranes, but there was a clear correspondence between the transition found in the membrane and gel → liquid-crystalline transition of the extracted lipids dispersed in water. An important point to note is that in the case of animal cell membranes, transitions were either not observed or occurred at temperatures far beyond the physiological: that is either below 20°C or above 50°C. The latter transition probably represents denaturation of membrane proteins.

Thermotropic lipid phase transitions have also been reported for a variety of membranes by use of probe techniques (for example, lipid spin labels [7], fluorescent probes [8]), X-ray diffraction [9,10], as well as light scattering and dilatometry [11]. Although there are certain inconsistencies between data obtained by the probe approach and other methods (due to the perturbing effects of the various probes) all experiments indicate that thermotropic phase transitions do occur in the membranes of various microorganisms. The data are somewhat less simple in the case of the membranes of eukaryotic cells. Thus, recent information obtained by freeze-cleave electron microscopy, indicates that such transitions produce thermotropic lipid–protein phase-segregation in the nuclear membranes, and possibly certain other intracellular membranes of eukaryotic cells, including lymphocytes, but not in the plasma membranes [12–14]. However, Raman-spectroscopic analyses [14a] show that some of the phospholipids in animal-cell plasma membranes undergo cooperative lipid state transitions within *microdomains*.

A number of investigators have established a relationship between certain membrane functions (transport; enzyme activity) and membrane lipid 'fluidity'. The basis of these correlations is the observation that the temperature activity curves of the membrane functions show discontinuities which correlate with the transition temperatures of the membrane lipids. The most convincing data have been reported for *E. Coli* (for example [11,15,16]). Analogous data exist for the membranes of *Acholeplasma laidlawii* [17].

Data on animal cells are fewer and less impressive. However, Grisham and Barnett [18] report that the characteristic temperature for the sodium-potassium ATPase activity of a plasma membrane isolate from kidney correlates with a lipid phase change detected by a fatty acid spin probe. Eletr et al. [19] show a similar correlation for the glucose-6-phosphatase and UDP-glucuronyl transferase in liver microsomal membranes. Unfortunately, the evaluation of such spin-label data is complicated by the fact that many spin probes are membrane perturbants [7,20], which tend to generate a locally fluid environment. We have already treated the report by Barnett and associates [21] arguing that the membranes of neoplastically transformed cells have a different fluidity from those of the non-transformed parental cells (Chapter 1; p. 55).

When an aqueous dispersion of mixed lipids (excluding cholesterol for the moment) is cooled down through the gel → liquid-crystalline transition temperature of any one of the components, segregation of the lipids may occur; this process requires the lateral translation of the lipids in the liquid-crystalline phase [22]. Whether phase separation occurs or not depends very much on the character of the lipids; for example, the transition temperature of dioleoyl phosphatidylcholine does not vary dramatically in the presence of equi-molar amounts of saturated phosphatidylcholines with hydrocarbon chain lengths between 14 and 22 carbons, although the transitions of the saturated components broaden and occur at lower temperatures [23]. In other words, the two components behave as separate phases. In contrast, when the hydrocarbon moieties of the two components are very similar, co-crystallization tends to occur. However, the hydrocarbon chains are not the only important entity; for example, equi-molar mixtures of dimyristoylphosphatidylcholine and dimyristoylphosphatidylethanolamine yield a very broad thermotropic transition in contrast to the sharp transitions of the individual lipids [24]. Similarly, admixture of increasing proportions of egg phosphatidylcholine with ox brain cerebroside, progressively drops the broad cerebroside transition temperature to lower temperatures.

Lipid phase separations have also been visualized electron microscopically by the freeze-fracture technique [25]. Bilayers prepared from synthetic phosphatidylcholines exhibit smooth fracture phases, when quick frozen from above the transition temperature, whereas below the transition temperature, clearcut 'band patterns' are observed. When a mixture of dioleoyl- and distearoylphosphatidylcholine is quick frozen from between a transition temperature of the individual lipids, one observes segregation into smooth areas, indicating separation of a gel phase and a liquid-crystalline phase.

Turning to biological membranes, Engelman [26] observed that the X-ray diffraction characteristics of *Acholeplasma laidlawii* membranes exhibit the characteristic of the gel state at low temperatures. As the temperature is increased through the phase transition temperature, the diffraction patterns become typical of fluid state. Similar data have been obtained for *E. Coli* membranes [10]. Correlation of these data with the temperature-dependence of proline accumulation and succinic dehydrogenase activity yields some interesting results. The Arrhenius plots of both biological activities exhibit discontinuities at temperatures which vary with the fatty acid composition of the membranes. However, the transition temperatures of the biologic processes occur at *lower* measured temperatures than the phase transition of the lipids as detected by X-ray diffraction: also, they are not identical for the two processes. It is suggested that the lipids associated with the two functional mechanisms are in a more fluid state than the bulk lipids in the membranes. The same has been proposed for the β-galactosidase and β-glucoside transport systems of *E. Coli* [16,27,28].

In this connection, the experiments of Tsukagoshi and Fox [29] are extremely pertinent. These authors determined the variation of the characteristic temperature of the lactose transport system in an unsaturated fatty-acid auxotroph of *E. Coli* as a function of the fatty acid supplement. In membranes supplemented with a single fatty acid, the characteristic temperature is 13 °C for oleate and 30 °C for elaidate. When transport is induced at 37 °C for a short period after shifting from growth with an oleate supplement to growth with an elaidic acid supplement, a single transition temperature is detected, indicating the transport is influenced primarily by the average fatty acid composition of the membrane lipids. However, when transport is induced at 25 °C, two discontinuities are observed in the transport; Arrhenius plots of these lie near to the characteristic temperatures of the individual oleic acid- and elaidic acid-derived lipids. If the cells are subsequently incubated for 10 min at 37°C, the system again assumes the characteristics of the mixed lipid state. These authors suggest that the transport enzymes are inserted in association with freshly synthesized lipids. When the lipids are in a gel-stage at the induction temperature, intermixing with the pre-existing lipids is prevented and the transport system exhibits the characteristic temperature appropriate to the new lipid. However, if induction takes place at a temperature where both the new lipid and the pre-existing lipid are liquid-crystalline, intermixing by lateral diffusion can occur and the characteristic temperature then reflects only the average composition. This can also be achieved by incubating at high temperature for a period of time after incubation of the low temperature.

Presumptive phase separation of membrane lipids has also been demonstrated by freeze-fracture electron microscopy in certain biomembranes, including the nuclear membranes of lymphoid cells, both normal and neoplastic ([14]: Chapter 1).

Above the critical temperature, the fracture phases of the nuclear membranes exhibit numerous, uniformly distributed intramembranous particles of varying size. When the cells are cooled prior to fixing for electron microscopy, one observes a rather sharp discontinuity in fracture phase appearance; this which occurs at ~18°C in the case of *Tetrahymena pyriformis* and at ~22°C in the case of mouse lymphocytes. At this temperature and below, the fracture faces become populated with smooth, particle-free areas, whose extent depends on the degree of cooling and also on the rate of cooling. In the case of *Tetrahymena*, the margins of these smooth areas are surrounded by regions of aggregated intramembranous particles, whereas in the case of lymphocyte nuclear membranes, no such aggregation is observed. These morphologic changes can be reasonably attributed to a liquid-crystalline → gel transition with extrusion of penetrating membrane proteins (the intramembranous particles) from the crystalline phase. This extrusion can occur either laterally, crowding the intramembranous particles into a liquid-crystalline area or perpendicular to the plane of the membrane, forcing the particles out of the fracture face. Although this morphological phenomenon has also been observed in the surface membranes of *Acholeplasma laidlawii* and *E. Coli*, it has definitely not been detectable in the plasma membrane of eukaryotic cells [12–14], although Raman spectroscopy clearly shows that cooperative gel → liquid-crystalline phase transitions do occur within some of these membranes, apparently in very small domains [14a].

3.3. Phospholipid biosynthesis

3.3.1. Fatty acid biosynthesis

The complete *de novo* biosynthesis of the fatty acid chains of phospholipids and glycolipids is catalyzed by two enzyme systems, acetyl-CoA-carboxylase and fatty acid synthetase, the latter lying in the cytoplasm. Acetyl-CoA-carboxylase which, in liver, is localized on the membranes of endoplasmic reticulum [30] catalyzes the carboxylation of cytoplasmic acetyl-CoA with CO_2. The fatty acid synthetase system of the soluble cytoplasm catalyzes the condensation of one molecule of acetyl-CoA and 7 molecules of malonyl-CoA to make a molecule of hexadecanoic acid (palmitic acid). The palmitic acid can be further elongated by enzymes located in the microsomes and mitochondria. The saturated fatty acid chains can also be desaturated by enzymes located in the endoplasmic reticulum.

Acetyl-CoA carboxylase acts as the rate-limiting regulating enzyme in fatty acid biosynthesis. Its properties have been studied extensively and have been well reviewed by Lane et al. [31]. The acetyl-CoA carboxylases purified from various animal tissues exist as enzymatically active polymeric filaments (4000 Å $\times$ 100 Å, molecular weight about 4 $\times$ 10^6). A single hepatocyte contains about 50,000 such filaments. The filamentous polymers of acetyl-CoA carboxylase can dissociate into weight-homogenous, high-molecular-weight protomers, which are themselves complex, multi-subunit structures. These protomers are enzymatically inactive. Their particle weight is about 400,000 and they are made up of four smaller subunits with molecular weights of about 100,000 each. Each 400,000 dalton protomer has one binding site for acetyl-CoA and one binding site for the positive modulators, citrate or isocitrate. Each 400,000 dalton subunit also contains one molecule of covalently bound biotin. In the presence of citrate or isocitrate the protomers aggregate to form the enzymatically active filamentous polymer. Citrate thus acts as an allosteric activator.

The source of all the carbon atoms of cytoplasmically synthesized fatty acids is cytoplasmic acetyl-CoA. This in turn derives from intramitochondrial acetyl-CoA which is formed by the oxidative decarboxylation pyruvate and the β-oxidation of long chain fatty acids. However, mitochondrial acetyl-CoA cannot pass into the cytoplasm as such because the mitochondrial membrane is impermeable to acyl-CoA derivatives. Rather the acetyl group of the mitochondrial acetyl-CoA is first transferred to carnitine by a transferase located on the inner mitochondrial membrane [32]. The acetyl carnitine thus formed passes through the mitochondrial membrane to the cytoplasm, where the acetyl group can be transferred to cytoplasmic CoA (Fig. 3.5). Mitochondrial acetyl groups can also be transferred to the cytoplasm in the form of citrate, the product of the condensation of acetyl-CoA and oxaloacetate. Citrate readily permeates through the mitochondrial membrane to the cytoplasm via a carrier system. In the cytoplasm, citrate undergoes cleavage by the ATP-dependent citrate cleaving enzyme with the formation of acetyl-CoA. The mitochondrial membrane thus acts as a regulator for transfer of acetyl-CoA from mitochondria to the cytoplasm, with carnitine and citrate acting as carriers.

The transfer of fatty acids such as palmitic acid between the cytoplasmic space

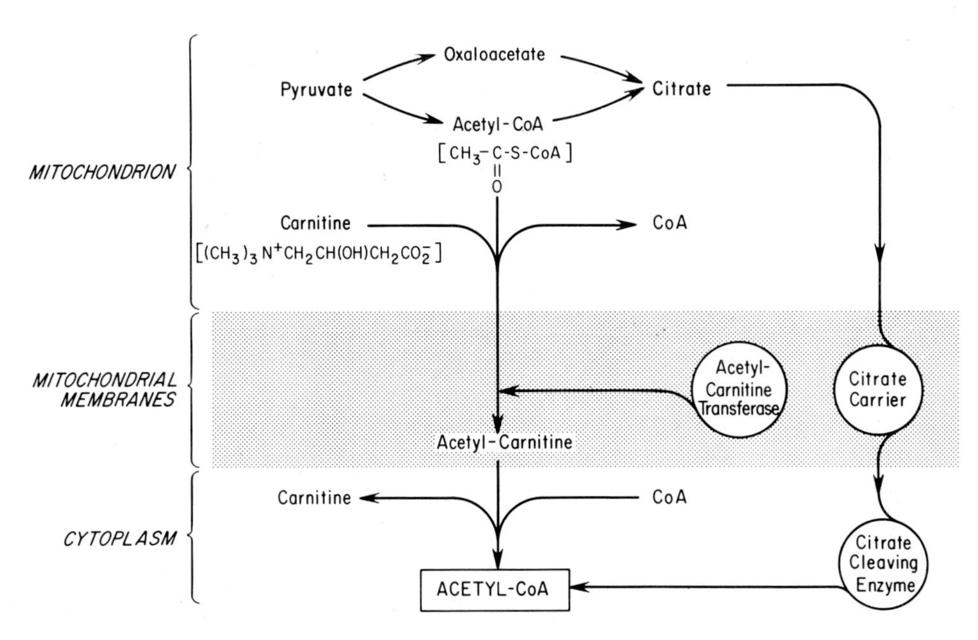

Fig. 3.5. Transfer of acetyl-CoA equivalents across the mitochondrial membrane.

and the mitochondrial compartment is mediated and regulated by two carnitine fatty acyltransferases associated with the inner mitochondrial membrane [33]. The sequence of reactions involved is depicted in Fig. 3.6. Starting with, for example, palmitic acid generated by cytoplasmic fatty acid synthetase, this is converted into palmitoyl-CoA by a fatty acyl-CoA synthetase located in the outer mitochondrial membrane. Palmitoyl-CoA cannot penetrate through the inner mitochondrial membrane. However, the CoA moiety of the palmitoyl-CoA is replaced by L-carnitine to form L-palmitoyl carnitine in a reaction mediated by a carnitine acyltransferase located on the outer surface of the inner mitochondrial membrane. L-Palmitoyl carnitine is then transferred into the intramitochondrial space through the action of a vectorially oriented carnitine-palmitoyltransferase B, which in the presence of mitochondrial-CoA, generates intramitochondrial palmitoyl-CoA.

The rate of fatty acid biosynthesis is regulated through the activity of acetyl-CoA-carboxylase in at least two ways: Long term regulation depends on the biosynthesis of the enzyme, which decreases with fat feeding [34]. Short term regulation (e.g. [35]) occurs through the action of positive modulators such as citrate, as well as through feedback inhibition.

Acetyl-CoA carboxylase is inhibited by a number of compounds, most important of which are long-chain fatty acyl-CoA derivatives, such as palmitoyl-CoA,

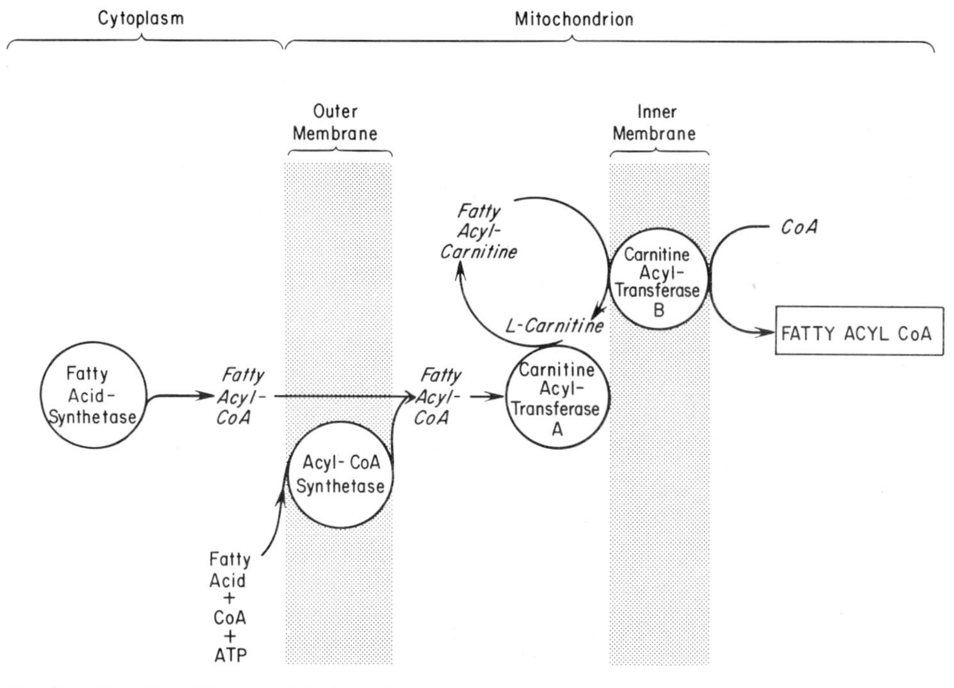

Fig. 3.6. Transfer of fatty acyl CoA equivalents across the mitochondrial membrane.

steroyl-CoA and oleoyl-CoA. These fatty acid derivatives are effective at extremely low concentrations (3 to 8 $\times$ 10^{-7} M), whereas the corresponding free fatty acids are not inhibitory at similar levels. Since the fatty acyl-CoA derivatives are the end products of cytoplasmic fatty acid biosynthesis, that is the products of the fatty acid synthetase complex, their action constitutes end product inhibition of an early, rate-controlling step. This role is consistent with the fact that all processes which depress hepatic fatty acid synthesis, lead to increased long-chain fatty acyl-CoA concentrations in liver.

The inhibitory action of fatty acyl-CoA derivatives probably involves a shift of the protomer $\rightleftharpoons$ polymer equilibrium towards the protomer state, that is, the opposite to the effect of citrate. Importantly, the inhibition of acetyl-CoA carboxylase by long-chain fatty acyl-CoA derivatives can be largely reversed by (+) palmitoyl carnitine (see for example [32,36]). This activating effect is not specific with respect to (+) palmitoyl carnitine, since acetyltrimethyl ammonium ion is also effective [37]. What is important here however, is not only the fact that fatty acyl carnitine can in itself reverse the inhibition of acetyl-CoA carboxylase by fatty acyl-CoA derivatives, but also that these derivatives can be transferred from the cytoplasmic space into the mitochondrial space (and vice versa) by their conversion to fatty acyl carnitines through the action of acyl-carnitine transferases, in a fashion similar to the translocation of acetyl-residues [32].

3.3.2. Phospholipid assembly

The principal membrane phospholipids in animal cells are phosphatidylcholine and phosphatidyethanolamine. As detailed in a recent comprehensive review [38], most, if not all, *de novo* synthesis of these phosphatides, as well as of serine- and inositol-phosphoglycerides and sphingomyelin, occurs in/on the membranes of the endoplasmic reticulum. Indeed, it was established already in 1963 by Wilgram and Kennedy [39] that the two final enzymes involved in the *de novo* biosynthesis of phosphatidylcholine, namely CDP-choline-1,2, diglyceridecholine phosphotransferase and CTP-choline phosphate cytidyltransferase, are located primarily in the microsomal membrane fraction of liver, whereas simultaneously isolated mitochondrial fractions contain very little of these enzyme activities. Similar information is now available for the biosynthesis of phosphatidylserine and phosphatidylinositol and sphingomyelin.

The most comprehensive study on the subcellular localization of phosphatide biosynthesis has been published by van Golde et al. [40]. Rat livers were fractionated and their nuclei, mitochondria, rough and smooth microsomes, Golgi-complexes and plasma membrane fragments isolated and characterized using specific marker enzymes. The various subcellular fractions were assayed for biosynthesis of sphingomyelin, CDP-diglycerides, phosphatidylinositol, phosphatidylserine, decarboxylation of phosphatidylserine, formation of lecithin via *N*-methylation, and activation of palmitic and octanoic acids. None of these processes are found to be present in the Golgi complex. Concordant with previous data, the endoplasmic reticulum appears to be the principal cellular site for the synthesis of sphingomyelin, CDP-diglycerides, phosphatidyl-inositol, phosphatidylserine and lecithin. The biosynthesis of phosphatidylserine appears four times more active in rough than in smooth microsomal particles, suggesting a ribosomal participation in this process.

Except for synthesis of CDP-diglycerides, mitochondria do not exhibit any of the biosynthetic activities assayed. However, these membranes are the only particles which could decarboxylate phosphatidylserine. This activity is localized within the inner mitochondrial membrane.

Activation of palmitate was localized predominantly in the endoplasmic reticulum and mitochondria. The plasma membrane fragments also contain some palmitoyl-CoA synthetase activity.

Thus, the evidence now available indicates that, in animal cells, mitochondrial membranes are the only membrane system other than endoplasmic reticulum capable of *de novo* biosynthesis of phosphatides. This, however, is restricted to phosphatidic acid and diphosphatidyl glycerol [41–47]. These acidic phospholipids, of course, play a specific functional role in mitochondria.

In rat liver, phosphatidylserine decarboxylase activity is also located predominantly in the mitochondria, although the biosynthesis of this phosphatide takes place in the endoplasmic reticulum. It is possible that phosphatidylserine is transported to the mitochondria and decarboxylated there to yield mitochondrial phosphatidylethanolamine [48].

3.4. Degradation of phosphatides

Phospholipids can be degraded by a number of phospholipases as illustrated in Fig. 3.7. Phospholipases A_1 and A_2 are esterases which cleave the ester linkages between glycerol and the fatty acids at the 1 and 2 positions. Phospholipase C is a phosphoesterase which splits off the phosphorylated head groups of glycerol phosphatides and sphingomyelin, leaving either diglycerides or ceramide. Phospholipase D is also a phosphoesterase which removes the unphosphorylated head groups of phospholipids. Phospholipase B acts on lysophosphatides generated by the action of phospholipase A and cleaves off the remaining fatty acid.

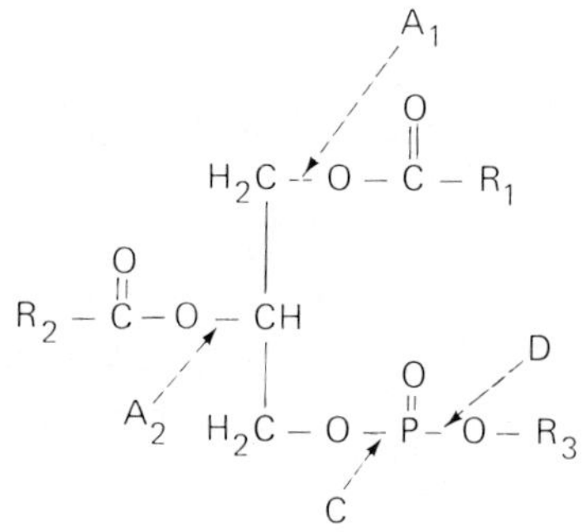

A_1

$H_2C-O-C(=O)-R_1$

$R_2-C(=O)-O-CH$

A_2 $H_2C-O-P(=O)-O-R_3$ D

C

Fig. 3.7. Actions of phospholipases.

The phospholipases of the A type are very prevalent in various tissues and could, without other mechanisms, lead to accumulation of toxic lysophosphatides. However, this eventuality is prevented by two mechanisms, namely (a) the action of lysophospholipases and (b) reacylation (Fig. 3.8).

In the latter metabolic pathway, lysophosphatides are reacylated by fatty acyl-CoA through the actions of various acyltransferases. There are different acyltransferases for saturated and unsaturated fatty acids. This process not only prevents accumulation of lytic levels of lysophosphatides, but also serves for *in situ* exchange of phospholipid fatty-acids. This may serve in three major roles:

(a) The exchange of phospholipids between diverse lipid compartments may occur far more readily in the form of lysophosphatides because of their high water solubility. This item is discussed further on.

(b) A change of the functional state of a cell may require a different fatty acid composition of the membrane lipids, which could be brought about by a

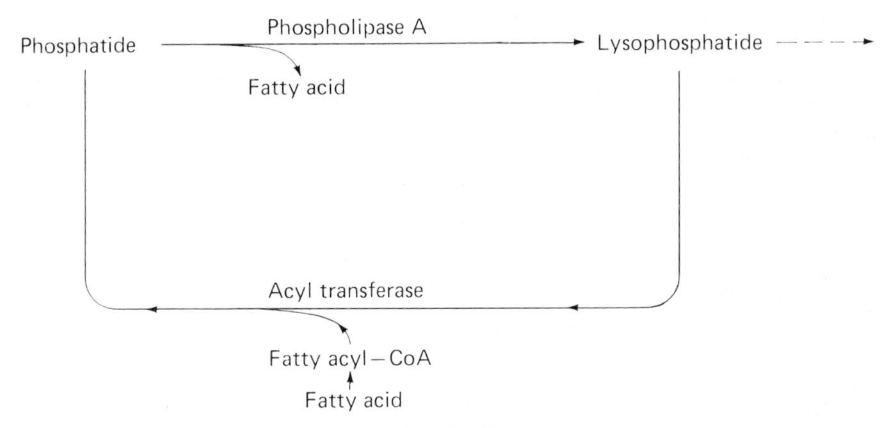

Fig. 3.8. Decacylation–reacylation cycle of phosphatides.

deacylation–reacylation cycle. An example of this occurs in the mitogenic stimulation of lymphocytes which is accompanied by a very marked enhancement of acyltransferase activity.

(c) The cycle allows for the repair of phospholipids whose fatty acid moieties have been oxidized. Lipid peroxidation is a rather common process, which occurs in all cellular membranes including plasma membranes. Peroxidation is thought to be initiated by mechanisms involving catalysis by heme groups, Fe^{2+} and certain other agents, in the presence of molecular oxygen. It then continues via a free radical chain reaction. Lipid peroxidation is markedly accelerated by a number of mechanisms, for example, excessive oxygen tension, ionizing radiation, and various toxic agents [49]. Lipid peroxidation can be inhibited by lipid-soluble anti-oxidants such as vitamin E and is often accelerated by vitamin E deficiency.

3.5. Phospholipid exchange

3.5.1. Phospholipid composition of various cellular membrane compartments

Table 3.1 gives the relative phospholipid complement in the major membrane fractions of cultured hamster fibroblasts (BHK21) and of rat liver [50,51]. These are the only cell types for which such comprehensive data exist but it is known that overall phospholipid composition varies extensively from one cell type to the next and from one species to another. Nevertheless the data in Table 3.1 reveal compositional trends which appear to apply to all normal animal cells but which do not necessarily hold for tumor cells.

All membranes contain phospholipids although these substances are not syn-

TABLE 3.1

Phospholipid composition of subcellular membranes from BHK21 fibroblasts and rat liver.

Phospholipid class	Phospholipid content[1]									
	BHK21[2]					Rat liver[3]				
	H	N	M	ER	PM	H	N	M	Ms	PM
Diphosphatidylglycerol	3.3	1.7	15.6	2.0	0.4	3.9	—	14.6	—	1.0
Phosphatidylethanolamine	23	20	27	14	18	22.4	27.6	35.3	26.9	24 7
Phosphatidylserine	6.6	2.4	1.9	3.6	8.9	3.1	7.3	1.3	2.0	4.2
Phosphatidylinositol	5.7	3.9	3.0	4.3	2.3	5.3	4.1	0.5	8.8	6.7
Phosphatidylcholine	50	68	48	62	44	52.8	52.5	48.1	54.3	46.1
Sphingomyelin	6.9	2.5	4.2	10	18	8.8	4.9	—	7.2	16.8
Phosphatidylcholine + Sphingomyelin	56.9	70.5	52.2	72	62	61.6	57.4	48.1	62.5	52.9

[1] In %. H = homogenate, N = nuclei, M = mitochondria, Ms = microsomes, ER = endoplasmic reticulum, PM = plasma membrane.
[2] From Renkonen et al. [50].
[3] From Bergelson et al. [51].

thesized in all membranes. As might be expected, diphosphatidylglycerol is most prevalent in mitochondria–it is synthesized there–but this phosphatide is also found in other membranes (cross contamination has been excluded). Phosphatidylcholine and phosphatidylethanolamine, both synthesized in endoplasmic reticulum, are the most common phospholipids in all membranes. Sphingomyelin, which is also synthesized in endoplasmic reticulum, is typically most prevalent in the plasma membrane, whereas nuclear membranes and mitochondria contain rather low levels of this phospholipid.

Two important general conclusions can be drawn: (a) There must be mechanisms of transporting these lipids from the sites of biosynthesis to the various cellular membranes. (b) The various cellular membranes differ in their avidities for specific phospholipids.

3.5.2. Exchange in vivo

Several studies show convincingly that upon intravenous or intraperitoneal injection of [^{32}P]phosphate or [^{14}C]choline into rats, phosphatidylcholine, phosphatidylethanolamine and phosphatidylinositol in liver microsomal membranes label more rapidly than the mitochondrial fractions [52]. This is not the case for phosphatidic acid, which can be synthesized by the mitochondria. Subfractionation studies show that the specific radioactivity sequence for phospholipids other than phosphatidic acid is: microsomes > outer mitochondrial membranes > inner mitochondrial membranes. These data are consistent with the idea that phospholipid synthesis takes place in the endoplasmic reticulum and is followed by an exchange first with the outer mitochondrial membrane succeeded by a further exchange into the inner mitochondrial membrane. Direct evidence for such a process has been obtained by pulse-labelling of isolated hepatocytes with [^{32}P]phosphate, [Me-^{14}C]choline, [2-^{14}C]ethanolamine and [U-^{14}C]inositol, following this with a 'cold chase' of the radioactive precursors and evaluation of the subsequent distribution of radioactivity in microsomal and mitochondrial phospholipids [53]. The 'cold chase' was invariably followed by a drop in specific radioactivity of labelled phospholipid in microsomal membranes and a continued rise for some time for the same phospholipids in the mitochondrial fraction. The data indicate a system moving towards isotopic equilibrium and suggest a transfer of intact phospholipid molecules between endoplasmic reticulum and mitochondria in the intact cells. This experiment is all the more convincing because hepatocyte preparations such as those employed in these studies are not in optimal metabolic state, and might be rather 'leaky'.

Wirtz and Zilversmit [54] have also established that the specific radioactivity of mitochondrial phospholipids changes in parallel with that of microsomal phospholipids following treatment of rats with phenobarbital or carbon tetrachloride; both agents interfere with the phospholipid metabolism of the endoplasmic reticulum. Phenobarbital treatment lowers the specific activities of microsomal and mitochondrial phosphatidylcholine and phosphatidylethanolamine while carbon tetrachloride administration to rats increases the specific activity of microsomal and mitochondrial phosphatidylethanolamine proportionally.

Concordant data have been obtained for plasma membrane isolates [55]. Most of the plasma membrane phospholipids showed a half-life comparable to that of the microsomal phospholipids. However, the sphingomyelin of plasma membranes has a longer half-life than that of the microsomes.

One possible interpretation of the label incorporation studies is that phospholipid exchange between membranes occurs only during biosynthesis and that 'once in place', the phosphatides remain fixed to their membrane structures. This, however, is not the case and it appears much more likely that the phospholipids of various membranes and even of various tissues are in dynamic state of equilibrium. For example, Le Kim et al. [56] have shown that, following intravenous administration of doubly labelled phosphatidylcholine to rats, the complete phosphatide molecules are incorporated into a large number of different tissues. Subsequent cell fractionation studies showed that the label locates in various intracellular membrane structures such as mitochondria and microsomal membranes. The authors concluded that the pattern of incorporation reflects a continuous interchange of phospholipids between the various tissues and that interchange is mediated by serum lipoproteins.

Tissue culture studies support this contention. For example, Peterson and

Rubin [57] found that phospholipids exchange freely between cultured chick embryo fibroblasts and the serum in the growth medium. Similar results were obtained by Illingworth et al. [58] who measured the synthesis and extracellular release of phospholipids by cells of a human prostatic adenoma grown in culture. The cells, previously pulse-labelled with [2-^{3}H]glycerol and [Me-^{14}C]choline, were incubated in a medium containing 0.3% albumin with and without unlabelled plasma lipoproteins. Under these conditions labelled lecithin, sphingomyelin, phosphatidylinositol and phosphatidylethanolamine transfer from the cells to the medium. The 'accretion' of labelled phospholipids is significantly greater with the lipoproteins than without, indicating that both mass transfer and exchange of phospholipids occur concurrently.

The authors argue that, since the biosynthesis of these phosphatides occurs in the endoplasmic reticulum and since the cell phospholipids were labelled via *de novo* synthesis, the exchange of labelled phospholipids between the cells and the lipoproteins in the surrounding medium must represent a secondary exchange, occurring after exchange of the phospholipids from the endoplasmic reticulum into the plasma membrane.

The role of the plasma membrane as an intermediate in this exchange was studied for the case of liver from squirrel monkeys, previously injected with [Me-^{14}C]choline and [^{3}H]leucine. These studies showed very clearly that both lecithin and sphingomyelin exchange *in vitro* between the purified plasma membrane isolates and unlabelled low or high density lipoproteins from the same species.

By far the largest amount of information in this area comes from studies on exchange of phospholipids between erythrocytes and plasma (for example [59–64]. In the case of man and dog, the *in vitro* exchange rates between plasma and erythrocytes were about 13% in 12 hr for phosphatidylcholine and 14% for sphingomyelin. However, incubation of erythrocytes with isolated ^{32}P-labelled plasma lipoprotein leads to a rapid, time-independent transfer of label to the erythrocytes [63]. The predominant component in this rapid exchange appears to be lysophosphatidylcholine. (Presumably, this exchange might involve a reacylation process within the erythrocyte membrane, as described below.)

The phospholipids and plasma lipoproteins can also exchange with the phospholipids in various isolated cellular membranes. For example, a rapid exchange of choline-labelled phospholipids occurs between plasma lipoproteins and mitochondria [65]. Similarly, one can achieve a net transfer of labelled phospholipids from microsomes to plasma lipoproteins [66], as well as an exchange of phospholipid, particularly in the phosphatidylcholine fraction.

3.5.3. Mechanisms of exchange of phospholipids between membranes

3.5.3.1. *General considerations*

Phospholipids such as occur in biomembranes exhibit very low monomer solubilities, i.e., they have exceedingly low critical micelle concentrations. In the case of dipalmitoylphosphatidylcholine, the critical micelle concentration is 4.6×10^{-10} molar at 20°C [67]. The critical micelle concentrations of the more unsaturated phospholipids found in biomembranes are likely to be somewhat greater. Nevertheless, transfer of phospholipids between membranes in the form of monomers diffusing through an aqueous phase would occur at a negligible rate. Evidence supporting this contention comes from the experiment of Kornberg and McConnell [68] indicating very slow exchange of spin labelled phosphatidylcholine from spin labelled vesicles to unlabelled vesicles.

Concordant results have been obtained by Ehnholm and Zilversmit [69]. These investigators prepared liposomes from [^{32}P]phosphatidylcholine, cholesterol, dicetyl phosphate and Forssman antigen in molar ratios of 1.0, 0.75, 0.11, 0.01. They also prepared similar liposomes, lacking the glycolipid. They showed that the antigen-containing liposomes could be quantitatively precipitated by use of specific antibody. Using the antibody to separate the Forssman-containing liposomes from the other liposomes, the authors show that little or no phosphatidylcholine exchange takes place under ordinary conditions. However, the addition of partially purified phospholipid exchange protein (p. 164) markedly promotes phosphatidylcholine exchange. Such studies indicate that although collisions must be taking place at a high frequency between the liposomes, these collisions do not bring about phospholipid exchange. However, the investigations are somewhat non-conclusive with regard to biomembranes, since these are far more complex in terms of phosphatide composition and since the presence of small amounts of lysophosphatide can readily induce fusion of phospholipid vesicles while cholesterol tends to hinder this process. For example, Papahadjopoulos, et al. [70] show that neutral phosphatidylcholine vesicles have only a limited capacity to fuse. The extent of fusion depends on the degree of unsaturation of the acyl chains and, in general, fusion occurs more readily when the membrane phospholipids are in a liquid-crystalline state than when they are in a solid, ordered state. Fusion occurs more readily between negatively charged phospholipid vesicles. However, incorporation of equal molar amounts of cholesterol into phosphatide vesicles which, in the absence of cholesterol, are in a liquid-crystalline state at a given temperature, suppresses the ability of these vesicles to fuse.

Again, one cannot extend such data very readily to biomembranes, where the role of membrane proteins may be dominant. However, reasoning from the behavior of the lipid *per se*, membranes such as plasma membranes, with a high proportion of cholesterol, should be more resistant to fusion than membranes with a low proportion of cholesterol such as mitochondrial membranes. At the moment it is not possible to extend the information very simply to considerations of lipid *exchange* (or membrane fusion) between diverse biomembranes.

3.5.3.2. *Transmembrane exchange*

Before turning to mechanisms of exchange of phospholipids between membranes, we must briefly discuss the exchange of phospholipid molecules between various sites of a given membrane.

As we discuss in more detail in the section on membrane structure (Chapter 1), there is considerable evidence that biomembranes contain regions of phospholipid bilayer. In certain membranes, some of the phospholipid molecules can undergo rather rapid lateral diffusion within the two monolayers of the bilayer. Other phosphatide molecules appear tightly associated with membrane proteins and thus exhibit more restricted diffusion.

It is generally assumed that the two monolayers of the bilayer domains are discrete domains with little or no intermixing of phospholipids, although the evidence for this concentration is less than conclusive. Indeed, much thinking in this area represents extrapolations from the experiment of Kornberg and McConnell [71] showing that the exchange of spin-labelled phosphatidylcholine between the inside and outside of artificial phosphatidylcholine liposome membranes ('flip-flop') occurs with a half-time of about 6.5 hr at 30 °C. Assuming that the transmembrane exchange of the spin-labelled phosphatidylcholine analog is representative of that of the natural phospholipid, which is by no means certain, this exchange rate would be too slow to account for rapid membrane processes such as membrane transport. However, a half-time of 6.5 hr is by no means slow compared with many other biologic processes. It seems unjustified, therefore, to use this information as an explanation for the apparent asymmetry of phospholipid distribution across certain biomembranes, for example, the erythrocyte membrane [72]. Moreover, recent studies on the rate of 'flip-flop' of spin-labelled phosphatidylcholine in *biomembranes* yield half-times in the order of minutes.

Thus, McNamee and McConnell [73] have succeeded in incorporating dioleoylphosphatidyltempoylcholine (OPTC), a spin-label phosphatidylcholine analog, into excitable membrane vesicles derived from the electric organ of *Electrophorus electricus* by incubating the spin-labelled vesicles with the biomembrane vesicle at 30 °C. They then measured the rate of inside-outside translocation of the spin-labelled phospholipid and found a half-time of 3.8 to 7 min at 15 °C. This is an order of magnitude more rapid than the 'flip-flop' rate found for OPTC in egg phosphatidylcholine vesicles. The difference is not due to the difference in the properties of the two spin-labels, because the 'flip-flop' rate of

OPTC in egg phosphatidylcholine vesicles is comparable to the rate measured for dipalmitolphosphatidyltempoylcholine.

Also, Huestis and McConnell [74] find a half-time of 20 to 30 min for 'flip-flop' in red blood cells at 37 °C. Finally, the data of Grant and McConnell [75] suggest that the rate of 'flip-flop' in the membranes of *Acholeplasma laidlawii* is in the order of less than 1 min at 0 °C.

The available data thus indicate that the rate of transmembrane transfer of phosphatide molecules in biologic membranes may be rapid compared with artificial membranes. This could be due to a number of mechanisms, for example, discontinuities in the membrane phospholipid bilayer, possibly in the vicinity of membrane proteins, or even architectural modulations of membrane proteins. In any case, the rapid 'flip-flop' in biological membranes implies that the distribution of phospholipids in these membranes will always represent either an equilibrium or a steady state, depending on other factors, and will not reflect the mode of incorporation of the lipids into the biomembranes.

Any asymmetric distribution of phospholipids observed across biomembranes might thus be reasonably attributed to an asymmetric distribution of membrane proteins. An alternative, but not exclusive possibility might be that the distribution of phospholipids across a biomembrane depends on the different properties of the compartments separated by the membrane.

3.5.3.3. *Transfer in the form of lysophospholipids*

Lysophosphatides are vastly more soluble than phosphatides and are known to exchange readily. Glycerophospholipids with given head group might therefore exchange in their lyso forms. After exchange these could be acylated by appropriate acyl transferases, via suitable fatty acyl-CoA derivatives. The presence of such acyl transferases has been documented for a large variety of biomembranes [76]. Moreover, numerous biomembranes contain the phospholipases (A_1 and/or A_2) necessary to generate lysophosphatides. This includes endoplasmic reticulum and mitochondria [77,78]. Even erythrocyte membranes contain some phospholipase activity [76] and can reacylate lysophosphatides with exogenously supplied fatty acid acyl-CoA derivatives [76]. Finally, deacylation and reacylation of phosphatides has been well documented in lymphoid cells [76]. Both processes are markedly activated upon mitogenic stimulation of these cells.

3.5.3.4. *Membrane flow*

It is possible that phospholipids never leave the membrane systems of their cells and that they pass from one membrane to another, possibly as lipoprotein units, until they reach their 'ultimate' localization. The morphologic continuity necessary for such a process exists between the outer membrane of the nuclear envelope and the endoplasmic reticulum and between the endoplasmic reticulum and the Golgi apparatus. Some electronmicroscopic evidence also suggests that the outer mitochondrial membrane and the endoplasmic reticulum are in continuity [79]. Lateral diffusion of phospholipids within such membrane continua might explain the redistribution of biosynthetically incorporated phosphatide label within intercellular membranes. At present there is no conclusive evidence either for or against this hypothesis.

3.5.3.5. *Extracellular, protein-mediated exchange*

It has long been known that phospholipids can exchange between the various classes of plasma lipoproteins, and it has been suggested that this exchange involves the formation of collision complexes between the lipoproteins [80]. If phospholipids could diffuse within such collision complexes, an exchange of phospholipids would follow dissociation of the complexes. It now appears that the situation may be more complicated, since recent studies show that some of the lipoprotein subunits can also exchange [81]. Specifically, apolipoprotein-glutamic acid and apolipoprotein-alanine, both low-molecular weight apolipoproteins, readily transfer *in vitro* from very low-density lipoproteins to other lipoproteins. Their transfer to high-density lipoproteins always exceeds that to low-density lipoproteins and is proportional to the concentration of lipoproteins present in the incubation mixture. A similar transfer occurs *in vivo*. Moreover, the process of transfer between very low-density and high-density lipoproteins is bidirectional and represents a real exchange phenomenon.

It now appears that this kind of exchange process may not not limited to soluble lipoproteins. Thus, Langdon [82] has presented provocative evidence that both apolipoprotein-glutamic acid and apolipoprotein-alanine are present in human erythrocyte membranes.

It is conceivable therefore, that the exchange processes in both plasma lipoproteins and possibly membranes involve transfer of whole phospholipid-lipoprotein apoprotein subunits, in which process the protein functions as a carrier of the phospholipid.

3.5.3.6. *Intracellular, protein-mediated exchange*

A number of workers have demonstrated that exchange of phospholipids between microsomal membranes and mitochondrial membranes is markedly accelerated by soluble, phospholipid-exchange proteins which are found in a large variety of cells. This topic has been recently reviewed by Wirtz [83]. Present data indicate the existence of a series of phospholipid-exchange proteins, which have different activities with respect to the transfer of individual classes of phospholipid. The specificity appears to be directed primarily at the head group and, at least in liver, there appear to be specific proteins facilitating the exchange of phosphatidylcholine, phosphatidylinositol and phosphatidylethanolamine.

The phosphatidylcholine-exchange protein of liver has been most extensively characterized [84,85]. This protein, as well as other phospholipid exchange-proteins isolated from beef liver, heart, and brain all have molecular weights between 20,000 and 30,000 and isoelectric points between 4.7 and 5.8. Amino acid analyses of the phosphatidylcholine exchange-protein indicate 190 amino acids per mole, with an N-terminal glutamic acid. The average hydrophobicity is 1.109 kcal per residue, a value not very far from the hydrophobicity exhibited by some soluble proteins which have no specific or general lipid associations.

Demel et al. [85] have studied the mode of action of phosphatidylcholine-exchange protein, making use of monolayer techniques. In this, they prepared monomolecular films of ^{14}C-labeled phosphatidylcholine at air–water interfaces and monitored the surface radioactivity and surface pressure before and after injection of the exchange protein under the monomolecular film. Injection of the protein reduces surface radioactivity, although the surface pressure remains constant. This indicates that the phosphatidylcholine bound to protein can exchange with phosphatidylcholine located at the interface. Support for this notion comes from the demonstration that the exchange protein can transfer radiolabelled phosphatidylcholine between two monolayers connected only through a common subphase. Finally, it is shown that the exchange protein can facilitate the transfer of phosphatidylcholine between labelled monolayers and phosphatidylcholine liposomes.

It is suggested that in the exchange process, the exchange protein collides with either the monolayers, or the halves of a bilayer in the case of liposomes, to form a transient collision complex, in which protein bound-phosphatide is replaced by monolayer- or liposome-phosphatide. It is proposed that, during the interaction of the exchange protein with the monolayer of a bilayer, the protein penetrates into the hydrophobic region of these layers. During this period the protein-bound phosphatide mixes freely with the liposome or monolayer associated phosphatide.

It is clear from these data that the phospholipid composition of a given biomembrane might be regulated to some extent by the exchange proteins in its environment. There is information that *the reverse may also be true.* Thus, Kagawa, et al. [86] have shown that a phospholipid-exchange protein isolated from beef heart transfers phosphatidylcholine from phosphatidylcholine liposomes, to vesicles reconstituted from mitochondrial hydrophobic protein, cardiolipin and phosphatidylethanolamine. This is in contrast to the behavior of high-density lipoprotein apoproteins [83] which, although they strongly bind phospholipids, do not accept phosphatidylcholine from rat liver microsomes in the presence of phosphatidylcholine-exchange protein.

3.6. Phospholipid turnover

Pasternak [87] has reviewed work from his laboratory and others on the phospholipid turnover in normal and neoplastic cells. The turnover studies have involved radioactively labelled inositol and choline as specific precursors for phosphatidylinositol and phosphatidylcholine. The turnover pattern of inositol- and choline-phospholipids exhibits a 'fast' component and a 'slow' component. This biphasic pattern has been observed with human lymphocytes, primary chick fibroblasts, and P815Y mastocytoma cells and does not depend on cell cycle.

The turnover is not due to exchange of the choline or inositol head groups [88]. In the case of choline phospholipids, phosphatidylcholine accounts for most of the 'fast' component of turnover, whereas the 'slow' component is enriched in sphingomyelin. The turnover of total phosphatidylcholine is equally high in exponentially-growing as in resting-cells, occurs throughout the cell cycle and is maximal in the S-phase, where there is net synthesis of phospholipids and other macromolecules [87].

Pasternak [89] has also studied the effect of serum addition to chick fibroblast cultures which have been prelabelled with [^{3}H]inositol and [^{3}H]choline. Careful analysis of the distribution of label between phospholipid and the acid soluble phosphate pool of cells shows that serum causes a stimulation of label incorporation into phospholipid and that this stimulation is about equivalent to the rate of elimination of the phospholipid by turnover. This effect is also seen with hepatoma cells in culture and is somewhat dependent on serum concentration. Pasternak relates these findings to the serum-mediated 'release of contact inhibition'. However, he does not consider the fact that serum contains lipoproteins and that, depending on the nature of the serum, the exchange between membrane and lipoprotein may lead to an altered apparent turnover. The data can thus not be interpreted unambiguously.

A number of speculations have been presented over the years as to the significance of physiological turnover of phospholipids. One idea that is often raised is that there is continuous movement of membranes both internally between endoplasmic reticulum, nuclear membrane, plasma membranes, lysosomes, and also externally during endocytosis and exocytosis. Another possibility is that membrane phospholipids may be altered by degradative enzymes or peroxidation and that they will, therefore, have to be replaced to maintain normal membrane function.

3.7. Phospholipid anomalies in tumors

3.7.1. Composition

3.7.1.1. Whole-cell analyses

Several laboratories have compared the phospholipid patterns of unfractionated tumor cells with those of non-neoplastic cells of the same type. As is evident from Bergelson's recent comprehensive review [90], these studies have not proven very revealing. This is in part because the analyses are for the most part rather incomplete and lack adequate biological controls. Moreover, whole tissue analyses are not likely to be informative because of the very different (and often opposite) phospholipid compositions of various membranes (Table 3.1). We will therefore only deal with the few comprehensive comparisons of normal and neoplastic cells published recently.

Cunningham [91] has examined rates of incorporation of [^{32}P]orthophosphate into the phospholipids of 3T3 cells under various conditions. He finds that, at isotopic equilibrium, the proportions of diverse labelled phospholipids is the same in non-confluent as in confluent 3T3 cells and their non-confluent and confluent, polyoma-derived, neoplastic variants. The total phospholipid composition of the 3T3 cells (Table 3.2) is rather similar to that of BHK21 cells (Table 3.1) but differs from the values reported for normal Syrian hamster fibroblasts (Table 3.3) [92].

TABLE 3.2

Phospholipid composition of normal and polyoma-transformed mouse 3T3 fibroblasts at low- and high-growth densities[1,2]

Phospholipid	Proportions (%)			
	3T3		Py 3T3	
	Non-confluent	confluent	Non-confluent	confluent
Diphosphatidylglycerol + phosphatidic acid	2.7	2.1	4.8	2.4
Phosphatidylethanolamine	27	29	25	26
Phosphatidylserine	6	7	7	6
Phosphatidylinositol	3.1	3.2	3.4	3.9
Phosphatidylcholine	52	54	52	53
Sphingomyelin	7	3	6	7
Lysophosphatidylcholine	2.2	1.7	1.8	1.7

[1] From Cunnigham [91].
[2] According to ^{32}P-labelling at isotopic equilibrium.

TABLE 3.3

Phospholipid composition of syrian hamster fibroblasts, spontaneously-transformed and polyoma-transformed variants *in vitro* and their *in vivo* tumors[1]

Phospholipid	Phospholipid content (%)				
	Culture			Tumor	
	Normal cells	Polyoma transformed	Spontaneously transformed	Polyoma	Spontaneous
Phosphatidylethanolamine[2]	23.1	25.7	15.5	38.7	34.5
Phosphatidylserine	13.7	11.6	17.1	12.9	13.5
Phosphatidylinositol	2.5	5.2	2.6	—	—
Phosphatidylcholine	45.4	39.7	41.7	22.5	24.6
Sphingomyelin	15.6	17.5	22.9	25.8	27.1
Choline phosphatide[3]	60.4	57.2	64.6	48.3	51.7

1 From Chao et al. [92].
2 Includes diphosphatidylglycerol.
3 Phosphatidylcholine + sphingomyelin

However, upon reaching confluency, the normal cells exhibit a 2-fold increase of incorporation of label into phosphatidylcholine while the rates of incorporation into phosphatidylethanolamine and phosphatidylserine decrease 7- and 2-fold, respectively, under the same conditions. Concordantly, alleviation of density dependent growth inhibition by addition of fresh serum leads to an increase in orthophosphate incorporation into all phospholipids within 2 hr of serum addition.

Importantly, the polyoma variants show no change in the rate of phospholipid incorporation upon reaching high densities or after addition of fresh serum.

Cunningham's results contrast with those of Chao et al. [92]. These workers have examined the overall lipid composition of Syrian hamster fibroblasts in secondary culture, as well as their spontaneous- and polyoma-transformants, and tumors derived therefrom (Table 3.3). They report some interesting features.

First, each cell class (normal cells, neoplastic variants in culture and the tumors derived therefrom) appears to have its own phospholipid individuality, except for phosphatidylserine, which does not vary significantly in proportion. Importantly, however, the neoplastic cells, whether isolated from initial culture or from the tumors, show a drastic increase in the proportion of sphingomyelin. This is particularly remarkable in the case of the tumors.

Another interesting feature is that normal and neoplastic cells grown *in vitro* have a cholesterol/phospholipid ratio of 0.3, whereas this value is 1.0 for both solid tumors. Similarly, the cholesterol/sphingomyelin molar ratio of the transformed cells grown *in vitro* does not differ significantly from that of the normal cells (1.5 to 2.0) whereas the values are 3.43 and 3.6 for the tumors caused by polyoma transformed- and spontaneously-transformed cells, respectively. A possible explanation for these findings is that the cells cultured *in vitro* were artificially depleted of cholesterol by the culturing conditions, a situation which does not arise *in vivo*.

Howard et al. [93] have compared the lipid composition and metabolism of the human diploid cell line WI-38 and SV40 transformant thereof, WI 38VA13. They did not carry out any specific membrane analysis, but their overall compositional data are revealing, nevertheless (Table 3.4).

TABLE 3.4

Phospholipid composition of normal and SV40-transformed WI38 Cells[1]

Phospholipid	Phospholipid composition[2] (%)	
	WI38	WI38VA13
Phosphatidylethanolamine	12	13
Phosphatidylserine	12	4
Phosphatidylinositol	13	10
Phosphatidylcholine	57	57
Sphingomyelin	7.4	13
Lysophosphatidylcholine	2.9	2.0

[1] From Howard et al. [93].
[2] Molar ratios cholesterol/phospholipid are 0.29 and 0.48 in WI38 and WI38VA respectively, whereas the molar ratios of cholesterol/sphingomyelin were essentially identical.

The major significant difference in phospholipid composition is in the greater sphingomyelin content of the transformed cells. This observation is in accord with the data of Chao et al. [92] and observations on hepatomas to be discussed below. The SV40-transformed cells also show an overall cholesterol/phospholipid molar ratio of 0.48, 1.6-fold greater than the value of 0.29 in the WI38. The cholesterol/sphingomyelin molar ratios are essentially identical in the two cell lines (WI38 = 3.9; WI38VA13 = 3.7) because of the greater proportion of sphingomyelin in the transformed cells.

3.7.1.2. *Membrane analyses*

Bergelson et al. [51] have presented an extensive analysis of the phospholipid composition of homogenates, mitochondria, nuclei, microsomes and, in some instances, plasma membrane isolates, of rat and mouse liver, various hepatomas, and the Jensen sarcoma. The most complete data comparing liver with an ascites hepatoma are shown in Table 3.5.

TABLE 3.5

Phospholipid composition of various membrane fractions from rat liver, an ascites hepatoma and the Jensen rat sarcoma[1]

Phospholipid	Phospholipid content[2]													
	Liver					Zajdela ascites hepatoma					Jensen sarcoma			
	H	N	Mit	Ms	PM	H	N	Mit	Ms	PM	H	N	Mit	Ms
Diphosphatidylglycerol	3.9	—	14.6	—	1.0	4.1	3.6	6.3	5.2	3.3	2.4	3.9	8.2	7.7
Phosphatidylethanolamine	22.4	27.6	35.3	26.9	24.7	24 9	21.5	19.5	19.0	24.9	27.6	29.2	26.0	22.0
Phosphatidylserine	3.1	7.3	1.3	2.0	4.2	13.1	10.9	11.9	13.2	10.4	13.6	8.2	9.8	19.9
Phosphatidylinositol	5.3	4.1	0.5	8.8	6.7	6.6	5.4	6.1	7.3	5.3	6.8	6.0	4.2	4.5
Phosphatidylcholine	52.8	52.5	48.1	54.3	46.1	34.4	38.3	40.9	37.3	37.7	31.7	29.2	26.0	22.0
Sphingomyelin	8.8	4.9	—	7.2	16.8	11.3	15.4	10.6	12.3	14.5	15.1	10.2	8.9	10.8
Lysophosphatidylcholine	3.7	3.6	—	—	0.5	5.6	5.7	4.7	5.7	3.9	2.8	4.8	3.6	3.7
Phosphatidylcholine + sphingomyelin	61.6	57.4	48.1	62.5	52.9	45.7	53.7	51.5	49.6	51.2	46.8	39.5	34.9	32.8

[1] From Bergelson et al. [51].

[2] In %. H = homogenate; N = nuclei; Mit = mitochondria, Ms = microsomes, PM = plasma membrane, — = not detected.

In this system, one observes a significant difference between the overall phospholipid composition of liver and the tumor; the proportion of phosphatidylserine and sphingomyelin is greater in the hepatoma and the proportion of phosphatidylcholine is lower. Moreover, the proportion of total choline phosphatide (phosphatidylcholine plus sphingomyelin) is 46% in the case of the hepatoma compared with 62% in normal liver.

Much more dramatic differences emerge when one looks at the subcellular fractions. Particularly remarkable is the fact that the microsomal fraction of the hepatoma contains a very substantial level of diphosphatidylglycerol, whereas this phosphatide is absent in the microsomes from normal liver. In contrast, the diphosphatidylglycerol complement of the hepatoma mitochondria is about half that of normal liver. An equally significant difference is the markedly greater sphingomyelin complement of all hepatoma fractions. This is particularly dramatic in the case of the mitochondria which contain 11% sphingomyelin compared with none in the normal liver.

Finally, all of the hepatoma membrane fractions contain a significantly greater proportion of phosphatidylserine than the corresponding liver fractions and a much lower complement of phosphatidylcholine. The proportion of total choline phosphatide (phosphatidylcholine plus sphingomyelin) is also significantly less in the hepatoma fractions particularly in the microsomes and plasma membranes.

These data are in line with those of Feo et al. [94] who investigated the mitochondria and microsomes from the minimal deviation type hepatoma Morris 5123. Although these authors do not specifically comment on it, their Figure 1 also shows an increased proportion of sphingomyelin in the hepatoma mitochondria.

Bergelson et al. [95] have recently extended their earlier studies to a comparison of the phospholipid patterns in the mitochondria and microsomes from slow and fast growing hepatomas, as well as from regenerating liver (Table 3.6). Diphosphatidylglycerol was more prevalent in the microsomes of both hepatomas (compared to liver) and depleted in the mitochondria. In contrast, sphingomyelin is increased in the mitochondria of both hepatomas and also in the microsomes of the highly malignant hepatoma 22. Importantly, the phospholipid patterns in mitochondria and microsomes isolated from liver and varying stages of regeneration showed no deviation from the pattern seen in particles from normal liver. Accordingly, the biochemical abnormalities of phospholipid composition observed in the tumors cannot be readily attributed to rapid cell growth. The possibility that the anomalous phospholipid patterns in the mitochondrial and microsomal membranes of hepatoma cells might represent cross contamination or exchange during fractionation has been explored and excluded by Bergelson et al. [51,96].

Bergelson et al. [51] have also studied the phospholipid distribution in homogenates, nuclear fractions, mitochondria and microsomes derived from the Jensen

TABLE 3.6

Phospholipid composition of mitochondria and microsomes from normal C_3HA mouse liver and two Hepatomas[1]

Phospholipid	Mitochondria			Microsomes		
	Liver	Hepatoma 48	Hepatoma 22	Liver	Hepatoma 48	Hepatoma 22
Diphosphatidylglycerol	18.3	12.8	10.2	—	2.1	6.3
Phosphatidylethanolamine	33.4	36.8	26.7	24.9	20.2	30.2
Phosphatidylserine	—	—	6.4	2.9	4.5	7.1
Phosphatidylinositol	4.4	4.7	11.0	6.5	11.1	6.3
Phosphatidylcholine	43.9	43.5	37.4	58.0	56.0	39.5
Sphingomyelin	—	2.2	7.7	6.3	5.4	10.1
Lysolecithin	—	—	0.6	1.4	0.7	0.5
Phosphatidylcholine + sphingomyelin	43.9	45.7	45.1	64.3	61.4	49.6

[1] From Bergelson et al. [95]; — = not detected.

sarcoma (Table 3.5). In this case, it is of course difficult to use a reference tissue. Nevertheless, the distribution pattern is very similar to that observed in the hepatomas, with an unusually low complement of phosphatidylcholine and phosphatidylcholine plus sphingomyelin, unusually high sphingomyelin levels, particularly in the mitochondria, unusually high phosphatidylserine levels, particularly in the microsomal fractions, and anomalous levels of diphosphatidylglycerol, particularly in the microsomal fraction. The phosphatide pattern is also totally different from that observed for the membrane fraction of normal BHK21 fibroblasts (Table 3.1).

The sphingomyelin levels reported by Bergelson et al. [51] for nuclei isolated from the Zajdela ascites hepatoma and from Jensen sarcoma are remarkable when compared with the values published for nuclei isolated from normal tissues. For example, Kleinig [97] reports a value of 3.2 % for the sphingomyelin content of pig liver nuclear membranes and Renkonen et al. [50] give a value of 2.5 % for these organelles isolated from BHK21 cells. On the other hand, the sphingomyelin levels reported by Bergelson et al. [51,95] for microsomes isolated from Zajdela ascites hepatoma, Jensen sarcoma and hepatoma 22 all fall in the range of the values reported as typical for such isolates (about 10 %).

3.7.1.3. *Bile-front membranes*

Three laboratories have published phospholipid analyses of bile-front plasma membrane isolates derived from various hepatomas and from normal liver [51,97,98]. The data, summarized in Table 3.7 show no consistent pattern. Indeed there are major differences between the analyses of normal liver reported by the

TABLE 3.7

Phospholipid composition of plasma membrane isolates derived from normal liver and several hepatomas[1]

Phospholipid	Composition (%)									
	Rat							Mouse		
	Liver[1]	Liver[2]	Liver[3]	484A[1]	3924A[3]	H35[3]	Zajdela[2]	Liver[1]	147042[1]	143060[1]
Diphosphatidylglycerol + phosphatidic acid	2.3	1.0	—	2.6	—	—	3.3	1.0	5.2	2.2
Phosphatidylethanolamine	19.0	24.7	25.6	22.1	17.1	14.2	24.9	20.2	12.7	19.1
Phosphatidylserine	15.6	4.2	6.8	11.3	6.1	7.0	10.4	13.6	14.4	17.5
Phosphatidylinositol	6.0	6.7	6.9	3.0	7.0	8.0	5.3	3.8	6.3	3.7
Phosphatidylcholine	30.5	46.1	41.4	24.0	52.8	54.1	37.7	27.2	34.6	28.6
Sphingomyelin	22.5	16.8	9.1	24.5	21.2	8.3	14.5	22.8	14.2	17.8
Lysophosphatides	5.2	0.5	6.1	12.3	1.4	—	3.9	11.2	11.8	10.8

[1] From Van Hoeven and Emmelot [98].
[2] From Bergelson et al. [45].
[3] From Selkirk et al. [99].

three laboratories. We attribute these inconsistencies to the difficulties involved in obtaining comparable plasma membrane fractions from tissues differing in surface architecture [98]. The problem is particularly severe in the liver-hepatoma system, for the following reasons:

(a) No satisfactory system has been developed for the isolation of *all* regions of the hepatocyte surface; most methods deal only with the highly-specialized 'bile-front' domain.

(b) Bile-front isolation depends on the unusual surface architecture of the bile-canalicular region. But the surface topology of hepatoma cells deviates from that of normal hepatocytes, and highly deviant hepatomas have primitive if any bile-canalicular specialization.

(c) The diverse methods used for the isolation of liver plasma membranes can yield very different products. This is true also for the hepatoma studies (Table 3.7) and we suspect that the plasma membrane isolates analyzed are not reliably comparable.

Nevertheless, all of the plasma membrane isolates contain a very high level of sphingomyelin compared with the values recorded for other membrane fractions. This appears to be a general quality of plasma membranes and, at least in the tumors where information is available, appears to be maintained even in the neoplastic state. Current plasma membrane analyses thus do not reflect the rather dramatic alterations of sphingomyelin proportion which are observed in nuclear membranes, mitochondrial membranes and microsomal membranes [51,95]. One should note, however, that the decreased proportion of sphingomyelin found in plasma membrane isolates of the mouse hepatoma 147042 and 143060 (Table 3.7), as well as in the Zajdela ascites hepatoma of the rat, when compared with the values obtained in the same laboratory for normal liver, might be consistent with the Bergelson hypothesis of 'membrane dedifferentiation'. However, the membranes of hepatoma 3924A, a highly dedifferentiated tumor, exhibit a very high sphingomyelin content. Clearly, much more careful membrane fractionation is in order to clarify this situation.

The compositional studies can be summarized as follows:

(a) All whole-cell analyses but one [91] show an increased proportion of *sphingomyelin* in neoplastic cells versus the appropriate non-neoplastic cell type.

(b) Data on intracellular membranes are limited but dramatic. *First*, all studies demonstrate very high *sphingomyelin* levels in the mitochondria and nuclei. Microsomes from tumor cells also show increased sphingomyelin levels. *Second*, tumor microsomes are enriched in diphosphatidylglycerol whereas the tumor mitochondria exhibit unusually low levels of this phosphatide.

(c) Data on plasma membranes isolates do not show any consistent pattern, presumably because of unresolved technical difficulties in isolating these membranes.

To explain the altered phospholipid patterns observed, Bergelson et al. [51,95] propose that neoplasia involves a state of membrane 'dedifferentiation', that is, the distinctive lipid patterns, characteristic of normal cellular membranes, are lost. However, we lack clues as to how this might come about; in fact we do not know the mechanisms responsible for the patterns characteristic of normal cells. Critical among these must be the protein content and organization of the membranes.

The observation that tumor mitochondria are enriched in sphingomyelin, as well as cholesterol is intriguing from a physico-chemical point of view. Thus, a variety of physico-chemical studies show that the gel → liquid-crystal transition of sphingomyelin lies at 40 °C; much above the values observed for most phospholipids. This high transition temperature can be attributed to the generally low degree of unsaturation found in sphingomyelin [3], and implies that membranes rich in sphingomyelin might be generally less fluid at physiological temperatures than other membranes, or might contain non-fluid domains. However, cholesterol tends to abolish gel → liquid-crystalline transition of sphingomyelin. As noted before, this is because cholesterol fluidizes the hydrocarbon chains below the transition temperature and 'defluidizes' above the transition temperature. Cholesterol thus tends to 'fluidize' sphingomyelin at physiologic temperatures. Moreover, in mixed lipid systems, cholesterol would tend to convert what might be a mixture of separated solid and fluid-phases into a generally liquid-crystalline phase.

An increase in the proportion of sphingomyelin, particularly the mitochondria membranes, would be expected to interfere with their normal physical and biological properties. However, this deleterious effect might be largely compensated by the presence of higher proportions of cholesterol. One can thus imagine that the elevated cholesterol levels in the membranes, particularly mitochondria of the hepatoma cells, represents somewhat of a 'compensatory mechanism'.

The altered sphingomyelin content of some tumor membranes should be viewed in the light of the known close association of this lipid with some proteins. For example, the species-difference in the sphingomyelin content of certain erythrocytes is clearly related to the affinity of the membrane proteins for this phosphatide [101]. Also, the 5′-nucleotidase of rat liver is very closely associated with sphingomyelin [102].

3.7.2. Phospholipid metabolism

3.7.2.1. *Fatty acid content and regulation of fatty acid biosynthesis in hepatomas*

It appears that the fatty acid patterns found in tumors are strongly dependent on the host. As reviewed by Bergelson [90] the fatty acid composition of hepatoma is in general similar to that of normal liver. The same is true for various leukemic cells compared with normal white blood cells. However, all minimum deviation hepatomas carried *in vivo* exhibit defective regulation of *de novo* fatty acid biosynthesis, although this control is functioning normally in host livers [104–107].

Majerus et al. [107] have demonstrated that the levels of acyl-CoA carboxylase and fatty acid synthetase in hepatomas do those of respond to fasting and fat-refeeding of tumor bearing animals, as do the host livers. However, the acyl-CoA carboxylases purified from several hepatomas appear to be identical to those of the host livers in terms of kinetic and other properties. Moreover, the levels of carboxylase activity in tumors parallel the content of immuno-chemically reactive protein and there is no change in the rate of enzyme synthesis in tumors of animals switched to a fat-free diet. Finally, both acyl-CoA carboxylase and fatty acid synthetase in tumors are inhibited by palmitoyl-CoA to the same extent as normal liver enzymes. These results are interpreted to indicate that the defective control of fatty acid biosynthesis is due to the failure of the tumors to regulate the quantity of the acyl-CoA carboxylase rather than to production of an altered enzyme.

The possibility that the altered or defective regulation of fatty acid biosynthesis in hepatomas is due to an anomalous blood supply has been excluded by the experiments of Bartley and Abraham [108]. These authors transplanted hepatic tissue to a subcutaneous environment where it received no portal blood supply. However, the autografts exhibited the same levels of acyl-CoA carboxylase and fatty acid synthetase as normal liver and these enzymes varied as in normal liver in response to fasting and refeeding of the animals.

It appears, therefore, that the lack of adequate control of fatty acid biosynthesis in hepatomas does not constitute an example of deleted feedback inhibition. However, lack of regulation of acyl-CoA carboxylase biosynthesis is also not the only explanation for the phenomenon. Other feasible explanations include a greater stability of the enzyme in the tumor compared with normal hepatocytes as well as the lack of production of some unknown feedback inhibitors. We also consider another possibility, namely that defective regulation may be due to an alteration of membrane mediated control. Thus, regulation of *de novo* fatty acid synthesis depends on the steady state concentration of cytoplasmic fatty acyl-CoA. The levels of these substances depend on the activity of the acyl-carnitine transferases of the mitochondrial membranes. Enhanced activities of these enzymes (possibly related to the altered phospholipid or cholesterol composition of mitochondrial membranes) could lead to accelerated removal of fatty-acyl-CoA from the cytoplasmic space. Moreover, increased levels of fatty-acyl-carnitine would decrease the inhibitory effects of residual fatty-acyl-CoA.

3.7.2.2. *Phospholipid turnover*

There is very little experimentation on the comparative phospholipid turnover of normal and neoplastic cells. What information there is deals only with overall phospholipid metabolism and does not deal with the turnover, etc. of the phospholipid components of specific cellular membranes.

Thus, Cunningham [91] has measured the incorporation of ^{32}P-labelled orthophosphate into the phospholipids of confluent and non-confluent 3T3 fibroblasts and their polyoma transformants. Correcting for differences in phosphorus uptake by fibroblast under different culture conditions [109], he finds that addition of fresh serum (to a level of 10%) induces further cell division in density-inhibited 3T3 cells and produces an increased turnover of phospholipids, particularly phosphatidylethanolamine and phosphatidylinositiol. Importantly, polyoma virus transformed 3T3 cells exhibit an invariant rate of turnover, which is the same as that of non-confluent 3T3 cells. Knox (cited by Pasternak [87]) has also observed that the turnover of ^{3}H-choline labelled phospholipids in SV40 transformed 3T3 cells is greater than in the normal variants.

These data suggest an increase of phospholipid turnover during release of contact inhibition and a greater level of turnover in certain virally transformed cells compared with their normal variants. However, the data are too sparse to arrive at firm conclusions.

3.8. *Exchange*

We know of only one study dealing with the important matter of phospholipid exchange. In this, Anghileri [110] has compared the fate of ^{32}P-labelled liver phospholipids administered intraperitoneally to normal rats, as well as to animals bearing subcutaneously transplanted Jensen sarcoma. The label distribution in the injected phospholipid was 54.5% phosphatidylcholine, 30.3% phosphatidylethanolamine, 5.6% phosphatidylserine, 6.3% sphingomyelin and 3.1% lysophosphatidylcholine. Anghileri observed that the blood concentration of phosphatide label decreases more sharply in tumor-bearing animals than in normal animals. Also, the radioactivity in livers of tumor-bearing rats decreases after 8 hr, whereas in control animals it is steady for prolonged periods. Analysis of the tumors showed a steady increase in phospholipid labelling between 3 and 24 hr. The incorporation of label from radioactive orthophosphate, in contrast, was much lower than from the injected phospholipids, indicating that the accumulation of label in the tumors could not be due to *de novo* synthesis of phospholipid by the tumor. The tumor-bearing animals also exhibited a faster excretion of injected labelled phospholipids as well as a mobilization of phospholipids from other tissues into the tumor.

This study suggests a much greater exchangeability of the phospholipids in the tumor membranes. A high exchangeability, if extended to intracellular membranes, might fit with the 'dedifferentiation' hypothesis of Bergelson et al. [51,95]. However, since the specific lipid pattern of a membrane must depend on the relative affinities of membrane proteins for various lipids, one must look to membrane proteins, as well as to phospholipid exchange proteins, for an ultimate explanation of any phospholipid alterations in tumor membranes. Much more extensive experimentation, based on the fundamental approaches detailed earlier is clearly indicated.

3.9. Conclusions

The information available to date indicates that no one phospholipid composition is characteristic of all tumors. Moreover, no tumor-specific phospholipid had been proven to exist, and there are no indications that the overall phospholipid metabolism of neoplastic cells is significantly different from that of normal cells. Overall phospholipid analyses have not proven very revealing because of the cellular heterogeneity of the tissues studied, variability among different tissues, both normal and neoplastic, the difficulty of finding suitable controls for neoplastic cell populations, and the different and often opposite phospholipid patterns of diverse intracellular membranes.

Despite these obstacles, an important general trend has been revealed through comparisons of the phospholipid composition of diverse cellular membranes from normal and neoplastic tissues. The phospholipid distribution characteristic of specific subcellular membranes, particularly those of the mitochondria and endoplasmic reticulum, tend to disappear in neoplastic cells; that is, the phospholipid patterns tend to become the same for all subcellular membranes. Recent data suggest that this trend is less pronounced in minimum deviation tumors and more pronounced in highly malignant tumors. Moreover, it has been demonstrated that this 'membrane dedifferentiation', at least in hepatomas, is not related to cell growth but to the neoplastic process.

In view of the extensive information pointing to membrane alterations in the neoplastic process, the relative deficiency of solid experimentation on possible phospholipid alterations in the neoplastic process is rather surprising. This is particularly so, since there is a large amount of very solid information about the chemistry, physics and molecular biology of phospholipid in model systems and also in simple cells. However, this information has not been applied to the tumor field in a systematic or regular fashion. Although extensive analytical data are available, there are very few comprehensive analyses on systems where one can make adequate comparisons between the normal and the neoplastic state. Another topic which has been totally neglected but which is very critical to the whole field is the matter of phospholipid exchange. Thus, while anomalous phospholipid patterns have been observed in the mitochondria and endoplasmic reticulum of hepatoma cells, organelles which are known to exchange phospholipids via phospholipid exchange proteins, exchange mechanisms have not been investigated at all in neoplastic systems.

To sum, one can say that, although there are very many investigations on the phospholipid composition and phospholipid metabolism of neoplastic cells, most of the experimentation that has been performed is excessively self-contained. Moreover, many very important experimental approaches have not been applied to the problem. We feel that the field deserves further experimentation using modern techniques, modern information on the chemistry, physics and biology of phospholipids, correlation with possible alterations of glycolipid and cholesterol metabolism, etc., and careful, rigorous biological controls.

References

[1] Shipley, G. in *Biological Membranes*, D. Chapman and D.F.H. Wallace, eds., Academic Press, London, Vol. II, 1973, p. 1.
[2] Singer, S.J. and Nicolson, G.L., *Science*, *175*, 720 (1972).
[3] Chapman, D. in *Biological Membranes*, D. Chapman and D.F.H. Wallach, eds., Academic Press, London, Vol. II, 1973, p. 91.
[4] Träuble, H. and Eibl, H., *Proc. Natl. Acad. Sci. U.S.*, *71*, 214 (1974).
[5] Steim. J.M., Tourtellotte, M.E., Reinert, J.C., McElhaney, R.M. and Rader, R.L., *Proc. Natl. Acad. Sci. U.S.*, *63*, 104 (1969).
[6] Blazyk, J.F. and Steim, J.M., *Biochim. Biophys. Acta*, *266*, 737 (1972).
[7] Keith, A.D., Sharnoff, M. and Cohn, G.E., *Biochim. Biophys. Acta*, *300*, 329 (1973).
[8] Radda, G.K. and Vanderkooi, J., *Biochim. Biophys. Acta*, *265*, 509 (1972).
[9] Engelman, D.M., *Chem. Phys. Lipids*, *8*, 298 (1972).
[10] Esfahani, M., Limbrick, A.R., Knutton, S., Oka, T. and Wakil, S.J., *Proc. Natl. Acad. Sci. U.S.*, *68*, 3180 (1971).
[11] Overath, P. and Träuble, H., *Biochemistry*, *12*, 2625 (1973).
[12] Speth, V. and Wunderlich, F., *Biochim. Biophys. Acta*, *291*, 621 (1973).
[13] Wunderlich, F., Batz, V., Speth, V. and Wallach, D.F.H., *J. Cell Biol.*, *61*, 633 (1974).
[14] Wunderlich, F., Speth, V. and Wallach, D.F.H., *Biochim. Biophys. Acta*, *373*, 34 (1974).
[14a] Wallach, D.F.H., Bieri, V., Verma, S.P. and Schmidt-Ullrich, R., *Ann. N. Y. Acad. Sci.*, 1975, in press.
[15] Overath, P., Schairer, H.U. and Stoffel, W., *Proc. Natl. Acad. Sci. U.S.*, *67*, 606 (1970).
[16] Wilson, G. and Fox, C.F., *J. Mol. Biol.*, *55*, 49 (1971).
[17] de Kruyff, B., van Dijck, P.W.M., Goldbach, R.W., Demel, R.A. and van Deenen, L.L.M., *Biochim. Biophys. Acta*, *330*, 269 (1973).
[18] Grisham, C.M. and Barnett, R.E., *Biochemistry*, *12*, 2635 (1973).
[19] Eletr, S., Zakim, D. and Vessey, D.A., *J. Mol. Biol.*, *78*, 351 (1973).
[20] Bieri, V., Wallach, D.F.H. and Lin, P.-S., *Proc. Natl. Acad. Sci. U.S.*, *71*, 4797 (1974).
[21] Barnett, R.E., Furcht, L.T. and Scott, R.E., *Proc. Natl. Acad. Sci. U.S.*, *71*, 1992 (1974).
[22] Shimshick, E.J. and McConnell, H.M., *Biochemistry*, *12*, 2351 (1973).
[23] Phillips, M.C., Ldbrooke, V.D. and Chapman, D., *Biochim. Biophys. Acta*, *196*, 35 (1970).
[24] Oldfield, E. and Chapman, D., *FEBS Letters*, *23*, 285 (1972).
[25] Ververgaert, P.H.J., Verkleij, A.J., Elbers, P.F. and van Deenen, L.L.M., *Biochim. Biophys. Acta*, *311*, 320 (1973).
[26] Engelman, D.M., *J. Mol. Biol.*, *58*, 153 (1971).
[27] Wilson, G., Rose, S.P. and Fox, C.F., *Biochem. Biophys. Res. Commun.*, *38*, 617 (1970).
[28] Linden, C.D., Wright, K.L., McConnell, H.N. and Fox, C.F., *Proc. Natl. Acad. Sci. U.S.*, *70*, 2271 (1973).
[29] Tsukagoshi, N. and Fox, C.F., *Biochemistry*, *12*, 2822 (1973).
[30] Yates, R.D., Higgins, J.A. and Barrett, R.J., *J. Histochem. Cytochem.*, *17*, 379 (1969).

[31] Lane, D.M., Moss., J. and Polakis, S.E. in *Current Topics in Cellular Regulations*, B.L. Horecker and E.R. Stadman, eds., Academic Press, New York, Vol. 8, 1974, p. 139.
[32] Fritz, I.B., *Perspect. Biol. Med.*, *10*, 643 (1967).
[33] Hoppel, C.L. and Tomec, R.J., *J. Biol. Chem.*, *247*, 852 (1972).
[34] Nakanishi, S. and Numa, S., *Eur. J. Biochem.*, *16*, 161 (1970).
[35] Jacobs, R.A. and Majerus, P.W., *J. Biol. Chem.*, *248*, 8392 (1973).
[36] Goodridge, A.G., *J. Biol. Chem.*, *247*, 6946 (1972).
[37] Greenspan, M.D. and Loewenstein, J.M., *J. Biol. Chem.*, *243*, 6273 (1968).
[38] van den Bosch, H., van Golde, L.M.G. and van Deenen, L.L.M., *Ergeb. Physiol.*, *66*, 13 (1972).
[39] Wilgram, G.F. and Kennedy, E.T., *J. Biol. Chem.*, *238*, 2615 (1963).
[40] van Golde, L.M.G., Raben, J., Batenbrug, J.J., Fleischer, B., Zambrano, F. and Fleischer, S., *Biochim. Biophys. Acta.*, *360*, 179 (1974).
[41] Shepard, E.H. and Huebscher, G., *Biochem. J.*, *113*, 429 (1969).
[42] Stanacev, N.Z., Stuhne-Sekalec, L., Brookes, K.D. and Davidson, J.B., *Biochim. Biophys. Acta*, *176*, 650 (1969).
[43] Daae, L.N.W. and Bremer, J., *Biochim. Biophys. Acta*, *210*, 92 (1970).
[44] Dawson, R.M.C., *Essays in Biochem.*, Vol. 2, P.N. Campbell and G.D. Greville, eds. Academic Press, London, 1966, p. 69.
[45] Hosteller, K.Y., van den Bosch, H. and van Deenen, L.L.M., *Biochim. Biophys. Acta*, *239*, 113 (1971).
[46] Sauner, M.-T. and Levy, M., *J. Lipid Res.*, *12*, 71 (1971).
[47] Miller, E.K. and Dawson, R.M.C., *Biochem. J.*, *126*, 805 (1972).
[48] Dennis, E.A. and Kennedy, E.P., *J. Lipid Res.*, *13*, 263 (1972).
[49] Slater, T.F. in *Free Radical Mechanisms in Tissue Injury*, Pion, London, 1972.
[50] Renkonen, O., Luukkonen, A., Brotherus, J. and Kääriäinen, L. in *Control of Proliferation in Animal Cells*, B. Clarkson and R. Baserga, eds., Cold Spring Harbor Laboratory, 1974, p. 495.
[51] Bergelson, L.D., Dyatlovitskaya, E.V. and Torkhovskaya, T.I., *Biochim. Biophys. Acta*, *210*, 287 (1970).
[52] McMurray, W.C. and Dawson, R.M.C., *Biochem. J.*, *112*, 91 (1969).
[53] Jungalwala, F.B. and Dawson, R.M.C., *Biochem. J.*, *117*, 481 (1970).
[54] Wirtz, K.W.A. and Zilversmit, D.B., *Biochim. Biophys. Acta*, *187*, 468 (1969).
[55] Lee, T.C., Stephens, N., Moehl, A. and Snijeer, F., *Biochim. Biophys. Acta*, *291*, 86 (1973).
[56] LeKim, D., Betzing, H. and Stoffel, W., *Hoppe-Seyler's, Z., Physiol. Chem.*, *353*, 949 (1972).
[57] Peterson, J.A. and Rubin, H., *Exp. Cell Res.*, *58*, 365 (1969).
[58] Illingworth, D.R., Portman, O.W., Robertson, A.L. and Magyar, W.A., *Biochim. Biophys. Acta*, *306*, 422 (1973).
[59] Lovelock, J., James, A. and Rowe, C., *Biochem. J.*, *74*, 137 (1960).
[60] Polonovski, J. and Paysant, M., *Bull. Soc. Chim. Biol.*, *45*, 339 (1963).
[61] Sakagami, T., Minari, O. and Orii, T., *Biochim. Biophys. Acta*, *98*, 111 (1965).
[62] Reed, C.F., *J. Clin. Invest.*, *47*, 749 (1968).
[63] Soula, G., Valdiguie, P., Douste-Blazy, L., *Bull. Soc. Chim. Biol.*, *49*, 1317 (1967).
[64] Shohet, S.B. and Nathan, D.G., *Biochim. Biophys. Acta*, *202*, 202 (1970).
[65] Tarlov, A., *Fed. Proc.*, *27*, 458 (1968).
[66] Zilversmit, D.B., *J. Lipid Res.*, *12*, 36 (1971).
[67] Smith, R. and Tanford, C., *J. Mol. Biol.*, *67*, 75 (1972).
[68] Kornberg, R.D. and McConnell, H.M., *Proc. Natl. Acad. Sci. U.S.*, *68*, 2564 (1971).
[69] Ehnholm, C. and Zilversmit, D.B., *Biochim. Biophys. Acta*, *274*, 652 (1972).
[70] Paphadjopoulos, D., Poste, G., Schaeffer, B.E. and Vail, W.J., *Biochim. Biophys. Acta*, *352*, 10 (1974).
[71] Kornberg, R.D. and McConnell, H.M., *Biochemistry*, *10*, 1111 (1971).
[72] Zwaal, R.F.A., Roelofsen, B. and Colley, C.M., *Biochim. Biophys. Acta*, *300*, 159 (1973).
[73] McNamee, M.G. and McConnell, H.M., *Biochemistry*, *12*, 2951 (1973).
[74] Huestis, W. and McConnell, H.M. quoted by McNamee and McConnell [73].
[75] Grant, C.W.W. and McConnell, H.M., *Proc. Natl. Acad. Sci. U.S.*, *70*, 1238 (1973).
[76] Ferber, E. in *Biological Membranes*, D. Chapman and D.F.H. Wallach, eds., Academic

Press, London, 1973, Vol. 2, p. 221.
[77] Victoria, E.J., van Golde, L.M.G., Hostetler, K.Y., Scherphof, G.L. and van Deenen, L.L.M., *Biochim. Biophys. Acta*, *239*, 443 (1971).
[78] Newkirk, J.D. and Waite, M., *Biochim. Biophys. Acta*, *225*, 224 (1971).
[79] Ghidoni, J.J. and Thomas, H., *Experientia*, *25*, 632 (1969).
[80] Gurd, F.R.N. in *Lipid Chemistry*, J.D. Hanahan, ed., Wiley, New York, 1960, p. 260.
[81] Eisenberg, S., Bilheimer, D.W. and Levy, R.I., *Biochim. Biophys. Acta*, *280*, 94 (1972).
[82] Langdon, R.G., *Biochim. Biophys. Acta*, *342*, 213 (1974).
[83] Wirtz, K.W.A., *Biochim. Biophys. Acta*, *344*, 95 (1974).
[84] Kamp, H.H., Wirtz, K.W.A. and van Deenen, L.L.M., *Biochim. Biophys. Acta*, *318*, 313 (1973).
[85] Demel, R.A., Wirtz, K.W.A., Kamp, H.H., Geurtsvan Kessel, W.S.N. and van Deenen, L.L.M., *Nature*, *New Biol.*, *246*, 102 (1973).
[86] Kagawa, Y., Johnson, L.W. and Racker, E., Biochem. *Biophys. Res. Commun.*, *50*, 245 (1973).
[87] Pasternak, C.A. in *Tumor Lipids*, R. Wood, ed., American Oil Chemist Society, Champagne, Ill., 1973, p. 66.
[88] Pasternak, C.A. and Bergeron, J.J.M., *Biochem. J.*, *119*, 473 (1970).
[89] Pasternak, C.A., *J. Cell Biol.*, *53*, 231 (1972).
[90] Bergelson, L.D., *Prog. Chem. Fats Other Lipids*, *13*, 1 (1972).
[91] Cunningham, D.D., *J. Biol. Chem.*, *247*, 2464 (1972).
[92] Chao, F.-C., Eng, L.F. and Griffin, A., *Biochem. Biophys. Acta*, *260*, 197 (1972).
[93] Howard, B.V., Butler, J.D. and Bailey, J.M. in *Tumor Lipids*, R. Wood, ed., American Oil Chemists Society Press, Champaign, Ill., 1973, p. 200.
[94] Feo, F., Canuto, R.A., Bertone, G., Garcea, R. and Pani, P., *FEBS Letter*, *33*, 229 (1973).
[95] Bergelson, L.D., Dyatlovitskaya, E.V., Sorokina, I.B. and Gorkova, N.P., *Biochim. Biophys. Acta*, *360*, 361 (1974).
[96] Dyatlovitskaya, E.V., Trusova, V.M., Greshnykh, K.P., Gorkova, M.T. and Bergelson, L.D., *Biokhimiya*, *37*, 607 (1972).
[97] Kleinig, H., *J. Cell Biol.*, *46*, 396 (1870).
[98] van Hoeven, R.P. and Emmelot, P., *J. Memb. Biol.*, *9*, 105 (1972).
[99] Selkirk, J.K., Elwood, J.C. and Morris, H.P., *Cancer Res.*, *31*, 27 (1971).
[100] Wallach, D.F.H. and Lin, P.-S., *Biochim. Biophys. Acta*, *300*, 211 (1973).
[101] Kramer, R.K., Schlatterer, Ch. and Zahler, P., *Biochim. Biophys. Acta*, *282*, 146 (1972).
[102] Widnell, C.C. and Unkeless, J.C., *Proc. Natl. Acad. Sci. U.S.*, *61*, 1050 (1968).
[103] Sabine, J.R., Abraham, S. and Chaikoff, I.L., *Biochim. Biophys. Acta*, *116*, 407 (1966).
[104] Sabine, J.R., Abraham, S. and Chaikoff, I.L., *Cancer Res.*, *27*, 793 (1967).
[105] Sabine, J.R., Abraham, S. and Morris, H.P., *Cancer Res.*, *28*, 46 (1968).
[106] Elwood, J.C. and Morris, H.P., *J. Lipid Res.*, *9*, 337 (1968).
[107] Majerus, P.W., Jacobs, R. and Smith, M.B., *J. Biol. Chem.*, *243*, 3588 (1968).
[108] Bartley, J.C. and Abraham, S., *Biochim. Biophys. Acta*, *260*, 169 (1972).
[109] Cunningham, D.D. and Pardee, A.B., *Proc. Natl. Acad. Sci. U.S.*, *64*, 1049 (1969).
[110] Anghileri, L.J., *Z. Krebsforsch.*, *78*, 326 (1972).

4

Glycolipids

Extensive experimentation during the past five years indicates that the metabolism of acidic and neutral glycolipids is often deranged in neoplasia, leading to an anomalous glycolipid composition. This topic has been analyzed in greatest detail by comparing diverse cultured rodent fibroblasts with transformants thereof, induced by oncogenic DNA and RNA viruses.

4.1. General characteristics of glycolipids

4.1.1. Metabolism

The important glycolipids in animal cells are sphingoglycolipids. These consist of a lipid moiety, ceramide, varying only slightly in fatty amide composition and a variable, polar segment generally containing from one to seven monosaccharide units (Figs. 4.1–4.3).

The lipid moiety, ceramide, is synthesized in the endoplasmic reticulum by the same enzymatic processes as are involved in the biosynthesis of the ceramide portion of sphingomyelin. In contrast, the elaboration of the carbohydrate moiety occurs by sequential addition of monosaccharides from sugar-nucleotide donors to the growing carbohydrate 'head-groups'. This progressive glycosylation is catalyzed by a series of specific glycosyltransferases (Fig. 4.4).

The biosynthesis of the glycolipids is unusual in the sense that the various gangliosides and neutral glycolipids can act both as 'end-products' (in membranes) and as substrates for subsequent glycosylations.

Older studies on the subcellular localization of the glycosyltransferases involved in ganglioside biosynthesis have been reviewed by Roseman [1]. In

The general structure for *sphingoglycolipids* is:

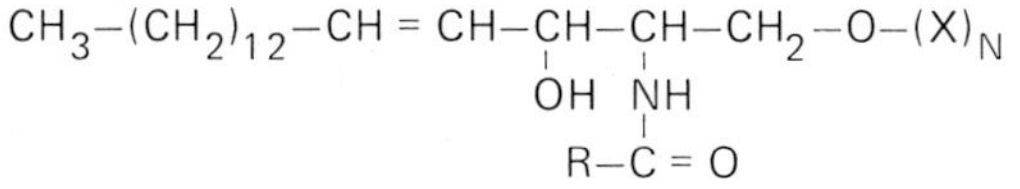

$$CH_3{-}(CH_2)_{12}{-}CH = CH{-}\underset{OH}{CH}{-}\underset{\substack{NH \\ | \\ R-C=O}}{CH}{-}CH_2{-}O{-}(X)_N$$

In *ceramides* (X) = H

Cerebrosides are *ceramide monohexosides* where X = *galactose* or *glucose* and N = 1. In *ceramide oligohexosides* N = 1, but there are equivalent molar amounts of sphingosine, hexose and fatty acid (R); the only nitrogen is in the sphingosine. In *gangliosides* additional nitrogens occur in the sugar moiety. Then X = (hexose(s) ± hexosamine(s) ± neuraminic acid).

The structure of a typical (galactosyl) cerebroside monohexoside (Gal→ceramide) is:

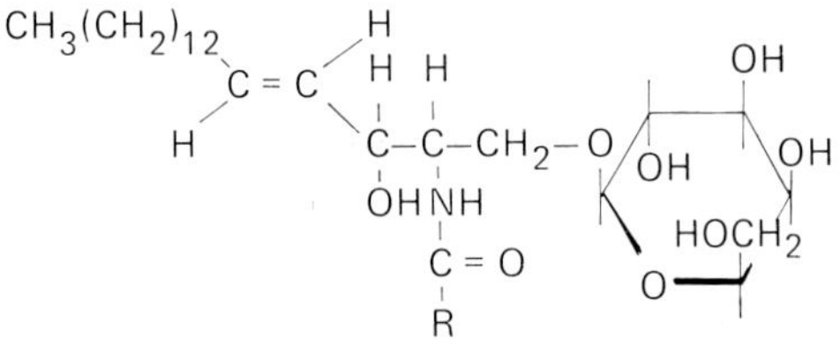

where

CRO = lignoceryl ($C_{24}H_{47}O$)	in kerasin
cerebronyl (2–hydroxylignoceryl)	in phrenosin
nervonyl ($C_{24}H_{45}O$)	in nervone
2–hydroxynervonyl	in oxynervone

Fig. 4.1. Sphingoglycolipids.

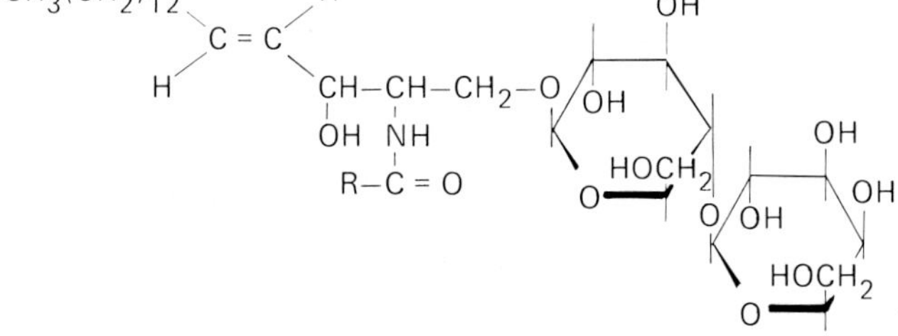

Fig. 4.2. A ceramide oligohexoside (ceramide lactoside; Gal1 → 4Glc → ceramide).

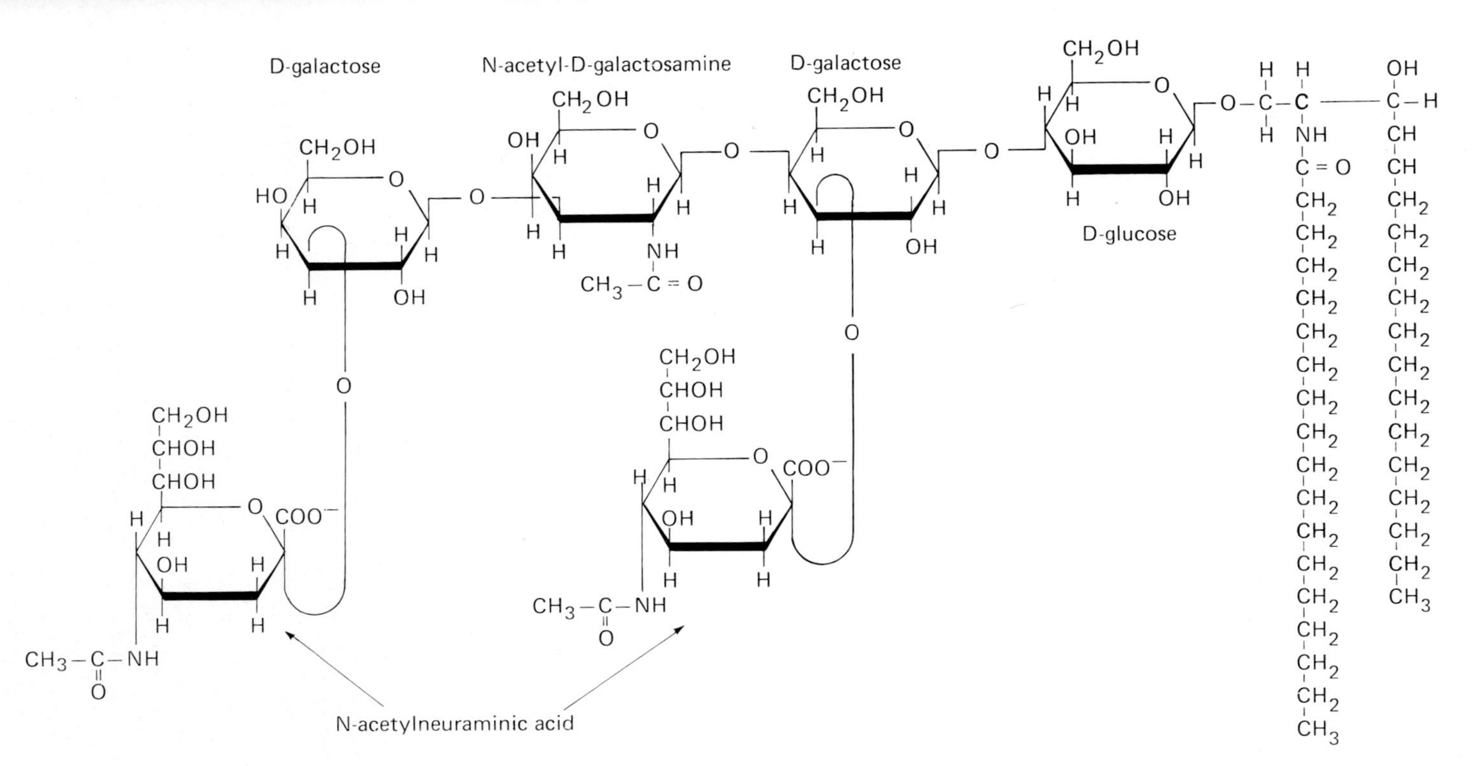

Fig. 4.3. An acid glycolipid or ganglioside [disialosyl tetrahexosyl ceramide (G_{D1a})].

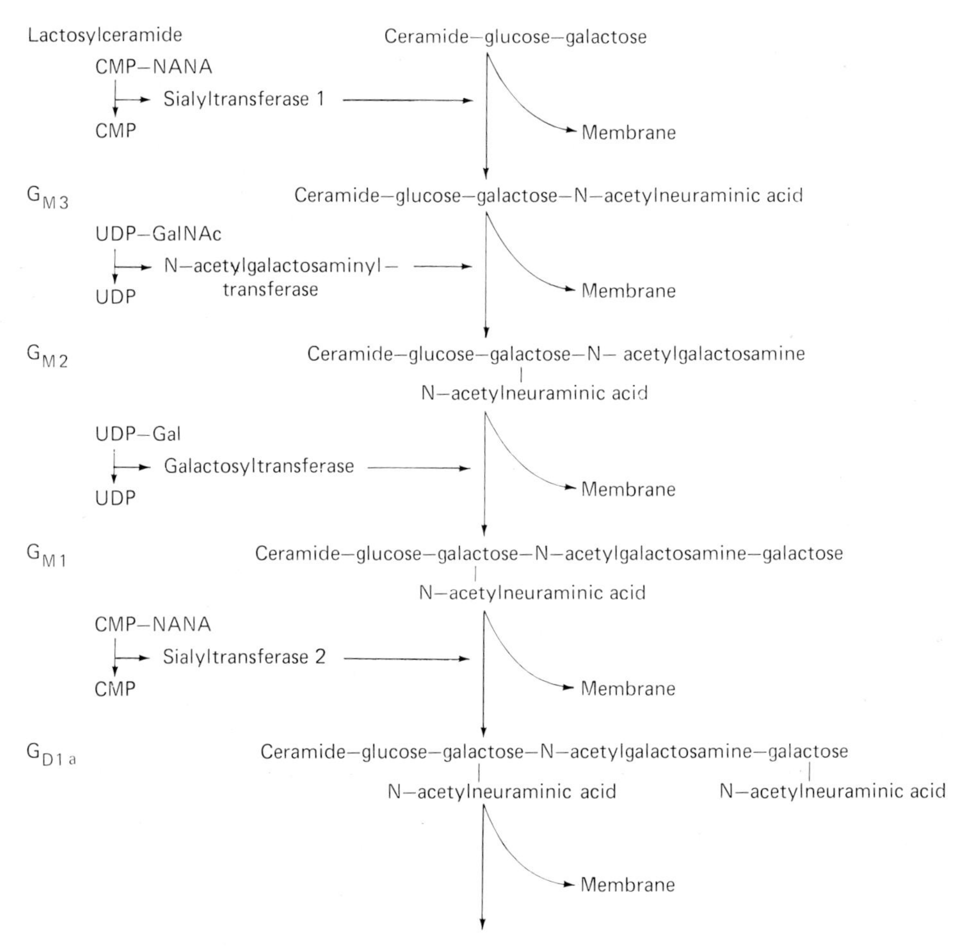

Fig. 4.4. Biosynthetic route for the principal gangliosides found in normal mouse fibroblasts.

embryonic chick brain two types of *N*-acetylneuraminyltransferase, a galactosyltransferase and an *N*-acetylgalactosaminyltransferase appear located primarily at synaptic nerve endings, but in liver these enzymes appear concentrated in the Golgi apparatus. As detailed in Chapter 8, recent studies on a variety of cells, including cultured fibroblasts suggest the presence of glycosyltransferases on external cell surfaces.

The location of glycosyltransferases in Golgi membranes presents an interesting topological problem. This is because in liver the *glycoproteins* destined for secretion are contained within the cisternal spaces of the Golgi apparatus, whereas the small substrates for the glycosyltransferases, that is, the sugar-nucleotides, are generated within the cytoplasmic space and the enzymes themselves are bound to, or firmly embedded within the membrane. Accordingly, the enzymes must utilize sugar nucleotides supplied from one side of the membrane and transfer the glycose unit to the acceptors on the other side. This would require that the enzymes either penetrate the membrane or that the sugar nucleotides somehow permeate the membrane. The topologic problem is similar for glycolipids unless one can invoke rapid, sequential 'flip-flop' for these substances during their biosynthesis. The listed considerations apply also to cell-surface-located glycosyltransferases. To glycosylate extracellular, high-molecular weight substrates the enzymes must utilize sugar-nucleotides generated intracellularly.

Experimentation documenting fluctuations of glycolipid synthesis during the mitotic cycle [2,3] provide evidence that there exist mechanisms regulating glycolipid biosynthesis. Such regulatory mechanisms may be rather intricate since all of the gangliosides can, in principle, serve not only as end products, i.e., membrane components, but also as substrates for a subsequent glycosylation (Fig. 4.4). This is obviously most true for G_{M3}. Patterns of glycolipid composition and biosynthesis might thus be regulated by (a) the number and nature of the membrane receptors for a given glycolipid and (b) the activities of the various glycosyltransferases. Moreover, regulation at the level of the glycosyltransferases could, as in the case of other enzymes, occur via allosteric mechanisms or through control of enzyme synthesis. Unfortunately, we lack significant information about these crucial topics.

The principal neutral glycolipids found in rodent cells are given in Table 4.1.

TABLE 4.1

Structure of major neutral glycolipids

Symbol	Name	Structure
Gl-1	Glucosylceramide	Glc $\rightarrow$ ceramide
Gl-2	Ceramidelactoside:	e.g., Gal1 $\xrightarrow{\beta}$ 4Glc $\rightarrow$ ceramide
Gl-3	Ceramidetrihexoside:	Gal1 $\xrightarrow{\alpha}$ 4Gal1 $\xrightarrow{\beta}$ 4Glc $\rightarrow$ ceramide
Gl-4	Ceramidetetrahexoside (globoside)	GalNAc1 $\xrightarrow{\beta}$ 3Gal1 $\xrightarrow{\alpha}$ 4Gal1 $\xrightarrow{\beta}$ 4Glc $\rightarrow$ ceramide
Gl-5	Ceramidepentasaccharide; Forssman hapten:	GalNAc1 $\xrightarrow{\alpha}$ 3GalNAc1 $\xrightarrow{\beta}$ 3Gal1 $\xrightarrow{\alpha}$ 4Gal1 $\xrightarrow{\beta}$ 4Glc $\rightarrow$ ceramide

The acid sphingoglycolipids or gangliosides bear one or more sialic acid residue on their head-groups, which are therefore negatively charged at neutral pH ($pK_a \sim 2.6$). The major gangliosides found in rodent cells are listed in Table 4.2.

The catabolic degradation of glycolipid head-groups occurs in stepwise fashion through the action of various glycosidases (e.g. neuraminidase), which are principally of lysosomal origin [4–6].

TABLE 4.2

Structures of some major gangliosides in animal cells

Symbol	Structure[1]	Name
G_{M3}	(Cer → 1Glc4 $\xrightarrow{\beta}$ 1Gal3 $\xrightarrow{\beta}$ 2NANA)	Monosialohematoside
G_{M3}	(Cer → 1Glc4 $\xrightarrow{\beta}$ 1Gal3 $\xrightarrow{\beta}$ 2N-glycolyl-NA)	
G_{M2}	(Cer → 1Glc4 $\xrightarrow{\beta}$ 1Gal4 $\xrightarrow{\beta}$ 1GalNAc) 3 ↑α 2NANA	Tay-Sachs ganglioside
G_{M1}	(Cer → 1Glc4 $\xrightarrow{\beta}$ 1Gal4 $\xrightarrow{\beta}$ 1GalNAc3 $\xrightarrow{\beta}$ 1Gal) 3 ↑α 2NANA	Receptor for cholera enterotoxin
G_{D1a}	(Cer → 1Glc4 $\xrightarrow{\beta}$ 1Gal4 $\xrightarrow{\beta}$ 1GalNAc3 $\xrightarrow{\beta}$ 1Gal3 3 ↑α ↑α 2NANA 2NANA	
G_{T1}	Cer → 1Glc4 $\xrightarrow{\beta}$ 1Gal4 $\xrightarrow{\beta}$ 1GalNAc3 $\xrightarrow{\beta}$ 1Gal3 3 ↑α ↑α 2NANA 2NANA8 $\xrightarrow{\alpha}$ 2NANA	Receptor for tetanus toxin

[1] Cer, ceramide; Glc, glucose; Gal, galactose; GalNAc, *N*-acetylgalactosamine; NANA, *N*-acetylneuraminic acid.

4.1.2. Physical properties

The glycolipids, like the phospholipids, are amphipathic substances. In aqueous solutions they tend to arrange in micellar or bilayer arrays with their carbohydrate moieties in the aqueous phase and the hydrocarbon chains of the ceramide moiety oriented into the apolar core of the micelle. Early evidence [7,8] indicates that the gangliosides (monomeric molecular weight in organic solvents of about 1500) form micelles in aqueous solutions with a particle weight of about 2×10^5 (see also [9]).

As reviewed by Shipley [10], X-ray studies on neutral sphingoglycolipids reveal a liquid-crystalline phase only at temperatures exceeding ~70°C. At a composition of 80% cerebroside, a bimolecular lipid layer is formed with a thickness of 47.5 Å and a surface area per molecule in contact with water of 55.8 Å^2.

Analyses of cerebrosides using nuclear magnetic resonance and differential scanning calorimetry reveal a gel → liquid-crystalline phase transition centered at 65°C [11]. This is abolished in the presence of cholesterol (1 : 1 molar ratio); it appears that cholesterol tends to 'fluidize' the acyl chains of cerebrosides below their transition temperature and to 'defluidize' above the transition temperature, maintaining an intermediate fluidity. It also appears that, in mixed phosphatide-cerebroside (ganglioside) systems, cholesterol tends to generate a generally liquid-crystalline phase from what would otherwise be a mixture of gel and liquid-crystal systems. Increasing proportions of egg-lecithin lower the transition temperature of ox-brain cerebroside; at a 1 : 1 molar ratio it is 20°C [12].

Bimolecular films of egg-lecithin/ox-brain cerebroside exhibit a greater electrical resistance, lower A.C. capacitance and greater breakdown voltage than lecithin films [13]. These data, together with reflectivity measurements and comparisons with sphingomyelin, indicate that the hydrocarbon chains of the cerebrosides do not interdigitate in the apolar core of the bilayer. This makes for a greater thickness of this region. The data also indicate that the polar headgroups of the glycolipids do not extend out from the membrane surface but lie parallel to it (Fig. 4.5).

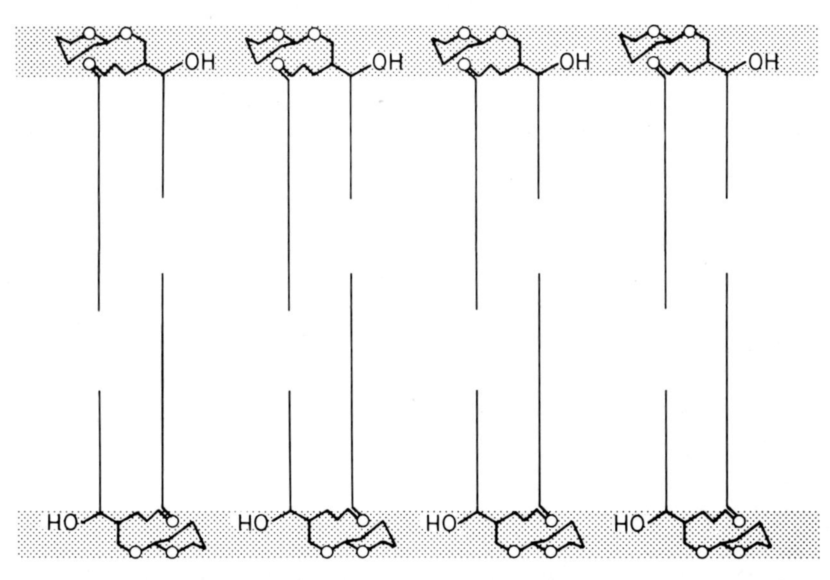

Fig. 4.5. Bilayer orientation of cerebrosides. After Clowes et al. [13].

Clowes et al. [13] have also studied bilayer films of egg-lecithin + gangliosides. These films are 79 ± 2 Å thick.

The gangliosides exhibit much greater solubilities than phospholipids. For example, whereas dipalmitoylphosphatidylcholine has a critical micelle concentration (CMC) of approximately 5×10^{-10} M at 20°C, CMCs of various gangliosides range in the order of 10^{-4} M [14]. The CMC increases with sialic acid substitution; thus, the CMC for monosialoganglioside is 8.5×10^{-5} M, that of disialoganglioside is 9.5×10^{-5} M and that of trisialoganglioside is 1×10^{-4} M.

One can therefore expect that the transfer of glycolipids from their sites of biosynthesis to their 'ultimate' membrane locations could occur by simple diffusion of the monomeric form. Teleologically speaking, no carrier system like the 'sterol-carrier proteins' or the 'phospholipid-exchange proteins should be necessary'. This does not exclude the existence of such proteins.

4.1.3. Exchange

The physical properties of the glycolipids suggest that these substances might exchange rather readily between membranes–barring, of course, specific, high-affinity interactions with other membrane components. As pointed out in Chapters 3 and 5, there is considerable evidence that membrane phospholipids, membrane cholesterol and possibly even membrane proteins can exchange both between membranes and between the plasma membrane and the extracellular milieu.

The possibility of glycolipid exchange has been much less investigated, but there is solid information that it can occur. For example, Marcus and Cass [15] have shown that glycosphingolipid Lewis antigens can be readily absorbed by intact erythrocytes to alter the Lewis specificity of these cells. The carbohydrate moiety of the Lewis antigen (Fig. 4.6) may also be attached to *glycoproteins* and, in this form, cannot be transferred to Lewis-negative red cells. Also, Laine and

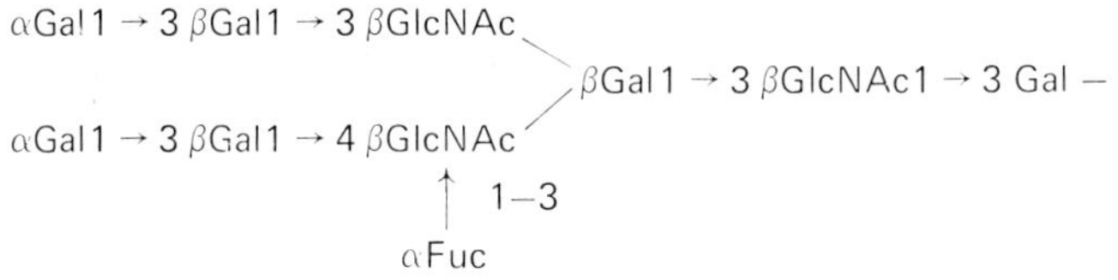

αGal1 → 3 βGal1 → 3 βGlcNAc
βGal1 → 3 βGlcNAc1 → 3 Gal –
αGal1 → 3 βGal1 → 4 βGlcNAc
↑ 1–3
αFuc

Fig. 4.6. Lewis antigen.

Hakomori [16] have very recently shown that exogenous glycosphingolipids may be taken up by the plasma membrane of cultured hamster fibroblasts (NIL), therein changing these cells growth behavior. In this study, the authors followed transfer of a ceramide tetrasaccharide with the structure GalNAcβ1 → 3Galα1 → 4Galβ1 → 4Glc → ceramide whose galactose residues had been tritiated by the galactose-oxidase-borohydride method [17]. They found that the exposure of NIL cells (and polyoma transformed NIL cells) to 8×10^{-4} molar ceramide tetrahexoside (probably above the CMC) increases cellular ceramide tetrahexoside by 60% to 100%, compared with the levels in the same cells, cultured at the same population density, in media lacking added glycolipid. There was also an associated slight increase in ceramide trihexoside, but no significant change in the content of ceramide dihexoside and an actual slight decrease in the content of Forssman glycolipid.

Absorption of the ceramide tetrasaccharide resulted in a markedly delayed growth rate due to an extended lag of the prereplicative phases of the cell cycle, particularly the G1 phase. The reduced growth rate was also associated with a reduced saturation density.

Bacterial glycolipids can also bind to animal cells and in some cases alter their growth characteristics. Thus, glycolipid mR595 ($[KDO]_3$-lipid A, where KDO = 2-keto-3-deoxyoctonic acid and lipid A = glucosaminyl-β-1,6-glucosamine with fatty acid substituted at hydroxyl and amino groups) is taken up by normal and SV40 transformed rat fibroblasts [18]. mR595 inhibits the growth of transformed cells and stops it at confluency. A preliminary report [19] suggests changes in membrane transport and architecture but no alterations of membrane gangliosides or glycoprotein.

Gahmberg and Hakomori [17] have attempted to localize glycolipid adsorbed to NIL cells by fractionating the cells and isolating their plasma membranes. They could localize more than 80% of the labelled glycolipid taken up in a plasma membrane isolate. Unfortunately, this fraction is not well characterized. It apparently contains most of the plasma membranes of the cells, but there is no assurance that the fraction is not also contaminated by other subcellular fragments. The data can thus only be taken to show that glycolipids can be transferred from the extracellular milieu into the cell.

Transfer of gangliosides has been demonstrated by Cuatrecasas [20]. His studies exploit the high affinity of G_{M1} ganglioside for cholera enterotoxin (see below) and demonstrate that a substantial but finite number of ganglioside molecules can be rapidly transferred from an incubation medium to the plasma membranes of hepatocytes or adipocytes even at 4 °C. This ganglioside incorporation leads to increased binding of cholera toxin.

Sphingoglycolipids and sphingomyelin possess the same apolar moieties and physical studies show that glycosphingolipids interact with the apolar regions of other lipids in much the same way as does sphingomyelin. Accordingly, membranes with a 'preference' for sphingomyelin, for example, plasma membranes, might show some affinity also for sphingoglycolipids. However, while sphingomyelin exchanges poorly, due to its very low critical micelle concentration, the sphingoglycolipids appear to exchange very readily.

But there are more critical questions to be answered, namely: (a) are the glycolipid compositions of various membranes and diverse tissues determined solely by 'extramembranous' factors, for example, biosynthesis, catabolism, solubility, or (b) do other membrane components, proteins in particular influence membrane glycolipid content and turnover. Unfortunately we lack sufficient information to address these questions adequately. However, if diverse membranes (or tissues) exhibit binding selectivities for certain glycolipids, one must suspect these to involve selective interactions between some proteins in the membranes and the specific sugar headgroup of the preferred glycolipid.

The evidence that glycolipids can transfer from extracellular fluids to cells raises another critical question: How representative is the *in vitro* glycolipid composition of cells of the *in vivo* state. If glycolipids can exchange one would observe cellular glycolipid specificities only if diverse cells differ in their glycolipid binding affinities. As already mentioned, this would probably require differences in membrane protein composition or organization.

4.1.4. Subcellular localization of gangliosides and glycolipids

It is often stated that glycolipids and gangliosides are more concentrated in the plasma membranes of various cells than in intercellular membranes, but the data available is somewhat ambiguous. For example, Dodd and Gray [21] find that approximately 7% of the total lipids in rat liver bile-front plasma membrane isolates is made up by neutral and acidic sphingoglycolipids, whereas no such lipids were detected in the mitochondria (which are also depleted in sphingomyelin). However, this analysis does not tell how much of the total cellular glycolipid is in the plasma membranes. Another study on liver plasma membrane isolates [22] reports only 3-fold enrichment of gangliosides over whole cells, although plasma membrane-specific markers were enriched 50-fold.

In another study, Renkonen [23] compared the ganglioside content of plasma membrane fragments isolated from BHK fibroblasts with that of endoplasmic reticulum. They found that the plasma membrane fragments contained between 8.1 and 13 μg of ganglioside *N*-acetylneuraminic acid per milligram of protein, compared with 1.2 to 2.6 μg/mg of protein in endoplasmic reticulum. However, they point out that plasma membrane ganglioside would account for only perhaps 5% of the total ganglioside of the cells.

Evidence reviewed by Klenk [24] on the glycolipid composition of enveloped viruses indicates that these substances constitute a very significant proportion of the total polar lipids in the viral envelopes. Moreover, the viral glycolipids are clearly host specific and exhibit specific antigenicities also found in the host plasma membrane. Analyses have been carried out on arbor viruses, which bud from the plasma membrane of the host cells and also from cytoplasmic vesicles, and on rhabdo-viruses which also bud from the plasma membrane. All of these virus membranes are characterized by rather high proportions of sphingomyelin, as well as cholesterol/phospholipid ratios close to one. These are features which are generally characteristic of plasma membranes and one can argue that the high glycolipid content of virus envelopes reflects a higher glycolipid concentration of a plasma membrane. Indeed, a study by Renkonen and associates [25] shows significant differences between the lipid patterns of Semliki Forest virus envelopes and those of the whole cells or endoplasmic reticulum; the lipids of the envelopes closely resemble those of the purified plasma membranes of the host cells.

An enrichment of ganglioside content has also been reported for a plasma membrane fraction isolated from 3T3 cells [26]. The plasma membrane fraction contained ~3 μg of lipid bound *N*-acetylneuraminic acid per mg protein; the value of whole cells was 1.3 μg/mg. The specific ganglioside pattern in the plasma membrane fraction appeared similar to that of whole cells. This may not be a meaningful finding, however, since the plasma membrane isolate was only 3× purified over whole cells in terms of specific markers.

Weinstein et al. [27] have carried out an analyses of the glycolipids in mouse L-cells. Four glycolipids account for ~0.7 % of the total cellular lipid. Surface membranes isolated by the fluorescein mercuric acetate method also contain about 0.7 % of the total lipid in the form of glycolipids. However, the total amount of glycolipid in the plasma membrane fraction is very small. Thus, the plasma membrane isolates account for 0.7 % of the total membrane protein and only approximately 3 % of total cellular glycolipid and of total cellular gangliosides. In these cells also, most of the cellular glycolipids, both neutral and acidic must therefore be located in membranes other than the plasma membrane. Cuatrecasas' studies on the binding of cholera toxin to cells also indicate that intracellular membranes contain G_{M1} ganglioside [18].

Even though most of the cellular glycolipid may be located intracellularly, that exposed at the cell surface can be labelled by treating whole cells first with galactose oxidase and then reducing with borohydride [17]. By this procedure accessible galactose residues become labelled.

Gahmberg and Hakomori [17,29] have used this approach to show that ceramide penta, tetra- and tri-saccharides of NIL cells label differently during various stages of the mitotic cycle. All of these glycolipids label maximally during the G1-phase and minimally during the S-phase, although the total glycolipid complement of the cells remain constant throughout the mitotic cycle. The implication of the study is significant, since the galactose-oxidase procedure labels only carbohydrates exposed at the external surface of the plasma membrane under the condition employed. What is involved apparently is an altered exposure of the galactose entities of the ceramides at different phases of the mitotic cycle.

4.2. *The role of gangliosides in the binding of cholera toxin and tetanus toxin to biomembranes*

Certain gangliosides can act as high affinity binding sites for tetanus toxin and cholera enterotoxin. The tetanus toxin consists of two dimers of molecular weight 73,000 linked noncovalently. Each dimer consists of a light chain of molecular weight 21,000 and a heavy chain of molecular weight 53,000 linked together by disulphide linkages. The cholera toxin is a protein with a molecular weight of 84,000. It is probably composed of six subunits and has 5 to 6 intrachain disulphide bridges.

The binding specificities of various gangliosides for these toxins are shown in Table 4.3. The ganglioside that reacts with cholera toxin appears to be G_{M1}.

TABLE 4.3

Reactivities of various gangliosides for cholera and tetanus toxins[1]

	Reactivity	
Ganglioside Structure	Cholera Toxin	Tetanus Toxin
Cer–Glc–Gal–GalNAc–Gal | NANA	+	–
Cer–Glc–Gal–GalNAc–Gal | ⋮ NANA NANA	–	–
Cer–Glc–Gal–GalNAc–Gal | NANA ⋮ NANA	–	+
Cer–Glc–Gal–GalNAc–Gal | ⋮ NANA NANA ⋮ NANA	–	+

[1] From van Heyningen [30].
Cer = ceramide; Glc = glucose; Gal — galactose; GalNAc = *N*-acetylgalactosamine, NANA = *N*-acetylneuraminic acid; --- = neuraminidase-susceptible linkage.

The terminal Gal residue is essential for binding of both toxins. Substitution of the NANA on the proximal Gal-residue with another molecule of NANA linked by a neuraminidase-sensitive bond, interferes with the capacity of the ganglioside to bind cholera toxin, but makes it a highly efficient ligand for tetanus toxin. Trisialoganglioside is also an efficient ligand for tetanus toxin but does not bind cholera toxin. The sialosyl residue bound to the terminal Gal is not required for tetanus toxin binding but interferes with the binding of cholera toxin. The binding ratios at saturation are 1 :2 (toxin: ganglioside) [30].

Despite the importance of these multiple sialic residues for tetanus binding, binding is apparently not a simple electrostatic process. This has been shown by Clowes et al. [13] in the study of the interaction of egg lecithin-ganglioside bilayers with tetanus toxin. Without the toxin, the bilayers have a thickness of approximately 79 Å. Addition of tetanus toxin leads to an increase in the thickness of the films to 119 Å, together with a decrease of electrical resistance.

Nevertheless, there appears to be no modification of the bilayer structure and no penetration of the toxin into the bilayer region. A study of the effects of ionic strength and pH on the interaction show that the process could not occur exclusively through electrostatic interactions.

Very extensive studies on the interaction of cholera enterotoxin with cellular membranes and also with isolated gangliosides, have been published by Cutracasas [20,28,31,32]. These studies have involved the use of biologically-active, ^{125}I-labelled cholera toxin. The toxin binds very tightly to the plasma membranes of adipocytes and hepatocytes. The dissociation constants for the formation of the initial toxin–membrane complexes lie in the vicinity of 5×10^{-10} to 1×10^{-9} M, respectively. The binding can be attributed to the membrane gangliosides and, consistent with observations of van Heyningen [30], the monosialoganglioside, G_{M1} appears to be the principal receptor.

It would appear that cholera toxin and tetanus toxin can serve to probe the states of their ganglioside receptors in normal and neoplastic cells. One approach of obvious appeal would be the use of ferritin conjugated toxin for electron-microscopic exploration of ganglioside distribution on cellular membranes.

4.3. Alterations of ganglioside content and metabolism in neoplastically transformed cells

4.3.1. Mouse cell lines

As shown in Table 4.4, normal cultured mouse fibroblasts contain a substantial proportion of gangliosides 'higher' than G_{M3}. However, one can observe substantial variations between strains and between different clones of the same strain [38]. When various strains of mouse fibroblasts are transformed with SV40 or polyoma virus (Table 4.4) one often observes a striking depletion but not elimination of gangliosides larger than G_{M3} [39–41]. However, Yogeeswaran et al. [26] report that, whereas one polyoma-transformed clone exhibited the described drastic 'simplification' of ganglioside pattern, two other SV40-transformed clones had similar ganglioside proportions as normal 3T3 fibroblasts.

A simplified ganglioside pattern was also observed when Swiss 3T3 cells are transformed with the Moloney isolate of the murine sarcoma virus (MSV) together with murine leukemia virus (MLV) as 'helper'. In contrast, cells treated with the non-transforming virus, namely (MLV), did not show an altered ganglioside composition [35].

An analogous but slightly different pattern is observed when Balb/c.3T3 cells are transformed with the Kirsten strain of murine sarcoma virus (Table 4.4), [36]. The major difference in this case is that the transformed cells contain very

TABLE 4.4

Ganglioside content and glycosyltransferase activity of normal mouse fibroblasts and transformants thereof induced by various DNA and RNA viruses[1]

Cell type[2]	Ganglioside content (% of total)				Glycosyltransferase activity (% of control)[3]			
	G_{M3}	G_{M2}	G_{M1}	G_{D1a}	ST_1	NAcGalT	GalT	ST_2
AL/N – normal	13	17	32	38	100	100	100	100
– SV40-transformed	79	5	9	7	56	15	654	184
– Py-transformed	78	9	6	4	142	13	168	192
Swiss 3T3 – normal	37	16	24	22	100	100	100	100
– SV40-transformed	91	1.4	5.7	3	104	12	6	36
– Py-transformed	95	1.4	2.7	1	110	10	11	82
Swiss 3T3 – Mock infected	63	3	3	30	100	100	100	100
– MSV infected	91	4	5	5	117	20	276	85
Balb/c 3T3 – normal	17	46	2	35	ND	100	100	ND
– Ki MSV	13	85	<1.5	1	ND	102	<1	ND

[1] From Brady and associates [34–38].
[2] SV40 = Simian Virus; Py — polyoma; MSV = murine sarcoma virus; KiMSV — Kirsten strain of murine sarcoma virus.
[3] ST_1 = sialyltransferase 1, NAcGalT = *N*-acetylgalactosaminyltransferase; GalT – galactosyltransferase; ST_2 = sialyltransferase 2; ND – not determined.

appreciable amounts of G_{M2}, whereas this ganglioside is markedly depleted in the strains transformed by DNA viruses and by the Moloney isolate of MSV.

To elucidate the altered ganglioside patterns following transformation, Brady and associates [35–38] have compared the glycosyltransferase activities of the various normal cell lines and their viral transformants, using glycolipids as acceptors. They assayed the four enzyme systems shown in Fig. 4.4. For measurement of the sialyltransferase 1, the glycolipid acceptor is ceramide–glucose–galactose, and the sugar donor is CMP-NANA. For measurement of sialyltransferase 2, the glycolipid acceptor is G_{M1}, and the sugar donor is again, CMP-NANA. For the measurement of *N*-acetylgalactosaminyltransferase, the acceptor is G_{M3}, and the sugar donor is GDP-GalNAC. Galactosyltransferase was measured using G_{M2} as acceptor and UDP-Gal as donor. In all cases, the assays are carried out on cell homogenates in the presence of small amounts of non-ionic detergent.

All of the transformed cells showed some alteration in the activities of glycosyltransferases participating in ganglioside biosynthesis, but the patterns are not very consistent (Table 4.4). In the case of transformation by SV40 and polyoma, as well as in the case of Moloney sarcoma virus, a dramatic decrease in the activity of *N*-acetylgalactosaminyltransferase was observed (Table 4.5). According to the scheme in Fig. 4.4, a decrease of *N*-acetylgalactosaminyltransferase activity

TABLE 4.5

UDP-*N*-Acetylgalactosaminyltransferase activity in normal and transformed mouse fibroblasts[1]

Cell line	UDP-GalNAc Transferase activity (% of control)
Swiss 3T3	
Normal	100
SV40-transformed	12
Polyoma-transformed	16
MSV-transformed	20
MLV-infected	80
N AL/N	
Normal	100
SV40-transformed	15
Polyoma-transformed	13
Balb/c 3T3	
Normal	100
Ki MSV-transformed	103

[1] From Cumar et al. [37] and Fishman et al. [36].

would lead to the observed depletion of G_{M2}, G_{M1} and G_{D1a}, provided the enzyme activity is actually rate limiting in the sequence.

However, in the case of Moloney leukemia virus-infected 3T3 cells and in Balb/c.3T3 cells transformed with the Kirsten strain of MSV virus, one finds different patterns. Indeed, with the Kirsten virus one observes a drastic reduction in the activity of G_{M2}: UDP-galactosyltransferase activity; this is in accordance with the altered ganglioside composition of these RNA virus-transformed cells.

As far as other glycosyl transferases are concerned, sialyltransferase 1 is unchanged in transformed Swiss 3T3 cells, slightly reduced in an SV40-transformed AL/N line and actually elevated in the polyoma transformed AL/N line. Sialyltransferase 2 is unaffected in the murine sarcoma virus-infected cells and polyoma-transformed Swiss 3T3 cells, elevated in the transformed AL/N lines and reduced in the SV40-transformed 3T3 lines. Finally, galactosyltransferase is elevated in all transformed Swiss lines and decreased in the transformed AL/N line.

Importantly, when Swiss 3T3 cells are exposed to a very high multiplicity of polyoma virus, yielding lytically infected cells, no change in the activity G_{M3}; UDP-*N*-acetylgalactosaminyltransferase is observed [41]. Also, revertants of SV40-transformed cells possess essentially the same ganglioside complement as the normal parental 3T3 cells and actually have a somewhat higher specific activity of G_{M3}: UDP-*N*-acetylgalactosaminyltransferase [41]. These revertants exhibit similar culture characteristics as non-transformed cells, are poorly tumorigenic in animals, but still contain the full SV40 genome. These data indicate that the simple expression of the viral genome is not adequate to produce the altered patterns of ganglioside composition and synthesis observed in mouse fibroblasts. Rather, the glycolipid changes appear to be a consequence of tumorigenic transformation. The situation thus seems analogous to the changes in sugar and amino acid transfer induced by oncogenic viruses (Chapter 7); these do not appear to be due to expression of the viral genome but to an altered expression of the host genome.

Brady et al. [34,42,43] have observed a drastic decrease in G_{M3}: UDP-*N*-acetylgalactosaminyltransferase activity in 24 of 26 different DNA virus-transformed mouse-cell lines. Moreover, the ganglioside content in these 24 lines fits the impairment of biosynthetic activity. They have also shown that the defect was not dependent on growth conditions, such as medium and pH, did not generally vary with the number of passages, and did not depend on cell density in culture. Thus, none of the four glycosyltransferases investigated varied in specific activity within Swiss 3T3 cells over a 10-fold range of cell density [38]. Similarly, the transfer of UDP-*N*-acetylgalactosamine to G_{M3} occurred at a similar specific rate in sparse and confluent cultures of Balb/c 3T3 cells. Finally, Mora and coworkers [35,41] have shown convincingly that cells transformed by oncogenic viruses do not generate or accumulate a diffusible inhibitor of amino sugar transfer.

Recent experimentation [43] shows that ganglioside anomalies such as documented for virally-transformed fibroblasts can also accompany non-viral neoplastic conversion. Thus, chemically- or X-ray-transformed Balb/c 3T3 mouse fibroblasts exhibit marked reductions of G_{M1} and G_{D1a} and an increase above normal of G_{M2}. This effect can be attributed to a marked decrease ($>85\ \%$) in the activity of UDP : G_{M2} galactosyltransferase, without alterations in other transferases.

4.3.2. Hamster cell lines

One also observes altered ganglioside composition and metabolism in other cell types consequent to transformation, but the information available is often conflicting; for example, Hakomori and Murakami [44] report that baby hamster kidney cells (BHK) which contain only one principal ganglioside, namely G_{M3}, become depleted in this ganglioside upon spontaneous neoplastic transformation or transformation by polyoma virus. A similar alteration has been reported for BHK cells transformed by two strains of Rous sarcoma virus [45]. This alteration was attributed to a decrease in the activity of lactosylceramide: CMP-*N*-acetylneuraminic acid transferase 1. However, Brady and Fishman [33] cite experiments on normal and polyoma virus transformed BHK cells which showed no difference in G_{M3} content.

As demonstrated by Nigam et al. [46] the situation can get rather complicated as far as alterations of a specific biosynthetic path are concerned. These authors studied an SV40 transformant of BHK cells, a revertant thereof, showing normal growth characteristics in culture and a back revertant showing abnormal growth characteristics in culture (Table 4.6). In contrast to the parental BHK cell which contains primarily G_{M3}, the revertant showed approximately equal proportions of G_{M3} and G_{M2}.

TABLE 4.6

Ganglioside content and *N*-acetylgalactosaminyltransferase activity in hamster fibroblasts with normal and neoplastic culture characteristics[1]

Cell Type[2]	Ganglioside Content (% of G_{M3} + G_{M2})		*N*-acetylgalactosaminyl-transferase activity (%)
	G_{M3}	G_{M2}	
Transformant	91	9	16–31
Revertant	54	46	100
Back-revertant	67	33	16–49

[1] From Nigam et al. [46].
[2] The transformant is a clone of SV40 converted BHK cells. The revertant is a clone of the transformed cells which has reverted back to normal growth. The back revertant is a clone of the revertant which has returned to abnormal growth.

Transformation of BHK cells with hamster sarcoma virus [47] or an Adeno-7-SV40 hybrid did not affect the levels of this ganglioside. Moreover, Sakiyama et al. [48] found a ten-fold variation in the amount of palmitate labelled G_{M3} in 6 NIL clones obtained from a single cell source. Two of these clones showed a density dependent increase in the incorporation of G_{M3}, two were density independent, and two showed a decrease in labelling incorporation with increasing cell density. When the clones exhibiting the strongest cell-density-dependent increase of palmitate incorporation in G_{M3} were transformed with polyoma or hamster-sarcoma virus, the clones lost this density-dependence. One should note, however, that the incorporation of palmitate into G_{M3} does not give a reliable estimate of the content of this ganglioside [48,49].

4.3.3. Rat kidney fibroblasts

Brady and Fishman [33] report a change in ganglioside metabolism when newborn rat kidney (NRK) cells are transformed by a temperature-sensitive mutant of Kirsten sarcoma virus, but only at the temperature (31 °C) permissive for transformation. As in the case of BHK cells, the major ganglioside of NRK cells is G_{M3}. Infected cells have more G_{M3} than control cells at the non-permissive temperature, whereas transformed cells (i.e., at the permissive temperature) contained substantially less G_{M3} than the controls. The changes in G_{M3} fit the levels of lactosylceramide: CMP-n-acetylneuraminic acid transferase, which are enhanced at 39 °C but depressed at 31 °C (transformation). The pattern seen in the transformed cells thus corresponds to the original observations by Hakomori and Murakami [44] on BHK fibroblasts.

4.3.4. Human fibroblasts

Hakomori [50] has reported a decrease of 'higher' gangliosides and G_{M3} when human diploid fibroblasts (8166) are transformed by SV40 virus. Brady and Fishman [33] report a decrease of 'higher' gangliosides also when human lung fibroblasts (WI-38) are transformed with SV40, but they do not observe a loss of G_{M3}. Their preliminary experiments suggest a deficiency in galactosyltransferase.

4.3.5. Rat liver and rat hepatomas

Information on the ganglioside content of rat liver and rat hepatomas is sparse and contradictory. Thus, Brady et al. [51] report that 63% of adult rat liver ganglioside is G_{M3}, 11% G_{M1}, and 25% G_{D1A}, whereas Siddiqui and Hakomori [52] give values of 36%, 36% and 23% respectively. The data on hepatoma 5123 are even more disparate: Brady et al. [51] gives values of 90%, 9%, and 1% for G_{M3}, G_{M1}, and G_{D1a} respectively, while Siddiqui and Hakomori [52] report values of 10%, 38%, and 52% respectively. The last results appear consistent with the qualitative results of Cheema et al. [53] suggesting that hepatomas 5123 and 7777 contain G_{M3}, G_{M1} and G_{D1a}, with the latter predominating. Similar patterns are also reported for hepatomas 5123 and 7800 [52]; in all instances tested, there appears to be a depletion of trisialoganglioside.

A recent preliminary report by Dnistrian et al. [54] demonstrates substantial differences in ganglioside content between 'plasma membranes' isolated from normal rat liver and similar fractions purified from hepatoma 5123. The hepatoma membranes are typified by an 8-fold increase in total ganglioside content over the levels in liver membranes. This is due to a 4-fold increase in G_{M3}, and 8-fold increase in G_{M2}, and a 22-fold increase in disialoganglioside. No trisialoganglioside was detected, in contrast to membranes from normal livers. The results are in accord with the whole-cell studies of Siddiqui and Hakomori [52].

The data available allow no conclusions as to the specific nature of the ganglioside anomalies in rat hepatomas. This is regrettable, since the rat-liver hepatoma system is a most important *in vivo* tumor model. The data on normal rat liver all showing high levels of G_{D1a} and some also suggesting a high G_{M1}

content, also, do not fit the analyses of NRK cells [33] which appear to contain predominantly G_{M3}. One wonders whether the difference has to do with cell type or the disparity between the *in vivo* and *in vitro* situations.

4.3.6. Lymphoid cells

We know of only one study on the ganglioside metabolism and composition of lymphoid cells [55]. This compared the acid sphingoglycolipids of chicken lymphoid leucosis cells with their parental lymphocytes (bursal cells). The leukemic cells lacked the trisialoganglioside present in normal cells as well as G_{M1} and G_{M2}, but showed an accumulation of G_{M3}. The accumulation of G_{M3} and the depletion of trisialoganglioside could be ascribed to blocked conversion of disialoganglioside to trisialoganglioside. Indeed the leukemic cells show a marked depletion of sialyltransferase activity. However, the low levels of G_{M2} are not easily explained since the cells show high levels of NAcGal transferase. The low levels of G_{M1} can be correlated with reduced UDP-Gal : G_{M2} galactosyltransferase activity. Certainly no single enzyme defect can explain the anomalous ganglioside patterns of the leukemic cells. In view of the extensive information on the biology and biochemistry of lymphocytes further studies on the glycolipids of these cells appear desirable and would indubitably reveal significant information.

4.3.7. Binding of cholera toxin and its effect on cell growth

Cuatrecasas and associates [56,57] have demonstrated that the binding of ^{125}I-labelled cholera toxin to several neoplastically-transformed mouse cell lines follows the ganglioside composition of the cells. As noted above, G_{M1} is the principal natural membrane receptor for cholera toxin. TAL/N cells of early passage, which contain G_{M1} as well as G_{D1a}, G_{M3} and G_{M2}, bind cholera toxin avidly and are most sensitive to its stimulatory action on adenylate cyclase. SV40-transformed AL/N cells, which lack G_{M1}, G_{D1a} and G_{M2}, bind little cholera toxin and are only slightly susceptible to its metabolic effects. Late passage TAL/N cells contain either no G_{M1} and G_{D1a}, or too little to be detectable and are intermediate in their affinity for and susceptibility to cholera toxin.

The described use of cholera toxin constitutes an important new approach. Particularly important in the present context is the fact that the toxin interacts specifically with *plasma membrane gangliosides*. In addition the toxin binding assays allow detection of very minute amounts of G_{M1}, i.e. $\sim 10^3$ molecules/cell.

4.4. Neutral glycolipids

The major simple, neutral glycolipids are listed in Table 4.1, and the structures of some important blood group substances and blood group glycolipids are depicted in Figs. 4.7 and 4.8. As in the gangliosides, the carbohydrate moieties of the neutral glycolipids are built up by the sequential addition of single sugars to the existing carbohydrate through the action of specific glycosyltransferases.

```
B      αGal1 → 3βGal1 → 3βGlcNAc1 → 3βGal1 → 4Glc → Cer
Le^b     βGal1 → 3βGlcNAc1 → 3Gal1 → 4Glc → Cer
           2            4
           ↑            ↑
           1            1
        αL – Fuc     αL – Fuc
'X – hapten'   βGal1 → 4βGlcNAc1 → 3βGal1 → Glc → Cer
                   2
                   ↑
                   1
               α – L – Fuc
Le^a  βGal1 → βGlcNAc1 → 3Gal1 → 4Glc → Cer
                  4
                  ↑
                  1
              α – L – Fuc
```

Fig. 4.7. Blood group glycolipids.

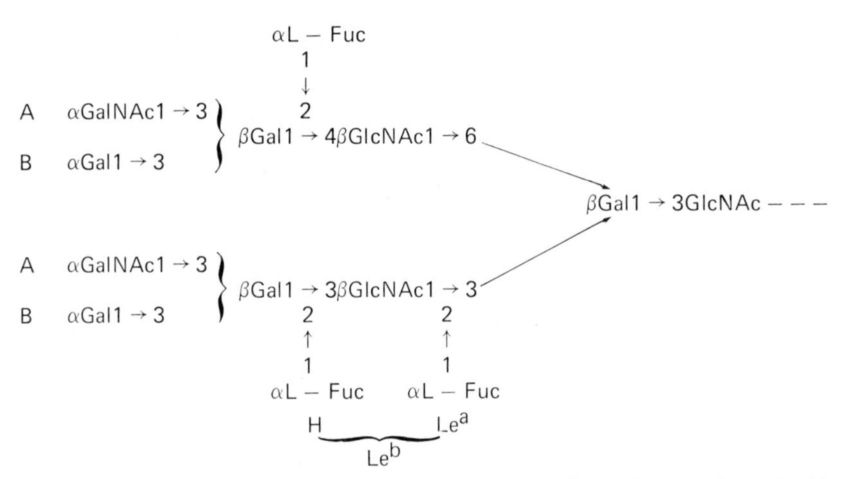

Fig. 4.8. Possible composite structure of A-, B-, H-, Le^a- and Le^b- specific megalosaccharides. The structure shown is a composite and in individual blood groups certain sugar residues will be lacking, that is, H determinants lack terminal αGalNAc or Gal residues, and Le^a determinants do not have the L-Fuc linked 1 → 2 to subterminal Gal. Also the determining sugars may sit on incomplete and/or unbranched chains. The megalosaccharides can be linked to either ceramide or protein.

4.4.1. Variations in glycolipid synthesis with cell density in culture

There are considerable differences in glycolipid composition between diverse lines of cultured cells; for example, some lines contain predominantly neutral glycolipids such as G1-3, G1-4, and G1-5, whereas others contain gangliosides primarily. In addition, there is substantial variation in the glycolipid patterns found for different clones of the same strain of 3T3 or BHK cells [50]. These facts must be born in mind when evaluating the glycolipid contents of transformed clones. Also, many cultured cell lines exhibit density-dependent patterns of glycolipid composition and synthesis.

Chemical analysis showed that the G1-3 content of cultured hamster BHK and NIL fibroblast increases sharply as the cultures become confluent. Various BHK clones differ in their density dependence of growth, and glycolipid analysis on such clones indicates that the less 'contact-inhibited' clones contain relatively low amounts of G1-3 and show no important variation in the level of this glycolipid upon attaining confluence [50]. In the case of BHK cells, the alteration of glycolipid composition occurring near confluence can be related to the enhanced incorporation of [^{14}C]glucosamine into G1-3. This is accompanied by a clear increase in the activity of UDP-galactose/lactosylceramide-α-galactosyltransferase (the synthetase of G1-3). Indeed, this enzyme activity is several times higher in confluent cells compared with the same cells at very low population densities [59]. In contrast, the β-galactosyltransferase (UDP-galactose/glucosylceramide-β-galactosyltransferase) does not change in activity with cell density.

In the case of NIL cells, there is a dramatic increase in the production of G1-3, G1-4 and G1-5 when the cells become confluent [48,59]. However, this effect again becomes less striking in NIL clones that show a lesser growth dependence on cell density. Kijimoto and Hakomori [60] have measured the rate of [^{14}C]galactose assimilation into the neutral glycolipids of NIL cells and observed that the incorporation of the label increases most dramatically during the early stage of confluence rather than later on.

These data have led Hakomori [61] to suggest that cell contact leads to an extension of the carbohydrate chains of membrane glycolipids ('contact extension') and accumulation of glycolipids containing longer carbohydrate chains ('contact accumulation'). He further suggests that these phenomena induce 'contact inhibition' and that defective 'contact extension' and 'contact accumulation' account for the impaired 'contact inhibition' of neoplastic cells in culture. Hakomori's basic argument is that glycosyl extension somehow acts as a signal for a cell to become contact inhibited; in transformed cells which lack such glycolipids, the extension response is not forthcoming and cells are not contact inhibited. However, the picture appears more complicated. Thus, when dense cultures of NIL-2 cells are trypsinized and replated at low cell densities, synthesis of G1-3 continues at the rate characteristic of the dense cultures for up to 10 hr after subculturing [62]. In addition, when the growth of NIL-2 cells in low density cultures is restricted by use of media with low concentrations of glutamine (or serum), one observes a significant increase in the synthesis of neutral glycolipids compared with that found in sparse, rapidly-growing cultures [62]. Finally, the addition of cyclic AMP to cultured fibroblasts does not alter glycolipid biosynthesis (in either normal or transformed cells) even though growth rates, density dependence and culture morphology change [63]. These data suggest that cultured cells can increase glycolipid synthesis in the absence of cell–cell contact and at low culture densities.

4.4.2. *Altered composition and metabolism in neoplastic cells*

4.4.2.1. *Human tumors*

Evidence indicating that tumors might be abnormal in their content and/or metabolism of glycolipids dates back to a time when the actual nature of the glycolipids was not fully understood. For example, in 1949, Ohuti [64] reported a depletion of Blood Group A and Blood Group B activities in blood group extracts from human gastric cancer (see also [65]). The specificities of the blood group substances reside in their carbohydrate moieties which may be attached to either protein or ceramides (Figs. 4.7 and 4.8).

A number of studies carried out since then have documented frequent anomalies in the content of blood group haptens in human cancers. This topic has been reviewed by Hakomori [61].

The systematic characterization of glycolipids in human tumor tissues came somewhat more recently. Thus, Hakomori and Jeanloz [66] demonstrated an accumulation of fucose-containing glycolipids in human adenocarcinomas. Subsequently, Hakomori et al. [67] characterized these glycolipids structurally and showed them to consist of Le^a glycolipids and a positional isomer thereof, termed the 'X-hapten' (Fig. 4.7) [68]. A small proportion of Le^b glycolipid was also isolated [69]. The abnormal fucose glycolipids could be found also in normal tissues, but only in trace quantities. Very large amounts of Le^a-active glycolipids were found also in tumors of Le^b individuals [69]. Hakomori and associates early postulated that the blood group anomalies observed might be due to deficiencies in certain galactosyltransferases and their recent data suggest that this is actually the case in some instances of human gastrointestinal carcinoma [70]. Tumors deficient in Blood Group A (determinant = αGalNAc) were found deficient in NAcGal-transferase and tumors depleted in Blood Group B (determinant = αGal) were deficient in galactosyltransferase activity.

4.4.2.2. *Glycolipid alterations of cells transformed* in vitro *hamster cells*

In their initial studies on the transformation of BHK fibroblasts by polyoma virus, Hakomori and Murakami [44] reported an increase in the amount of G1-2 in the neoplastically transformed cells. A similar increase in G1-2 was reported subsequently in BHK cells transformed by two strains of Rous sarcoma virus [45]. However, in both of these studies G1-2 was the only neutral glycolipid detected in either normal or transformed BHK lines. More recently, Hakomori [50] reported an appreciable amount of G1-3 in untransformed BHK cells, but not in cells transformed spontaneously or by polyoma virus. As noted before, the G1-3 levels of the untransformed cells increase as the cultures become confluent. The data of the Hakomori group are at variance with work cited by Brady and Fishman in their review [33]. These authors describe experiments carried out by them on normal BHK cells and a series of polyoma transformed clones thereof. They found identical neutral-glycolipid patterns in all of the transformants. Moreover, two clones transformed by temperature-sensitive polyoma mutants showed no significant change of neutral glycolipid composition when switched from the permissive- to the non-permissive temperatures of the mutant viruses.

The composition and metabolism of neutral glycolipids has also been studied in another hamster cell line, NIL cells, both normal and transformed by various oncogenic viruses [47,48,62]. When glycolipid composition and metabolism were assayed by measuring the incorporation of [1-^{14}C]palmitate from the tissue culture medium, it was observed that the 'higher' neutral glycolipids, namely G1-3, G1-4 and G1-5, became more highly labelled in high density cultures of untransformed NIL cells than in low density cultures. Also, these glycolipids were markedly depleted in the virally transformed cells (Table 4.7).

TABLE 4.7

Incorporation of [1 − ^{14}C]-labelled into neutral glycolipids of normal and neoplastic NIL cells in sparse and dense culture[1]

Cell Type	Incorporation of [1 − ^{14}C]palmitate into[2]				
	Gl-1	Gl-2	Gl-3	Gl-4	Gl-5
NIL-2					
Sparse	15.6	7.9	2.5	2.0	5.4
Dense	5.5	6.7	18.2	7.6	11.3
NIL-2/HSV[3]					
Sparse	26.5	17.6	0	0	0
Dense	4.0	8.6	0	0	0

[1] From Critchley and MacPerson [62].
[2] Relative weighted values.
[3] Transformed by hamster sarcoma virus.

The virally transformed cells also did not show increased biosynthesis of these glycolipids at high-culture densities. Studies of various NIL clones transformed with hamster sarcoma virus generally reveal a depletion of the higher 'glycolipids' [48,62], but this is not universally true [63]. Moreover, of six revertant clones selected from HSV transformed NIL cells, none showed a restoration of the neutral glycolipids [62].

Sakiyama and Robbins [71] have attempted to determine whether there is a correlation between the composition and biosynthesis of neutral glycolipids and tumorigenicity. They find very extensive variation in glycolipid synthesis among individual tumors, but most tumor cells, if not all, show some density-dependent glycolipid effect, mostly in G1-5. Also, their *least* tumorigenic clones lacked all of the density-dependent neutral glycolipids. Critchley and MacPherson [62] also noted that all of eleven tumor lines derived from transformed NIL cells contain G1-5, which exhibits a density-dependent increase in biosynthesis. However, 10 of the 11 tumor lines exhibited reduced G1-3 and G1-4 synthesis.

The problems of clonal variations in hamster cell lines complicate evaluation of the above results. Thus, while Critchley and MacPherson [63] report very minor differences between the glycolipid content of various clones, Sakiyama et al. [48] describe extensive clonal variation. These workers could also observe no clear correlation between glycolipid composition and the degree of cell density in culture. Indeed, their most 'contact inhibited' clone lacked all three of the higher neutral glycolipids.

Rat cells. Brady and Fishman [33] cite studies from their laboratories on the neutral-glycolipid composition of newborn rat kidney fibroblasts in culture. In these cells, which have a rather similar glycolipid composition to BHK cells, G1-3 apparently increases as the normal, contact-inhibited fibroblasts reach confluence. However, after transformation of the newborn-rat kidney cells with Kirsten murine sarcoma virus, the G1-3 content of the cells decreased. Moreover, when the cells are transformed with a temperature-sensitive mutant of murine sarcoma virus, the G1-3 content varies with temperature. Cells grown at the transforming temperature (31 °C) have less G1-3 than cells grown at the non-transforming temperature (39 °C). The authors propose that the lowered G1-3 content derives from reduced α-galactosyltransferase activity; hence, reduced conversion of G1-2 to G1-3.

Steiner et al. [72] report that rat embryo fibroblasts contain series of up to six different fucosylglycolipids. When the cells are transformed by a murine sarcoma-murine leukemia virus complex, the incorporation of radioactive fucose into the various fucolipids changes. The studies indicate a decrease in the amount of fucosylglycolipid with complex polysaccharide chains. The metabolic basis for this alteration has not been established at this writing.

4.4.2.3. *The Forssman paradox*

Evaluation of neutral glycolipid patterns in transformed cells is complicated by the fact that various hamster cells transformed by polyoma virus react with antibodies directed against Forssman Hapten (G1-5), whereas the parental cells do not [13-15]. Moreover, several studies indicate that treatment of normal fibroblasts with proteases make them reactive to Forssman antibody [76]. These data are difficult to reconcile with Hakomori's hypothesis that transformed cells show impaired 'extension' responses. The only mechanism known for the production of G1-5 is the stepwise addition of monosaccharide units to the 'lower' glycolipids (G1-1, G1-2, G1-3, G1-4). If α-galactosyltransferase activity is impaired in transformed cells, as proposed by Hakomori [61], the transformed cells should also show a reduction of G1-5 in addition to G1-4 and G1-3. Also, if higher glycolipids such as G1-5 participate in a cell–cell recognition process they should not be in a 'cryptic' state.

4.4.2.4. *The organizational of glycolipids in plasma membranes and changes thereof following malignant transformation*

The head groups of glycolipids are antigenic and one can produce antibodies specific for their various head groups. The use of such antibodies allows conclusions as to the degree of exposure of the head groups at membrane surfaces. It has now become apparent that the activity of various cells with antibodies directed against their membrane glycolipids is not necessarily proportional to the content of the glycolipid present. For example, as shown by Koscielak et al. [77], G1-4 is the major glycolipid component of human erythrocyte membranes, but anti-serum against this glycolipid does not react with intact human membranes. However, the reactivity of the red cells with the antibody could be increased 10-fold by treatment of the cells with various proteases and also with sialidase. This situation is very analogous to the exposure of G1-5 by protease treatment and to the increased reactivity of various lectin receptors upon protease treatment (Chapter 10). However, it does not always hold. Thus, Kijimoto and Hakomori [60] report that NIL cells from sparse cultures react with anti-Forssman antibody even without protease treatment, but this reactivity decreases markedly when the cells reach confluence. In contrast, polyoma-transformed cells do not exhibit significant reactivity with anti-Forssman antibody, whether at high or low-culture density.

Gahmberg and Hakomori [17] have developed an elegant approach for the study of the surface topology of galactolipids. In this they treat intact cells with galactose oxidase and then reduce the oxidized sugars with tritiated borohydride to label them with ^{3}H. Since galactose oxidase cannot penetrate into the interior of intact cells, the method only labels those galactosyl residues accessible to the enzyme on the external surface of the plasma membrane. All of the glycolipids other than glucosylceramide and lactosylceramide are labelled by this procedure.

Gahmberg and Hakomori [29,78] then combined their surface labelling procedure with analytical techniques to determine how glycolipids might be inserted into the plasma membranes of normal NIL fibroblasts and a polyoma transformant thereof (Table 4.8). They found that the normal fibroblasts contain a glucosylceramide, a lactosylceramide (βGal1 → 4Glc → Cer); a digalactosylceramide (αGal1 → 4βGal → Cer); a trihexosylceramide (αGal1 → 4βGal1 → 4Glc → Cer); two kinds of ceramide tetrasaccharides (a. αGalNAc1 → 3βGal NAc1 → 3αGal → 4βGal1 → Cer, a new type of Forssman active glycolipid; b. globoside, namely βGalNAc1 → 3αGal1 → 4βGal1 → 4βGlc → Cer), and a ceramide pentasaccharide having the classical structure of Forssman antigen (αGalNAc1 → 3βGalNAc1 → 3αGal1 → 4βGal1 → 4βGlc → Cer).

The polyoma transformant of the NIL-2K cells (NILpy) is unique in that it contains an additional ceramide tetrasaccharide which cannot be detected in the normal cells. This glycolipid is lacto-*N*-neotetraosylceramide (βGal1 → 4βGlcNAcl → 3βGal1 → 4Glc → Cer). This unique ceramide tetrasaccharide is found not only in cells cultured *in vitro* but also in tumors produced in hamsters by the *in vitro* transformed cells. However, the transformed cells contain considerably lower amounts of tetra- and pentasaccharides per cell than the normal fibroblasts.

Although the higher glycolipids were much less prevalent in the transformants, they were much more accessible to labelling with the galactose oxidase-borohydride method; that is, the specific activities of the label in the glycolipids of the transformed cells were invariably greater than in the parental cells (Table 4.8).

In the parental NIL cells, the surface labelling of the glycolipids is cell-cycle dependent. The surface exposure appears highest in the G−1 phase, decreases significantly during the S-phase and increases again after mitosis. In NIL cells, the surface exposure is also low after trypsinization. However, the chemical quantity of the ceramide pentasaccharide and ceramide tetrasaccharide is constant during the varying phases of the cell cycle, although the ceramide di- and tri-hexosides increase significantly at the G−1 phase. Accordingly, since large variations of surface labelling of ceramide pentasaccharide and tetrasaccharide occur during cell cycle, the observed changes should depend on the exposure of these glycolipids on the cell surface.

In contrast to the normal NIL cells, the polyoma transformant does not show a significant change in the exposure of the various glycolipids during the cell cycle, suggesting that these cells' membranes possess a distinctly different geometry from that of the parental cells. The exposure of the galactosyl residues of the ceramide tetrasaccharide fraction in the case of NILpy cells could be attributed to the presence of the unique lacto-*N*-neotetraosylceramide. Whereas the surface

TABLE 4.8

Struc' es, chemical quantities and degrees of 'surface exposure' of neutral glycosphingolipids found in NIL-2K hamster fibroblasts and their polyoma-induced transformant[1]

Glycolipid	Structure	Chemical quantity μmole lipid/100 mg dry cell residue[2]		Surface exposure[3]	
		NIL-2K	NILpy	NIL-2K	NILpy
Gl-1	Glc → Ceramide	ND	ND	ND	ND
Gl-2	βGal1 → 4Glc → Ceramide	0.55	0.965	36	44
	αGal1 → 4Gal → Ceramide				
Gl-3	αGal1 → 4βGal1 → 4Glc → Ceramide	0.30	0.159	543	1811
Gl-4	βGalNAc1 → 3αGal1 → 4βGal1 → 4Glc → Ceramide	0.29	0.075	297	3265
	αGalNAc1 → 3βGalNAcl → 3αGal1 → 4Gal(orGlc) → Ceramide	0.37	0.072		
	βGal1 → 4βGlcNAcl → 3βGal1 → 4Glc → Ceramide				
Gl-5	αGalNAcl → 3βGalNAcl → 3αGal1 → 4βGal1 → 4Glc → Ceramide	0.34	0.036	261	44

[2] From Gahmberg and Hakomori [29].
[2] Calculated on the basis of galactose content. ND = not determined.
[3] Labelling by treatment of whole cells with galactose oxidase and borotritride (cmp/mole glycolipid).

labelling of NIL cells glycolipids is greatly enhanced by castor-bean lectin at low concentration, this is not the case for the polyoma transformant.

It is of major interest that, although the chemical quantity of ceramide pentasaccharides and ceramide tetrasaccharides is less in NILpy cells compared with the non-transformed parental cells (see also refs. [48,60,62,71,78], possibly due to blocked synthesis of ceramide trihexoside (cf. [59]), their reactivities in the polyoma transformants with extracellular galactose oxidase is nevertheless vastly greater than in the case of the non-transformed cells. The authors have considered two explanations: *First*, it may be possible that the glycolipids on normal NIL cells are shielded by various types of membrane proteins. Deletion of these proteins or their structural modification could thus make the glycolipids more accessible. Indeed, there have been several reports suggesting that some external surface proteins on tumor cell surfaces are eliminated or deleted [78-83]; *Second*, it is conceivable that the glycolipid components of the transformed cells are more reactive with external galactose oxidase due to some increase of their mobility within the plane of the tumor membranes. There is no evidence for this proposal.

4.5. Polypeptide–glycosphingolipid complexes

In 1960, Wallach et al. [85] observed that the lipid extracts of Ehrlich ascites carcinoma had a higher molecular ratio of nitrogen to phosphorus than would be expected from their phospholipid composition. They suggested that some amino groups of phosphatidylethanolamine were associated in amide linkages to amino acids or polypeptides. They were not, however able to isolate specific phospholipid–peptide complexes. More recently, Diatlovitskaya et al. [86] isolated a specific phospholipid–peptide complex from the Jensen rat sarcoma. In addition, Petering et al. [87] reported the presence in tumors of a heterogenous fraction which included some phospholipid–glutamic acid complexes, with a high content of long-chain fatty acids (C20, C22 and C24).

More rigorous investigations in this direction have been carried out by Skipski and associates [88]. These workers have isolated novel polypeptide–glycosphingolipid complexes – *neoproteolipids* – which exist in high concentrations in various animal tumors, but can also be detected in non-neoplastic tissues. The neoproteolipids are soluble in chloroform–methanol mixtures and insoluble in water and, in this property, resemble the proteolipids which have been isolated from nervous tissues [89]. However, the new proteolipids described by Skipski and associates differ chemically, physico–chemically, and in distribution from the classical proteolipids. Skipski and associates have developed several chromatographic systems allowing separation of the neoproteolipid from other components in lipid extracts and also distinction between these complexes from different tissues.

The neoproteolipid isolated from the Walker carcinosarcoma 256 has been most extensively characterized. It accounts for about 1% of the lipid in this tumor, and its overall composition is given in Table 4.9. The data are interpreted to indicate that the neoproteolipid is a polypeptidylglycosphingolipid complex, bearing small amounts of cholesterol and phospholipid.

TABLE 4.9

Composition of neoproteolipid from Walker carcinosarcoma[1]

Component	Wt. %
Phosphorus	0.4–0.5
Fatty acids[2]	25.0–31.0
Sphingosine	10.0–14.0
Cholesterol	1.0–1.5
Carbohydrate[3]	40.0–45.0
Peptide	8.0–11.0

[1] From Skipski [88].
[2] 58 % C_{24} or longer; ~ 60 % saturated.
[3] Calculated as monosaccharide. The principal sugars are galactose, glucose, fucose and galactosamine.

Neoproteolipids are not tumor-specific substances; rather, they are concentrated in some tumors. They are also found in significant amounts in normal rat embryos and in the spleens of normal adult rats, although never at the levels found in tumor tissues. Traces are also found in some other tissues, but not in liver, brain, kidney, plasma, muscle or heart of normal animals, However, in tumor-bearing animals, the levels of neoproteolipid in the blood plasma rise and the amounts in erythrocytes and spleen increase substantially. It is suggested that the neoproteolipids are produced at a higher rate and in larger amounts in tumor tissue and that they exchange into the extracellular fluid and plasma of the tumor-bearing animals and thence into cells such as erythrocytes.

Some studies comparing the sera of normal human individuals with those of cancer patients suggest that measurable quantities of neoproteolipid is present in the high density lipoproteins of the tumor-bearing individuals.

The data presently available are provocative, but far too incomplete to allow conclusions as to the specific nature of the 'neoproteolipids' or their relationship to neoplasia. It is conceivable that they are in the category of carcino-embryonic substances. It is also possible that they constitute a feature of the often anomalous glycolipid metabolism in tumor cells.

4.6. Coda

The glycolipid composition and metabolism of neoplastic cells has been extensively investigated. *In vitro* experiments comparing cultured fibroblasts with their neoplastic transformants have received particular attention.

The data indicate that oncogenic conversion is often, but not universally, associated with altered ganglioside and/or cerebroside compositions. When deviations do occur, they are often correlated with altered activity of one or more of the glycosyltransferases participating in the assembly of the glycolipid head-groups. However, no consistent pattern has emerged and there is no evidence that neoplastic conversion can lead to suppression of the genes coding for specific glycosyltransferases.

The glycolipid anomalies observed in neoplastic cells have given rise to numerous speculations about the relevance of these anomalies to the neoplastic state. However, we consider any generalizations in this area premature for the following reasons:

(a) There is no apparent consistency in the anomalies shown by neoplastic cells.

(b) Glycolipids serve as both substrates and products. Accumulation of a given glycolipid thus depends not only upon the glycosyltransferases participating in its metabolism, but also upon the affinities of diverse membranes for this lipid; this will depend upon the membrane proteins.

(c) It is not known which enzymes in a glycosyltransferase are rate limiting and what factors regulate glycosyltransferase activity.

(d) Glycosyltransferases appear to be membrane-bound enzymes, and it is not known how membrane changes, unrelated to glycolipid composition, might affect glycosyltransferase activity.

Relevant to this issue is the observations by Zakim and Vessey [90] that UDPGlc, UDPMan and UDPXyl inhibit the UDP-glucuronyltransferase of guinea pig liver microsomes below 16 °C but *not* at higher temperatures. The microsomal lipids undergo a gel → liquid-crystalline transition when the temperature is increased above 16 °C and it appears that the liquid-crystalline state correlates with the normal substrate specificity and allosteric behavior of this transferase. The phospholipid composition and cholesterol content of microsomes is commonly altered in neoplasia (Chapters 3 and 5) and these changes are expected to modify the state of the membrane lipids. It is significant in this context that the UDPGal: G_{M2} galactosyltransferase of neoplastic 3T3 mouse fibroblasts, although depressed in activity, shows the same kinetic properties as the enzymes of normal cells [43].

(e) There is no adequate information about the glycolipid content of various subcellular membranes of normal and neoplastic cells, nor about glycolipid exchange.

Resolution of these issues is imperative, particularly since recent data indicate that the disposition and/or exposure of plasma membrane glycolipids varies with cell cycle and changes after neo-plastic transformation.

References

[1] Roseman, S., *Chem. Phys. Lipids.*, *5*, 270 (1970).
[2] Bosman, H.B. and Winston, R.A., *J. Cell Biol.*, *45*, 23 (1970).
[3] Chatterjee, S., Sweeley, C.C. and Velicer, L.F., *Biochem. Biophys. Res. Commun.*, *54*, 585 (1973).
[4] Kolodny, E.H., Kanfer, J.N., Quirk, J.M. and Brady, R.O., *J. Biol. Chem.*, *246*, 1426 (1971).
[5] Tallman, J.F. and Brady, R.O., *J. Biol. Chem.*, *247*, 7570 (1972).
[6] Tallman, J.F. and Brady, R.O., *Biochim. Biophys. Acta*, *293*, 434 (1973).
[7] Klenk, E. and Gilen, W., *Hoppe-Seyler's Z. Physiol. Chem.*, *319*, 283 (1960).
[8] Albers, R.W. and Koval, G.J., *Biochim. Biophys. Acta*, *60*, 359 (1962).
[9] Trams, E.G. and Lauter, C.J., *Biochim. Biophys. Acta*, *60*, 350 (1962).
[10] Shipley, G.G. in *Biological Membranes*, D. Chapman and D.F.H. Wallach, eds., Academic Press, London, Vol. II, 1973, p. 1.
[11] Oldfield, E. and Chapman, D., *FEBS Letters*, *21*, 303 (1972).
[12] Clowes, A.W., Cherry, R.J. and Chapman, D., *Biochim. Biophys. Acta*, *249*, 301 (1971).
[13] Clowes, A.W., Cherry, R.J. and Chapman, D., *J. Mol. Biol.*, *67*, 49 (1972).
[14] Yoshe, H.C. and Rosenberg, A., *Chem. Phys. Lipids*, *9*, 279 (1972).
[15] Marcus, D.M. and Cass, L., *Science*, *164*, 553 (1969).
[16] Laine, R.A. and Hakomori, S.-I., *Biochem. Biophys. Res. Commun.*, *54*, 1039 (1973).
[17] Gahmberg, C.G. and Hakomori, S.-I., *J. Biol. Chem.*, *248*, 4311 (1973).
[18] Brailovsky, C., Trudel, M., Lallier, R. and Nigam, V.N., *J. Cell. Biol.*, *57*, 124 (1973).
[19] Brailovsky, C., Lallier, R. and Nigam, V.N., *J. Cell. Biol.*, *63*, 34a (1974).
[20] Cuatrecasas, P., *Biochemistry*, *12*, 3558 (1973).
[21] Dodd, B.J. and Gray, G.M., *Biochim. Biophys. Acta*, *150*, 397 (1968).
[22] Dnistrian, A.M., Skipski, V.P., Barclay, M., Essner, E.S. and Stock, C.C., *J. Cell. Biol.*, *63*, 86a (1974).
[23] Renkonen, O., Gahmberg, C.G., Simons, K. and Kääriäinen, L., *Acta Chem. Scand.*, *24*, 733 (1970).
[24] Klenk, H.D. in *Biological Membranes*, D. Chapman and D.F.H. Wallach, eds., Academic Press, Vol. II, 1973, p. 145.
[25] Renkonen, O., Kääriäinen, L., Simons, K. and Gahmberg, C.G., *Virology*, *46*, 318 (1971).
[26] Yogeeswaran, G., Sheinin, R. and Wherrett, G.R., *J. Biol. Chem.*, *247*, 5146 (1972).
[27] Weinstein, D.B., Marsh, J., Glick, M.C. and Warren, L., *J. Biol. Chem.*, *245*, 3928 (1970).
[28] Cuatrecasas, P., *Biochemistry*, *12*, 3547 (1973).
[29] Gahmberg, C.G. and Hakomori, S.-I., *Biochem. Biophys. Res. Commun.*, *59*, 283 (1974).
[30] van Heyningen, W.E., *Naunyn-Schmiedebergs Arch. Pharm.*, *276*, 289 (1973).
[31] Cuatrecasas, P., *Biochemistry*, *12*, 3567 (1973).
[32] Cuatrecasas, P., *Biochemistry*, *12*, 3577 (1973).
[33] Brady, R.O. and Fishman, P.H., *Biochim. Biophys. Acta*, *3-5*, 121 (1974).
[34] Brady, R.O., Fishman, P.H. and Mora, P.T., *Fed. Proc.*, *32*, 102 (1973).
[35] Mora, P.T., Fishman, P.H., Bassin, R.H., Brady, R.D. and McFarland, V.W., *Nature New Biol.*, *245*, 226 (1973).
[36] Fishman, P.H., Brady, R.O., Bradley, R.M., Aaronson, S.A. and Todaro, G.J., *Proc. Natl. Acad. Sci. U.S.*, *71*, 298 (1974).
[37] Cumar, F.A., Brady, R.O., Kolodny, E.H., MacFarland, V.W. and Mora, P.T., *Proc. Natl. Acad. Sci. U.S.*, *67*, 757 (1970).
[38] Fishman, P.H., McFarland, V.W., Mora, P.T. and Brady, R.O., *Biochem. Biophys. Res. Commun.*, *48*, 48 (1971).

[39] Mora, P.T., Brady, R.O., Bradley, R.M. and McFarland, V.N., *Proc. Natl. Acad. Sci. U.S.*, *63*, 1290 (1969).
[40] Brady, R.O. and Mora, P.T., *Biochim. Biophys. Acta*, *218*, 308 (1970).
[41] Mora, P.T., Cumar, F.A. and Brady, R.O., *Virology*, *46*, 60 (1971).
[42] Brady, R.O., Fishman, P.H. and Mora, P.T. in *Advances in Enzyme Regulation*, G. Weber, ed., Vol. 11, 1973, p. 231.
[43] Coleman, P.L., Fishman, P.H., Brady, R.O. and Todaro, G.J., *J. Biol. Chem.*, *250*, 55 (1975).
[44] Hakomori, S.-I. and Murakami, W.T., *Proc. Natl. Acad. Sci. U.S.*, *59*, 254 (1968).
[45] Hakomori, S.-I., Teather, C. and Andrews, H., *Biochem. Biophys. Res. Commun.*, *33*, 563 (1968).
[46] Nigam, V.N., Lallier, R. and Brailovsky, C., *J. Cell Biol.*, *58*, 307 (1973).
[47] Robbins, P.W. and MacPherson, I.A., *Nature*, *229*, 569 (1971).
[48] Sakiyama, H., Gross, S.K. and Robbins, P.W., *Proc. Natl. Acad. Sci. U.S.*, *69*, 872 (1972).
[49] Hakomori, S.-I., Saito, T. and Vogt, P.K., *Virology*, *44*, 609 (1971).
[50] Hakomori, S.-I., *Proc. Natl. Acad. Sci. U.S.*, *67*, 1741 (1970).
[51] Brady, R.O., Borek, C. and Bradley, R.M., *J. Biol. Chem.*, *244*, 6552 (1969).
[52] Siddiqui, B. and Hakomori, S.-I., *Cancer Res.*, *30*, 2930 (1970).
[53] Cheema, P., Yogeeswaran, G., Morris, H.P. and Murray, R.K., *FEBS Letters*, *11*, 181 (1970).
[54] Dnistrian, A.M., Skipski, V.P., Barclay, M., Essner, E.S. and Stock, C.C., *J. Cell Biol.*, *63*, 86a (1974).
[55] Keenan, T.W. and Doak, R.L., *FEBS Letters*, *37*, 124 (1973).
[56] Hollenberg, M.D., Fishman, P.H., Bennett, V. and Cuatrecasas, P., *Proc. Natl. Acad. Sci. U.S.*, *71*, 4224 (1974).
[57] Siegel, M.I. and Cuatrecasas, P. in *Cellular Membranes and Tumor Cell Behavior*, Walborg, E.E. ed., Williams and Wilkins, Baltimore, in press (1975).
[58] Lloyd, K.O., Kabat, E.A. and Licerio, E., *Biochemistry*, *7*, 2976 (1968).
[59] Kijimoto, S. and Hakomori, S.-I., *Biochem. Biophys. Res. Commun.*, *44*, 557 (1971).
[60] Kijimoto, S. and Hakomori, S.-I., *FEBS Letters*, *25*, 38 (1972).
[61] Hakomori, S.-I., *Tumor Lipids*, R. Wood, ed., Oil Chemists Society Press, Champaign, Ill., 1973, p. 269.
[62] Critchley, D.R. and MacPherson, I., *Biochim. Biophys. Acta*, *296*, 145 (1973).
[63] Sakiyama, H. and Robbins, P.W., *Arch. Biochem. Biophys.*, *154*, 407 (1973).
[64] Ohuti, K., Tohoku, J., *Exp. Med.*, *51*, 297 (1949).
[65] Masamune, H., Yosizawa, T., Masukawa, A. and Tohoku, J., *Exp. Med.*, *58*, 381 (1953).
[66] Hakomori, S.-I. and Jeanloz, R.W., *J. Biol. Chem.*, *239*, 3606 (1964).
[67] Hakomori, S.-I., Koscielak, J., Bloch, K.J. and Jeanloz, R.W., *J. Immunol.*, *98*, 31 (1967).
[68] Yang, H. and Hakomori, S.-I., *J. Biol. Chem.*, *246*, 1192 (1971).
[69] Hakomori, S.-I. and Andrews, H., *Biochim. Biophys. Acta*, *202*, 225 (1970).
[70] Stellner, K., Hakomori, S.-I. and Warner, G.S., *Biochem. Biophys. Res. Commun.*, *55*, 439 (1973).
[71] Sakiyama, H. and Robbins, P.W., *Fed. Proc.*, *32*, 86 (1973).
[72] Steiner, S., Brennan, P.J., Melnick, J.L., *Nature, New Biol.*, *245*, 19 (1973).
[73] Fogel, M. and Sachs, L., *J. Nat. Cancer Inst.*, *29*, 239 (1962).
[74] Fogel, M. and Sachs, L., *Exp. Cell Res.*, *34*, 448 (1964).
[75] O'Neill, C.H., *J. Cell Sci.*, *3*, 405 (1968).
[76] Makita, A. and Seyama, Y., *Biochim. Biophys. Acta*, *241*, 403 (1971).
[77] Koscielak, J., Hakomori, S.-I. and Jeanloz, R., *Immuno Chem.*, *5*, 441 (1968).
[78] Gahmberg, C.G. and Hakomori, S.-I., *J. Biol. Chem.* *250*, 2438, 2447 (1975).
[79] Robbins, P.W. and MacPherson, I., *Proc. Royal Soc. London*, *177*, 49 (1971).
[80] Gahmberg, C.G., Kiehn, B. and Hakomori, S., *Nature*, *248*, 413 (1974).
[81] Gahmberg, C.G. and Hakomori, S., *Proc. Natl. Acad. Sci. U.S.* *70*, 3329 (1973).
[82] Hynes, R.O., *Proc. Natl. Acad. Sci. U.S.*, *70*, 3170 (1973).
[83] Stone, K.R., Smith, R.E., Joklik, W.K., *Virology*, *58*, 86 (1974).
[84] Wickus, G.G., Branton, P.E. and Robbins, P.W. in *Control of Proliferation in Animal Cells*, B. Clarkson and R. Baserga, eds., Cold Spring Harbor Laboratory, 1974, p. 541.
[85] Wallach, D.F.H., Soderberg, J. and Bricker, L., *Cancer Res.*, *20*, 397 (1960).

[86] Diatlovitskaia, E.V., Torkhovskaia, T.I. and Bergelson, L.D., *Biokhimiia*, *34*, 177 (1969).
[87] Petering, H.G., Van Giessen, G.J., Buskirk, H.H., Crim, J.A., Evans, J.S. and Musser, E.A., *Cancer Res.*, *27*, 7 (1967).
[88] Skipski, V. in *Tumor Lipids*, R. Wood, ed., American Oil Chemistry Society Press, Champaign, Ill., 1973, p. 225.
[89] Folch-Pi, J. in *Brain Lipids and Lipoproteins and the Leucodystrophies*, J. Folch-Pi and H. Bauer, eds., Elsevier, Amsterdam, 1963, p. 18.
[90] Zakim, D. and Vessey, D.A., *J. Biol. Chem.*, *250*, 343 (1975).

5

Cholesterol

5.1. Introduction

Animal cells, unlike most microbial organisms, contain an important amount of cholesterol (0.5-2.4 mg/100 mg dry wt). Most of the sterol (Fig. 5.1), is associated with cellular membranes; and most of it is normally in the unesterified form.

Cholesterol exhibits an exceptionally low monomer solubility in aqueous solvents [1]. It undergoes a reversible monomer–micelle equilibrium with a very low critical micelle concentration – 25–40 $\times$ 10^{-9} M at 25 °C. The maximal aqueous solubility of cholesterol in micellar form is 4.5 $\times$ 10^{-6} M. Cholesterol micelles are stabilized by strong intermolecular forces, as well as by solvent exclusion. The micelles are probably rod shaped, 1000 Å in length and 20 Å in diameter, with molecular weights near 2 $\times$ 10^5.

High energies of self association tend to prevent direct cholesterol–protein

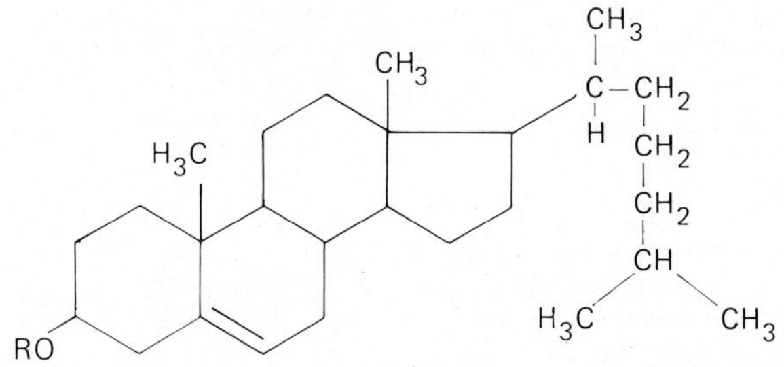

Fig. 5.1. Cholesterol and its esters. In *cholesterol*: R = H; in *cholesterol esters*: R = $-\overset{\overset{O}{\|}}{C}-R$, with C_{12}–C_{20} saturated and unsaturated acyl groups.

associations. Thus, unlike many other sterols, cholesterol does not bind to serum albumin or apolipoproteins. The ready association of cholesterol with phospholipid monolayers or bilayers, as well as with some biomembranes, probably results from high-affinity interactions with phosphatides.

Diverse membranes differ in their cholesterol content (Table 5.1). The greatest concentrations of cholesterol occur in plasma membranes. These exhibit molar cholesterol phospholipid ratios near unity. In contrast, the membranes of mitochondria and nuclei contain little cholesterol, whereas endoplasmic reticulum occupies an intermediate position.

As noted, cholesterol does not dissolve readily in aqueous solvents and tends to self-associate in aqueous systems. However the sterol readily associates with phosphatides. Incorporation of cholesterol into artificial phosphatide bilayers markedly alters their properties [10]. The steroid nucleus interacts with the fatty acid chains of the phosphatides and changes their molecular motions. The motions of the phosphatide methylene groups are not affected until the gel → liquid-crystalline transition temperature is reached; above this temperature their

TABLE 5.1

Cholesterol/phospholipid ratios in various cellular membranes

Membrane class	Cholesterol/phospholipid (mole/mole)
Plasma Membranes	
a. Erythrocytes (various species)[1]	0.8–1.0
b. Myelin[2]	1.0
c. Lymphocytes[3,4]	0.9–1.0
d. Liver Bile Fronts[5]	0.43–0.90
Nuclear Membranes	
a. Pig liver[2]	0.10
b. Rat liver[6]	0.23
c. Mouse liver[6]	0.17
d. Lymphocytes[7]	0.25
Microsomes[6]	
a. Rat liver	0.16
b. Mouse liver	0.23
Mitochondria[2]	
a. Rat liver	0.015–0.046
b. Rat intestine	0.036–0.123
Lysosomes[5]	0.6

[1] From Rouser et al. [2].
[2] From Graham and Green [3], and Parson and Yano [4].
[3] From Ferber et al. [5].
[4] From Schmidt-Ullrich et al. [6].
[5] From Thines-Sempoux [7].
[6] From Franke et al.[8].
[7] From Jarasch et al. [9].

mobilities are hindered. However, cholesterol also interferes with the crystalline packing of the fatty acids, which ordinarily occurs below the transition temperature. Cholesterol further gives the glycerol moieties of phosphatides greater motional freedom in pure phosphatide systems. Accordingly, the gel → liquid-crystal transitions then occur over a wide range of temperature and in a poorly-cooperative fashion. Speaking in biologic terms, the presence of cholesterol tends to 'fluidize' the fatty acid chains of membrane phospholipids which, in the absence of the sterol, would be in an ordered crystalline state at physiologic temperatures. The sterol also tends to '*defluidize*' the acyl chains of phospholipids which would otherwise be in a disordered (liquid-crystalline) state at physiologic temperatures. An important effect of the sterol, therefore, is to '*buffer*' the *fluidity* of the hydrocarbon chains of membrane phosphatides. Equally significant may be the influence of cholesterol on the *cooperativity* of membrane lipid *state-transitions* (Chapter 11).

Finally, cholesterol may play a critical biological role in creating a generally fluid, liquid-crystalline phase in lipid mixtures which, in the absence of cholesterol, would form separated crystalline and liquid-crystalline domains. Cholesterol may thus play an essential role in the insertion and assimilation of unusual functional lipids, such as glycosphingolipids, into a generally phospholipid region.

The precise role of cholesterol in a biomembrane cannot be readily assessed without experimentation far more extensive than has been applied to any biological system to date. Thus, the action of this sterol depends not only on temperature, but also on the degree of unsaturation of the phosphatide hydrocarbon chains and on the net charge on the phosphatide head groups [11]. Moreover, according to data derived from simple cholesterol–phospholipid systems, the 'buffering effect' of the sterol is not a simple function of cholesterol/phospholipid ratio; for a saturated phospholipid, the major changes occur between a cholesterol/phospholipid molar ratio of 0.2 (such as is found in many intracellular membranes) and a ratio of >0.5 (such as is more typical for plasma membranes). These estimates [10] constitute average values for phospholipid–cholesterol systems and do not take into account the fact recently documented [12, 13, 13a] that the interaction of cholesterol and phospholipids may not occur in a random fashion, but also in *localized clusters* with unique physical properties.

Another problem to be considered is the possible differential interaction of cholesterol with phosphatides on the inner and outer surfaces of a membrane. This problem is documented in a model study presented by Huang et al. [14]. Their data suggest that up to cholesterol/phospholipid ratios of 0.3, the sterol is symmetrically distributed between the inner and outer surfaces of artificial cholesterol–phospholipid bilayers of small radius. However, with cholesterol/phospholipid ratios greater than 0.3, the sterol is distributed preferentially on the inner face of a lecithin–cholesterol vesicle. As one might anticipate, this effect depends on the degree of unsaturation of the phosphatide hydrocarbon chains.

As a functional consequence of cholesterol inclusion into phosphatide bilayers, one observes marked reductions in the permeabilities to anions [15], cations [16] and non-electrolytes [17]. Insertion of cholesterol into phosphatide bilayers also interferes with the enhanced cation transport through these membranes normally induced by valinomycin [18]. Moreover, the presence of cholesterol interferes with the fusion of phosphatide vesicles [19].

5.2. Cholesterol content of biomembranes

Importantly, cholesterol enrichment of guinea pig erythrocytes [20] as well as the membranes of *Acholeplasma laidlawii* [21] impair active and passive potassium permeation. How this relates to the inhibition of the phospholipid activation of rabbit kidney cortex Na^+/K^+-ATPase (e.g. [22]) is not clear. However, Kimelberg and Papahadjopoulos [22] propose that cholesterol influences the activity of membrane-bound enzymes by regulating the 'fluidity' of their phospholipid environment. They suggest the term 'viscotropy' to describe the effects of lipid fluidity (viscosity) on the activity of membrane enzymes. Moreover, Papahadjopoulos [23] speculates that adenylate cyclase – a critical mediator between extracellular information and cellular metabolism – also depends on the 'fluidity' of its environment and that this fluidity varies with cholesterol/phosphatide ratio. Certainly cyclase activity depends critically on phosphatides [24].

The role of cholesterol cannot be dissociated from that of membrane phospholipids, including the unsaturation of their fatty acids. Thus, incorporation of cholesterol into dilinoleyllecithin bilayers produces *no* permeability change, in contrast to the effects observed with more saturated phosphatides [25]. Cholesterol also condenses monolayers of unsaturated phosphatides to a greater extent than it does saturated phospholipid films [25]. Furthermore, the phospholipid activation of purified Na^+/K^+-ATPase is less susceptible to cholesterol inhibition, if the phosphatides are highly unsaturated [22]. Finally, Bloj et al. [26] argue that cholesterol-related modification of erythrocyte membrane Na^+/K^+-ATPase depends on subtle modifications of cholesterol/phospholipid ratios in the membrane domains that contain the enzyme.

Many authors imply that there is a simple correlation between cholesterol content and membrane 'fluidity' in many biomembranes, but this conclusion is not justified. *First*, the sterol exchanges readily and its level in membrane cannot depend on membrane lipid alone. *Second*, plasma membranes contain so much cholesterol that small changes in the cholesterol/phospholipid ratio *per se* (increases in particular) are not likely to alter membrane 'fluidity' remarkably. *Third*, in biomembranes, where proteins comprise $>60\%$ of the membrane mass, membrane 'fluidity' must depend critically on the behavior of membrane proteins; the nature of this dependence is largely unknown. *Fourth*, the cholesterol distribution within a membrane is not likely to be uniform in view of the clustering tendency of this molecule.

5.3. Cholesterol biosynthesis

5.3.1. Overview

Most animal cells contain the machinery to synthesize cholesterol. Cholesterol biosynthesis is a complicated process and the conversion of acetyl-CoA to cholesterol requires 26 well-defined enzymatic reactions [27]. A limiting step in cholesterol biosynthesis (Fig. 5.2) is the conversion of β-hydroxyl-β-methylglutaryl-CoA to mevalonate. This step is catalyzed by the enzyme β-hydroxyl-β-methylglutaryl-CoA reductase (HMG-CoA reductase). This enzyme is associated with the microsomal membranes of hepatocytes [28] and other cells.

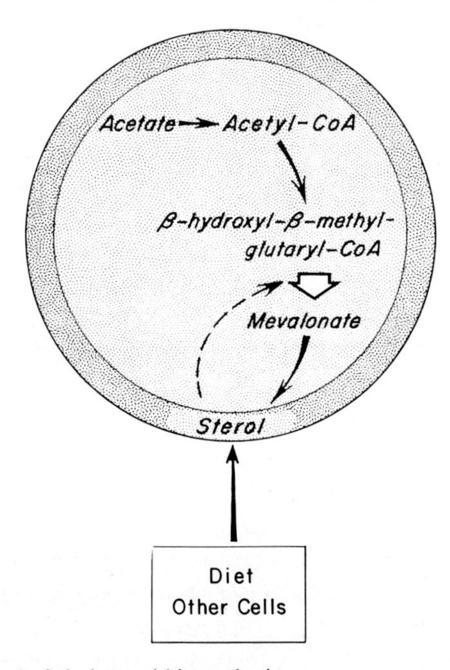

Fig. 5.2. Feedback control of cholesterol biosynthesis.

5.3.2. Feedback inhibition of cholesterol biosynthesis

HMG-CoA reductase is subject to at least three independently-operating negative feedback mechanisms. These are (a) endogenous feedback, (b) dietary feedback (intermittent feeding and fasting) and (c) diurnal rhythm (cf. review by Sabine

[29]). Until recently the initiating agent for the feedback control of liver-HMG-CoA reductase was thought to be cholesterol, whether endogenous or dietary.

The *in vivo* feedback control of sterol biosynthesis is not restricted to hepatocytes. Thus, cholesterol synthesis by bone marrow and diverse leukocytes is inhibited by cholesterol feeding [30, 31]. Also, cholesterol biosynthesis in rat granulocytes (but not lymphocytes or platelets) decreases 90% upon cholesterol feeding; this decrease correlates with an 87% drop in the activity of microsomal HMG-CoA reductase [32]. Furthermore, although normal mouse lymphocytes exhibit very low rates of cholesterol synthesis, but production is further inhibited by feeding cholesterol for three days or more [33].

Feedback regulation of cholesterol biosynthesis can also be demonstrated *in vitro*. Thus, Avigan et al. [34] have examined the regulation of cholesterol biosynthesis in normal cultured (primary) human fibroblasts, as well as in several established cell lines, including HTC (hepatoma 7288C). They show that cholesterol biosynthesis is reduced 100-fold when fibroblasts are first incubated with lipid-poor serum and then treated with cholesterol-enriched medium. Cholesterol added to the medium in codispersion with lecithin has the same effect as cholesterol-rich serum. The authors attribute their findings to feedback inhibition of sterol biosynthesis. Established cell lines showed lower levels of steroid biosynthesis when the protein in the culture medium is cholesterol-depleted. Cholesterol addition reduces synthesis by factors of approximately 6, 4, 2 and 2 in the cases of hepatoma 7288C, Hela L3 cells, L2071 cells and BRL 62 cells respectively. Clearly, the feedback mechanisms are present in the established lines but to a smaller extent than *in vivo*.

A number of experimenters have investigated sterol biosynthesis in L-cells. These fibroblasts do not manufacture cholesterol but desmosterol. Sterol biosynthesis in L-cells can be inhibited *in vitro* by a number of sterols through depression of HMG-CoA-reductase activity (cf. [35–39]). Thus, Kandutsch and Chen [36, 37] have recently demonstrated that the rates of acetate incorporation into the sterols of cultured L-cell fibroblasts (and mouse fetal liver cells) can be inhibited by various sterols, but, importantly, *not* by pure cholesterol or several metabolically related sterols. (Inhibitory sterols could be extracted from impure cholesterol samples.) Derivatives of cholesterol oxygenated at the 7, 20, 22, or 25 positions are highly inhibitory at concentrations of 5×10^{-7} to 5×10^{-8} M. Moreover, inhibitory sterols specifically depress the activity of the key regulatory enzyme in the sterol biosynthetic pathway (3-hydroxy-3-methylglutaryl-CoA reductase) within 6 hr.

It is also found that longer incubation of L-cell fibroblasts with inhibitory steroids, diminishes the concentration of cellular desmosterol and results in the reduction or cessation of growth, unless an appropriate cholesterol analogue is supplied in the medium [39]. A sterol precursor which can enter the biosynthetic path after the feedback control step can also alleviate growth inhibition.

The most effective inhibitor of cell growth is 25-hydroxycholesterol, followed in order of decreasing activity by 20α-hydroxycholesterol, 7-ketocholesterol and 7α-hydroxycholesterol. The first three of these sterols inhibit growth by more than 90%. The relative effectiveness of the sterols in inhibiting cell replication correlates well with their capacity to inhibit sterol biosynthesis and to depress the levels of HMG-CoA-reductase activity. However, the levels of sterol required to arrest growth were several times higher than those necessary to inhibit sterol synthesis. The authors consider the possibility that the inhibitory sterols compete with desmosterol for sites within the cellular membranes, with resulting impairment of membrane function. (The depression of HMG-CoA-reductase activity might then be a distal event.) However, it is argued that the inhibitory sterols depress growth by virtue of their restriction of sterol synthesis. Since it is well known, for example, that cholesterol depletion of erythrocytes and other cells results in increased membrane fragility, the observed growth retardation may simply be a reflection of membrane instability and damage. However, an alternate possibility is that the effect is more subtle and is mediated via the effect of cholesterol on the structure of other membrane lipids.

Bates and Rothblat [38] demonstrate that addition of human serum to L-cells induces a rapid sterol flux, accompanied by a drastic reduction in sterol synthesis by the cells. Serum lecithin–cholesterol acyltransferase activity is not involved and both influx and efflux values approach saturation within 8 hr at 37°C. Because L-cells lack the ability to produce cholesterol and synthesize desmosterol as the end product of sterol biosynthesis, one can differentiate between cellular sterol (desmosterol) and exogenous serum sterol (cholesterol).

Bates and Rothblat show that the influx of cholesterol was essentially the same after addition of high-density lipoprotein and low-density lipoprotein at equivalent cholesterol levels, whereas the efflux of desmosterol from the cells was 3 to 6 times greater when high-density lipoprotein was added to the medium than when low-density lipoprotein was used. The high-density lipoprotein could reduce the desmosterol content of the cells by as much as 50%, whereas low-density lipoprotein caused only a 4% reduction. Both classes of serum lipoprotein, as well as serum, produce an increase in total cellular sterol, but, surprisingly, high-density lipoproteins do not appreciably reduce cholesterol biosynthesis. The data indicate that sterol synthesis is not simply related to cellular sterol content.

In the case of human fibroblasts [40], two low-density fractions (d. 1.006–1.019 and 1.019–1.063 g/ml) of human lipoprotein inhibit cholesterol biosynthesis, but the fraction with d. 1.063–1.21 g/ml, as well as egg-yolk lipoprotein, is ineffective at equivalent cholesterol concentrations. On the other hand, Watson [41] shows that cholesterol, presented in mixed lecithin–cholesterol vesicles, effectively inhibits cholesterol biosynthesis by cultured HTC cells. In contrast, cholesterol-free lecithin dispersions or cholesterol-poor plasma lipoproteins, allow free sterol production by the cultivated hepatoma cells.

Current evidence indicates that the feedback inhibition of HMG-CoA reductase is not an allosteric effect. Thus, the liver enzyme has a half life of only about 3 hr [42]. Moreover, studies on human fibroblasts using metabolic inhibitors suggest that the regulation of HMG-CoA-reductase activity depends on unhampered protein synthesis [40]. However, Higgins et al. [42a], using immunological techniques together with biochemical assays, show that HMG-CoA reductase can exist in states of different activity, both when isolated and *in vivo*. Moreover, they document that HMG-CoA-reductase *activity* declines *immediately* after cholesterol is supplied, whereas the synthesis of the enzyme declines only at a later stage.

The mechanisms for cholesterol exchange discussed below are clearly relevant to the precise mechanisms of feedback regulation. However, as already noted, the ultimate feedback regulator may not be cholesterol (cf. also [34,43,44]).

5.4. Exchange

The irregular cholesterol distribution among various cellular membrane compartments is remarkable. Unlike phospholipids, the other major membrane lipids, and despite its very low solubility, cholesterol associates rather 'loosely' with other membrane components. It elutes easily, even with organic solvents that do not denature membrane proteins. It also exchanges rapidly between plasma membranes and lipoproteins or phosphatide dispersions.

Because of its extremely low monomer solubility, cholesterol is unlikely to exchange in this form. Exchange via the micellar state also appears unfavorable because cholesterol micelles are large and also rather insoluble. The most appealing proposal for the interchange between plasma membranes and extracellular carriers (lipoproteins) is that of Rothblat [45], Fig. 5.3. He suggests that exchange occurs during 'adsorption' of the carrier onto the membrane. Cholesterol is then incorporated into, or released from the plasma membrane. An analogous mechanism might be necessary for the transfer of cholesterol from its site of synthesis (in the endoplasmic reticulum) to other membranes. This transfer could proceed via a soluble, intracellular lipoprotein carrier, during transient membrane fusion or through transient membrane adhesion.

The exchangeability of erythrocyte membrane cholesterol has been extensively explored by Bruckdorfer and colleagues. They first showed [46] that cholesterol exchanges rapidly between rat erythrocyte ghosts and human low-density serum lipoprotein. The exchange rate is minimal at pH 5, but does not vary markedly over a wide range of ionic conditions or temperature. Similar exchange patterns occur with cholesterol-lecithin dispersions (1/1; mole/mole), [47]. Also, using lecithin dispersions containing various proportions of the sterol, cholesterol can be extracted from, or added to the ghosts. With an excess of a cholesterol-free

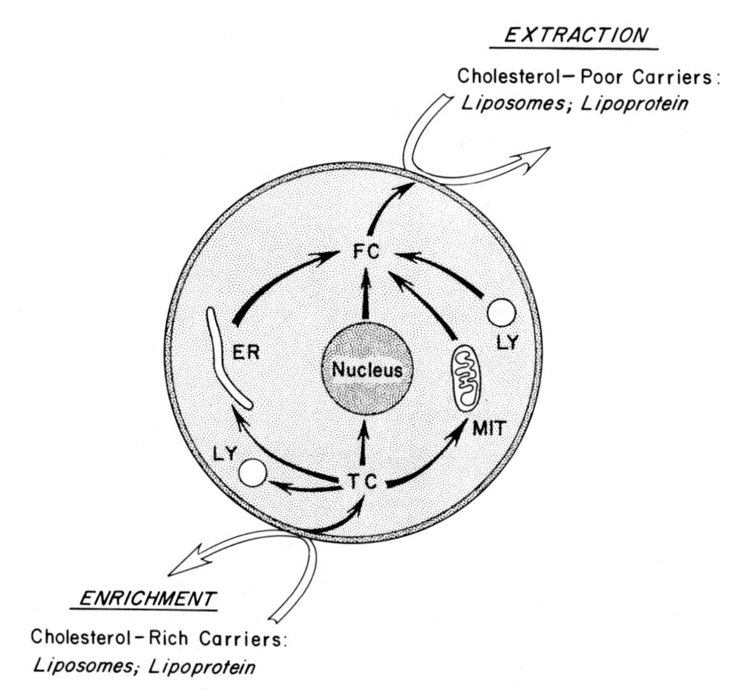

Fig. 5.3. Cholesterol exchange.

lecithin dispersion, up to 60% of the ghost cholesterol can be extracted. That 40% of the sterol cannot be eluted indicates the existence in the membranes of 'cholesterol-binding domains', with higher affinities than those of simple lipid bilayers.

Use of the dispersed mixtures of lecithin and various cholesterol analogues allows facile modification of the steroid content in the erythrocyte membranes [48]. Thus, cholesta-4,6-dien-3-one, cholesta-4-en-3-one, 7-dehydro-cholesterol and sterophenol exchange readily and can replace cholesterol. Indeed cholesta-4,6-dien-3-one exchanges more rapidly than cholesterol. Ergosterol, in contrast, cannot substitute for cholesterol.

The sterol in the membranes of *intact* human erythrocytes can also be modified by incubating washed cells with sonicated lecithin or lecithin-sterol suspension [49]. Cholesterol-free lecithin dispersions extract cholesterol from the cells and thereby increase osmotic fragility. Removal of 48% of the membrane cholesterol induces hemolysis. Cholesterol depletion also augments glycerol permeation. Interestingly, 25% replacement of membrane cholesterol with cholesta-4,6-dien-3-one produces no significant increase in osmotic fragility, and 26% exchange with lathosterol actually reduces fragility.

The high cholesterol/phospholipid ratio in erythrocyte membranes might be attributed to equilibration with the plasma lipoprotein pool. However, this argument does not apply to tissue cells which do not interact with plasma lipoproteins. Also, even in erythrocytes the situation is not straightforward. Thus, exchange between erythrocytes and lipoproteins does not depend on the levels of free cholesterol in these compartments, but on the cholesterol/phospholipid ratios [50]. Moreover, there are at least two cholesterol 'pools' in the membranes [51]. One contains molecules with substantial membrane affinity, but which can nevertheless exchange. The other consists of sterols that are weakly associated and exchange very freely. One might suppose that the latter corresponds to the fraction of cholesterol that can be extracted by lecithin dispersions [47]. It may also be this component that accounts for the cholesterol enrichment of erythrocyte (and platelet) membranes in 'spur cell' anemias [52], and for the abnormal erythrocytes found in guinea pigs on high cholesterol diets [53].

Direct measurements of cholesterol exchange have been reported for the system Ehrlich ascites carcinoma–serum low-density lipoprotein. Abell et al. [54] found a *net* uptake of 7–12% of the steroid available in the lipoprotein. However, studies with radioactively-tagged lipoprotein, show 50–60% exchange [50, 55]. Taken together, the data indicate a two-way exchange.

A number of studies on cultured cells [45, 56–58] give evidence of cholesterol exchange between the cells and the medium. This exchange is much accelerated by adding lipid-poor serum or cholesterol-free phosphatide dispersions to the medium. Also, by differentially radio-labelling both the unesterified cholesterol and the protein of α-lipoprotein, Bailey and Butler [58] showed that the transfer of sterol to L-cells was not due to uptake of the whole lipoprotein complex. These workers also suggest that many cultured cells derive their cholesterol from the culture medium.

Unesterified cholesterol. added to the culture medium, can also be taken up by malignant mouse lymphoblasts [57]. Moreover, both L5187Y cells and L-cells that have taken up labelled unesterified cholesterol can transfer approximately 50% of this free sterol to serum when the cells are moved to fresh medium [57]. Also, through the activity of a cholesterol esterase in L-cells, approximately 50% of the cholesterol ester taken up by these cells can be released into the medium as the free sterol; cholesterol ester itself is 'excreted' much less readily [57].

Cholesterol exchange has also been studied with macrophages [59–61]. When macrophages are incubated with fetal calf serum containing ^{3}H-labelled cholesterol, the fall in specific activity in the serum is accompanied by an equivalent rise in the macrophage (and vice versa when the label was originally in the cell). Eighty percent of the total cell cholesterol is exchangeable in 4 hr. The time course of the exchange showes two phases: Werb and Cohn [59,60] propose that the rapidly-equilibrating cell component, accounting for 60–70% of the total cell sterol, represents the fast exchange between lipoprotein and the surface membrane, while the slowly-equilibrating component represents the slow exchange between the surface membrane and internal membranes, particularly lysosomes. When the lysosomal compartment is expanded by adding dextran sulphate to the medium, the contribution of the slowly-equilibrating component increases [60]. Cells containing few lysosomes (e.g., L-cells) show only a small slowly-equilibrating component. Werb and Cohn [60] have eliminated endocytosis as a means of cholesterol transfer to the lysosomes and it seems that the appearance of label in these organelles is due to a true exchange from the surface membrane, probably mediated by some form of carrier molecule which itself may be a lipoprotein or apolipoprotein.

Werb and Cohn [61] have also shown that when excess cholesterol is introduced into macrophage lysosomes via phagocytosis, this is transferred out of the cell via the plasma membrane, eventually reaching the external serum lipoproteins.

It appears that the transfer of free sterol between a cell (enriched in this component) and the culture medium is similar to that which occurs between erythrocytes and cholesterol-depleted lipoproteins. Indeed, Rothblat and Kritchevsky [57] observed an increased transfer when medium-serum was not heat-inactivated: under these conditions the lecithin/cholesterol acyltransferase enzyme would be active and hence deplete the lipoproteins of some of their unesterified cholesterol; this would be partly replenished by transfer from the cell membrane. This type of mechanism has been frequently proposed (e.g. [62]).

The most important exchanging components in serum are the low-density lipoproteins. However, Stein and Stein [63] have shown that Landschütz Ascites cells lose approximately 4% of their unesterified cholesterol to a high-density apolipoprotein during a 2 hr incubation. Albumin did not substitute as an acceptor. The complete high-density lipoprotein was ~7 times more efficient than the apoprotein. This may reflect the denaturing properties of the delipidation technique, although co-dispersions of the apolipoprotein, phosphatidylcholine and sphingomyelin accepted 4 times as much sterol as did the apolipoprotein alone. It would appear therefore that the high-density lipoprotein can be at least partially reconstituted by this technique and that the phospholipid is important for efficient transfer of cholesterol. During the incubation with the apolipoprotein alone [63] some phosphatidylcholine is released from the cell to the protein: this may be necessary to effect an efficient transfer of sterol.

Studies with isolated membranes show that cholesterol exchange occurs equally rapidly through the inner and outer surfaces of erythrocyte membranes [64]. Also, exchange can be demonstrated in many membrane types [3,65]. Thus, incubation of low-density plasma lipoproteins with liver bile-fronts, intestinal microvillous membranes, muscle sarcolemma, kidney and liver microsomal membranes, and mitochondria from various sources, yields 80–96 % equilibration of the cholesterol pools in 24 hr. In the process, mitochondrial cholesterol can be enriched by as much as 95 % (10–95 %), yielding cholesterol/phospholipid ratios of 0.061 and 0.207 for the inner and outer membranes, respectively [3]. This has the following functional consequences (Table 5.2):

TABLE 5.2

Effects of cholesterol enrichment on some enzymic properties of rat-liver mitochondria[1]

Enzyme	Activity	
	Control	Cholesterol-enriched
Succinate-cytochrome *c* reductase[2]	8.4	19.2
NADH oxidase[2]	0.45	0.50
Rotenone-insensitive, NADH-cytochrome *c* reductase[2]	0.18	0.20
Fatty acid oxidase[2]	0.12	0.13
Monoamine oxidase[3]	0.11	0.11
ATPase[2] – dinitrophenol	3.7	3.6
ATPase[2] + dinitrophenol	11.2	13.0

[1] From Graham and Green [3].
[2] μmoles/mg protein/hr.
[3] Increase of absorbance at 250 nm/mg protein/hr.

(a) Cholesterol-enriched mitochondria are resistant to large amplitude swelling.
(b) Activities of NADH-oxidase, rotenone-insensitive NADH-cytochrome *c* reductase, fatty acid oxidase and monoamine oxidase do not change significantly with cholesterol enrichment.
(c) Cholesterol enrichment does not appear to affect respiratory control.
(d) Succinate-cytochrome *c* reductase activity increases upon cholesterol enrichment. This effect is reversible.

The recent study of Smith and Green [66] provides an interesting model approach to the study of cholesterol exchange. They used a fluorescent sterol, sterophenol,

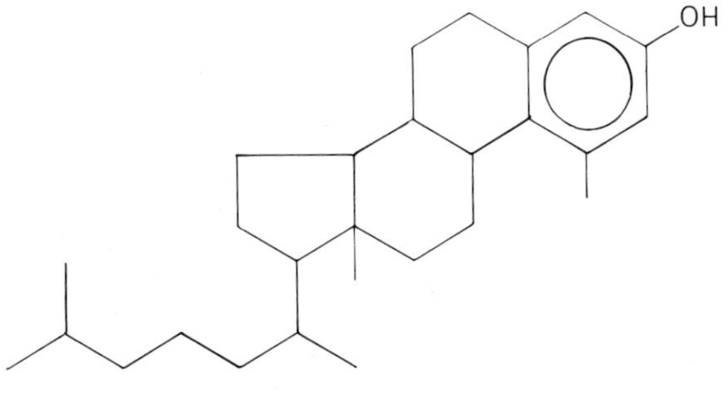

as a cholesterol analogue, and utilized the quenching of this sterol's fluorescence by I^- to measure the rate of its transmembrane exchange. Their approach is analogous to the use of spin-label techniques to estimate phospholipid 'flip-flop' in lipid bilayers [67]. They observed an average half-time for sterophenol 'flip-flop' of 72 min. This is much less than the value of 6.5 hr measured for spin-labelled phospholipids in liposomes but slower than the values obtained for phosphatide spin-labels in various biomembranes (Chapter 3). The possibility exists that the two pools proposed by Cooper et al. [51] represent the inside and outside portions of erythrocyte membranes.

No adequate, general mechanism has been proposed for the cholesterol 'selectivity' of cellular membranes and one aspect of cholesterol exchangeability in subcellular membranes stands out particularly; namely, some normal intracellular membranes will not accumulate cholesterol even when pushed. Thus, mitochondrial inner membranes normally have cholesterol/phospholipid ratios <0.05 (Table 5.1). However, even a 95 % enrichment, induced by equilibration with plasma lipoproteins, [56] yields a ratio of <0.1, much lower than that of plasma membranes. This suggests that these membranes normally discriminate against cholesterol; this appears in contradistinction to the *tumor* situation (see below).

Haberland and Reynolds [1] argue, with some reason, that membrane proteins are improbable cholesterol ligands because of the favorable self-association energies of cholesterol. Indeed, no specific cholesterol 'carrier' proteins have been identified. However, 'sterol carrier proteins' [68] may serve this function. In any case, some cholesterol-metabolizing enzymes must have substantial, specific affinities for cholesterol; but these enzymes could hardly account for the different cholesterol contents of intracellular membranes. If phosphatide is determining, as Haberland and Reynolds [1] have suggested, one must invoke different phospholipid compositions and/or arrays in diversr membrane systems; extensive experimentation indicates that these do exist. Thus, De Kruyff et al. [12] demonstrate in model systems that cholesterol preferentially interacts with phosphatidyl choline species showing the lowest transition temperatures in the case of phosphatide mixtures which differ by 4C-atoms or more. Accordingly, one might observe non-random distributions of cholesterol in biomembranes with heterogeneous fatty acid compositions. On the other hand, as pointed out by Oldfield and Chapman [69], in a mixed lipid system the presence of cholesterol would tend to convert liquid mixed gel and liquid-crystalline domains into an overall liquid-crystalline system.

In a given cell, with a given machinery for phosphatide biosynthesis, the different phospholipid compositions of various membrane systems must arise from the relative phosphatide affinities of membrane proteins. We reason, therefore, that the different cholesterol compositions of the various membrane systems in a given cell also reflect differences in membrane protein composition, even though the membrane cholesterol may not be primarily protein associated.

5.5. Altered cholesterol content and biosynthesis in tumors

5.5.1. Cholesterol content

Several hepatomas contain much more cholesterol than normal livers. Thus, Snyder et al. [70] obtain cholesterol values of 3.8–4.1 wt % for normal livers but 9.6–10.4 and 7.1–7.5 wt % for hepatoma 7777 and hepatoma 7794A respectively. Sterol ester levels in the hepatomas are also elevated (6.6–7.3 and 28.0–28.2 wt % in 7794A and 7777 respectively, vs. 3.8–4.4 wt % in normal livers).

Rats bearing hepatoma 7777 exhibit a 50 % increase in plasma free cholesterol and a virtual doubling of cholesterol esters [71]. This is remarkable because the *livers* of the *tumor*-bearing animals take up *dietary* cholesterol far more efficiently than the livers of normal rats and vastly more efficiently than the hepatomas do (Table 5.3). This low uptake by the tumor is typical of hepatomas and is a significant feature of the defective feedback control of cholesterol biosynthesis seen in liver tumors, including hepatoma 7777 (see below). It thus appears that the high cholesterol content of hepatoma 7777 and the high levels of plasma cholesterol seen in the tumor-bearing animals arise from excessive sterol synthesis by the tumor.

TABLE 5.3

Uptake of dietary cholesterol by hepatoma 7777, host livers and livers of normal rats

Tissue	Cholesterol content[1] (wt %)	Cholesterol uptake[2]
Livers of control animals	3.8–4.1	77
Livers of animals bearing hepatoma 7777		130
Hepatoma 7777	9.6–10.4	1.2
Hepatoma 7794A	7.1–7.5	—

[1] From Snyder et al. [70].
[2] Calculated from (Grigor et al. [71]). Uptake in cpm cholesterol-7-^{3}H per gm tissue 1 hr after i.v. injection of 8 μCi of labelled sterol.

Van Hoeven and Emmelot [72] have carried out lipid analysis on plasma membranes of normal and neoplastic hepatocytes and report higher cholesterol contents and higher cholesterol/phospholipid ratios in the plasma membranes of one rat and two mouse hepatomas than in membranes from normal cells (Table 5.4). This is very relevant to the generally defective dietary feedback control

TABLE 5.4

Cholesterol content and cholesterol/phospholipid ratios in plasma membranes isolated from liver and hepatomas[1]

Tissue	Cholesterol (μmoles/mg membrane protein)		Cholesterol/phospholipid ratio (mole/mole)
	Free	Esterified	
Rat liver	0.223	0.003	0.65 ± 0.06
Rat hepatoma 48A	—	—	0.89 ± 0.02
Rat hepatoma 484A	0.584	0.061	0.89 ± 0.02
Mouse liver	0.282	0.008	0.80 ± 0.01
Mouse hepatoma 147042	0.374	0.021	1.00 ± 0.06
Mouse hepatoma 143066	0.338	0.032	1.08 ± 0.06
Mouse hepatoma 41896	—	—	1.00 ± 0.006

[1] From van Hoeven and Emmelot [72].

of cholesterol biosynthesis in tumors, but these tumors were not characterized in this regard. The authors also find unusually high-cholesterol/phospholipid ratios in hepatoma microsomes (0.52 vs. 0.12 in normal livers). This may be due to the tendency of hepatoma plasma membrane fragments to co-sediment with microsomes.

Recent studies [73,74] also show a striking difference in the cholesterol/phospholipid molar ratios of microsomal and mitochondrial membranes from normal liver and hepatoma 5123 (Table 5.5). Consistent with the data of van Hoeven and Emmelot [72], the hepatoma microsomes exhibit high cholesterol/phospholipid ratios. A very dramatic cholesterol-enrichment was also observed in the hepatoma microsomes.

TABLE 5.5

Cholesterol in mitochondrial and microsomal membranes of normal rat liver and hepatoma 5123[1]

Source	Cholesterol (μg/mg protein)	Cholesterol/phospholipid ratio (mole/mole)
Liver		
Mitochondria	12.3	0.12
Microsomes	25.9	0.17
Hepatoma 5123		
Mitochondria	40.8	0.86
Microsomes	25.8	0.37

[1] From Feo et al. [74].

Several reports deal with the cholesterol content of neoplastic cells of other than liver origin. Thus, Chao et al. [75] found that the cholesterol/phospholipid molar ratios of embryonic fibroblasts, a polyoma transformant therefore and a spontaneous transformant *in vitro* were 0.30, 0.30 and 0.34 respectively. When the two tumor types were grown *in vivo*, ratios ranged from 1.0–1.05. Reculture again produced lower cholesterol/phosphatide ratios. It is conceivable that the *in vivo* ratio represents a real pattern and that the low *in vitro* values are due to cholesterol depleting media. Unfortunately, such tissue analyses may be not very meaningful in view of the large differences in the cholesterol content of various intracellular membranes.

Inbar and Schinitzky [76, 76a] claim that the plasma membranes of leukemic lymphocytes are markedly depleted in cholesterol. They report values of 31 μg/mg cell lipid for a Moloney-virus lymphoma compared to 54 μg/mg cell lipid in normal lymphocytes. A subsequent report cites 53.5 μg cholesterol/mg of cell lipid for rat lymphocytes and 54.7 μg/mg for mouse lymphocytes, to be compared with 51.8 μg/mg and 51.6 μg/mg for two types of leukemic mouse lymphocytes. The corresponding cholesterol/phospholipid molar ratios were 0.25 and 0.26 for rat lymphocytes and mouse lymphocytes and 0.19 and 0.17 for D2 mouse leukemias. On the basis on these data, it is concluded that a decrease in the molar ratio of cholesterol to phospholipids leads to a 'higher fluidity' of the lipids in the surface membranes of neoplastic lymphocytes. However, the authors do not provide separate cholesterol or phospholipid analyses of various subcellular membranes. This is regrettable since cytoplasmic membranes have lower cholesterol values than plasma membranes. One can accordingly not draw firm conclusions from whole cell analyses. Moreover, the role of cholesterol in membranes depends not only on the cholesterol/phosphatide ratio, but also on phosphatides unsaturation. In fact, as pointed out already, cholesterol tends to increase the fluidity of phospholipids *below* their transition temperature and fosters the intermixing of lipids which, in the absence of cholesterol, form separate phases. Accordingly, conclusions about plasma membrane 'fluidity', based on whole cell analyses and whole cell experiments with fluorescent probes of the type which might localize in all cellular membranes, appear rather overextended. Finally, the low cholesterol values recorded by Inbar and Schinitzky [76] also appear unusual, because they can be increased by equilibration of the cells with cholesterol-lecithin liposomes, but cannot be decreased by equilibrating the cells with cholesterol-free lecithin liposomes.

The data of Inbar and Schinitzky [76a] can also not be reconciled with the cholesterol values reported for purified plasma membranes isolated from leukemic mouse [78] or human [78a] lymphocytes. The membranes from mouse cells contain 164 μg cholesterol/mg membrane protein and their cholesterol/phospholipid ratio of 0.9 is in excellent agreement with the values of 0.99 to 1.03 reported for most lymphoid cells [5, 6, 79]. The membranes from human leukemic cells [78a] contained 185 μg/mg protein and the low cholesterol phospholipid ratio observed (0.4) is due to unusually *high* phospholipid levels in these membrane isolates. There is no cholesterol 'depletion'.

Perdue et al. [80–82] have compared the composition of plasma membranes from normal chick embryo fibroblasts with membranes from RNA virus-induced chicken sarcomas grown *in vivo* and *in vitro*. The plasma membrane isolates from 12-day chick embryo cells contained 1998 μg phospholipid/mg membrane protein and 613 μg cholesterol/mg. The values for the virus converted cells were 1979 μg/mg and 619 μg/mg respectively. However, the cholesterol/phospholipid ratios were 0.66 in both normal and neoplastic cells. When the transformed cells were grown *in vivo*, the plasma membrane isolates gave maximal phospholipid values of 1134 μg/mg and cholesterol levels of 273 μg/mg protein. The cholesterol/phospholipid molar ratio, 0.64, does not differ significantly from the 0.66 obtained for normal or neoplastic cells propagated *in vitro*. However, all of the ratios are below those generally found in plasma membranes ($\sim$1.0) This may be due to contamination of the plasma membrane fraction by other membranous components some of which (e.g., Fractions D and E) exhibit cholesterol/phospholipid ratios of <0.5 and correspond to $\sim$10 times the membrane mass of the plasma membrane fractions.

Summarizing, there is evidence that plasma membranes, as well as some intracellular membranes of some hepatomas are enriched in cholesterol compared to the membranes of normal hepatocytes. The data now available show no comparable phenomenon in neoplastic fibroblasts or lymphocytes. There is no evidence for depletion of cholesterol in neoplastic lymphocytes.

The high cholesterol/phospholipid ratios reported for hepatoma mitochondria and microsomes appear highly significant since they occur in a range where changes in cholesterol proportion most profoundly influence the physical state of membrane lipids. From the purely lipid point of view one would expect the hepatoma membranes to be much less fluid than those of normal cells. This could have major functional consequences in terms of membrane permeability and the function of membrane-associated enzymes, including those participating in cholesterol biosynthesis.

5.5.2. Cholesterol biosynthesis and its regulation in neoplasia

5.5.2.1. Cholesterol biosynthesis in leukemias

The leukemic granulocytes of a spontaneous rat ascites leukemia exhibit the same rate of cholesterol biosynthesis as normal granulocytes [30]. However, unlike in normal cells, the rate in leukemic leucocytes is not subject to dietary feedback inhibition. Siperstein [30] reports that the L_2C leukemia of guinea pigs is capable of very rapid cholesterol synthesis and that invasion of various tissues of the body by the leukemic cells leads to marked increase in cholesterol synthesis in organs such as the liver. While lymphocytes of normal lymph nodes consistently show definite feedback inhibition of cholesterol synthesis, the L_2C cells were completely unaffected by feeding of exogenous cholesterol (Table 5.6)*.

TABLE 5.6

Loss of feedback control of cholesterol biosynthesis in guinea pig L_2C leukemia[1]

Tissue	Cholesterol biosynthesis (nmoles acetate-2-^{14}C incorporated per g tissue or per 10^9 cells per hr)	
	Normal Diet	5 % Cholesterol
Normal liver	5.3	0.3
Leukemia liver	47.5	38.3
Normal lymph nodes	9.9	0.7
L_2C cells	55.0	50.0

[1] From Siperstein [30].

Leukemic mouse lymphocytes of thymic origin isolated from spleen, lymph nodes and thymuses show 10-*fold higher rates of cholesterol biosynthesis* than normal cells [33]. This increase correlates well with these cells' high HMG-CoA-reductase levels. Leukemic lymphocytes from peripheral blood also exhibit enhanced cholesterol biosynthesis [83]. The rate of sterol production by the neoplastic cells is inhibited about 50% by cholesterol feeding, but only after 7 days; even then the inhibited values are still 5–6 times higher than those of normal cells. The increased rate of sterol synthesis does not produce raised plasma cholesterol levels, such as found in animals bearing hepatoma 7777. Also, fatty acid biosynthesis is not much greater than it is in normal cells. The high rates of intrinsic sterol biosynthesis point to defective endogenous feedback control. Dietary feedback control, although detectable, is also clearly deficient**.

5.5.2.2. Cholesterol biosynthesis in hepatomas

5.5.2.2.1. Endogenous control

Ample evidence indicates that hepatomas synthesize cholesterol at an excessive rate *in vivo* and that the tumors [70, 71] and their membranes [72] are enriched

* Our own experiments (Phillipot and Wallach, to be published) show that the cholesterol biosynthesis of L_2C cells *can* be inhibited *in vitro* by 25-hydroxycholesterol *and* by a nitroxide derivate of androstane.

** Recent data (Kandusch, A.A. personal communication) show that exposure of leukemic lymphocytes to small levels of inhibitory sterols (e.g., 25 OH cholesterol) markedly lowers cholesterol biosynthesis by leukemic lymphocytes, although not to the low levels found in inhibited normal cells.

in cholesterol. This must be due to defective endogenous control of cholesterol biosynthesis.

5.5.2.2.2. Dietary feedback control

In 1961, Siperstein and La Marr [84] reported that the feeding of a high cholesterol diet to C57L/J mice blocked cholesterol biosynthesis in their livers, as expected, *but not* in BW 7756 hepatomas carried as subcutaneous transplants in the same animals. The detailed experiments [85,86] were soon confirmed and extended [87,88] and defective feedback inhibition of cholesterol biosynthesis by dietary cholesterol has now been reported for all hepatomas tested *in vivo*. These include three mouse tumors, 19 rat hepatomas (both 'minimal deviation' and poorly differentiated), several aflatoxin-induced trout hepatomas [89] and two human hepatomas [90]. Moreover, the livers of trout and rats lose the feedback control mechanism shortly after treatment with the hepatocarcinogen aflatoxin, months or years before development of tumor [91]. It is known that aflatoxin interferes with the association of polysomes and the membranes of endoplasmic reticulum, but the effect occurs also with other hepatocarcinogens as shown by Horton and associates (92–95). This information is summarized in Table 2 of Sabine's review [29]. The phenomenon does not appear to be related to hepatotoxicity [96]. However, Horton and associates [93,94] show that although carcinogen treatment consistently abolishes dietary feedback control of sterol biosynthesis, control is often regained, even after continued exposure to carcinogen.

We must here stress that impaired feedback of cholesterol biosynthesis in neoplastic cells has been *consistently* demonstrated only *in vivo*, that is, in the only situation properly representative of the malignant state. Thus, Howard et al. [97] report that while SV40-transformed, human WI38 fibroblasts exhibit 4-fold greater *in vitro* biosynthesis of cholesterol from [^{14}C]acetate than the parental WI38 cells, cholesterol biosynthesis is turned off in both cell types within 1–2 days upon transfer to cholesterol-enriched media. Also, HTC cells, which descend from hepatoma 7288 C, but become adapted to culture *in vitro*, show marked depression of sterol biosynthesis when cholesterol is added to the culture medium [98,41]. Yet hepatoma 7288 C, propagated *in vivo*, does not respond to cholesterol feeding [99].

What mechanisms might account for the lack of dietary feedback control of cholesterol biosynthesis in hepatomas or other cells? Sabine's review [29] suggests three possibilities:

(a) *First*, tumors might synthesize cholesterol by pathways that do not involve the HMG-CoA reductase step. However, several workers have proven that in hepatomas the defect does lie at the HMG-CoA-reductase level [100–102] or very close to it. This is illustrated in Table 5.7. (However, we caution that levels of HGM-CoA reductase *activity* cannot be equated *a priori* with levels of enzyme *protein*.)

TABLE 5.7

Effect of cholesterol feeding on cholesterol biosynthesis and activity of HMG-CoA reductase in livers and hepatomas of ACI/j MAI rats[1]

Tissue[2]	Cholesterol synthesis[3]	HMG-CoA reductase[4]
Liver (control)[5]	90.0	0.77 ± 0.26
Liver (high cholesterol)[5]	1.6	0.007 ± 0.01
Hepatoma 9121 (control)[5]	91.0	1.34 ± 0.32
Hepatoma 9121 (high cholesterol)[5]	192.0	1.69 ± 0.38

[1] From Siperstein et al. [100].
[2] Tumors grown intramuscularly.
[3] 2-[^{14}C]acetate incorporation into cholesterol (nmoles/gm liver slices).
[4] nmoles [^{14}C]HMG-CoA incorporated into mevalonate (per microsomes from 800 mg liver per hour).
[5] Control = low cholesterol diet; high cholesterol = 5 % cholesterol diet.

(b) *Second*, tumors might contain an altered HMG-CoA reductase. However, Kandutsch and Hancock [102] find no difference between normal and tumor enzyme extracts in terms of K_M and heat stability. Moreover, Brown et al. [40] have solubilized and purified HMG-CoA reductase from liver and hepatoma 9121 and find no feedback suppression of the hepatoma enzyme. Brown et al. [40] also obtain equivalent enzyme yields from tumors and liver. The two enzyme preparations resemble each other with respect to the following: both gave molecular weights near 200,000 and heat inactivation between 72 °C and 75 °C; both showed inactivation at 4 °C with first order kinetics, a half-life of 6 min and protection by 4 M KCl; both yielded identical K_M values for both NADPH and HMG-CoA. Such data indicates that hepatomas and liver synthesize identical or closely related enzymes. One must therefore determine whether the lack of feedback suppression in hepatomas might not arise from the failure of cholesterol (or the immediate regulator) to suppress the synthesis of the normal enzyme.

(c) A *third* mechanism to be considered is a possible 'impaired access' of regulatory sterol to the tumor cells, or inadequate intracellular concentrations of the HMG-CoA reductase regulator within tumors. There is some support for this alternative. Thus, after pulse feeding of [^{14}C]cholesterol, the [^{14}C]-cholesterol levels in hepatomas lie well below those of livers [103]. Moreover, even after prolonged cholesterol feeding, transplantable hepatomas exhibit less cholesterol uptake from plasma than normal livers [44].

The results of the pulse experiments might be attributed to the lack of a portal blood supply in hepatomas. However, in the long-term feeding studies the tumors should be adequately exposed to dietary cholesterol via the plasma. The same applies to leukemic granulocytes grown intraperitoneally. Also the livers of rats treated with hepatocarcinogens develop loss of feedback control of sterol biosynthesis, as well as impaired cholesterol uptake, long before there are changes in vascularization [96]. Moreover, the last experiments show no correlation between impairment of sterol uptake and loss of feedback control. The same holds for hyperplastic liver nodules and primary hepatomas induced by chemical carcinogens [104]. These take up dietary cholesterol at only slightly lower rates than normal livers and to a much greater extent than transplantable hepatomas. Both the nodules and primary hepatomas show impaired inhibition of cholesterol biosynthesis. Since they show no significant impairment of cholesterol uptake, the lesion must lie at an intracellular locus. Finally, the studies of Snyder et al. [70], Grigor et al. [71] and van Hoeven and Emmelot [72] clearly demonstrate abnormally high cholesterol in hepatomas. Accordingly, defective control of cholesterol biosynthesis, whether endogenous or dietary cannot be easily related to overall cellular cholesterol levels. Moreover, as already discussed, cholesterol associates into micelles at concentrations $>5 \times 10^{-9}$ M and the micelles become unstable at equivalent cholesterol concentrations $>5 \times 10^{-6}$ M, that is, within the range of cholesterol levels found in cells (moles/liter cell water). Cellular cholesterol content thus gives little information about cholesterol 'activity'.

It is also difficult to invoke defective transit of the information carried by high 'extracellular cholesterol' to the control locus for HMG-CoA-reductase biosynthesis because (a) this does not explain defective endogenous control and (b) HTC cells in culture show adequate feedback regulation [41] whereas cells of the parent 7288 C hepatoma carried *in vivo* do not.

We must now consider the implications of experiments [36, 37, 39] indicating that several different, oxidized derivatives of cholesterol, as well as certain nitroxide-sterol derivatives, act as potent regulators or sterol biosynthesis, while cholesterol itself is ineffective. This discovery is perhaps not totally surprising, since the physical properties of cholesterol are such as to complicate its potential role as a regulatory substance. In all probability the diverse regulatory steroids are substantially more water soluble and may therefore achieve biologically-significant monomer concentrations.

What is remarkable, however, is the evidence that several sterols of widely divergent structure are equally potent as inhibitors of sterol biosynthesis, HMG-CoA-reductase activity and cell growth. This circumstance leads us to suspect that sterol-control of sterol biosynthesis may depend less on molecular specificity than on the effects of the regulatory sterols on membrane structure.

Indeed, since the sterols which act as effective inhibitors of sterol biosynthesis are considerably more polar than cholesterol, one can suggest that they will behave as membrane perturbants in the same fashion as documented for certain nitroxide analogues of cholesterol [105]. The controlling element in HMG-CoA reductase activity (biosynthesis) might then be the physical state of critical membrane domains. In these terms a defective control of cholesterol biosynthesis might derive from:

(a) Impaired generation of regulatory steroids. (b) A *membrane 'state'* which is not responsive to the perturbing action of regulatory sterols. A possible reason for such a postulated 'unresponsive membrane state' could be the high cholesterol content of the endoplasmic reticulum in hepatomas. It is thus conceivable that the high cholesterol/phospholipid ratios of hepatoma membranes constitute a 'proximal' response to a critical membrane alteration and that impaired regulation of sterol synthesis is a more remote effect.

Apart from these considerations one should explore the possible role of sterol carrier proteins (SCP). As reviewed by Scallen et al. [68] there exist at least two soluble liver proteins which participate in the conversion of squalene to cholesterol by enzymes in liver endoplasmic reticulum (SCP_1 and SCP_2). These carrier proteins appear to derive from the endoplasmic reticulum. They bear their substrates (e.g. squalene) in hydrophobic, non-covalent association. It appears possible that the substrate remains attached to the carrier throughout the various enzymatic steps leading finally to cholesterol. They might, in this fashion, function in the feedback control of cholesterol biosynthesis. There are some indications that SCPs exist also in tissues other than liver, but we know of no report on neoplastic cells of any kind. Clearly, the study of SCPs in normal and neoplastic cells could provide important insights into the mechanism in normal feedback control of sterol biosynthesis and the defective process observed in many neoplasms.

5.6. Exchange

The importance and large extent of cholesterol exchange by tumors *in vivo* has been well documented in the recent study on cholesterol metabolism of Ehrlich ascites carcinoma [106]. It is shown that only about 3% of the cholesterol accumulation of Ehrlich ascites carcinoma cells grown in mouse peritoneal cavity is due to *de novo* cholesterol synthesis. The remainder of the cellular cholesterol is apparently derived from lipoproteins, particularly the very low-density lipoproteins in the ascites fluid. In terms of the whole organism, therefore, there is a flow and exchange of cholesterol in the direction of the tumor. This pattern is very similar to what is observed for phospholipids injected intravenously into tumor-bearing animals (Chapter 3). However, while this flow may appear to be unidirectional *in vivo*, *in vitro* experiments show that Ehrlich ascites tumor cells can also very readily transfer cellular cholesterol to very low density lipoprotein.

5.7. Summation

Cholesterol is a major component in the membranes of animal cells. It is extremely insoluble in water and its membrane affinity derives from membrane phosphatides. Diverse membranes differ considerably in cholesterol content, with plasma membranes at the high end of the scale and mitochondria at the low. When present at molar ratios cholesterol/phospholipid >0.3 the sterol acts to buffer the fluidity of membrane lipids and allows the intermixing of lipids with very diverse hydrocarbon moieties. Cholesterol biosynthesis requires a number of membrane-bound enzymes and feedback control of cholesterol biosynthesis by sterols occurs at the level of HMG-CoA reductase, a membrane-bound enzyme. It appears that cholesterol is not the final regulator. The feedback control mechanism depends on protein biosynthesis.

Many tumors show increased cholesterol/phospholipid ratios in various membranes. The changes seen in mitochondria and microsomes are in a range where one might observe significant differences in the physical state of the membrane lipids. *In vivo*, feedback control of HMG-CoA-reductase activity – hence cholesterol biosynthesis – does not occur in all hepatomas tested so far. Feedback control is also lacking in precancerous livers and in some leukemias. However, at least one hepatoma, which lacks *in vivo* feedback suppression, shows this in cells adapted to culture. *In vivo* feedback suppression is conjectured to involve cellular membranes; the mechanisms involved remain to be established.

References

[1] Haberland, M.E. and Reynolds, J.A., *Proc. Natl. Acad. Sci. U.S.*, *70*, 2313 (1973).
[2] Rouser, G., Nelson, G.J., Fleischer, S. and Simon, G. in *Biological Membranes*, D. Chapman, ed., Academic Press, New York, 1968, p. 6.
[3] Graham, J.M. and Green, C., *Eur. J. Biochem.*, *12*, 58 (1970).
[4] Parson, D.F. and Yano, Y., *Biochim. Biophys. Acta*, *135*, 362 (1967).
[5] Ferber, E., Resch, K., Wallach, D.F.H. and Imm, W., *Biochim. Biophys. Acta*, *266*, 494 (1972).
[6] Schmidt-Ullrich, R., Ferber, E., Knüfermann, H. and Wallach, D.F.H., *Biochim. Biophys. Acta*, *332*, 175 (1974).
[7] Thines-Sempoux, D. in *Lysosomes in Biology and Pathology*, J.T. Dingle, ed., North Holland, Amsterdam, 1973, p. 278.
[8] Franke, W.W., Deumling, B., Zentgraf, H., Falk, H., Rae, P.M.M., *Exp. Cell. Res.*, *81*, 365 (1973).
[9] Jarasch, E.D., Reilly, C.E., Comes, P., Kartenbeck, J. and Franke, W.W., *Hoppe-Seyler's Z., Physiol. Chem.*, *354*, 974 (1973).
[10] Chapman, D. in *Biological Membranes*, D. Chapman and D.F.H. Wallach, eds., Academic Press, London, Vol. 2, 1973, p. 91.
[11] Träuble, H. and Eibl, H., *Proc. Natl. Acad. Sci. U.S.*, *71*, 214 (1974).
[12] de Kruyff, B., van Dijck, P.W.M., Demel, R.A., Schnijff, A., Brants, F. and van Deenen, L.L.M., *Biochim. Biophys. Acta*, *356*, 1 (1974).
[13] Phillips, M.C. and Finer, E.G., *Biochim. Biophys. Acta*, *356*, 199 (1974).
[13a] Wallach, D.F.H., Bieri, V., Verma, S.P. and Schmidt-Ullrich, *Ann. N.Y. Acad. Sci.*, 1975, in press.
[14] Huang, C.-H., Sipe, J.P., Chow, S.T. and Martin, R.B., *Proc. Natl. Acad. Sci. U.S.*, *71*, 359 (1974).
[15] Papahadjopoulos, D. and Watkins, J.C., *Biochim. Biophys. Acta*, *135*, 639 (1967).
[16] Papahadjopoulos, D., Nir, S. and Ohki, S., *Biochim. Biophys. Acta*, *266*, 561 (1972).

[17] deGier, J., Mandersloot, J.G. and van Deenen, L.L.M., *Biochim. Biophys. Acta*, *173*, 143 (1969).
[18] deGier, J., Haest, C.W.M., Mandersloot, J.G. and van Deenen, L.L.M., *Biochim. Biophys.* *211*, 373 (1970).
[19] Papahadjopoulos, D., Poste, G. and Schaeffer, B.E., *Biochim. Biophys. Acta*, *323*, 23 (1973).
[20] Kroes, J., Ostwald, R. and Keith, A.D., *Biochim. Biophys. Acta*, *274*, 71 (1972).
[21] de Kruyff, B., Demel, R.A. and van Deenen, L.L.M., *Biochim. Biophys. Acta*, *225*, 331 (1972).
[22] Kimelberg, H.K. and Papahadjopoulos, D., *Biochim. Biophys. Acta*, *282*, 277 (1972).
[23] Papahadjopoulos, D., *J. Theoret. Biol.*, *43*, 329 (1974).
[24] Rethy, A., Tomasi, V., Trevisani, A. and Barnabei, O., *Biochim. Biophys. Acta*, *290*, 58 (1972). (1972).
[25] Demel, R.A., Geurts van Kessel, W.S.M. and van Deenen, L.L.M., *Biochim. Biophys. Acta*, *266*, 26 (1972).
[26] Bloj, B., Morero, R.D. and Farias, R.N., *FEBS Letters*, *38*, 101 (1973).
[27] Dempsey, M.E., *Annu. Rev. Biochem.*, *43*, 967 (1974).
[28] Gaylor, J.L., *Adv. Lipid Res.*, *10*, 89 (1972).
[29] Sabine, J.R. in *Tumor Lipids*, Randall Wood, ed., American Oil Chemists Society Press, Champaign, Ill., 1973, p. 21.
[30] Siperstein, M.D. in *Current Topics in Cellular Regulation*, Academic Press, New York, Vol. 2, 1970, p. 65.
[31] Takeuchi, N., Iritani, N., Wells, W.W. and Westerman, M.P., *Biochim. Biophys. Acta*, *125*, 375 (1966).
[32] Polsky, F.I., Brown, M.S. and Siperstein, M.D., *J. Clin. Invest.*, *52*, 65a (1973).
[33] Chen, H.W., Kandutsch, A.A., Heiniger, H.J. and Meier, H., *Cancer Res.*, *33*, 2774 (1973).
[34] Avigan, J., Williams, C.D. and Blass, J.P., *Biochim. Biophys. Acta*, *218*, 381 (1970).
[35] Rothblat, G.H. and Burns, C.H., *J. Lipid. Res.*, *12*, 653 (1971).
[36] Kandutsch, A.A. and Chen, H.W., *J. Biol. Chem.*, *248*, 8404 (1973).
[37] Kandutsch, A.A. and Chen, H.W., *J. Biol. Chem.*, *249*, 6057 (1974).
[38] Bates, S.R. and Rothblat, G.H., *Biochim. Biophys. Acta*, *360*, 38 (1974).
[39] Chen, H.W., Kandutsch, A.A., Waymouth, C., *Nature*, *251*, 419 (1974).
[40] Brown, M.S., Goldstein, J.L. and Siperstein, M.D., *Fed. Proc.*, *32*, 2168 (1973).
[41] Watson, J.A. in *Tumor Lipids*, R. Wood, ed., American Oil Chemists Society Press, Champaign, Ill., 1973, p. 34.
[42] Hickman, P.E., Horton, B.J. and Sabine, J.R., *J. Lipid Res.*, *13*, 17 (1972).
[42a] Higgins, M.J.P., Brady, D. and Rudney, H., *Arch. Biochem. Biophys.*, *163*, 271 (1974).
[43] Horton, B.J. and Sabine, J.R., *Eur. J. Cancer*, *9,1*, 11 (1973).
[44] Harry, D.S., Morris, H.P. and McIntyre, N., *J. Lipid Res.*, *12*, 313 (1971).
[45] Rothblat, G.H., *Adv. Lipid Res.*, *7*, 135 (1969).
[46] Bruckdorfer, K.R. and Green, C., *Biochem. J.*, *104*, 270 (1967).
[47] Bruckdorfer, K.R., Edwards, P.A. and Green, C., *Eur. J. Biochem.*, *4*, 506 (1968).
[48] Bruckdorfer, K.R., Graham, J.M. and Green, C., *Eur. J. Biochem.*, *4*, 512 (1968).
[49] Bruckdorfer, K.R., Demel, R.A., deGier, J. and van Deenen, L.L.M., *Biochim. Biophys. Acta*, *183*, 334 (1969).
[50] Quarfordt, S.H. and Hilderman, H.L., *J. Lipid Res.*, *11*, 528 (1970).
[51] Cooper, R.A., Dilroy-Puray, M., Lando, P. and Greenberg, M.S., *J. Clin. Invest.*, *51*, 3182 (1972).
[52] Cooper, R.A., *J. Clin. Invest.*, *48*, 1820 (1969).
[53] Kroes, J. and Ostwald, R., *Biochim. Biophys. Acta*, *249*, 647 (1971).
[54] Abell, L.L., Levy, B.B., Brodie, B.B. and Kendall, F.E., *J. Biol. Chem.*, *195*, 357 (1952).
[55] Bell, F.P. and Schwartz, C.J., *Biochim. Biophys. Acta*, *231*, 553 (1971).
[56] Bailey, J.M., *Exp. Cell. Res.*, *37*, 175 (1965).
[57] Rothblat, G.H. and Kritchevsky, D., *Biochim. Biophys. Acta*, *144*, 423 (1967).
[58] Bailey, J.M. and Butler, J., *Arch. Biochem. Biophys.*, *159*, 580 (1973).
[59] Werb, Z. and Cohn, Z.A., *J. Exptl. Med.*, *134*, 1545 (1971).
[60] Werb, Z. and Cohn, Z.A., *J. Exptl. Med.*, *134*, 1570 (1971).
[61] Werb, Z. and Cohn, Z.A., *J. Exptl. Med.*, *135*, 21 (1972).

[62] Glomset, J.A., *J. Lipid Res.*, *9*, 155 (1968).
[63] Stein, O. and Stein, Y., *Biochim. Biophys. Acta*, *326*, 232 (1973).
[64] Bruckdorfer, K.R., personal communication.
[65] Graham, J.M. and Green, C., *Biochem. J.*, *103*, 16C (1967).
[66] Smith, R.J.M. and Green, C., *FEBS Letters*, *42*, 108 (1974).
[67] Kornberg, R.D. and McConnell, H.M., *Proc. Natl. Acad. Sci. U.S.*, *68*, 2564 (1971).
[68] Scallen, T.J., Srikantaiah, M.V., Seetharam, B., Hansbury, E. and Gavey, K.L., *Fed. Proc.*, *33*, 1733 (1974).
[69] Oldfield, E. and Chapman, D., *FEBS Letters*, *23*, 285 (1972).
[70] Snyder, F., Blank, M.L. and Morris, H.P., *Biochim. Biophys. Acta*, *176*, 502 (1969).
[71] Grigor, M.R., Blank, M.L. and Snyder, F., *Cancer Res.*, *33*, 1870 (1973).
[72] Van Hoeven, R.P. and Emmelot, P. in *J. Memb. Biol.*, *9*, 105 (1972).
[73] Filipek-Wender, H., Karon, H. and Torlinski, L., *Acta Biol. Med. Germ.*, *29*, 823 (1972).
[74] Feo, F., Canuto, R.A., Bertone, G., Garcea, R. and Pani, P., *FEBS Letters*, *33*, 229 (1973).
[75] Chao, F.C., Eng., L.F. and Griffin, A., *Biochim. Biophys. Acta*, *260*, 197 (1972).
[76] Inbar, M. and Shinitzky, M., *Proc. Natl. Acad. Sci. U.S.*, *71*, 2128 (1974).
[76a] Inbar, M. and Shinitzky, M., *Proc. Natl. Acad. Sci. U.S.*, *71*, 4229 (1974).
[77] Vlodavsky, I., Sachs, F., *Nature*, *250*, 67 (1974).
[78] Warley, A. and Cook, G.M.W., *Biochim. Biophys. Acta*, *323*, 55 (1973).
[78a] Marique, D. and Hildebrand, J., *Cancer Res.*, *33*, 2761 (1973).
[79] Allan, D. and Crumpton, M.J., *Biochem. J.*, *120*, 133 (1970).
[80] Perdue, J.F., Kletzien, R. and Miller, K., *Biochim. Biophys. Acta*, *249*, 419 (1971).
[81] Perdue, J.F., Kletzien, R. and Wray, V.L., *Biochim. Biophys. Acta*, *266*, 505 (1972).
[82] Perdue, J.F. and Miller, K., *Biochim. Biophys. Acta*, *298*, 817 (1973).
[83] Chen, H.W. and Heiniger, H.J., *Cancer Res.*, *34*, 1304 (1974).
[84] Siperstein, M.D. and La Marr, D.V., *Proc. 5th Internatl. Congr. Biochem.*, *9*, 420 (1961).
[85] Siperstein, M.D. and Fagan, V.M., *Clin. Res.*, *12*, 67 (1964).
[86] Denton, M.D. and Siperstein, M.D., *Clin. Res.*, *12*, 265 (1964).
[87] Sabine, J.R., Abraham, S. and Chaikoff, I.L., *Biochim. Biophys. Acta*, *116*, 407 (1966).
[88] Sabine, J.R., Abraham, S. and Chaikoff, I.L., *Cancer Res.*, *27*, 793 (1967).
[89] Siperstein, M.D. and Luby, L.J. in *Fish in Research*, Neuhaus, O., W. and Halner, J.E., eds., Academic Press, New York, 1969, p. 87.
[90] Siperstein, M.D. and Fagan, V.M., *Cancer Res.*, *24*, 1108 (1964).
[91] Siperstein, M.D., *J. Clin. Invest.*, *45*, 1073 (1966).
[92] Horton, B.J., Horton, J.D. and Sabine, J.R., *Biochim. Biophys. Acta*, *239*, 475 (1971).
[93] Horton, B.J. and Sabine, J.R., *Eur. J. Cancer*, *7*, 459 (1971).
[94] Horton, B.J., Horton, J.D. and Sabine, J.R., *Eur. J. Cancer*, *8*, 437 (1972).
[95] Horton, B.J. and Sabine, J.R., *Eur. J. Cancer*, *9*, 1 (1973).
[96] Horton, B.J., Horton, J.D. and Pitot, H.C., *Cancer Res.*, *33*, 1301 (1973).
[97] Howard, B.V., Butler, J.D., Bailey, J.M. in *Tumor Lipids*, R. Wood, ed., American Oil Chemists Society, Champaign, Ill., 1973, p. 200.
[98] Watson, J.A., *Lipids*, *7*, 146 (1972).
[99] Siperstein, M.D., Fagan, V.M. and Morris, H.P., *Cancer Res.*, *26*, 7 (1966).
[100] Siperstein, M.D., Clyde, A.M. and Morris, H.P., *Proc. Natl. Acad. Sci. U.S.*, *68*, 315 (1971).
[101] Goldfarb, S. and Pitot, H.C., *Cancer Res.*, *31*, 1879 (1971).
[102] Kandutsch, A.A. and Hancock, R.L., *Cancer Res.*, *31*, 1396 (1971).
[103] Sabine, J.R., Horton, B.J. and Tan, C.S., *Proc. 10th Internat. Cancer Conf.*, *282* (1970).
[104] Horton, B.J., Mott, G.E., Pitot, H.C. and Goldfarb, S., *Cancer Res.*, *33*, 460 (1973).
[105] Bieri, V., Lin, P.-S. and Wallach, D.F.H., *Proc. Natl. Acad. Sci. U.S.*, *71*, 4797 (1974).
[106] Brenneman, D.E., McGee, R. and Spector, A.A., *Cancer Res.*, *34*, 2605 (1974).

6

Enzymes and enzyme regulation

6.1. Introduction

Neoplastic cells very commonly exhibit a quantitatively abnormal composition of various enzymes or a different distribution than normal of various isozymes. This includes also a large number of membrane associated enzymes, many of which are treated, in connection with specific membrane functions.

In this chapter, we will first present an introductory exploration of the topic. We will then address three important regulatory systems, which involve the interaction of a large number of different enzymes, some of them membrane associated and some of them depending on some membrane function for their interaction.

Weber [1] has proposed a unifying hypothesis, the *Molecular Correlation Concept*, to account for the multiplicity of enzyme alterations in neoplasia. This hypothesis derives primarily from studies on hepatomas and proposes the following:

(1) Neoplasia involves an ordered pattern of gene expression which is reflected as an imbalance of 'opposing and competing key enzymes' and of overall metabolic patterns in both anabolic and catabolic pathways.

(2) The metabolic imbalance in neoplasia produces a shift in isozyme distributions towards the emergence of enzymes with low K_M-values and depletion of the high K_M enzymes that are subject to physiological control loops.

(3) The isozyme shifts and metabolic anomalies vary directly with *growth rate* suggesting that the rate of cell proliferation and development of metabolic imbalance represent a common pattern of gene expression. However, the correlation with growth holds only for neoplasia and not for fetal differentiating or regenerating liver.

(4) Tumor cells gain a selective advantage over normal cells by a progressive predominance of anabolic metabolism vs. catabolism.

The Molecular Correlation Concept is an important hypothesis, which, as we shall show, has also been extended to certain membrane enzymes [1]. The hypothesis will be difficult to test but it has obvious heuristic significance.

Numerous authors have compared the enzyme composition of plasma membrane preparations isolated from normal rat livers and from hepatomas (e.g., Morris hepatomas). As we have noted before (Chapters 2 and 3) such membranes do not behave identically upon subcellular fractionation. For example, 5′-nucleotidase is particularly enriched in the bile canaliculi which, when fragmented, tend to sediment not with the bile-front preparations but with the microsomes. However, it is well recognized that one of the early manifestations of neoplastic transformation in liver is a loss of bile canalicular specialization.

Graham [2] has compared the Na^+/K^+-Mg^{2+} ATPase of microsomes from normal BHK fibroblasts and polyoma-transformed BHK cells. The cells were disrupted by nitrogen cavitation (see Chapter 2) and the microsomes were isolated by zonal centrifugation from the 'large-particle' supernatant using a sucrose step gradient of a density between 1.02–1.10. The resulting low density fraction, containing the microsomal membranes, was then subfractionated in a 5–25% (w/w) dextran (mol. wt. 40,000) gradient. Both, normal and transformed cells yielded five bands of which two contained most of the ATPase activity, suggesting a concentration of plasma membrane material. In the transformed cells the specific Na^+/K^+-Mg^{2+}-ATPase activity was only ~50% of that found in the membranes of normal cells. However, the K_M for ATP of the Na^+/K^+-Mg^{2+} ATPase, estimated from Lineweaver–Burk plots, was calculated to be 0.028 mM in the membranes of transformed cells, compared with 0.25 mM for the membranes of normal cells. This 10-fold increase in affinity for ATP is remarkable, particularly if one considers the likelihood that the ATPase measured probably reflects transport activities, including amino acid transport, and that polyoma-converted BHK fibroblasts exhibit enhanced amino acid transport (Chapter 7).

Barclay and Terebus Kekish [3] have found that 'plasma membrane' preparations isolated from Morris hepatoma 5123 tc exhibit much lower (by a factor of 30) specific Na^+/K^+-ATPase-activities than membranes from normal cells. This is also true for various subfractions obtained by ultracentrifugation in sucrose density gradients. 5′-Nucleotidase also is diminished in hepatoma membranes compared with membranes isolated from normal liver.

When the plasma membrane fragments were extracted with 0.15 M NaCl [3] 25% of the protein was removed from the membranes of normal hepatocytes, compared with 14% for the tumor membranes. After sonication, the maximum 5′-nucleotidase- and Na^+/K^+-ATPase-activities were found in a low density fraction (density ~1.125) with both normal and tumor material. The authors suggest that the low ATPase and 5′-nucleotidase levels observed indicate impaired purine–pyrimidine and nucleotide metabolism in the hepatomas, but before this conclusion can be accepted, it is necessary to establish the kinetic properties of the tumor enzymes as well as to check for fractionation artifacts.

Plasma membrane preparations isolated from rat hepatoma 484 A (4-dimethylaminoacylbenzene-induced) exhibit more than 3 times the specific alkaline glycerolphosphatase activity found in membranes from normal cells [4]. However, in contrast to the case of normal cells, the tumor enzymes are not influenced by male sex hormones (e.g., testosterone). Moreover, the alkaline glycerolphosphatase activity can be increased by treatment with neuraminidase. This also occurs in the case of HeLa cells [5]. The increase of activity is some 20% and correlates to a release of neuraminic acid of about 70%.

Raftell and Blomberg [6, 7] and Blomberg and Raftell [8] have analyzed the activities of commonly used plasma membrane- and endoplasmic reticulum-marker enzymes in the membranes isolated from normal rat hepatocytes and rat hepatoma cells. The tumor was originally induced by 4-dimethylaminoazobenzene. The authors found by far the highest specific activity of nucleoside monophosphate hydrolase (presumably 5′-nucleotidase) in the plasma membrane fraction. They noted that the enzyme is decreased to a remarkable extent in the tumor membranes as well as in the homogenates and all the microsomal fractions. This basically excludes the possibility that the enzyme-bearing fragments sediment in a different fashion in the two cell types. However, since 5′-nucleotidase is particularly concentrated in bile canalicular regions and these tend to dedifferentiate in hepatocarcinogensis, it is not surprising to find a decrease of the enzyme in material isolated from neoplastic cells.

A different fractionation behavior is also probable because the relative distribution of whatever 5′-nucleotidase is present, showed an increase in activity in the so-called smooth microsomal membranes in the case of the tumors (probably fragments of plasma membranes). The hydrolysis of ADP and UDP were also concentrated in the smooth microsomal membranes from tumors relative to the comparable fractions from normal liver. Moreover, tumor homogenates showed decreased activities of ADPase and UDPase in comparison to homogenates of normal liver. The highest specific activities of UDPase and ADPase occurred in the rough microsomal membranes, from normal hepatocytes, whereas these enzyme activities are significantly enriched in the smooth microsomal membrane fraction from hepatoma cells. The endoplasmic reticulum enzyme, glucose-6-phosphatase, is rather low in hepatoma microsomes compared with microsomes from normal hepatocytes [4, 5, 6–8].

Blomberg and Raftell [8] have investigated various enzyme-active membranes in their liver-hepatoma systems using immunoprecipitation techniques. Membrane antigens were extracted with 1% Na-deoxycholate (DOC) and 0.5% Lubrol W. They found two antigens with esterase activity in both tumor and normal liver plasma membrane preparations. However, the tumor plasma membranes contained one immunologically distinct esterase and this was precipitable only with anti-tumor serum. When they analyzed the glycosidase activity in the plasma membrane preparations of various cell types, they only found β-D-glucuronidase in membranes isolated from normal tissues, whereas tumor membranes contained three glycosidases, namely β-glucuronidase, β-galactosidase and *N*-acetyl-β-glucosaminidase. These enzymes were not restricted to the plasma membrane preparations but were also found in other subcellular fractions. It is conceivable that they relate to the altered glycoprotein and glycolipid composition of tumors discussed in Chapters 2 and 4.

In any event, the results of the Raftell group seems to support data earlier reported by Bosmann [9] who, using a similar immunological method, found a decrease in 5′-nucleotidase in tumor plasma membranes. However, Bosmann [9] suggested that, because the immunoprecipitates from hepatoma membranes differed from those of normal liver, the enzyme involved might be a different enzyme, possibly an isozyme.

This matter has been investigated in a more recent study by Raftell et al. [10] on hepatomas D23 and D33 (induced by 4-dimethylaminoazobenzene). The authors analyzed multi-enzyme complexes released by use of 1 % DOC, 0.5 % Lubrol W, using two-dimensional immunorelectrophoresis. In the case of both hepatomas, only one out of nine immunologically different esterase-active antigens was also detectable in adult liver microsomes. The multi-enzyme complexes of all hepatoma membranes showed reduced activities of nucleoside di- and tri-phosphate, hydrolase, acid phosphatase and NADH-oxidoreductase. The only difference between the D23 and the D33 tumor, was that the former did not contain a γ-L-glutamyl-β-naphtylaminidase antigen, whereas four immunoprecipitates active for this enzyme were found in material extracted from the D33 tumor. The authors found that the tumors behave rather similarly with regard to the deletion and preservation of certain enzyme active antigens. Moreover, one ADP-ATPase-active antigen typical of fetal liver was also found in both hepatomas. The approach employed by these authors is obviously very promising, particularly with regard to the characterization of membrane isozymes.

We now want to turn to three topics of capital interest: *First*, the role of mitochondrial membranes in the integration of respiration with glycolysis and in the calcium metabolism of cells, *second*, the role of cyclic nucleotides in the control of cellular events and *third*, the role of endoplasmic reticulum membranes in stabilizing mRNA templates. Of these topics, the mitochondrial enzyme behavior is the best understood because of very extensive activity in numerous excellent laboratories on the properties of normal mitochondria. Amazingly, however, the behavior of tumor mitochondria has been neglected of late. As far as the role of cyclic nucleotides are concerned, this is an area which is receiving very extensive attention; it is also a field in very major flux. It is therefore, not possible to give a comprehensive overview, but we can present a picture of the state of the art at the present time. The same comment may be applied to the role of endoplasmic reticulum membranes in regulation of messenger RNA translation. This is obviously a topic of major importance in terms of the synthesis of both structural and enzymatically active proteins and has lead to the development of an elegant unifying hypothesis concerning these matters, the '*membron hypothesis*'.

6.2. Integration of respiration and glycolysis

6.2.1. Introduction

One of the most common metabolic anomalies is tumors in their propensity to produce large amounts of lactic acid from glucose [1].

Glucose is an important energy source for animal cells, which can extract energy by the oxidation of the sugar through two major pathways, namely, *glycolysis*, i.e.,

$$\text{Glucose} \rightarrow 2\ \text{lactate} + 2\ H^+$$

and *respiration*, i.e.,

$$\text{Glucose} + 6\ O_2 \rightarrow 6\ CO_2 + 6\ H_2O$$

The glycolytic pathway yields a net of two molecules of ATP per molecule of glucose, whereas complete, respiratory oxidation of glucose, coupled to oxidative phosphorylation produces 36 molecules of ATP.

Clearly, the glycolytic pathway is far less efficient ($\sim 5\%$) than respiration at the cellular level, but in the intact animal, this potential wastage of nutrient energy does not occur, since lactate produced by glycolysing cells can be efficiently reconverted to glucose in several tissues, the liver in particular.

A consequence of glycolysis that may be more relevant to tumor behavior is the fact that the conversion of each glucose molecule yields $2\ H^+$ ions, since at neutral pH, lactic acid (pK = 3.86), is $>99\%$ dissociated; this means 1 mole of H^+ per mole of ATP.

In contrast, respiratory oxidation of glucose yields a *neutral* molecule, CO_2, which becomes an acid only after its hydration to H_2CO_3 and subsequent dissociation of H_2CO_3 to HCO_3^- and H^+. But this hydration process occurs very *slowly*, unless catalyzed by carbonic anhydrase. Most cells lack significant carbonic anhydrase activity and in cells, other than erythrocytes, that contain this enzyme, its activity is usually coupled to specific proton-translocation mechanisms (e.g., in the kidney). In general, therefore, CO_2 becomes hydrated and produces H^+ only in the blood which, because of the special properties of hemoglobin, can accommodate the H^+ without major changes of pH. In the lungs, bicarbonate recombines with H^+ to form H_2CO_3; this is rapidly dehydrated and the resulting CO_2 eliminated in gaseous form.

The importance of the problem can be illustrated by a numerical example (Table 6.1). Lo et al. [12] have shown that homogenates of hepatoma 3683 produce 108×10^{-6} moles of lactate $\cdot$ sec$^{-1} \cdot$ L^{-1} under aerobic conditions, more than twice the 43×10^{-6} moles $\cdot$ sec$^{-1} \cdot$ L^{-1} generated by homogenates

TABLE 6.1

Respiration and aerobic lactate production by combined subcellular fractions from normal liver and two highly glycolysing tumors[1]

System[2]		Oxygen uptake[3] (μatoms $\cdot$ sec$^{-1} \cdot$ L^{-1})	Lactate (H^+) production (μmoles $\cdot$ sec$^{-1} \cdot$ L^{-1})[3]
Particles from	Soluble fraction from		
Liver	Liver	74	45
H3683	H3683	27	113
Novikoff	Novikoff	5	109
H3683	Liver	29	132
Novikoff	Liver	26	100
Liver	H3683	68	52
Liver	Novikoff	65	60

1 From [12].
2 Particles = material sedimented from homogenates at 6×10^6 g $\cdot$ min; soluble fraction = supernatant fluid from 6×10^6 g $\cdot$ min centrifugation. H3683 = rapidly growing Morris hepatoma. Novikoff = Novikoff hepatoma.
3 Data per L of tissue.

of normal liver. This corresponds to an excess, acid production of 65×10^{-6} moles $\cdot$ sec$^{-1} \cdot$ L^{-1}, an *enormous* load, especially if one considers the fact that the physiological hydrogen ion concentration lies near 10^{-7} M! That such tumor cells can survive at all *in vivo* is due largely to the exceedingly rapid long-range dissipation of H^+ gradients in aqueous media (due to the massive H-bonding of H_2O at $<50\,^{\circ}C$). However, fixed anion production must be dissipated by diffusion and, as shown in theoretical studies by von Ardenne and Rieger [13] and Jensen et al. [13a] high lactate production can generate dangerously low pericellular pH levels ($<$ pH 7). The general consequences of high intra- and peri-cellular acidity remain to be established. Apart from obvious alterations of enzyme activities and modification of cell contacts (Chapter 8), they would certainly include activation of certain lysosomal proteases (Chapter 7). In addition, very small deviations of pH can produce substantial modifications of lipid–protein interactions in the plasma membranes of some cells [14].

Cellular respiratory rate depends upon the availability of substrate, the oxygen concentration at the inner mitochondrial membrane and, since respiration (electron transport) is normally tightly coupled to the phosphorylation of ADP to ATP, on ADP. Because of the very high O_2-affinity of cytochrome oxidase, the terminal oxidase of the electron transport chain, electron transport is not oxygen-limited unless the oxygen concentration drops below $\sim 1\ \mu M$. Without substrate depletion, the major regulator of respiratory oxidation of substrate is therefore ADP and, at mitochondrial ADP levels $< 20\ \mu M$ [15] electron transport does not proceed at a maximum rate.

The major regulatory enzyme in the *glycolytic* sequence is phosphofructokinase (Fig. 6.1). This catalyzes the rate limiting reaction in glycolysis, namely the phosphorylation of fructose-6-phosphate to fructose-1,6-diphosphate. Phosphofructokinases are soluble enzymes, and those which have been well studied, namely the heart and skeletal muscle enzymes, are oligomeric proteins whose monomer–oligomer equilibria depend upon pH, protein concentration, the presence of substrate and various effectors. The active, regulatory species probably contains one catalytic subunit and three regulatory subunits, and is capable of binding one fructose-6-phosphate molecule and three inhibitory ATP molecules.

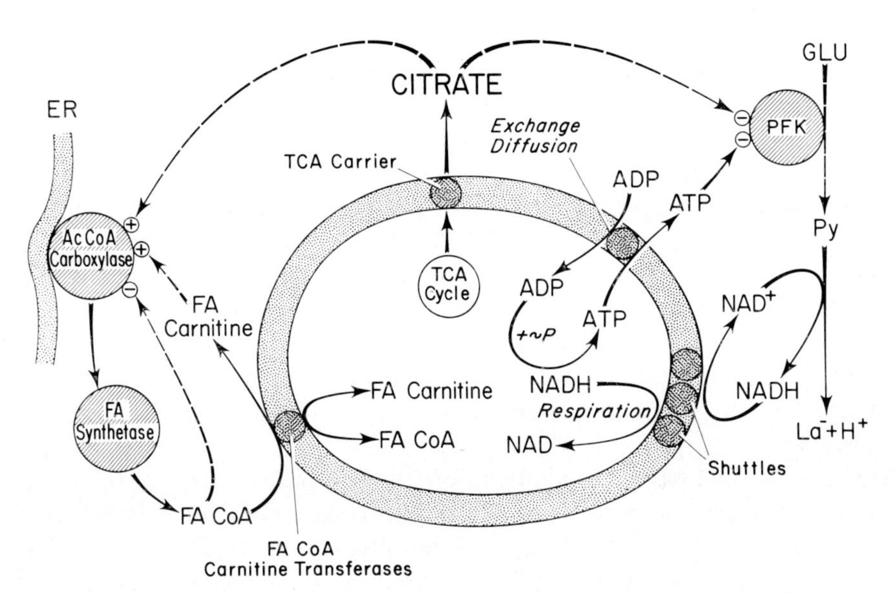

Fig. 6.1. Some mitochondrial membrane functions regulating cellular homeostasis.

Phosphofructokinase is *stimulated* by ADP and P_i and is *inhibited* by ATP and citrate. At high [ATP]/[ADP] ratios, enzyme activity is depressed; the reverse is true when the ratio is low. The high mitochondrial affinity for ADP during respiration-oxidative phosphorylation produces high cytoplasmic [ATP]/[ADP] ratios that inactivate or inhibit phosphofructokinase and therefore slow down glycolysis. Moreover, the normal intracellular concentrations of fructose-6-

phosphate and ATP are in a range where the phosphofructokinase reaction is extremely sensitive to cytoplasmic [ATP].

Citrate is also a negative modulator of the phosphofructokinase reaction; mitochondrially-generated citrate penetrates through the mitochondrial membrane via a citrate carrier system (Fig. 6.1). Thus, the active production of citrate by the tricarboxylic acid cycle during respiration can lead to inhibition of phosphofructokinase*.

The phosphofructokinase reaction is not the only control point in the glycolytic sequence. Hexokinase can act as a secondary regulator since it is strongly inhibited by its own product, glucose-6-phosphate. Moreover, the high affinity of mitochondria for ADP compared with the affinity exhibited by the ADP-requiring enzymes of the cytoplasm, namely phosphoglycerate kinase and pyruvate kinase, can cause an inhibition of glycolysis during active oxidative phosphorylation when intramitochondrial [ADP] becomes low, thus depressing cytoplasmic [ATP] production.

6.2.2. *The spatial segregation of respiration and glycolysis*

The general role of mitochondrial membranes in segregating various cellular metabolic processes and in regulating the interaction between these processes is beautifully described by Lehninger [16]. We will therefore only deal with processes directly pertinent to our central theme.

The glycolytic and respiratory systems utilize two sets of obligatory intermediates: ADP/ATP and NAD^+/NADH.

In glycolysis, ATP is converted to ADP produced in the hexokinase-catalyzed phosphorylation of glucose-1-phosphate to fructose-1,6-diphosphate. Four molecules of ATP are subsequently generated from 4 molecules of ADP by the conversion of 2 molecules of 1,3-diphosphoglycerate to pyruvate. In oxidative phosphorylation, ADP is also the universal acceptor of high-energy phosphate ($\sim$P; Fig. 6.1).

NAD^+ and NADH participate in two steps of the glycolytic sequence: (a) The oxidation and phosphorylation of glyceraldehyde-3-phosphate to 1,3-diphosphoglycerate, mediated by glyceraldehyde-3-phosphate dehydrogenase, requires conversion of NAD^+ to NADH (b) This NADH can be reoxidized to NAD^+ by glycolytic conversion of pyruvate to lactate. NAD^+ can, of course, also be regenerated via mitochondrial oxidative processes.

Since it is not permeable to ATP/ADP or NAD^+/NADH, the inner mitochondrial membrane spatially segregates respiration from glycolysis. The membrane is also not freely permeable to citrate, an important negative modulator of phosphofructokinase. The integration of glycolysis and respiration therefore

* Note that citrate is a positive modulator of acetyl-CoA carboxylase (Chapter 3).

depends on a set of substrate-specific carrier and shuttle systems that are part of the inner mitochondrial membrane. The carriers can be saturated as the substrate concentration is increased; they may also be inhibited by specific agents. In the mitochondria of rat liver specific carriers have been identified for ADP, ATP, inorganic phosphorus, Ca^{2+}, certain amino acids, certain intermediates of the tricarboxylic acid cycle, as well as some intermediates of fatty acid metabolism (Chapter 3).

6.2.2.1. ADP and ATP

ADP and ATP pass across the inner mitochondrial membrane by *exchange diffusion* in which one molecule of ATP moves out for every molecule of ADP entering (Fig. 6.1). The carriers involved exhibit a high affinity for ATP or ADP and are highly specific for these intermediates; GTP and GDP, or CTP and CDP are not transported. The carriers are inhibited by toxic glycosides, such as atractyloside, and the inhibition of oxidative phosphorylation induced by this agent derives from depletion of intramitochondrial ADP.

6.2.2.2. NADH and NAD^+

In anaerobic glycolysis, lactate acid is formed, from pyruvate at the expense of NADH.

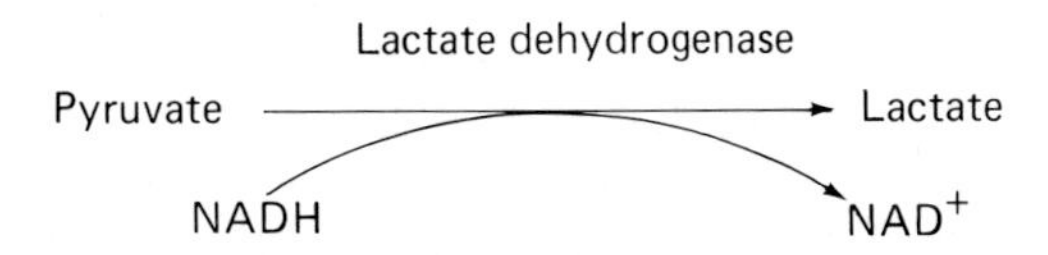

Unless NAD^+ is regenerated by this reaction, glycolysis cannot proceed beyond the triose stage. Since neither NADH nor NAD^+ can penetrate the mitochondrial membrane as such, their cytoplasmic levels are regulated by a set of 'shuttle systems' (Fig. 6.1) which transfer 'reducing equivalents' across the membrane. Important in these translocation reactions is the fact that the respiratory chain has a much higher affinity for NADH than lactate dehydrogenase.

The most important shuttle systems for NADH/NAD^+ are:

The α-glycerophosphate shuttle. In this shuttle, one of the intermediates of glycolysis, dihydroxyacetone-phosphate, is reduced by NADH to form L-glycerol-3-phosphate in a reaction mediated by a cytoplasmic enzyme, NAD^+-glycerol-phosphate-dehydrogenase:

$$\text{Dihydroxyacetone-phosphate} + \text{NADH} + \text{H}^+ \rightarrow \rightarrow \text{L-Glycerol-3-phosphate} + \text{NAD}^+.$$

The glycerol-3-phosphate thus generated can permeate through the mitochondrial membrane and, within the mitochondria, is oxidized back to dihydroxyacetone phosphate by a glycerol-3-phosphate dehydrogenase which is distinct from the cytoplasmic enzyme and bears a flavin prosthetic group (F),

$$\text{glycerol-3-phosphate} + \text{F} \rightarrow \text{dihydroxyacetone phosphate} + \text{FH}_2.$$

The reduced flavin protein now donates its reducing equivalent to the electron transport chain. The dihydroxyacetone can diffuse back into the cytoplasm where it can accept electrons from another molecule of, that is, extramitochondrial NADH.

The malate shuttle. In this system, extramitochondrial NADH is oxidized by oxaloacetate through the action of a cytoplasmic malate dehydrogenase, to yield malate (Fig. 6.1). This enters the mitochondria via the malate-succinate carrier system and is oxidized intramitochondrially to oxaloacetate by an intramitochondrial malate dehydrogenase to yield intramitochondrial NADH.

The malate shuttle, in contrast to the glycerophosphate shuttle, is bidirectional for the following reasons: The oxaloacetate inside the mitochondria can be converted to aspartate by transamination. This amino acid can pass through the mitochondrial membrane via a specific carrier, and can be converted back into oxaloacetate by a transaminase in the cytoplasm.

The acetoacetate shuttle. Another transfer system of possible importance is the *acetoacetate shuttle.* Here, extramitochondrial NADH is reduced acetoacetate to β-hydroxybutyrate which is reoxidized intramitochondrially by a β-hydroxybutyrate dehydrogenase. However, since β-hydroxybutyrate dehydrogenase is known to be bound to the mitochondrial membrane, its operation in a shuttle mechanism would require two orientations of the enzyme (one accessible from the cytoplasmic space and one oriented into the mitochondria).

6.2.2.3. NADPH and NADP$^+$

There is also a shuttle system allowing transfer of reducing equivalents of NADPH. This is *the α-oxoglutarate-(iso)citrate shuttle.*

In this system [17, 18] α-oxoglutarate is transferred by an exchange diffusion process tightly coupled to the translocation of malate and other dicarboxylates. α-Oxoglutarate is converted to isocitrate by isocitrate dehydrogenase as follows:

$$\alpha\text{-oxoglutarate} + \text{NADPH} + \text{H}^+ + \text{CO}_2 \rightleftarrows \text{isocitrate} + \text{NADP}^+.$$

This reaction can occur both within the cytoplasm and the mitochondria and thus allows transfer of 'reducing equivalents of *NADPH*'. The direction of transfer of reducing equivalents in this system, depends on the relative oxidation-reduction states of the mitochondrial and cytoplasmic compartments, as well as on the sign of the pH difference across the membranes.

6.2.3. The 'Pasteur Effect'

In 1861, Pasteur [19] discovered that, when yeast cells were transferred from an oxygen-free environment to an oxygen-containing environment, glucose consumption decreased and lactic acid production ceased. This phenomenon, known as the '*Pasteur Effect*', has now been documented for numerous *animal* cells. In fact, most animal cells under anaerobic conditions consume large amounts of glucose and produce enormous amounts of lactic acid in order to meet their energy requirements. However, when oxygen is admitted to most normal cells, the rate of glucose consumption declines dramatically and lactic acid production is reduced to zero. The oxygen concentration required for this switch is very low – just enough to activate mitochondrial electron transport.

The '*Pasteur Effect*' is not universal to all normal tissues. For example, various embryonic tissues produce lactic acid from glucose at very appreciable rates for prolonged periods of time and it has been suggested [20] that this metabolic feature may be an adaptation allowing survival during the anaerobic or hypoxic conditions which frequently occur during birth.

The high aerobic glycolysis of fetal tissues is not due to a small number of mitochondria per cell or due to abnormal functioning of oxidative phosphorylation. However, embryonic tissues have rather low levels of α-glycerol phosphate dehydrogenase activity; (for example, 6 pmoles per mg of protein per minute compared with 1650 pmoles per mg of protein per minute for adult liver [21]). The high aerobic glycolysis of embryonic tissue could thus be due to the limited ability of these tissues to transport reducing equivalents into the mitochondria via the glycerophosphate shuttle [22].

The inner *renal medulla* typically exhibits high rates of aerobic glycolysis. The reasons for this are not clear.

Polymorphonuclear leucocytes also lack a 'Pasteur Effect'. This is because these cells contain virtually no mitochondria and depend on glycolytic conversion of glucose to lactate for ATP production.

Mature mammalian *erythrocytes* lack mitochondria and any machinery for respiratory energy production. Erythrocytes thus depend solely on glycolysis to generate ATP. Without mitochondria, one might expect equivalent glycolysis whether oxygen is present or not. However, this is not the case and one observes enhancement of glycolysis under anaerobic conditions [23]. The 'Pasteur Effect' in erythrocytes is a different phenomenon from the one observed in cells capable of both glycolytic and respiratory ATP production. It very likely derives from the reversible binding of 2,3-diphosphoglycerate to hemoglobin [24]. In erythrocytes 2,3-diphosphoglycerate constitutes part of a pathway which bypasses glycolysis. It is produced from 1,3-diphosphoglycerate mutase and is then metabolized.

Most of the 2,3-diphosphoglycerate in erythrocytes (~5 mM) is bound to hemoglobin, but the affinity of *deoxy*hemoglobin is much greater than that of *oxy*hemoglobin. Therefore, 2,3-diphosphoglycerate is released and taken up during the cyclic oxygenation and deoxygenation of hemoglobin.

Since, 2,3-diphosphoglycerate inhibits hexokinase [26] and phosphofructokinase [26], both of which are essential for glycolysis, glycolysis is inhibited under aerobic conditions, that is when hemoglobin is highly oxygenated. Moreover, the lower affinity of oxyhemoglobin for 2,3-diphosphoglycerate tends to shift the metabolism of 1,3-diphosphoglycerate toward the production of 3-phosphoglycerate (phosphoglycerate kinase). This leads to the production of ATP, raises the ratio [ATP/ADP] and thus inhibits phosphofructokinase.

6.2.4. Altered glycolysis in tumors

6.2.4.1. Introduction

As already mentioned, Warburg [11] pointed out in 1926 that tumor cells generally exhibit high aerobic glycolysis. We will refer to this phenomenon as the '*Warburg phenomenon*'. Early studies by the Coris [27] and later work by Gullino et al. [28] have shown that this phenomenon is not characteristic only of tumor cells *in vitro*, but occurres also *in vivo*. Warburg's extensive studies on cancer led him to propose that cancer originates when a normal cell adopts an anaerobic metabolism as a means of survival after injury to its respiratory apparatus. In fact, he argued [29] that neoplastically converted cells would pass their defective energy metabolism on to their daughter cells. This prediction, made at a time when there was no knowledge that mitochondria have genetic material of their own and might thus determine their own destiny to some extent, has, in fact, not been disproved despite the fact that Warburg's suggestion cannot be generalized. Indeed it is possible though that a defect in the mitochondrial genome, once introduced, might be reproduced during cell division. (However, the mitochondrial genome does not code for all of the enzymes involved in respiration and oxidative phosphorylation [30].)

A necessary corollary to the Warburg hypothesis is that prolonged exposure of normal cells to anoxic condition should induce accelerated emergence of neoplastic cells. This issue has not received great attention. However, Goldblatt and associates [31,32] have attempted to relate the incidence of oncogenic transformation of rat embryo cells grown for long periods *in vitro* to the use of aerobic or anaerobic culture conditions. In these experiments cells were propagated in conventional tissue culture containers (i.e., gas-imperable, culture substrates), using conventional tissue culture media, with the gas phase either oxygenated or oxygen depleted. Under both conditions, atypical cells began to appear during repeated subculture and their incidence increased until most of the cells appeared atypical and yielded transplantable tumors upob subcutaneous inoculation into rats. However, this experimentation cannot be considered reliable, since recent experimental and theoretical studies [13a, 33] indicate that the cells cultured under the conditions employed by Goldblatt and associates [31, 32] would be hypoxic even with an oxygenated gas phase.

Goldblatt and Friedman [34] recognizing that ordinary culture conditions tend to produce pericellular hypoxia, even with an oxygenated gas phase, propagated mixed cultures of epithelial cells and fibroblasts, derived from primary cultures of rat embryo skin, in the presence or absence of 1% oxyhemoglobin (as oxygen carrier). These experiments have proven highly provocative.

Cells propagated without oxyhemoglobin yielded transplantable tumors after the 52nd passage, that is, after *13* months of frequently repeated subculture. In contrast, a set of subcultures derived from the 22nd passage of the original culture and propagated under the same conditions as the main-line culture (except for the addition of 1% oxyhemoglobin), failed to produce oncogenic cells after *23* months of repeated subculture. In other words, the presence of oxyhemoglobin appeared to 'keep malignancy in abeyance'. Unfortunately, the authors experiments do not allow one to relate their experiments unequivocally to the oxygenation of the cells; however, this could be done by propagating primary cells on gas-permeable plastics, allowing precise control of pericellular pO_2 [13a,33].

An important aspect of the experiments of Goldblatt and Friedman [34] is the fact that return to hemoglobin-free media, of cells that had been cultured for 23 months in the presence of 1% oxyhemoglobin and had failed to produce oncogenic cells, yielded tumorigenic cells in 4 months, that is, after 12 or more passages. Moreover, once the cells had become oncogenic, their cultures continued to produce transplantable tumors even when propagated in the presence of 1% oxyhemoglobin for up to 305 passages. The presence of oxyhemoglobin in the medium apparently cannot reverse a malignant transformation, once established. This finding, if proven to depend upon pericellular oxygen tension, would be consistent with Warburg's original hypothesis. The issue could be tested by use of gas-permeable film culture techniques [13a,33].

Although not a *universal* or *unique* aspect of cancer, high aerobic glycolysis is a very common feature of tumors growing *in vivo*, and one that increases with tumor anaplasia [36–38]. This point is illustrated in Table 6.2.

The same is true for the *in vitro* situation and the oncogenic conversion of various cultured cells is generally accompanied by an increase of aerobic glycolysis [39, 40]. However, as shown in Table 6.3, enhanced *in vitro* aerobic glycolysis cannot be considered a *necessary* feature of oncogenic conversion [41].

Indeed, the initial explantation of various tissues into cell culture generally leads to a decrease in oxygen consumption and an increase in glycolysis [42–45]. This fact is often ascribed to insufficient oxygenation of the cells in culture but this is not an adequate explanation; if the oxygen supply of the cells is diffusion limited, supply of nutrients such as glucose will be similarly restricted.

As first shown by Druckrey and associates [46] the aerobic glycolysis characteristic of malignant hepatocytes does not appear abruptly during hepato-

TABLE 6.2

Anaerobic and aerobic lactate production by host liver, a minimum deviation hepatoma (5123) and a highly dedifferentiated hepatoma (H35 TC)[1]

Tissue	Lactate production[2]		Ratio (Aerobic/anaerobic)
	Anaerobic	Aerobic	
Host liver	100	60	0.6
Hepatoma 5123	310	156	0.5
Hepatoma H35 TC	4240	3820	0.9

[1] Calculated from [36].
[2] The anaerobic lactate production of host liver has been set at 100.

carcinogenesis. On the contrary, glycolysis gradually increases during the long induction phase of chemical carcinogenesis and increases rapidly only in the last stages. The changes of glycolysis and respiration during hepatocarcinogenesis induced by diethylnitrosamine have been studied in more detail recently [47]. Hepatomas were induced in rats by daily feeding of 10 mg/kg of the carcinogen and hepatomas emerged between 95 and 120 days after treatment. Oxygen utilization per unit weight of liver diminished gradually during the period prior to the emergence of the tumor and then diminished abruptly thereafter. Aerobic glycolysis could not be detected until about 20 days after carcinogen treatment but increased markedly with appearance of tumors. Anaerobic glycolysis augmented gradually during the first 70 days and increased precipitously with tumor appearance.

The Warburg phenomenon might derive from anomalies in the glycolytic pathway, defects in respiration and/or oxidative phosphorylation, impaired integration of these processes or a combination of these possibilities.

TABLE 6.3

Glycolysis and respiration of several tumorigenic and non-tumorigenic mouse cell lines propagated *in vitro*[1]

Cell line[2]	O_2 Consumption[3]	Glucose consumption[3]	Lactate production[3]
C3H-M (6–8 mos)	0.43	0.32	0.83
C3H-E (6–8 mos)	0.19	0.43	0.41
C3H-L1 (6–8 mos)	0.22	0.43	0.67
L929 (25 y)	0.51	0.29	0.32

[1] From [41].
[2] C3H-M is non-tumorigenic; C3H-E is weakly tumorigenic; C3H-L1 and L929 are strongly tumorigenic.
[3] In μmoles/mg dry wt/hr.

6.2.4.2. Possible defects in the glycolytic system

Weinhouse [48] has recently reviewed information bearing on possible defects in the glycolytic systems of tumor cells. He points out that one cannot assume that phosphofructokinase is the only, or even the dominant regulator of glycolysis in neoplastic cells. He stresses that tumors, in general, and poorly differentiated hepatomas, in particular, tend to demonstrate isoenzyme compositions that resemble those of fetal liver. (These patterns are much less obvious in highly-differentiated 'minimum deviation' hepatomas.) One interesting finding is that pyruvate kinase II, which is under dietary and hormonal control and has kinetic properties that endow it with regulatory functions in gluconeogenesis and glucose utilization [49], diminishes in poorly differentiated hepatomas and is there replaced by large amounts of pyruvate kinase I; this is low in normal liver. This shift of isoenzyme complement is consistent with Weber's 'Molecular Correlation Concept' [1] and results in impaired regulation of glycolysis at the pyruvate kinase stage and possible competition of the glycolytic system for ADP. Indeed, Lo et al. [12] using isolated segments of the glycolytic sequence and mitochondria from well- and poorly-differentiated hepatomas, demonstrated that pyruvate kinase can compete with the mitochondrial respiratory system for ADP. In principle, therefore, high pyruvate kinase activity might be a significant determinant in the high aerobic glycolysis of tumors. However, Lo et al. [12] *also* showed that the high aerobic glycolysis in homogenates of many tumors could be *abolished by replacing the tumor mitochondria with mitochondria from tissues having a normal Pasteur Effect* (Table 6.1).

Hexokinase, the initial enzyme in glucolysis is generally considered a soluble cytoplasmic enzyme. However, several authors [50, 51] have reported significant amounts of hexokinase activity in tight association with plasma membrane isolates from rapidly growing rat hepatoma cells. These workers have suggested that the ability of rapidly growing rat hepatomas to efficiently utilize glucose relates to the presence of hexokinase in the plasma membranes of these cells. However, Bheatty and Hickie [52] could detect no hexokinase activity in plasma membrane isolates derived from rat hepatoma 5123 C. This hepatoma, although rather well differentiated and slow growing, still exhibits much greater aerobic glycolysis than normal liver [53, 54]. We suspect that some discrepancies arise from the very different fractionation properties of cells with varying degrees of plasma membrane differentiation (Chapter 2; [55]).

6.2.4.3. Mitochondrial defects

6.2.4.3.1. Mitochondrial number, size and enzyme content

An abundance of information indicates that tumor cells generally contain fewer mitochondria than their normal tissue equivalents [38, 56–58]. In addition, as pointed out by Hruban et al. [59] and Sordahl and associates [60, 61], tumors generally reveal a much more heterogeneous size distribution of mitochondria, than normal liver.

Much early experimentation has suggested a deficiency of cytochrome oxidase in tumor mitochondria [62]; these findings are supported by recent studies. Thus, Galeotti et al. [63] have shown that mitochondria from Walker carcinoma-sarcoma 256 contain only 10–20 % of cytochrome a, a_3, b and $c+c_1$, although the proportion of the cytochromes are as in normal liver. Also, Sato and Hagihari [64], who have measured the cytochrome content of several rat and mouse ascites hepatomas using spectrophotometric methods, find that all of the tumors contained low levels of cytochrome $a + a_3$, b, c_1, but essentially normal levels of cytochrome c. A subsequent study by Hagihari et al. [65] has extended these studies to a series of minimum deviation hepatomas (7794A, 7316B and 7793). These well-differentiated hepatomas also exhibit low contents of cytochrome $a + a_3$, b and c_1 and nearly normal levels of cytochrome c. However, the cytochrome composition of regenerating liver was similar to that of normal liver on a unit weight basis, although the growth rate of the tissue exceeded even that of the more rapidly growing hepatomas.

White and Tewari [66] in their study on the mitochondria from Novikoff hepatoma, observed that whereas the total activities of cytochrome oxidase and malate dehydrogenase activity per unit cell weight were less in the hepatoma than in normal liver, the *specific activities* of these two enzymes in the mitochondria were not significantly different. Their findings are consistent with those of Pedersen et al. [67], who could also discover no significant changes in the specific activities of cytochrome oxidase and malate dehydrogenase in the mitochondria from Morris hepatomas 9618A, 7800A and 3924A.

On the other hand, major changes have been observed in a number of enzymes associated with the *outer* membranes of the mitochondria [68,69], namely, monoamine-oxidase [66,70,70a], as well as rotenone-insensitive NADH-cytochrome c reductase [66]. This is shown in Table 6.4. In addition, there is a substantial decrease in the specific activities of several flavoproteins involved in electron transport. For example, NADH-cytochrome c reductase is decreased 4-fold in specific activity compared with host liver, NADH oxidase is reduced 8-fold and succinate dehydrogenase and succinate oxidase showed 4-fold and 5-fold decrease compared with host liver ([70]; Table 6.5).

White et al. [70,70a] have also compared mitochondria from hamster melanoma cells (CCL 49) with those from hamster NIL-2-fibroblasts and mitochondria from mouse 3T3 fibroblasts, with those from SV40 transformed cells. As shown in Table 6.5, the two hamster cell types have very comparable specific activities of cytochrome oxidase and malate dehydrogenase. Normal and transformed 3T3 cells also had comparable cytochrome oxidase levels.

Deficiencies in the enzyme complement of outer mitochondrial membranes thus appear to be a common feature of neoplastic cells.

TABLE 6.4

Activities of enzymes in outer mitochondrial membranes from some normal and neoplastic cells[1,2]

Enzyme	Specific Activity[3]					
	Normal hepatocyte	Novikoff hepatoma	Hamster NIL fibroblast	Hamster melanoma	Mouse 3T3 fibroblast	SV40 transformed 3T4 cells
Monoamine Oxidase	100	38	100	0	100	50
Rotenone-insensitive-NADH-cytochrome *c* reductase	100	43	100	36	—	—

[1,2] From [66, 70, 70a].
[3] Specific activity in % of normal cell.

6.2.4.3.2. *Sedimentation properties of tumor mitochondria*

Several experimenters have reported abnormal centrifugal fractionation properties for neoplastic hepatocytes [66, 70, 73]. Thus, mitochondria from rat hepatoma equilibrate isopycnically in *glycogen* gradients at density 1.07 [73] compared to 1.10 for normal liver mitochondria [74]. However, in *sucrose* gradients the mitochondria from both cell types (monitored by cytochrome oxidase activity) distributed in the density range 1.16–1.19 ([72], Fig. 1). The data suggest that the isolated tumor mitochondria were more sucrose-permeable than normal, possibly reflecting an *in vivo* state or an increased *in vitro fragility*.

White and associates [66, 70, 70a] have examined the different sedimentation

TABLE 6.5

Flavin enzymes of mitochondrial electron transport from some normal and neoplastic cells[1,2]

Enzyme	Specific activity[3]			
	Normal hepatocyte	Novikoff hepatoma	NIL fibroblast	Hamster melanoma
NADH-cytochrome *c* reductase	100	50	100	29
NADH-oxidase	100	13	100	18
Succinate oxidase	100	37	—	—
Succinate dehydrogenase	100	10	—	—

[1,2] From [66, 70, 70a].
[3] In % of normal.

properties of mitochondria isolated from normal and neoplastic cells in *sucrose* gradients and sucrose gradients containing 1% serum albumin.

They found that the mitochondria from normal liver invariably equilibrated as a discrete layer with a midpoint density of 1.187 g/ml. In contrast, the mitochondria from Novikoff hepatoma distributed between densities 1.144 and 1.161 g/ml. The authors then incorporated 1% bovine serum albumin (BSA) into the gradients, because of previous demonstrations that the presence of this protein could restore normal respiratory control by removal of free fatty acids [71,72]. This procedure altered the banding of the hepatoma mitochondria and produced a sharp layer with a mid point density of 1.171 g/ml. However, the density of the liver mitochondria remained at 1.187 g/ml. There was no significant difference in the specific enzyme activities of thc mitochondria isolated in the two gradient types.

The authors then carried out measurements on the binding of ^{14}C-labelled bovine serum albumin. They found that the hepatoma-mitochondria bound substantially more bovine serum albumin than those of normal liver. Indeed, in the case of the hepatoma mitochondria, 2% of the total mitochondrial protein consisted of bound bovine serum albumin, compared with only 0.76% for the liver mitochondria. Since the outer mitochondrial membrane is impermeable to BSA, the authors concluded that the excess BSA associated with the hepatoma mitochondria was bound to the outer mitochondrial membrane. The amount of outer membrane protein in the rat liver varied between 4% to 10% of the total mitochondrial protein, suggesting that about 8–20% of the outer membrane protein of liver mitochondria would be accounted for by bound BSA compared with 20–50% in the case of the hepatoma mitochondria. Further studies on sucrose gradients containing 1% BSA, indicated a heterogeneity of the hepatoma mitochondria which did not exist in liver mitochondria. In their subsequent study, White et al. [70] showed that the densities of hamster melanoma and hamster fibroblast mitochondria in linear sucrose gradients were both 1.165 g/ml, whether BSA was present or not, even though the melanoma mitochondria displayed similar enzyme deficiencies as found in hepatoma organelles. Mitochondria from normal and SV40-transformed 3T3 cells also showed comparable sedimentation characteristics [70a].

Cornbleet et al. [75] have shown that mitochondria from the well-differentiated, slow-growing Morris hepatoma 16, tend to be smaller than normal and thus to require unusually high centrifugal forces for sedimentation. In a subsequent study [76] the same group evaluated the distribution of mitochondrial marker enzymes in subcellular fractions derived from three well-differentiated Morris hepatomas ([21], R3B and 7794A), as well as from host livers.

The data (Table 6.6) indicate that mitochondria from hepatomas vary more in size and/or density than those from normal cells. The same patterns were obtained with markers for the inner mitochondrial membrane (cytochrome oxidase) and

TABLE 6.6

Distribution of a marker enzyme for inner mitochondrial membranes (cytochrome oxidase) and a marker for outer mitochondrial membranes (monoamine oxidase) in various subcellular fractions derived from host livers (HL) and hepatomas 21, R3B, 7794A[1,2]

Enzyme	Proportion of total recovered activity located in[3]				
	(N)	(Mit)	(Gr)	(Ms)	(Sol)
Cytochrome oxidase					
HL	28	68	4	0	0
21	72	26	2	0	0
R3B	41	54	5	0	0
7794A	26	60	14	0	0
Monoamine oxidase					
HL	26	63	7	4	0
21	60	34	4	2	0
R3B	31	58	6	6	0
7794A	23	56	18	2	0

[1] From the compilation in [76].
[2] For fractionation see [56].
[3] N = nuclear fraction; Mit = mitochondria; Gr = small granules (lysosomes, peroxisomes, small mitochondria); Ms = microsomal fraction (endoplasmic reticulum, small plasma membrane fragment, Golgi fragments); Sol = not sedimentable.

for the outer membrane (monoamine oxidase), but the 3 hepatomas show no consistent trend. Thus, hepatoma 21 contained the bulk of its mitochondrial membrane enzyme activity in the *nuclear* fraction, whereas hepatoma 7794A showed more mitochondrial material in fractions requiring high centrifugal forces for sedimentation. The mitochondria from hepatoma 21 isolated from the nuclear fraction exhibited similar (tight) coupling between respiration and phosphorylation as the particles in the mitochondrial fraction.

Present fractionation data provide further evidence for mitochondrial abnormalities in neoplasia, but the techniques employed do not allow conclusions as to underlying mechanisms. Isopycnic centrifugation in polymer density gradients [55, 74] would be much more informative in this respect than differential centrifugation and sucrose gradient centrifugation. However, available information reemphasizes the point, introduced in earlier chapters, that comparisons between the membranes of normal and neoplastic cells must take into account the possible different fractionation patterns of the two cell categories.

6.2.4.3.3. Coupling of phosphorylation to respiration: acceptor control ratio

As already noted, the rate of electron transport in normal, tightly-coupled mitochondria, depends critically on intramitochondrial ADP levels. When this is $<20\ \mu M$ (*state 4* respiration) electron transport proceeds at only 5–10% of the maximum observed at high ADP levels (*state 3* respiration). The acceptor control ratio is usually taken as the ratio of oxygen consumption in *state 3 versus* that in *state 4*.

It has been suggested that the Warburg phenomenon in tumors might derive from impaired coupling of oxidative phosphorylation to electron transport. Such a coupling defect would prevent an adequate yield of ATP via oxidative phosphorylation, forcing the cell to make up the energy deficit by glycolysis.

However, a whole series of studies (e.g. [12,60,67,76–79]) show that hepatoma mitochondria possess intrinsically normal acceptor control ratios, although tumor mitochondria are often more susceptible *in vitro* to the deleterious actions of fatty acids (Chapter 3) and certain divalent cations ([77]; Section 6.4.2.3.5.).

It appears unlikely, therefore, that the commonly high aerobic glycolysis of tumors is due to defects in the coupling of phosphorylation to mitochondrial electron transport.

6.2.4.3.4. Transfer of ADP and ATP across the mitochondrial membrane

As summarized earlier in this chapter, high aerobic glycolysis could arise from deficient transfer of ADP and ATP between the cytoplasmic and mitochondrial compartments. Indeed, this contention is consistent with the reconstitution experiments of Lo et al. ([12]; Table 6.1), showing that hepatoma mitochondria can induce high aerobic glycolysis in cytosol prepared from normal liver. On the other hand, a large number of experiments cited in Section 6.2.4.3.3. show that mitochondria from rapidly glycolysing tumors exhibit normal acceptor control *in vitro;* that is, ADP added externally, must have normal access to the intra-mitochondrial space of the isolated organelles.

It now appears that the situation involves *at least* one other important variable, that is, Ca^{2+}. Thus, when one adds Ca^{2+} to intact mitochondria isolated from normal tissues and respiring in a phosphate-containing medium, two Ca^{2+} ions are rapidly accumulated as an electron pair passes through each energy-conserving site of the electron transmitter chain [80,81]. At the same time, two H^{+} ions are ejected from the mitochondria for each electron pair passing through an energy conserving site. Sr^{2+} or Mn^{2+} can also be accumulated, but Mg^{2+} is not taken up. Moreover, inorganic phosphate is accumulated together with calcium in the same molar ratio that occurs in hydroxyapatite.

Significantly, under these conditions of calcium accumulation *no phosphorylation of ADP occurs*; that is, the energy of electron transport can *either* be transferred to ADP to form ATP, *or* can be used to accumulate calcium. It is clear that the calcium-accumulation capacity of mitochondria may serve as a mechanism for the control of Ca^{2+}-sensitive processes, for example, junction formation (Chapter 9), the metabolism of cyclic nucleotides and many other enzymatic mechanisms responsive to Mg^{2+}/Ca^{2+} ratio [82].

Ca^{2+} accumulation does not proceed indefinitely. Thus, in the case of mitochondria from normal tissues, addition of 100–150 nmoles Ca^{2+}/mg mitochondrial protein causes mitochondrial swelling, release of Ca^{2+} and loss of normal acceptor control; it also decreases coupling of phosphorylation to respiration and impairment of ADP/ATP exchange.

Recent studies indicate that the interaction of Ca^{2+} with mitochondria may be abnormal in certain tumors. First, Thorne and Bygrave [79,83] showed that mitochondria from Ehrlich ascites carcinoma can accumulate and retain Ca^{2+} in an ATP-supported system, whereas normal liver mitochondria cannot. Unlike the normal mitochondria, the tumor organelles do not swell or lyse and do not develop irreversible ATP hydrolysis. Moreover, whereas normal liver mitochondria, in the presence of substrate and inorganic phosphate maintain a continuous level of *state 3* respiration, tumor mitochondria are successively stimulated by repeated additions of Ca^{2+} (up to 350 nmoles/mg mitochondrial protein).

More recently [84,85] Thorne and Bygrave have demonstrated that the mitochondria of Ehrlich ascites carcinoma can accumulate very large levels of calcium, that ADP-ATP exchange in these organelles responds differently to calcium than in the case of normal mitochondria and that the interdependence of ADP-phosphorylation vs. calcium accumulation does not follow the pattern observed in normal liver or kidney mitochondria. For example, in the tumor mitochondria 100 nmoles of Ca^{2+}/mg protein (50 μM ATP) inhibits ADP and ATP exchange (50–75%), but stimulates this in liver mitochondria. Tumor mitochondria also do not exhibit the normal jump in electron transport following an ADP pulse. Thorne and Bygrave [86] have obtained similar results with Yoshida hepatoma 130, but not with Morris hepatoma 5123 (normal aerobic glycolysis). Thorne and Bygrave [86] have now established that similar anomalous patterns exist in Ehrlich ascites carcinoma, Yoshida HA 130 ascites hepatoma and Morris hepatoma 9618A; in contrast, mitochondria from regenerating liver behave as those from normal adult liver.

In summary, all of the tumor mitochondria showed (a) unusual capacity to take up Ca^{2+}, (b) calcium-induced inhibition of the normal ADP-induced jump of electron transport and (c) anomalous dependence of mitochondrial respiration upon the concentration of inorganic phosphate. An unusual capacity to retain Ca^{2+} and inorganic phosphate has also been observed in several other malignant tumors [87].

Reynafarje and Lehninger [88,89] have examined the relationship between electron transport and calcium accumulation in coupled mitochondria isolated from L1210 mouse lymphocytic leukemic cells and have discovered a most intriguing phenomenon, which, however, is not unique to tumors [90]. In their initial report [88] they showed that L1210 cells accumulate Ca^{2+} and eject H^+ during oxidation of succinate in the presence of Mg^{2+}, but that, as in the case of

Ehrlich carcinoma mitochondria, calcium uptake was greater than found in liver mitochondria. However, their studies revealed participation of two classes of calcium binding sites: One has a K_M of $\sim 8\ \mu M$, stimulates respiration and is saturated by ~ 40 ng-atoms Ca^{2+}/mg mitochondrial protein. The second has a $K_M \sim 120\ \mu M$, depends upon active electron-transport state and, upon calcium-binding, releases H^+ ions previously membrane-associated. The result is an apparent 'superstoichiometry'.

In a subsequent kinetic study, Reynafarje and Lehninger [89] demonstrated that within 5 sec after initiation of a Ca^{2+} pulse, L1210 mitochondria generate a massive, rapid burst of H^+ ejection ($H^+/2e^- > 20$) and calcium uptake ($Ca^{2+}/2e^- > 10$) both depending upon pre-existing and continuing electron transport. These rapid bursts are followed by a slower phase of H^+ ejection and Ca^{2+} uptake, both with normal stoichiometries ($H^+/2e^- \sim 4$–5; $Ca^{2+}/2e^- \sim 4$). The data suggest that L1210 cells bear Ca^{2+}-binding sites which are protonated during *state 4* respiration and which release these H^+ ions upon Ca^{2+}-binding. This accounts for the 'superstoichiometric pulse'. The slow phase of H^+-ejection and Ca^{2+}-accumulation represents the utilization of respiratory energy for Ca^{2+}-transport. Reynafarje and Lehninger [88] have coined the term 'membrane Bohr effect' for the superstoichiometric phase and compute that *state 4* mitochondria can bind 40 ng-atoms H^+/mg protein.

As documented in their most recent report [90], Reynafarje and Lehninger observe rapid, early, superstoichiometric Ca^{2+} uptake also in mitochondria from normal rat livers ($Ca^{2+}/2e^- = 4.2$ in first 6 sec). The phenomenon is thus not unique to neoplastic cells, but the effect is certainly more marked in L1210 mitochondria.

The Ca^{2+} metabolism of mitochondria is not autonomous, but appears to be regulated at least in part by cyclic nucleotides [62]. Thus, Borle [91] has demonstrated that the addition of cAMP to rat liver or kidney mitochondria produces a rapid burst of Ca^{2+} efflux, proportional to cAMP concentration between 10^{-7}–3×10^{-6} M. Cyclic GMP is active at only 10^{-5} M and then inhibits cAMP competitively. At low Ca^{2+} levels the extruded Ca^{2+} is rapidly reaccumulated. These results suggest that cAMP may participate in regulating cytoplasmic Ca^{2+} by its action on the efflux of calcium from mitochondria.

Recent information on membrane calcium-binding proteins is highly relevant here. Thus, Mikkelsen and Wallach [92] have demonstrated that the *high-affinity* calcium binding proteins of erythrocyte membranes are glycoproteins whose Ca^{2+} chelation apparently involves a *tyrosyl* residue and at least one *sialic acid* carboxyl. Moreover, Sottocasa et al. [93] and Gomez-Puyou et al. [94] have identified and purified a Ca^{2+}-binding sialoglycoprotein from inner mitochondrial membranes. The protein described in [93] was isolated from ox liver, contained $\sim 5\%$ carbohydrate, including sialic acid, and 30% phospholipid. Its SDS-PAGE molecular weight was $\sim 42{,}000$. It exhibited both high-affinity

($K_D \sim 10^{-7}$ M) and low-affinity ($K_D \sim 10^{-5}$ M) sites. The protein described in [94] was isolated from rat liver. Its apparent molecular weight was 67,000. It also exhibited a high-affinity Ca-binding site ($K_D \sim 10^{-7}$ M).

Highly significant to our main theme is the fact that Bosman et al. [95] have demonstrated that the neoplastic transformation of chick embryo fibroblasts by Rous sarcoma virus induces an increased biosynthesis of mitochondrial glycoproteins (as well as mitochondrial DNA, RNA and protein). This phenomenon (Table 6.7) can be attributed to transformation rather than infection, because it is not observed with a temperature-sensitive mutant (ts 68), which is permissive for transformation at 37 °C but not at 41 °C, although permissive for infection at both temperatures (Chapter 2).

TABLE 6.7

Mitochondrial glycoprotein biosynthesis in normal chick embryo fibroblasts and cells infected with Schmidt-Ruppin rous sarcoma virus under transforming and non-transforming conditions[1]

Cell system[2]	UDP-glucose[3] incorporation	GDP-mannose[3] incorporation
CEF (37 °C)	100	100
CEF (41 °C)	118	115
CEF-SR-RSV (37 °C)	186	201
CEF-SR-RSV (41 °C)	241	244
CEF ts 68 (37 °C)	186	210
CEF ts 68 (41 °C)	105	103

[1] Recalculated from the compilation in [95].
[2] CEF = check embryo fibroblasts; SR-RSV = Schmidt-Ruppin strain of Rous sarcoma virus; ts 68 = temperature sensitive mutant of SR-RSV, permissive for transformation at 37 °C.
[3] Incorporation of UDP-[^{14}C]glucose or GDP-[^{14}C]mannose into normal CEF at 37 °C set to 100.

Finally, we cannot over-emphasize the crucial role of mitochondria in regulating cytoplasmic levels of Ca^{2+}, and also suggest that Ca^{2+} can function as a 'cytoplasmic messenger', since intracellular Ca^{2+} levels $>10^7$ M 'decouple' intercellular junctions (Chapter 9) and the activities of plasma membrane adenyl- and guanyl-cyclases depend exquisitely on cytoplasmic Ca^{2+} levels.

We are thus faced with a fascinating chain of circular circumstantial evidence concerning mitochondrial abnormalities in neoplasia: (a) Most tumors exhibit high aerobic glycolysis; (b) high aerobic glycolysis can be explained by deficient communication across the mitochondrial membrane, including impaired exchange of ADP/ATP; (c) ATP/ADP exchange depends on Ca^{2+} binding/accumulation by mitochondrial membranes; (d) mitochondrial Ca^{2+} metabolism (binding) is abnormal in various tumors; (e) membrane Ca^{2+} binding (including that by mitochondria) is a function (not necessarily exclusive) of glycoprotein composition and (f) neoplastic transformation alters glycoprotein synthesis.

Clearly, this critical area of tumor biology demands elucidation.

6.2.4.3.5. Defects in the translocation of reducing equivalents from cytoplasm into tumor mitochondria

Weinhouse [96] as well as Boxer and Devlin [97], have proposed that the commonly high aerobic glycolysis of neoplastic cells may derive from deficiencies in the shuttle systems involved in the transfer of NAD(P)H reducing equivalents into the mitochondria.

Boxer and Devlin [97] documented a striking deficiency of cytoplasmic NAD^+ glycerophosphate-dehydrogenase activity in numerous tumors, resulting in a deficient α-glycerophosphate shuttle. They noted two exceptions: Morris hepatoma 5123, which is unusual in its low *aerobic* glycolysis and a hyperdiploid Lettré mutant of Ehrlich mouse ascites carcinoma (cf. also [98]). However, the Ehrlich-Lettré tumor exhibited much greater exchange diffusion of [^{14}C]malate with α-oxogluturate, suggesting that the deficiency in the α-glycerophosphate shuttle was compensated for by a shuttle involving α-oxogluturate [99]. It should be noted, however, that both the Ehrlich and Ehrlich–Lettré tumors retain the critical metabolic abnormality of high aerobic glycolysis. In any event, most strains of Ehrlich ascites carcinoma indicate a deficiency in the transfer of NAD(P)H reducing-equivalents from the cytoplasm into the mitochondria [100–102].

This matter has also been investigated for the Walker 256 carcinosarcoma [63, 103]. This tumor is deficient in all enzymes involved in the α-glycerophosphate and malate-asparatate redox shuttles, except for mitochondrial malate dehydrogenase, and reconstitution experiments show that the tumor mitochondria cannot oxidize externally added NADH. However, present data do not allow one to conclude that shuttle deficiencies account completely for the high aerobic glycolysis of tumors. The role of extramitochondrial, membrane-bound oxygenases (mixed function oxygenases; Section 6.2.4.3.6.) must also be considered [103].

Kovačević [104] has pointed out that the intracellular *aspartate* level may be rate-limiting for the *malate* shuttle because of the high K_M (5 mM) of aspartate-amino transferase for aspartate. He therefore, preincubated Ehrlich ascites carcinoma cells with glutamine, which is oxidized intracellularly to aspartate. He found that addition of glucose markedly reduced lactate production at a time after preincubation when the aspartate level is high. He also showed that aspartate accumulation within these cells in inhibited by glucose addition. His data suggest that the malate shuttle *does in fact* significantly influence aerobic lactate production, that the shuttle does not function (at least in his cell strain) because of glucose-limited aspartate accumulation.

Unfortunately, one cannot now properly evaluate the role of shuttle mechanisms in the high aerobic glycolysis of most tumors, since one does not know how much 'shuttling' is required. In this connection one must appreciate that glucose is not the only energy source for tumors; they survive very well on fatty acids for example (Chapter 3). The whole topic has clearly not received adequate attention.

6.2.4.3.6. Mixed function oxygenases

The membranes of the endoplasmic reticulum, the nuclear envelope and the Golgi apparatus contain a variety of electron transport systems which utilize molecular oxygen to oxidize specific substrates by direct insertion of O_2 or $\frac{1}{2}O_2$ into the product. The *mixed function* oxygenases utilize $\frac{1}{2}O_2$ for substrate oxidation and the other to oxidize NADPH, but also NADH.

These oxygenases constitute short, non-phosphorylating chains of cytochromes and flavoproteins. Their function is not blocked by agents which specifically influence mitochondrial electron transport or oxidative phosphorylation. They can account for a substantial proportion of cellular oxygen consumption (e.g., 15% in Walker 256 carcinosarcoma; [98]).

Comparisons of normal liver with Morris hepatoma 9618A-2, 7800, 7795 and 7787 reveal that the levels of mixed function oxygenases, including NADPH oxidase, NADPH-ferricyanide reductase, cytochrome *P*-450, cytochrome *P*-420 and cytochrome b_5 in the hepatomas are generally only 18 to 48% of those found in the membranes of normal liver [105,106].

In a highly discriminating study, Miyake et al. [107] have used difference-spectroscopy to measure the levels of cytochromes *P*-450, *P*-420 and b_5 in microsomes from rat liver, from the minimum deviation hepatoma 5123c and from the rapidly growing hepatoma 7777. In normal liver microsomes, cytochrome *P*-450 showed at least *three* spectroscopically identifiable components, whereas only two of these were apparent in hepatoma 5123c and only one in hepatoma 7777. Similarly, cytochrome *P*-420 exhibited three spectroscopically identifiable components in normal liver microsomes but only one in hepatoma 7777; also this was not converted to cytochrome *P*-450 by exposure to glycerol, as in the case in normal liver. Miyake et al. [107] also discovered a new oxygenase, cytochrome *P*-448, in the microsomal membranes of a methylcholanthrene-induced hepatoma.

The major results of this study are summarized in Table 6.8. Importantly, cytochrome *P*-450 is reduced in both hepatomas, more so in hepatoma 7777. At the same time, there is a marked increment in cytochrome *P*-420, and this increase is again particularly marked in most deviant hepatomas. Cytochrome b_5, an enzyme anchored to its membrane by an apolar 'tail' (Chapter 2), is diminished in both hepatomas. Cytochrome *P*-450 is also depleted in Yoshida ascites carcinoma [108].

TABLE 6.8

Activity of several 'mixed-function' oxygenases in the microsomes from rat liver and two rat hepatomas[1]

Cytochrome	Activity[2]		
	Liver	5123c	7777
P-450	100	41	2
P-420	100	190	675
b_5	100	29	17

[1] From Miyake et al. [107].
[2] In % of liver activity.

6.3. Cyclic nucleotides and cellular responses

6.3.1. Introduction

Within the past ten years it has become established that the response of numerous cell types to diverse *extracellular* regulator agents (hormones; neurotransmitters) occurs through the mediation of two cyclic nucleotides, cAMP and cGMP, which act as *intracellular* 'second messengers' and regulate processes as diverse as membrane transport (Na^+, K^+, Ca^{2+}, amino acid), enzyme activity (e.g., phosphorylases, protein kinases, glycogen synthetase, phosphofructokinase) and a host of cell-physiological mechanisms. The field of cyclic nucleotide research is a very active one and has been well reviewed from many points of view (e.g. [109–113]). Since cAMP and cGMP regulate numerous intracellular processes often deficient in neoplasia, extensive efforts are currently directed toward the assessment of cyclic nucleotide metabolism in tumor cells [114, 115]. Moreover, since the production of cyclic nucleotides in response to an extracellular regulator depends upon the binding of the regulator molecules to specific plasma membrane receptors and subsequent activation of membrane enzymes – adenylate cyclase and guanylate cyclase – studies on cyclic nucleotide metabolism represent an important aspect of biomembrane research.

6.3.2. Adenylate cyclase

Adenylate cyclase is a plasma membrane enzyme. It catalyzes the reaction

$$ATP \rightarrow 3',5'\text{-cAMP} + \text{pyrophosphate}.$$

Cyclic AMP acts as a diffusible, pleiotypic 'internal messenger' that can activate each cell type according to its phenotype. The intracellular level of cAMP is further regulated by phosphodiesterase, which catalyzes the reaction:

$$\text{cAMP} \rightarrow 5'\text{-AMP}$$

Phosphodiesterase is inhibited by several agents, prominently theophyllin.

In all cells which contain adenylate cyclase activity responsive to extracellular agents, the enzyme is stimulated by fluoride ion ($\sim$5–10 mM). The mechanisms underlying this phenomenon are not clear. In addition, adenylate cyclase requires Mg^{2+} or Mn^{2+}; these cations exert similar activating effects. In contrast, Ca^{2+} is generally inhibitory.

Current evidence indicates that the adenylate cyclase activation in diverse cells upon addition of hormones, fluoride, cholera toxin and certain detergents, represents the catalytic activity of a single class of enzyme molecules. This conclusion derives from the following facts: (a) The enzyme from diverse sources can always be stimulated by fluoride and exhibits similar kinetic properties, as well as cation responses [109-113], and an unusual time-dependent activation by cholera toxin [116]. (b) Membrane-bound adenylate cyclase from a homogeneous cell population can be stimulated by more than one hormone (in addition to fluoride) but maximal effective levels of two dissimilar agents do not produce additive responses [109–113]. (c) Propagation of S49 mouse lymphosarcoma cells in the presence of agents that stimulate adenylate cyclase, causes the death of most cells but allows the survival of a cell clone (1^{R}.1) which lacks an adenylate cyclase that can be activated by fluoride ion, hormones or cholera toxin [117].

As reviewed in [109–113] adenylate cyclase activity is extremely sensitive to lipid environment and some data indicate that the enzyme exhibits certain lipid preferences. Thus, Birnbaumer et al. [118] show that the glucagon (but not fluoride) activated enzyme of rat liver plasma membrane can be drastically inhibited by very limited phospholipase-A treatment and Pohl et al. [119] document that cyclase activity (and glucagon binding) can be partially restored by addition of certain exogenous phosphatides, notably phosphatidylserine.

The activation of adenylate cyclase occurs through a transduction process which occurs in the plasma membrane following binding of an extracellular regulator to its membrane receptor. Recent studies by Siegel and Cuatrecasas [120] on the activation of adenylate cyclase by cholera toxin give some insight into the mechanisms that might be involved. Cholera toxin binds very rapidly, primarily to a membrane ganglioside, G_{M1} (Chapter 4) but enzyme activation occurs only after a finite interval of time. The duration of this lag phase [120] decreases discontinuously with temperature (near 16 °C in mammalian cells). Using ^{125}I-labelled cholera toxin, Siegel and Cuatrecasas [120] have shown that the receptor-bound 'active' protein of cholera toxins reorients in the membrane to form a complex with the cyclase. This complex constitutes the enzymatically active form of the enzyme. The temperature-dependence suggests that the reorientation process depends on the state of membrane lipids (Chapters 3, 4 and 5). At a given temperature this state would also be responsive to local pH and ionic environment (Ca^{2+} in particular).

6.3.3. Guanylate cyclase

Recent experimentation shows that cyclic guanosine 3′,5′-monophosphate, cGMP, can also serve as an intracellular messenger. Cyclic GMP is generated by guanylate cyclases in the same manner as cAMP is produced from ATP by adenylate cyclase.

Most mammalian tissues contain both a soluble and a particulate guanylate cyclase, but there has been some dispute over the significance of the small amount of enzyme activity usually detected in particulate fractions (10–20%). This matter has been clarified through a recent study by Chrisman et al. [121] on the guanylate cyclase activity in rat lung. When homogenates of this tissue were centrifuged at 100,000 *g* for 1 hr, >90% of the activity in the homogenates was recovered in the supernatant fluid. However, addition of 1% Triton X-100 markedly increased the activity of the whole homogenate and the particles suggesting, that about one third of the enzyme activity is membrane-associated and masked in the absence of detergent. The soluble enzyme shows linear kinetics in double reciprocal plots, whereas the particulate enzyme exhibits cooperative behavior. Both the soluble and the particulate enzymes show a requirement for divalent cations, but the soluble enzyme is 10-fold more active in the presence of 3 mM Mn^{2+} than 3 mM Ca^{2+}, whereas the particulate enzyme is *inhibited* at $[Ca^{2+}] > 1$ mM. In this connection, we note that Ca^{2+} deprivation decreases the cGMP levels in a number of tissues, whereas cAMP is not affected [109], suggesting the possibility that *in vivo* cGMP and Ca^{2+} act as 'co-mediators'.

Recent evidence suggests that cGMP is a fundamental mediator of cholinergic receptor activity in diverse tissues [122] and that the effects of this nucleotide appear to be opposite to those of cAMP. This is also true in the case of lymphoid cells [123] and adipocytes [124]. However, in both of these cell types, insulin also increases cGMP levels in a process which apparently does *not* involve the cholinergic receptors.

An intriguing, unifying explanation for the opposite actions of cAMP and cGMP has been offered by Illiano et al. [124], who suggest that perhaps only one cyclase is involved. Depending on the ligand-receptor interaction occurring at the plasma membrane, the substrate specificity of the cyclase might change from ATP to GTP. The proportion of the hypothetical 'ATP- or GTP- utilizing' species would then depend upon the 'relative occupancy' of receptors that engender the 'ATP-form' versus the 'GTP-form'. If this hypothesis is correct, cell mutants deficient in adenylate cyclase activity [117] should also be deficient in guanylate cyclase.

In any event, present information suggests that intracellular cAMP/cGMP *ratios* may be more important in regulating cell metabolism than the concentrations of the two cyclic nucleotides.

6.3.4. *Variations of the levels of cAMP and cGMP during the cell cycle of normal cells: pleiotypic programs and pleiotypic control*

Sheppard and Prescott [125] have analyzed the levels of cAMP in synchronized Chinese hamster ovary fibroblasts during the cell cycle. Their data (Table 6.9) show that intracellular cAMP is depressed by 43 % during mitosis compared with the value during the S-phase. It increases by 57 % over the value of the S-phase during early G_1-phase and reaches about the same level in late G_1 as during S. The authors found that when the cells are blocked in the G_1-phase by isoleucine-deprivation of the medium, their cAMP level drops to about 15 pmoles/ml protein, which is about one third of the level observed in the presence of isoleucine. The depression observed suggests that the cAMP level somehow depends on protein synthesis.

Zeilig et al. [126] have also noted a fluctuation of cAMP levels during the cell cycle of HeLa cells synchronized by double thymidine blocking. They find that the lowest concentrations are reached during mitosis. Concordantly Milles et al. [127] have observed that addition of dibutyril-cAMP causes a delay in mitosis in synchronized human lymphoid cells.

Rudland et al. [128] have measured the intercellular cAMP and cGMP levels in a whole series of untransformed fibroblasts (mouse-Balb/3T3, Swiss 3T3-4A, mouse embryo, Swiss-3T6, hamster-BHK, monkey-BSC-1). They have demonstrated that the cAMP and cGMP levels rise and fall reciprocally during different growth states and conclude that the cAMP/cGMP ratio may be more important in the control of cellular events than the actual concentrations of the two cyclic nucleotides.

TABLE 6.9

Changes of cAMP Levels in synchronized Chinese hamster ovary cells at various stages of the cell cycle[1]

Phase of cell cycle	pmoles cAMP/mg protein
Mitosis	16 ± 2
Early G_1	44 ± 8
Late G_1	24 ± 4
Blocked in G_1 (isoleucine deprivation)	15 ± 3
S	28 ± 4

[1] From compilation by Sheppard and Prescott [125].

These authors analyzed three growth states, namely: (a) logarithmic growth, (b) resting phase, induced either by density inhibition of growth or exhaustion of the culture medium, and (c) reinitiation of growth by addition of fresh medium fortified with 20% serum; the last procedure induces a rather synchronized growth in previously quiescent cultures. The authors monitored the mitotic activity of the cells by [^{3}H]thymidine incorporation into the nuclei during a period of 24 hr. All fibroblasts showed a cAMP/cGMP ratio of 2 to 4 during logarithmic growth. The ratio rose to 12 to 30 in resting cells and dropped to 1 in activated cultures. This corresponds to an increase of 50–400% in the cAMP level during the resting phase in comparison with the logarithmic growth phase. In contrast, the cGMP level was lowest in the resting phase and increased between 2- and 4-fold in normal fibroblasts during the logarithmic growth phase and up to 10-fold during the state of activated cell growth. After reinitiation of growth the cAMP and cGMP concentrations dropped or rose, respectively to reach approximately equal values. This occurred within about 20 min after addition of fresh medium. The effects on both nucleotides were augmented by prostaglandin-E_1 (PGE_1).

The changes in cyclic nucleotide concentrations, particularly those of cGMP, are largely lost in cells that are transformed with SV40 or Py virus. In the neoplastic cells the values for cAMP remain between 6 and 9 nmoles/mg protein at all growth stages and between 5 and 4.3 nmoles/mg protein for cGMP (also at all growth stages), making for a cAMP/cGMP ratio between 1.2 and 1.9. Very likely the lack of reciprocal fluctuations of cAMP and cGMP levels in the transformed cells reflects the fact that these cells do not arrest in the G_1- or G_0-phases of the cell cycle.

Bannai and Sheppard [129] have shown that the cAMP level very sensitively indicates the growth stages of 3T3 mouse fibroblasts, as well as of L-cells. In both cell types the cAMP level in logarithmically growing cells is considerably lower than in quiescent, or density inhibited stages. These workers also found that the

TABLE 6.10

Effect of mitogenic doses of concanavalin A on the levels of cAMP and cGMP in human peripheral blood lymphocytes[1]

Conditions	Cyclic nucleotide[2]					
	cGMP			cAMP		
	0 min	10 min	20 min	0 min	10 min	20 min
Control	1.00	0.90	0.60	1.00	0.7	0.82
Concanavalin A	1.00	12.8	26.3	7.02	0.71	0.81

[1] From Hadden et al. [123].
[2] Control values at $t = 0$ set to 1.00.

increase of cAMP concentration closely followed the activity of adenylate cyclase; this increased as a function of culture density.

Hadden et al. [123] have found that, in the case of peripheral human blood lymphocytes, there is a 15–30-fold increase in intracellular cGMP concentration within 20 min after mitogenic stimulation with either concanavalin A or non-agglutinating phytohemagglutinin (PHA-NR 69). However, no changes in cAMP occur concurrently (Table 6.10). The fluctuations in the cGMP levels correlate very well with the rates of DNA synthesis. The authors accordingly conclude that the increase in the cGMP concentration represents a signal that can induce cell division, whereas the elevation of cAMP (or a high steady state concentration of cAMP) may actually inhibit initiation of cell division. It should be noted however that Smith et al. [130] have observed that PHA stimulation induces an increase in the intercellular levels of cAMP. Similar results were reported by Abell and Monahan [112].

Makman and Klein [131] have demonstrated that the adenylate cyclase activity of Chang liver cells in culture fluctuates with cell cycle after synchronization by single thymidine blockade. They found that the enzyme increases in specific activity and becomes more sensitive to epinephrine stimulation during mitosis. However, 17 to 20 hr after release of the cells from the thymidine block, both the basal- and epinephrine-stimulated adenylate cyclase activities decreased markedly (in the S-phase of the cell cycle). The activities then recovered, increased to a maximum at the peak of mitosis and decreased again as the cells entered the next period of DNA synthesis. This occurred in both stationary and suspension cultures.

The NaF-stimulated activity of adenylate cyclase (i.e., the intrinsic catalytic activity, independent of the activity of membrane receptors) showed a similar modulation, but recovery occurred already much earlier in the late S-phase. Less depression of NaF-stimulated activity was observed in stationary cultures, even in the earlier S-phase, compared with the basal- and epinephrine-stimulated activities. This suggests that the expression of the membrane receptors involved in hormonal control and basal activity might be regulated separately from the expression of catalytic activity. The authors propose that the receptors controlling adenylate cyclase are lost or hidden during S-phase and then replaced or uncovered during the G_1-phase.

As already noted by Bannai and Sheppard [129], the undulating intercellular levels of cAMP during the cell cycle strictly parallel the activity of plasma membrane adenylate cyclase; this increases with cell density. The plasma membrane antagonist of adenylate cyclase is phosphodiesterase, which forms 5′-AMP from cAMP. This enzyme has been characterized by Pastan and associates [132, 133].

D'Armiento et al. [132] demonstrated that contact inhibited 3T3 fibroblasts contain two forms of phosphodiesterase, with different kinetic characteristics.

One has a high affinity for the substrate (K_M = 2.5 μM); the second has a low affinity (K_M = 71 μM). The high-affinity enzyme increases in activity with cell density whereas the low affinity enzyme stays unchanged. An SV40 transformed 3T3 cell line which has low cAMP levels also exhibits low intracellular levels of the high affinity enzyme component.

The high affinity enzyme of either 3T3 cells or SV40-transformed 3T3 cells can be increased in activity by theophylline or by addition of dibutyryl-cAMP. This increase of enzyme activity is inhibited by actinomycin D and cycloheximide and it is therefore suggested that the enhancement of phosphodiesterase activity depends on *de novo* protein synthesis. In terms of K_M, the newly formed enzyme resembles the high affinity enzyme found in unstimulated cells. It is possible that accumulation of the enzyme is due to induction of its synthesis by cAMP.

More recently Russell and Pastan [133] have isolated a plasma membrane preparation from chick embryo fibroblast and have located the cAMP phosphodiesterase in this preparation. They show that the kinetic characteristics of this enzyme can be altered by trypsin treatment of the cells or the isolated fragments. They also demonstrate that prolonged homogenization of the cells releases increasing amounts of phosphodiesterase activity into the soluble fraction. DEAE cellulose chromatography of the soluble material resolves two enzymatically active peaks. The first hydrolyses both cGMP and cAMP (the latter according to normal Michealis–Menten kinetics). The second peak, in contrast, splits only cAMP and its kinetics resemble those determined for the membrane bound enzyme (not Michealis–Menten).

The authors suggest a negative-cooperativity model, with a cooperativity constant of 0.2, and argue that the enzyme might function to split cAMP in a negatively-cooperative fashion. It is also possible, however, that the membrane bound enzyme contains two different binding sites for cAMP and that its behavior does not necessarily represent cooperative behavior.

Trypsin treatment of isolated membranes alters the biphasic characteristics of the Lineweaver–Burke plots, reducing the K_M value of the high affinity site but not that of the low affinity site, and decreasing the V_{max} of both sites. It is interesting that, although the phosphodiesterase is sensitive to trypsinization of intact cells, it does not act significantly on externally-added cAMP.

Tomkins and colleagues [134, 135] have defined the coordinated metabolic patterns that occur under different conditions of cell growth as expressions of 'pleiotypic programs'. They further define the regulation of such coordinated cellular events as 'pleiotypic control'. They argue that, when growth is stimulated by some extracellular factor, the cells react with a 'positive pleiotypic response'. In contrast, when growth promoting agents are withdrawn a 'negative pleiotypic response' ensues. They further contend that cyclic nucleotides act as 'pleiotypic mediators'. In terms of present knowledge, cAMP appears to be principally involved in the 'negative pleiotypic response', that is, it tends to suppress various cellular metabolic processes involved in cellular replication, whereas cGMP might actually act as a 'positive pleiotypic mediator' [136]. This topic is obviously of major importance to our main theme, since neoplastic cells typically show a pleiomorphic deviation from the normal cell phenotype.

Kram et al. [135] have attempted to correlate the cAMP levels in normal mouse 3T3 fibroblasts (Balb/c) and their SV40 transformants with growth behavior. It was observed that, when both cell types were deprived of serum, intracellular cAMP increased by 60% after 18 hr in normal cells, but decreased by some 10% in transformed cells. Two hours after serum replenishment, the cAMP levels of normal cells returned to the values found before serum depletion; significantly, no such fluctuation of cAMP levels was observed in transformed cells. Concordantly, when replicating, untransformed mouse Balb/c-3T3 fibroblasts were transferred to a serum-deficient medium, the incorporation of leucine and uridine into macromolecules dropped and, after 20 hr of serum deprivation, reached values of 30% and 10% respectively compared to the controls. In contrast, thymidine incorporation into DNA was not affected for up to 5–6 hr, but then decreased sharply.

In another set of experiments the same group attempted to explore what cellular functions cAMP actually does control. For this they exposed the two cell types to 0.2 mM dibutyryl-cAMP (db-cAMP), a non-catabolized cAMP derivative (which can mimic some of the mediating effects of cAMP), with or without 1 mM theophylline (to block phosphodiesterase). It was observed that db-cAMP inhibited the uptake of uridine, leucine and 2-deoxyglucose by normal cells. Addition of theophylline produced further inhibition of these functions to the degree obtained with serum starvation. However, db-cAMP did not influence nutrient uptake by *transformed* cells. The uptake processes which are inhibited by db-cAMP are, in the absence of this reagent, also inhibited by PGE_1, presumably by activation of adenylate cyclase. Also cGMP can override the inhibitory effects of db-cAMP.

One of the major problems in the use of db-cAMP seems to be the fact that this nucleotide does not readily enter into cells and thus mimics primarily those effects of cAMP which occur at the cell surface [137]. Thus, db-cAMP, unlike cAMP, does not directly influence protein and RNA synthesis, or protein degradation.

Present evidence can be interpreted to signify that cAMP acts as a 'negative pleiotypic mediator', but the situation appears more complicated. Indeed, Goldberg and associates [138] suggest that cGMP actually acts as *the mediator* of 'positive pleiotypic responses' and the cGMP and cAMP act in a 'checks and balances' fashion (the 'Yin-Yang' hypothesis). Important to this argument is the fact that the cGMP levels in tissues are usually 10- to 50-fold *lower* than those of cAMP.

The first indications that agents such as epinephrine might act via increasing the level of cGMP were derived by Goldberg et al. [139] from their study on brain cells, of the effect of imidazole; this agent inhibits the phosphodiesterase-catalyzed hydrolysis of cGMP [139]. More recently, Otten et al. [140] have studied the cellular levels of cAMP in chick embryo fibroblasts at varying stages of cell proliferation. After 4 days of logarithmic growth these cells reach a stationary phase which persists for 2 days. Thereafter, the monolayers degenerate. During the first two days of logarithmic growth, the cGMP levels lie near 0.7 pmoles/μg DNA. As the cells growth became density inhibited, that is, in phases 3, 4 and 5, the values drop to 0.025, 0.09 and 0.035 pmoles/μg DNA, respectively, the fluctuations again appear to be reciprocal when related to the cAMP levels presented by Otten et al. [140]. (These cells inhibit an increase of cAMP during the attainment of the stationary phase.) Hadden et al. [123] have shown that within 20 min after stimulation of peripheral blood lymphocytes by highly purified PHA, there is a more than 10-fold increase in intracellular cGMP without any change in cAMP. This response is associated with enhanced incorporation of uridine into nuclear RNA and involves cGMP levels between 10^{-11} and 10^{-9} M. Since the cGMP decreased very rapidly after 20 min, Hadden and colleagues [123] propose that the initial cGMP pulse may act as an intracellular signal to induce proliferation. They consider this all the more likely since cAMP drops during the cGMP pulse.

In addition, Hadden et al. [123] have shown that acetylcholine has a *promoting* effect on the protein and RNA synthesis in human blood lymphocytes and it appears that the mitogen and acetylcholine have the same effect as far as the cyclic nucleotides are concerned, except that the latter act in a strong enough fashion to induce mitogenesis. Thus, acetylcholine increased the cGMP level only 2- to 3-fold, while the mitogens caused a 10-fold augmentation.

Hadden et al. [123] and Goldberg et al. [138] demonstrate that activation of cGMP production requires extracellular calcium, whereas this divalent cation inhibits the activation of adenylate cyclase in diverse tissues [122]. As already pointed out, however, the role of Ca^{2+} is complex and difficult to evaluate. The *intra-cellular* calcium concentration in particular is the critical factor and this depends markedly upon mitochondrial function.

Goldberg et al. [138] has found that addition of insulin (0.1–1000 munits/ml) raises the level of cGMP and drops the cAMP concentration [141] also in cultured mouse 3T3 cells; a 10–40-fold augmentation in the amount of cGMP per unit mass of DNA occurred within 20 min after hormone addition.

As we have already pointed out (p. 270), an analogous rapid and transient (2–10 min) stimulation of cGMP accumulation by insulin (0.1 munits/ml) has been observed in rat liver and rat adipocytes by Illiano et al. [124] and has led these authors to suggest that the cellular cGMP/cAMP balance depends on the relative proportion of 'GTP-vs. ATP-specific' forms of a single class of plasma membrane nucleotide cyclases.

The study of Illiano et al. [124] also illustrates the strong dependence of cyclic nucleotide responses on subtle variables, such as cell and species differences of plasma membrane receptors. Thus, rat-spleen and human-blood lymphocytes exhibit a clear rise in cGMP content following cholinergic stimulation no such effect could be demonstrated with insulin. This is consistent with other observations [142] showing a lack (or severe depletion) of insulin receptors on resting human lymphocytes; membrane preparations from resting lymphoid cells also exhibit no insulin-mediated inhibition of adenylate cyclase [143].

6.3.5. Cyclic nucleotide metabolism in transformed cells

6.3.5.1. Adenylate cyclase activity and cAMP levels

Makman [144] has reported comprehensively on the variation of the adenylcyclase activity of various cells as a function of cell growth and in response to various hormones. Makman [144] monitored adenylate cyclase activity by the hydrolysis of [^{32}P]ATP (followed by chromatographic isolation of labelled cAMP) in different clones of HeLa cells growing in suspension or stationary culture, in Chang liver cells, in rat hepatoma cells (HTC) and various lines of fibroblasts, including 3T3 cells and chemically transformed L929 cells. The external variables examined were NaF (8 mM), epinephrine (16 μM), glucagon (0.2 μg/ml in the presence of 20 mM GTP) and PGE_1 (1 μg/ml). The concentration of these agents were the ones which gave maximal effects in the various systems used.

The adenylate cyclase activity of HeLa S3 cells, adapted for growth and suspension culture, was stimulated by glucagon, PGE_1 and epinephrine. Both the basal and the stimulated activities increased after the cells had grown for 3 days in stationary culture. The most striking effects were obtained with NaF, which, by blocking energy metabolism, simulate a starvation situation. HeLa cells of the parental strain, grown normally in stationary culture, showed even greater basal adenylate cyclase activity and greater hormone responses.

HTC hepatoma cells exhibited no adenylate cyclase activity when grown in suspension culture and only very low activity when propagated in stationary culture. Cells grown in stationary culture gave an increase of activity upon addition of glucagon, epinephrine or postaglandin.

Chang liver cells exhibited high basal adenylate cyclase activities whether grown in suspension- or stationary-culture. This was responsive to epinephrine and NaF but only slightly to PGE_1; maximum stimulation was reached 2 to 3 days after transfer to stationary growth conditions. Chang cells showed a density dependent response only in stationary culture and it was small at that.

The variation of adenylate cyclase activity with culture density can also be demonstrated in fibroblasts. For example, the basal adenylate cyclase activity of 3T3 cells and L929 mouse fibroblast, increases with cell density in the presence of epinephrine and PGE_1 but does not respond to glucagon or NaF. Makman points out that, in general, cells with malignant growth characteristics exhibit less adenylate cyclase activity and he suggests that the cyclase activity may be stimulated by intercellular contact which might regulate or modulate the hormonal response of the cyclase in these cells.

Pastan and associates [132, 133] have carried out an extensive series of investigations on the differences in cAMP metabolism between various cultured fibroblasts and their transformed variants. As we have noted in Section 6.3.2., cAMP levels of cultured cells rise markedly during density inhibition of growth; this is not observed in a variety of transformants which lack density inhibition of cell replication.

In order to evaluate the mechanisms involved, Pastan and associates [145] have measured adenylate cyclase activities in normal chick embryo fibroblasts (CEF) and cells shortly after oncogenic conversion by the Schmidt–Ruppin (SR-RSV) or Bryan high titre (BH-RSV) strains of Rous sarcoma virus (RSV).

Both viruses caused a decrease of adenylate cyclase activity (Table 6.11) but in the case of BH-RSV, the reason appeared to be a decrease in the K_M for ATP whereas SR-RSV-transformed cells showed decrease in V_{max} (which could be overcome by fluoride) but no alteration in K_M. In view of difficulties involved in kinetic analyses of non-soluble systems (Chapter 7), the observed alterations in kinetic parameters cannot be simply interpreted. In particular, the increased K_M found in BH-RSV-transformed cells does not necessarily imply appearance of a qualitatively different cyclase. However, studies using various RSV mutants, that are temperature sensitive in producing the transformed phenotype, clearly show that the changes in cAMP levels and modifications of adenylate cyclase activity constitute an expression of the transformed state.

TABLE 6.11

Effect of oncogenic transformation by RNA tumor viruses on the adenylate cyclase activity of rat and chick fibroblasts[1]

Cell type[2]	Kinetic parameters of adenylate cyclase	
	K_M(ATP)[3]	V_{max}^3
CEF-Normal	0.23	40
CEF-BH-RSV transformed	1.08	38
CEF-SR-RSV transformed	0.28	21
NRK-Normal (low cell density)	0.24	13
NRK-Normal (high cell density)	0.21	48
NRK-Ki-RSV transformed (low-cell density)	0.23	17
NRK-Ki-RSV transformed (high-cell density)	0.25	22
NRK-MSV transformed (high-cell density)	0.21	11

[1] From [145, 146].
[2] Both transformed cells showed half the cAMP levels of normal cells.
[3] K_M in mM; V_{max} in pmoles · min^{-1} · mg protein^{-1} (in absence of fluoride).

Recently the Pastan group [146] has reported on the differing adenylate cyclase activity in crude membrane preparations isolated from normal rat kidney fibroblasts (NRK) and NRK cells transformed by the Kirsten strain of RSV (Ki-RSV) or Moloney sarcoma virus (MSV). The cyclase activity of transformed cells was at most 50 % of NRK cells and did not show the typical increase with cell density. The response to fluoride was also lower in the transformed cells. These effects could not be ascribed to a change of K_M for ATP which was ~0.24 mM for all cell types (Table 6.11). On the other hand the V_{max} was substantially lower in the transformants. Moreover the cyclase activity in the transformed cells showed a different dependence on Mg^{2+} than found in NRK cells. In the latter enzyme activity increased with Mg^{2+} concentration up to 15 mM; in the transformants activity became maximal at 1 mM.

Pastan and associates [132, 147] and others, have examined the effects of db-cAMP on various virally transformed rodent fibroblasts. These cells respond to db-cAMP by a decrease of motility and an increase in adhesion to the culture substrate through a mechanism not requiring new protein synthesis. Moreover, the transformed cells tend to assume the morphology of normal cells in the presence of db-cAMP.

Pastan et al. [147] propose that oncogenic viruses introduce different 'transforming genes' which modify adenylate cyclase activity in various ways either directly or indirectly by changing a critical plasma membrane component (protein, lipid or carbohydrate). As a consequence of this modification a whole cascade of pleiomorphic anomalies would result due to impairment of the 'pleiotypic mediation' system.

The adenylate cyclase activities in membrane preparations from normal and neoplastic liver have also been investigated. Thus Emmelot and Bos [4] have measured the adenylate cyclase activity in bilefront plasma membrane preparations (Chapter 2) isolated from normal rat hepatocytes, rat hepatoma 484A, mouse hepatocytes and mouse hepatoma 147042. In their assays, they inhibited phosphodiesterase activity with 10 mM theophyllin and measured adenylate cyclase by the conversion of [8-^{14}C]ATP. They also evaluated the influence of glucagon (10 μM), epinephrine (0.1 mM), bovine growth hormone (0.1 mg/mg of membrane protein), insulin (20 μM), and NaF (10 mM).

Some of the results are presented in Table 6.12. The bile-front preparations from normal mouse and rat liver showed a similar basal adenylate cyclase activity corresponding to about 2 μmoles cAMP formed per mg of protein per hour. In the membranes from both species, the basal activity was stimulated to a similar extent by various external agents, namely about 20-fold by glucagon, 5-fold by epinephrine and 10- to 15-fold by NaF. However, insulin and growth hormone had no significant effects.

The epinephrine receptor appeared to be very labile, because incubation of the membrane preparations for 5 to 10 min at 0 °C in the presence of the hormone,

TABLE 6.12

Adenylate cyclase activity in plasma membrane preparations isolated from normal mouse and rat liver, and from mouse hepatoma 147042 and rat hepatoma 484A[1]

Membrane source	Adenylate cyclase activity (μmoles cAMP formed mg protein$^{-1}\cdot$ hr^{-1})			
	Basal	+ Glucagon	+ Epinephrine	+ NaF
Mouse Liver	1.4–1.8	46	3.1–8.1	28–40
Hepatoma 147042	1.5–1.5	32–34	2.3–5.2	21–28
Rat Liver	1.1	27.5	4.5	20
Hepatoma 484A	0.8	7.4	1.1	9.2

[1] From Emmelot and Bos [4].

abolished the stimulatory effect nearly completely. This effect was not observed with glucagon and supports earlier observations by Bitensky et al. [148, 149] on rat liver homogenates.

The basal activities of adenylate cyclase were similar for the membrane preparation derived from normal and neoplastic hepatocytes, but these membranes showed different functional responses (Table 6.12). In general, the hepatoma membranes were stimulated less normally than by glucagon epinephrine and NaF. The greatest difference in responsiveness was noted in the case of glucagon and NaF.

Allen et al. [150] also found that the basal adenylate cyclase activities in homogenates of slow-, medium- and rapidly-growing Morris hepatomas were similar to that in liver homogenates, that all specimens responded with an 8–10-fold increase in activity but that the degree of response declined with tumor growth rate. However, regenerating liver behaved as does normal tissue in terms of adenylate cyclase activity. Rate of cell replication *per se* is thus not determining since the growth rate of regenerating liver is greater than that of most hepatomas.

Tomasi et al. [151], using a somewhat different method for membrane isolation, found a 3-fold higher basal adenylate cyclase activity in membrane preparations from normal rat hepatocytes. However, membranes isolated from the Yoshida ascites hepatoma exhibited only about 10 % of the basal specific adenylate cyclase activity found in normal rat hepatocytes. However, there is not reason to assume that the membranes isolated from the two cell types were identical. In fact, in the case of the tumor cells most of the adenylate cyclase activity was found in a supernatant fraction, whereas after fractionation of homogenates from normal rat hepatocytes most of the activity was found to be membrane-bound.

Tomasi et al. [151] do report some stimulation of the cyclase by glucagon ($\sim 10^{-5}$ M) epinephrine (2×10^{-6} to 10^{-5} M) and NaF (10 mM), but the activity of the soluble enzyme from the Yoshida ascites hepatoma was only doubled by glucagon. Also, the authors obtained only a 3-fold stimulation with glucagon with membranes isolated from normal hepatocytes compared with the 20-fold stimulation reported by Emmelot and Bos [4].

Chayoth et al. [152] have used a different approach to the problem. They analyzed the cAMP formed in rat liver tissue slices isolated from animals treated with the carcinogen, ethionine. They studied slices of uninvolved liver, slices of benign hyperplastic nodules and slices of hepatoma. All samples were incubated at 37 °C for 20 min in Krebs–Ringer buffer containing 0.1 % albumin, 0.1 % glucose, 10 mM theophylline, with or without PGE_1 (10 μg/ml) or glucagon (10 μg/ml). The reactions were terminated by boiling the slices and the cAMP determined by standard methods. The authors argue that in this way, the results obtained reflect more faithfully the *in vivo* situation.

Without incubation, the basal cAMP level in normal slices was 1.8 nmoles/g of protein. This level increased in the presence of theophylline and was further elevated by PGE_1 and glucagon. The increases in the cAMP levels could be correlated with an increase in adenylate cyclase activity. This was stimulated to 50 % by PGE_1 and 11-fold by glucagon. (However, intracellular cAMP rose only 2.5-fold in the presence of glucagon).

The cAMP levels, their responses, the adenylate cyclase activities and their responses in slices from ethionine treated animals were in general somewhat higher than in normal animals; however, PGE_1 produced a 5-fold greater increase in cAMP than normal. Even higher cAMP concentrations were found in the hyperplastic nodules and in the hepatomas and these basal cAMP levels could be increased 5- to 10-fold by PGE_1 and glucagon. However, there was no simple correlation between cAMP levels and adenylate cyclase activity; this was essentially uneffected by PGE_1 in the case of hyperplastic nodules (in contrast to normal liver material) and stimulated to a slightly lesser extent by glucagon in the case of hepatomas than in hyperplastic nodules. Chayoth et al. [152] also found a lower activity of phosphodiesterase in hepatomas than in benign nodules and, therefore, explain the elevated cAMP in hepatoma cells by the decrease in phosphodiesterase activity.

The results published by Chayoth et al. [152] fit other investigations, which also show that intracellular cAMP levels do not necessarily vary inversely with cell replication rate. Thus, Christofferson et al. [153] have analyzed the adenylate cyclase activity in a particulate fraction isolated from rat liver homogenates (material pelleted at 8.5×10^5 g · min) from rats feeds with 2-acetylamino-fluorene for several weeks. They compared particles isolated from normal liver, uninvolved liver from treated rats and particles from hepatomas. They found that despite an increase in basal adenylate cyclase activity going from normal liver, to livers from treated animals to hepatomas, the particles from all three cell categories gave an equivalent stimulation with both epinephrine (5×10^{-5} M) and NaF.

Further documentation of the very variable behavior of hepatoma cells has been presented by Butcher et al. [154] who studied a series of 15 Morris hepatomas propagated intramuscularly in rats. These authors found no consistent correlation between basal cAMP levels or changes induced by treatment of the animals with glucagon, theophylline, isoproterenol or epinephrine.

A basic problem in all of these studies is that they have focussed on the intracellular cAMP or adenylate cyclase activities and their responses to hormones such as glucagon, epinephrine, PGE_1. However, since the crucial issue may be the ratio of cAMP/cGMP rather than cAMP levels, it is difficult to arrive at extensive conclusions.

6.3.5.2. Phosphodiesterase and cAMP levels

Clark et al. [155] have measured the soluble phosphodiesterase activity in normal and regenerating rat liver, as well as in the slowly growing hepatoma 47-C and the rapidly growing hepatoma 3924A. All of the specimens apparently possess at least two enzymes, one with high affinity for cAMP ($K_M \sim 2\ \mu M$) and one with low affinity (K_M 100–500 μM).

The relative distributions of the two isozymes in the hepatomas differed significantly from the pattern in normal or regenerating liver (Table 6.13). Significantly, the 'high-K_M' activity was depressed in both hepatomas, while the 'low-K_M' phosphodiesterase activity was increased. In regenerating liver, the activities of both enzyme species were enhanced; that is, the isozyme shift is not simply a function of replication rate.

TABLE 6.13

Cyclic cAMP phosphodiesterase in normal and regenerating liver, and in two hepatomas[1]

Tissue	cAMP Phosphodiesterase activity 'High-K_M' enzyme	 'Low K_M' enzyme
Normal liver	100	100
Regenerating liver	154	114
Hepatoma 47-C	58	185
Hepatoma 3924A	13	265

[1] From Clark et al. [155].
[2] Activity in normal liver set to 100.

The authors suggest that the different isozyme distributions found in normal vs. neoplastic hepatocytes constitute an example of the patterns predicted by the 'Molecular Correlation Concept'. Indeed, this point is argued vigorously by Weber [1], but the matter appears more complicated. Thus, Tisdale and Phillips

[156] have developed a correlation between the degree of *malignancy* of certain cell types and intracellular cAMP level. They studied an L-cell derived line (A9H), the $A9T_4$ line obtained after four passages of A9H cells in mice, and a line of fast growing lymphomas cells (TLX5), as well as two hybrids, A9H/TLX (A) and $A9T_4$/TLX (A). Malignancy was characterized by tumor incidence after injection of 10^5 cells into mice and also by cell growth rate *in vitro*; both hybrid cell lines had a lower malignancy than the parent lines. There was a good correlation between cAMP levels and malignancy; both hybrid lines had significantly higher cAMP content, but there was no significant correlation between growth rate and cAMP level (Table 6.14).

TABLE 6.14

Concentrations of cAMP in TLX5 cells, A9H cells, $A9T_4$ cells and two hybrids[1]

Cell	Doubling time	pmoles cAMP/mg protein
TLX5	16	19.2
A9H	23	50
$A9T_4$	23	49
$A9T_4$/TLX	23	63
A9H/TLX	20	114

[1] From Tisdale and Phillips [156].

The cAMP phosphodiesterase activity in the cells was assayed in sonicates using [^{3}H]cAMP as substrate and identifying the 5′-AMP by thin layer chromatography. It was found that the cAMP-phosphodiesterase of each cell type had two K_M values, namely a low affinity value (370–1055 μM) and a high affinity value (K_M less than 5 μM) (Table 6.15).

TABLE 6.15

K_M and V_{max} values of low- and high-affinity cAMP-phosphodiesterases from several cell types[1]

Cell Line	Low-affinity enzyme		High-affinity enzyme	
	K_M^2	V_{max}^3	K_M^2	V_{max}^3
A9H/TLX	4.1	3.4	535	104
$A9T_4$/TLX	2.4	2.3	345	64
A9H	2.2	1.6	370	40
TLX5	4.4	0.3	1075	6

[1] From Tisdale and Phillips [156].
[2] In μM.
[3] μmoles · min^{-1} · mg protein^{-1}.

The enzymes from A9H and A9T_4/TLX cells, both of which replicate slowly, exhibited similar K_M values. However, the A9T_4/TLX and TLX cells yielded an enzyme with higher K_M values. The relative V_{max} value of either form of the enzyme rose 20-fold from the highly malignant TLX5 cells (low cAMP) to the less malignant A9H/TLX hybrid (high cAMP). The data suggest that the K_M values of both forms of enzymes increase with intercellular cAMP level and support the proposal of D'Armiento et al. [132] that the synthesis of the high affinity phosphodiesterase is regulated by cAMP.

6.3.6. Conclusion

The pace of cyclic nucleotide research as applied to tumor biology is so hot that we can neither provide a comprehensive review of the topic, nor present definitive conclusions. However, we feel that the following generalizations are reasonable:

(a) The cyclic nucleotide metabolism of neoplastic cells is commonly not responsive to external stimuli – for example, hormones, cell density.

(b) This unresponsiveness can be ascribed in part to a deficiency and/or altered distribution of specific receptors required for cyclase activation.

(c) The abnormal cyclic nucleotide metabolism may also be associated with, and derive in part from, modified cyclase activity and a shift in phosphodiesterase affinity for cyclic nucleotides.

(d) *In vitro*, the abnormal growth phenotype of virally transformed cells can often be suppressed by db-cAMP, a cAMP derivative, but it appears that the ratio cAMP/cGMP is more important in the regulation of intracellular metabolism than the concentration of either nucleotide.

(e) Major quantitative differences occur between various *in vitro* situations and between *in vitro* and *in vivo* responses.

In summary, it appears established that cAMP and cGMP act as 'pleiotypic mediators' and that the pleiotypic responses of many tumor cells are abnormal. However, the deviations of cyclic nucleotide metabolism appear *themselves pleiotypic*.

6.4. Membrane regulation of mRNA stability: the 'membron' hypothesis

The 'membron' hypothesis has been developed to account for the many-fold anomalies in protein (enzyme) biosynthesis that occur in neoplasms. Pitot and associates [157,158] have proposed that the differences of phenotype between normal and neoplastic cells, as well as between different tumors, might arise not only from differences in genome expression but also from alterations occurring at the translational level. Their basic argument, as reviewed in (e.g. [158,159]), is that the level and type of phenotypic expression in a cell depends upon the lifetime of various RNA messengers, that is, upon '*template lifetime*' or *template stability*. *Template lifetime* is defined as the period during which synthesis of a specific protein, once initiated, can proceed, without further synthesis of mRNA.

Evidence that the temporal stabilities of specific mRNAs can influence cellular enzyme phenotype was first presented by Pitot et al. [159] who used actinomycin-D to inhibit mRNA (template) synthesis in hepatic cells under conditions normally compatible with mRNA-directed protein synthesis. They found that the synthesis of diverse amino acid-metabolizing enzymes, induced in rat liver by feeding casein hydrolysates to normal rats can be blocked when actinomycin-D is administered to the animals during the initiation of induction. However, when actinomycin-D treatment is initiated 12 hr *after* feeding the casein hydrolysate, induction of enzyme synthesis is not inhibited. The length of time during which induction of enzyme synthesis remains actionmycin-D-insensitive, gives information about the 'stability' of the enzyme's mRNA template.

The template half-lives of enzymes induced by amino acid feeding vary from one enzyme to the next. Thus, in normal liver, the value is less than 3 hr for tyrosine transaminase, 6–8 hr for serine dehydrogenase and 18–24 hr for ornithine transaminase (Table 6.16). Significantly, the template lifetimes for these enzymes in various hepatocellular tumors differed substantially from those of normal liver; however, some tumors showed shorter half-lives than others for some given enzyme and longer half-lives for other enzymes (Table 6.16).

TABLE 6.16

Estimated mRNA-template lifetimes for various amino acid metabolizing enzymes of rat liver and several hepatocellular tumors[1]

Enzyle	Liver	Tumor
Serine dehydryogenase	6–8 hr	H-35 = <1 hr
		5123 = >2 wk
Ornithine aminotransferase	18–24 hr	5123 = >24 hr
Tryptophan oxygenase	>2 wk	H-35 = <12 hr
		5123, 7316, 7800 = <30 min
Tyrosine aminotransferase (induced)	2–3 hr	H-35 = >6 hr

[1] From Pitot et al. [157].

Different template lifetimes have also been observed for other enzyme categories and in other tumors. Thus, the lifetime of thymidine kinase in normal liver lies below 3 hr, but exceeds 12 hr in the H-35 hepatoma. Moreover, the template lifetimes of glucose-6-phosphate dehydrogenase and malic enzyme are both greater than 2 days in normal mammary gland [160] but exceed 2 days in mammary carcinoma R3230 AC. Unfortunately, no data exist for β-methyl-β-hydroxyl-glutaryl-CoA reductase in neoplasms. As noted in Chapter 5, this enzyme regulates cholesterol biosynthesis which in frequently deranged in tumors.

Pitot and associates argue that *template lifetime* is not invariant, but can be controlled by exogenous factors. Also their data (Table 6.16) suggest that the template of each enzyme is different. Finally, as shown in Table 6.16, the template lifetimes of diverse enzymes differ between normal liver and hepatocellular carcinomas, as well as between hepatomas. These differences could account for the wide variations of enzyme complement found in diverse hepatomas.

The polyribosomes of animal cells occur either bound to the membrane of endoplasmic reticulum (including outer membrane of the nuclear envelope) or free in the cytoplasm. The proportion of polysomes that are membrane bound depends on cell type and the physiological state of the cell.

Of the membrane associated polysomes, a small proportion are attached to the endoplasmic reticulum *via* the large ribosomal subunit. These polysomes appear to participate in the production of proteins that are 'secreted' into the cisternae of the endoplasmic reticulum [161]. However, even in non-secretory cells, 5–10% of the total polysome population fall into this category [162].

A second category of polysomes, is 'loosely' membrane associated and can be isolated with membrane fractions prepared (e.g., from HeLa cells) at low ionic strength [162]. This type of polysome can also be identified by thin section-electron microscopy [163].

Finally, Milcarek and Penman [164] have shown that the poly(A) mRNA segments of both loosely- and tightly-bound polysomes are attached directly to endoplasmic reticulum membranes (presumably to membrane proteins). The poly(A) segments of mRNA remain membrane bound even after degradation of the membrane bound polysomes. One third of the membrane bound poly(A) can be removed by EDTA treatment and the membrane affinity is eliminated by mild detergent treatment at 0 °C.

It is suggested that template lifetime depends upon the stability of the translational unit, or *membron* which is defined as a stable, functional complex between endoplasmic reticulum membranes and polysomes. The *membron* hypothesis derives from evidence indicating that stable templates can only develop through appropriate polysome membrane associations namely:

(a) Completely functional translational units (i.e., ones that can be released *in vitro* by puromycin) require association of relevant polysomes with endoplasmic reticulum [165].
(b) Polysomes bind preferentially to endoplasmic reticulum membranes, that have been recently stripped of polysomes [166].
(c) Binding of poly (GC) to stripped rough endoplasmic reticulum membrane inhibits polysome attachment.

(d) Modifications of rough endoplasmic reticulum, for example trypsinization, inhibit polysome binding [166].
(e) Polysome attachment to stripped endoplasmic reticulum produces functional units of protein synthesis equivalent to those of native rough endoplasmic reticulum [167].
(f) In comparison to free polysomes, polysome-membrane complexes resist degradation by ribonuclease.

6.4.1. Neoplastic cells

Several reports support the hypothesis that mRNA templates of hepatoma cells can differ in stability from those of normal hepatocytes. Thus, Lerman et al. [168] have demonstrated divergent translational control of protein synthesis in mouse livers vs. mouse hepatoma 22a. They isolated total homogenates, a nuclear fraction, a fraction composed of mitochondria + lysosomes and a fraction composed of polysome + ribosomes from both tumors and livers, 1 hr, 12 hr and 48 hr after injection of trace amounts of *N*-[^{14}C]methyl-*N*-nitrosurea ([^{14}C] MNU), an alkylating agent. The loss of radioactivity with time from DNA, RNA and protein in the various fractions was taken as a measure of the repair and/or replacement of these macromolecules.

Significantly, the radioactivity in all components from normal liver decreased rapidly up to 48 hr, while in hepatoma fractions, it increased during this interval. When the experiments were modified to monitor incorporation of [^{14}C]leucine into the proteins of the various fractions after a single dose of unlabelled MNU the authors observed a return to normal protein synthesis in normal liver within 15–24 hr. However, in the hepatomas protein synthesis remained depressed during this interval and declined even further up to at least 48 hr [168]. Moreover, incorporation of [^{14}C]leucine into nascent polypeptide chains was inhibited to the same extent as overall protein synthesis.

The data clearly indicate a significant difference in macromolecular biosynthesis in normal and neoplastic hepatocytes. The differential effects of MNU on [^{14}C]-leucine incorporation can be explained by two mechanisms: (a) The number of translating ribosomes is smaller and/or replaced more slowly in the hepatoma cells. (b) The rate of translation is decreased to a greater extent in the hepatomas. That this latter postulate may be the correct one, is suggested by the fact that addition of cell sap (containing tRNA) from normal livers did not repair the MNU-induced defect.

The results of Lerman et al. [168] and Abakumova et al. [169] fit those of Stewart et al. [170] who showed that a single dose of dimethylnitrosamine inhibited translation in rat liver *in vivo*. Moreover, Villa-Trevina [171] demonstrated already in 1967, that administration of dimethylnitrosamine to rats caused polysome breakdown in liver and accumulation of *nontranslating* 80 S ribosomes.

Decreased synthesis of mRNA was considered improbable in the experiments, since there was no reduction of [^{14}C]orotate incorporation into nuclear RNA at a stage where protein synthesis was reduced by 30%.

6.5. Coda

Neoplastic cells typically exhibit anomalies in several membrane-controlled cellular regulatory systems. These aberrations can account for many facets of tumor cell behavior, yet no coherent pattern has emerged. The plasma membrane, mitochondria and the membranes of endoplasmic reticulum all contain critical restriction points, but the interrelationships between these control sites, as well as their properties, remain obscure. This area of tumor biology, as well as others treated in this volume, urgently requires an integration of research strategies.

References

[1] Weber, G. in *The Role of Cyclic Nucleotide in Carcinogenesis*, J. Schultz and H.G. Gratzner, eds., Academic Press, New York, p. 57.
[2] Graham, J.M., *Biochem. J.*, *130*, 1113 (1972).
[3] Barclay, M. and Terebus-Kekish, O., *J. Natl. Cancer Inst.*, *51*, 1709 (1973).
[4] Emmelot, P. and Bos, C.J., *Biochim. Biophys. Acta*, *249*, 285 (1971).
[5] Miedema, E., *Exp. Cell Res.*, *53*, 488 (1968).
[6] Raftell, M. and Blomberg, F., *Biochim. Biophys. Acta*, *291*, 421 (1973).
[7] Raftell, M. and Blomberg, F., *Biochim. Biophys. Acta*, *291*, 442 (1973).
[8] Blomberg, F. and Raftell, M., *Biochim. Biophys. Acta*, *291*, 431 (1973).
[9] Bosmann, H.B., *Exp. Cell Res.*, *54*, 217 (1969).
[10] Raftell, M., Blomberg, F. and Perlmann, P., *Cancer Res.*, *34*, 2300 (1974).
[11] Warburg, O., *Über den Stoffwechsel der Tumoren*, Springer, Berlin, 1926; translated by Dickens, F.; *The Metabolism of Tumors*, R.N. Smith, New York, 1931.
[12] Lo, C.H., Cristofalo, V.J., Morris, H.P. and Weinhouse, S., *Cancer Res.*, *28*, 1 (1968).
[13] Von Ardenne, M. and Rieger, F., *Z. Naturforsch.*, *21b*, 472 (1965).
[13a] Jensen, M., Sherwood, P. and Wallach, D.F.H., *J. Theoret. Biol.*, 1975, in press.
[14] Wallach, D.F.H., Bieri, V., Verma, S.P. and Schmidt-Ullrich, R., *Ann. N. Y. Acad. Sci. U.S.*, 1975, in press.
[15] Chance, B. and Williams, W.R., *Advan. Enzymol.*, *17*, 65 (1965).
[16] Lehninger, A.L., *Biochemistry*, Worth Publishers, Inc., 1970, Ch. 18, 23.
[17] Quagliariello, E., Papa, S., Meijer, A.J. and Tager, J.M. in *Mitochondria Structure and Function*, eds. Ernster, L. and Drahota, Z., Academic Press, New York, 1969, p. 335.
[18] Papa, S., Lofrumento, N.E., Quagliariello, E., Meijer, A.J. and Tager, J.M., *J. Bioenerg.*, *1*, 287 (1971).
[19] Pasteur, L., *Comptes rend., Acad. Sci.*, *52*, 1260 (1861).
[20] Mott, J.C., *Br. Med. Bull.*, *17*, 144 (1961).
[21] Neubert, D., Peters, H., Teske, S., Kohler, E. and Barrach, H.J., *Naunyn-Schmiedebergs Arch. Pharmakol. Exp. Pathol.*, *268*, 235 (1971).
[22] Zebe, E., Delbrück, A. and Bucher, T.H., *Biochem. Z.*, *331*, 254 (1959).
[23] Asakura, T., Sato, Y., Minakami, S. and Yoshikawa, H., *J. Biochem.* (*Tokyo*), *59*, 524 (1966).
[24] Benesch, R., Benesch, R.E. and Yu, C.I., *Proc. Natl. Acad. Sci. U.S.*, *59*, 526 (1968).

[25] Dische, Z., *Trav. Soc. Chem. Biol.*, *23*, 1140 (1941).
[26] Tarui, S., Kono, N. and Uyeda, K., *J. Biol. Chem.*, *247*, 1138 (1972).
[27] Cori, C.F., Cori, G.T., *J. Biol. Chem.*, *65*, 397 (1925).
[28] Gullino, P.M., Clark, S.H. and Grantham, F.H., *Cancer Res.*, *24*, 780 (1964).
[29] Warburg, O., *Science*, *133*, 309 (1956).
[30] Kadenbach, B., *Eur. J. Biochem.*, *12*, 392 (1970).
[31] Goldblatt, H. and Cameron, G., *J. Exp. Med.*, *97*, 525 (1953).
[32] Goldblatt, H., Friedman, L. and Cechner, R.L., *Biochem. Med.*, *7*, 241 (1973).
[33] Jensen, M.L., Wallach, D.F.H. and Lin, P.S., *Exptl. Cell Res.*, *94*, 271 (1974).
[34] Goldblatt, H. and Friedman, L., *Proc. Natl. Acad. Sci. U.S.*, *71*, 1780 (1974).
[35] Aisenberg, A.C., *The Glycolysis and Respiration of Tumors*, Academic Press, New York, 1961.
[36] Burk, D., Woods, M. and Hunter, J., *J. Natl. Cancer Inst.*, *38*, 839 (1967).
[37] Potter, V.R., *Cancer Res.*, *24*, 1085 (1964).
[38] Wenner, C.E., *Advan. Enzymol.*, *29*, 321 (1967).
[39] Sanford, K.K., Barker, B.E., Woods, M.W., Parshad, R. and Law, L.W., *J. Natl. Cancer Inst.*, *39*, 705 (1967).
[40] Woods, M.W., Sanford, K.K., Burk, B. and Earle, W.R., *J. Natl. Cancer Inst.*, *23*, 1079 (1969).
[41] Kieler, J., Moore, J., Biczowa, B. and Radzikowski, C., *Acta Path. Microbiol. Scand.*, *79*, 529 (1970), series A.
[42] Paul, J., Pearsons, E.S., *Exp. Cell Res.*, *12*, 212 (1957).
[43] Paul, J., Pearsons, E.S., *Exp. Cell Res.*, *12*, 223 (1957).
[44] Warburg, O., Gawehn, K. and Geissler, A.W., *Z. Naturforsch. 15B*, 378 (1960).
[45] Warburg, O., Gawehn, K., Geissler, A.W. and Lorenz, S., *Hoppe Seyler's Z. Physiol. Chem.*, *321*, 252 (1960).
[46] Druckrey, H., Bresciani, F. and Schneider, H., *Z. Naturforsch 13B*, 516 (1958).
[47] Gottlieb, N., Heise, E. and Binder, N., *J. Natl. Cancer Inst.*, *35*, 413 (1965).
[48] Weinhouse, S., *Fed. Proc.*, *32*, 2162 (1973).
[49] Tanaka, T., Harano, Y., Sue, F. and Morinura, H., *J. Biochem.* (Tokyo), *62*, 71 (1967).
[50] Emmelot, P. and Bos, C.J., *Biochim. Biophys. Acta*, *121*, 434 (1966).
[51] Davidova, S.Y., Shapot, V.S. and Solowjeva, A.A., *Biochim. Biophys. Acta*, *158*, 303 (1968).
[52] Bheatty, R.S. and Hickie, R.A., *Biochem. Biophys. Res. Commun.*, *44*, 1443 (1971).
[53] Aisenberg, A.C. and Morris, H.P., *Nature*, *191*, 1314 (1961).
[54] Knox, W.E., Jamdar, S.C. and Davis, P.A., *Cancer Res.*, *30*, 2240 (1970).
[55] Wallach, D.F.H. and Winzler, R., *Evolving Strategies and Tactics in Biomembrane Research*, Springer-Verlag, New York, 1974, Ch. 2.
[56] Aisenberg, A.C., *Cancer Res.*, *21*, 295 (1961).
[57] Allard, C., Mathieu, R., DeLamirande, G. and Cantero, A., *Cancer Res.*, *12*, 407 (1952).
[58] Howatson, A.F. and Ham, A.W., *Cancer Res.*, *15*, 62 (1955).
[59] Hruban, Z., Mochizuki, Y., Slesers, A. and Morris, H.P., *Cancer Res.*, *32*, 853 (1972).
[60] Sordahl, L.A., Blailock, Z.R., Liebelt, A.G., Kraft, G.H. and Schwartz, A., *Cancer Res.*, *29*, 2002 (1969).
[61] Sordahl, L.A. and Schwartz, A. in *Methods of Cancer Research*, H. Busch, ed. Academic Press, New York, 1971, Vol. 6, p. 159.
[62] Schneider, W.C. and Potter, V.R., *Cancer Res.*, *3*, 353 (1943).
[63] Galeotti, T., Cittadini, A., Dionisi, O., Russo, M. and Terranova, M., *Biochim. Biophys. Acta*, *253*, 303 (1971).
[64] Sato, N. and Hagihara, B., *Cancer Res.*, *30*, 2061 (1970).
[65] Hagihari, B., Sato, N., Fukuhara, T., Tsutsumi, K. and Oyanagui, Y., *Cancer Res.*, *33*, 2947 (1973).
[66] White, M.T. and Tewari, K.K., *Cancer Res.*, *33*, 1645 (1973).
[67] Pedersen, P.L., Greenawalt, J.W., Chan, T.L. and Morris, H.P., *Cancer Res.*, *30*, 2620 (1970).
[68] Schnaitman, C., Erwin, V.G. and Greenawalt, J.W., *J. Cell Biol.*, *32*, 719 (1967).
[69] Sottocasa, G.L., Kuylenstierna, B., Ernster, L. and Berestrand, A., *J. Cell Biol.*, *32*, 415 (1967).
[70] White, M.T., Arya, D.V. and Tewari, K.K., *J. Natl. Cancer Inst.*, *53*, 553 (1974).

[70a] White, M.T., Arya, D.V. and Tewari, K.K., *J. Natl. Cancer Inst.*, *54*, 245 (1975).
[71] Emmelot, P., *Cancer Res.*, *22*, 38 (1962).
[72] Mehard, C.W., Packer, L. and Abraham, S., *Cancer Res.*, *31*, 2148 (1971).
[73] Wattiaux, R. and Wattiaux-de Coninck, S., *Eur. J. Cancer*, *4*, 201 (1968).
[74] Beaufay, H. and Berthet, J. in *Methods of Separation of Subcellular Structural Component*, Grant, J.K. ed., Biochem. Soc. Symp., 23, Cambridge University Press, 1963, p. 65.
[75] Cornbleet, P.J., Vorbeck, M.L., Lucas, F.V., Esterly, J.A., Morris, H.P. and Martin, A.P., *Cancer Res.*, *34*, 439 (1974).
[76] Martin, A.P., Cornbleet, P.J., Lucas, F.V., Morris, H.P. and Vorbeck, M.L., *Cancer Res.*, *34*, 850 (1974).
[77] Feo, F., *Life Sci.*, *6*, 2417 (1967).
[78] Feo, F., Canuto, R.A. and Garcea, R., *Eur. J. Cancer*, *9*, 203 (1973).
[79] Thorne, R.F.W. and Bygrave, F.L., *Cancer Res.*, *33*, 2562 (1973).
[80] Lehninger, A.L., *Biochem. J.*, *119*, 129 (1970).
[81] Carafoli, E. and Lehninger, A.L., *Biochem. J.*, *122*, 681 (1971).
[82] Spencer, T. and Bygrave, F.L., *J. Bioenerg.*, *4*, 347 (1973).
[83] Thorne, R.F.W. and Bygrave, F.L., *Biochem. Biophys. Res. Commun.*, *50*, 294 (1973).
[84] Thorne, R.F.W. and Bygrave, F.L., *Nature*, *248*, 348 (1974).
[85] Thorne, R.F.W. and Bygrave, F.L., *FEBS Letters*, *41*, 118 (1974).
[86] Thorne, R.F.W. and Bygrave, F.L., *Biochem. J.*, *144*, 551 (1974).
[87] Lehninger, A.L., personal communications.
[88] Reynafarje, B. and Lehninger, A.L., *Proc. Natl. Acad. Sci. U.S.*, *70*, 1744 (1973).
[89] Reynafarje, B. and Lehninger, A.L., *Biochem. Biophys. Res. Commun.*, *57*, 286 (1974).
[90] Reynafarje, B. and Lehninger, A.L., *J. Biol. Chem.*, *249*, 6067 (1974).
[91] Borle, A.B., *J. Memb. Biol.*, *16*, 221 (1974).
[92] Mikkelsen, R.B. and Wallach. D.F.H., *Biochim. Biophys. Acta*, in press.
[93] Sottocasa, G.L., Sandri, G., Panfili, E., deBernard, B., Gazzotti, P., Vasington, F.O. and Carafoli, E., *Biochem. Biophys. Res. Commun.*, *47*, 808 (1972).
[94] Gomez-Puyou, A., Tuena de Gomez-Puyou, M., Becker, G. and Lehninger, A.L., *Biochem. Biophys. Res. Commun.*, *47*, 814 (1972).
[95] Bosmann, H.B., Myers, M.W. and Morgan, H.R., *Biochem. Biophys. Res. Commun.*, *56*, 75 (1974).
[96] Weinhouse, S., *Science*, *124*, 267 (1956).
[97] Boxer, G.E. and Devlin, T.M., *Science*, *134*, 1495 (1961).
[98] Dionisi, O., Cittandini, G., Gelmuzzi, G., Galeotti, T. and Terranova, T., *Biochim. Biophys. Acta*, *216*, 71 (1970).
[99] Papa, S., Paradies, G., Galeotti, T., Dionisi, O. and Eboli, M.L., *Nature New Biol.*, *242*, 86 (1973).
[100] Borst, P., *Biochim. Biophys. Acta*, *57*, 270 (1962).
[101] McKee, R.W., Wong, W. and Landman, M., *Biochim. Biophys. Acta*, *105*, 410 (1965).
[102] Gordon, E.E., Ernster, L. and Dallner, G., *Cancer Res.*, *27*, 1372 (1967).
[103] Cittadini, A., Galeotti, T., Russo, M. and Terranova, T., *Biochim. Biophys. Acta*, *253*, 314 (1971).
[104] Kovačević, Z., *Eur. J. Biochem.*, *25*, 372 (1972).
[105] Brown, H.D., Chattopadhyay, S.K., Pennington, S.N., Spratt, J.S. and Morris, H.D., *Br. J. Cancer*, *25*, 135 (1971).
[106] Chattopadhyay, S.K., Brown, H.D. and Morris, H.P., *Br. J. Cancer*, *26*, 3 (1972).
[107] Miyake, Y., Gaylor, J.L. and Morris, H.P., *J. Biol. Chem.*, *249*, 1980 (1974).
[108] Sato, N. and Hagihara, B., *Cancer Res.*, *30*, 2061 (1970).
[109] Robison, G.A., Butcher, R.W. and Sutherland, E.W., *Cyclic AMP*, Academic Press, New York–London, 1971.
[110] Rodbell, M. in *Colloquium on the Role of Adenylate Cyclase and Cyclic 3',5'-AMP in Biological System*, P. Condliffe and M. Rodbell, eds., Fogarty International Center, Govt. Print. Off., Washington, D.C., 1971, p. 88.
[111] Perkins, J. in *Advan. in Cyclic Nucleotide Res.*, P. Greengard and G.A. Robison, eds., Raven Press, New York, 1973, Vol. 3, p. 1.

[112] Abell, C.W. and Monahan, T.M., *J. Cell Biol.*, *59*, 549 (1973).
[113] Birnbaumer, L., *Biochim. Biophys. Acta*, *300*, 129 (1973).
[114] Ryan, W.L. and Heidrick, M.L. in *Advan. Cyclic Nucleotide Res.*, P. Greengard and G.A. Robison, eds., Raven Press, New York, 1974, p. 81.
[115] *The Role of Cyclic Nucleotides in Cancer Research*, J. Schultz and H.G. Gratzner, eds., Academic Press, New York, 1973.
[116] Cuatrecasas, P., *Biochemistry*, *12*, 3547, 3558, 3567 (1973).
[117] Bourne, H.R., Coffino, P., Tomkins, G.M., *Science*, *197*, 750 (1975).
[118] Birnbaumer, L., Pohl, S.L. and Rodbell, M., *J. Biol. Chem.*, *246*, 1857 (1971).
[119] Pohl, S.L., Krans, H.M., Kozyreff, V., Birnbaumer, L. and Rodbell, M., *J. Biol. Chem.*, *246*, 4447 (1971).
[120] Siegel, M.I. and Cuatrecasas, P., in *Cellular Membranes and Tumor Cell Behavior*, ed. E.F. Walborg, Jr., Williams and Wilkins, Baltimore, in press, 1975.
[121] Chrisman, T.D., Garbers, D.L., Parks, M.A. and Hardman, J.G., *J. Biol. Chem.*, *250*, 374 (1975).
[122] Goldberg, N.D., O'Dea, R.F. and Haddox, M.K., *Advan. Cycl. Nucl. Res.*, *3*, 155 (1973).
[123] Hadden, J.W., Hadden, E.M., Haddox, M.K. and Goldberg, N.D., *Proc. Natl. Acad. Sci. U.S.*, *69*, 3024 (1972).
[124] Illiano, G., Tell, G.P.E., Siegel, M.E. and Cuatrecasas, P., *Proc. Natl. Acad. Sci. U.S.*, *70*, 2443 (1973).
[125] Sheppard, J.R. and Prescott, D.M., *Exptl. Cell Res.*, *75*, 293 (1972).
[126] Zeilig, C.E., Johnson, R.A., Friedman, D.L. and Sutherland, E.W., *J. Cell Biol.*, *55*, 296a (1972).
[127] Millis, A.J.T., Forrest, G. and Pious, D.A., *Biochem. Biophys. Res. Commun.*, *49*, 1645 (1972).
[128] Rudland, P.S., Seeley, M. and Seifert, W., *Nature*, *251*, 417 (1974).
[129] Bannai, S. and Sheppard, J.R., *Nature*, in press, 1974.
[130] Smith, J.W., Steiner, A.L., NewBerry, W.M. and Parker, C.W., *J. Clin. Invest.*, *50*, 432 (1971).
[131] Makman, M. and Klein, M.I., *Proc. Natl. Acad. Sci. U.S.*, *69*, 456 (1972).
[132] D'Armiento, M., Johnson, G.S. and Pastan, I., *Proc. Natl. Acad. Sci. U.S.*, *69*, 459 (1972).
[133] Russell, T. and Pastan, I., *J. Biol. Chem.*, *248*, 5835 (1973).
[134] Hershko, A., Mamont, P., Shield, R. and Tomkins, G.M., *Nature, New Biol.*, *232*, 206 (1971).
[135] Kram, R., Mamont, P. and Tomkins, G.M., *Proc. Natl. Acad. Sci. U.S.*, *70*, 1432 (1973).
[136] Sheppard, J.R. and Bannai, S., in *Control of Proliferation in Animal Cells*, Clarkson, B. and Baserga, R., eds.; Cold Spring Harbor Lab., New York, 1974.
[137] Ryan, W.L. and Durick, M.A., *Science*, *177*, 1002 (1972).
[138] Goldberg, N.D., Haddox, M.K., Dunham, E., Lopez, C. anf Hadden, J.W. in *Control of Proliferation in Animal Cells*, Clarkson, B. and Baserga, R., eds.; Cold Spring Harbor Laboratory, 1974, p. 609.
[139] Goldberg, N.D., Lust, W.D., O'Dea, R.F., Wei, S. and O'Toole, A.G., *Advances in Biochemical Psychopharmacology*, Costa, E. and Greengard, P., eds.; Vol. 3, p. 67, Raven Press, New York, 1970.
[140] Otten, J., Johnson, G.S. and Pastan, I., *Biochem. Biophys. Res. Comm.*, *44*, 1192 (1971).
[141] Sheppard, J.R., *Nature, New Biol.*, *236*, 14 (1972).
[142] Krug, U., Krug, F. and Cuatrecasas, P., *Proc. Natl. Acad. Sci. U.S.*, *69*, 2604 (1972).
[143] Tell, G.P.E., Cuatrecasas, P., Van Wyk, J.J. and Hintz, R.L., *Science*, *180*, 312 (1973).
[144] Makman, M.H., *Proc. Natl. Acad. Sci. U.S.*, *68*, 2127 (1971).
[145] Anderson, W.B., Johnson, G.S. and Pastan, I., *Proc. Natl. Acad. Sci. U.S.*, *70*, 1055 (1973).
[146] Anderson, W.B., Gallo, M. and Pastan, I., *J. Biol. Chem.*, *249*, 7041 (1974).
[147] Pastan, I., Anderson, W.B., Carchman, R.A., Willingham, M.L., Russell, T.R. and Johnson, G.S. in *Control of Proliferation in Animal Cells*, Clarkson, B. and Baserga, R., eds., Cold Spring Harbor Laboratory, New York, 1974, p. 563.
[148] Bitensky, M.W., Russell, V. and Robertson, W., *Biochem. Biophys. Res. Comm.*, *31*, 706 (1968).
[149] Bitensky, M.W., Russell, V. and Blanco, M., *Endocrinology*, *86*, 154 (1970).
[150] Allen, D.O., Munshower, J., Morris, H.P. and Weber, G., *Cancer Res.*, *31*, 557 (1971).
[151] Tomasi, V., Réthy, A. and Trevisani, A., *Life Sciences*, *12*, *Part II*, 145 (1973).

[152] Chayoth, R., Epstein, S.M., Field, J.B., *Cancer Res.*, *33*, 1970 (1973).
[153] Christoffersen, T., Morland, J., Osnes, J.B. and Eegjo, K., *Biochim. Biophys. Acta*, *279*, 363 (1972).
[154] Butcher, F.R., Scott, D.F., Potter, V.R. and Morris, H.P., *Cancer Res.*, *32*, 2135 (1972).
[155] Clark, J.F., Morris, H.P. and Weber, G., *Cancer Res.*, *33*, 356 (1973).
[156] Tisdale, M.J. and Phillips, B.J., *Exptl. Cell Res.*, *88*, 111 (1974).
[157] Pitot, H.C., Shires, T.K., Moyer, G. and Garrett in *The Molecular Biology of Cancer*, Busch, H., ed., Academic Press, New York, p. 523, 1974.
[158] Pitot, H.C. et al. in *Cellular Membranes and Tumor Cell Behavior*, Walborg, E.F., Jr., ed., Williams and Wilkins, Baltimore, Md., 1975, in press.
[159] Pitot, H.C., Peraino, C., Lamar, C., Jr. and Lesher, S., *Science*, *150*, 901 (1965).
[160] Hilf, R., *Cancer Res.*, *28*, 1888 (1968).
[161] Sabatini, D.D. and Blobel, G., *J. Cell Biol.*, *45*, 146 (1970).
[162] Rosbash, M. and Penman, S., *J. Mol. Biol.*, *59*, 227 (1971).
[163] Lee, S.Y., Krsmanovic, V. and Brawerman, G., *J. Cell Biol.*, *49*, 683, 1971.
[164] Milcarek, C. and Penman, S., *J. Mol. Biol.*, *89*, 327 (1974).
[165] Jost, J.P., Khairallah, E.A. and Pitot, H.C., *J. Biol. Chem.*, *243*, 3057 (1968).
[166] Shires, T.K., Narurkar, L. and Pitot, H.C., *Biochem. J.*, *125*, 67 (1971).
[167] Redman, C.M. and Cherian, M.G., *J. Cell Biol.*, *52*, 231 (1972).
[168] Lerman, M.I., Abakumova, O.Y., Kucenco, N.G., Gorbacheva, L.B. and Kukushkina, G.V., *Cancer Res.*, *34*, 1536 (1974).
[169] Abakumova, O.Y., Ugarova, T.Y., Gorbacheve, L.B., Kucenco, N.G., Philipenko, N.N., Sokolova, I.S. and Lerman, M.I., *Cancer Res.*, *34*, 1542 (1974).
[170] Stewart, B.W. and Magee, P.N., *Biochem. J.*, *125*, 943 (1971).
[171] Villa-Trevino, S., *Biochem. J.*, *105*, 625 (1967).

7

Membrane permeability

7.1. Transport

7.1.1. Introduction

Cells fixed in tissues obtain nutrients by diffusion from capillaries. Intensive studies, particularly on oxygen, indicate that the concentration of a given nutrient at points given in a tissue depends on the following:
(a) Rate of capillary blood flow.
(b) Intercapillary distance.
(c) The diffusion coefficient of the intercellular material for a given nutrient.
(d) Distance from the capillary wall.
(e) Distance from the arterial end of the capillary.
(f) Cellular nutrient uptake.
For a given tissue geometry, blood flow and diffusion coefficient cells most remote from the arterial end of the capillary are exposed to the lowest nutrient concentration. Such cells lead a precarious existence because many essential metabolites (e.g. oxygen) cannot reach them in excess.

Analyses of tissue diffusion processes show that harmonious coexistence of diverse cells in tissues can only occur when no cell type can accumulate essential nutrients much more efficiently than another. Should some process suddenly endow some cell population with an increased transport capacity for an essential metabolite, this clone could ultimately eliminate the normal cells. This clear principle of natural selection has led to the proposal that enhanced or uncontrolled membrane transport may play a significant role in the pathophysiology of cancer [1–3]. However, the field has received serious attention only very recently.

Before considering the recent data, it will be useful to consider the early model experiments of Shrivastava and Quastel [1]. These experimenters incubated rat brain cortex slices together with slices of kidney cortex, spleen, Walker carcinosarcoma, solid Ehrlich carcinoma, and suspensions of Ehrlich ascites carcinoma cells.

Their data show that co-incubation of brain slices with slices of kidney or spleen does not alter the glutamine levels of the brain cells. In contrast, co-incubation with the three tumor systems can lead to marked glutamine depletion in the brain cells. This effect depends on the glucose concentration of the medium: the glutamine depletion ranges from >80% at 1 mM initial glucose to 0% at 10 mM glucose. The authors attribute this phenomenon to the high glycolytic activity of the tumor cells, which deprives the normal tissue of the energy source essential for the net uptake of glutamine. At the same time, the glutamine is utilized by the tumor. The authors clearly relate their observations to the invasive process. They do not specifically raise the issue of membrane transport [2,3], and in their model, the competitive advantage of the tumor cells arises from enhanced glycolysis with attendant enhanced sugar uptake.

We will evaluate the possible relation of transport anomalies and related phenomena to the malignant process in Chapter 11. Here, however, we shall present the now extensive evidence demonstrating that many tumor cells show enhanced transport capacities for sugars and amino acids.

7.1.2. Definitions

7.1.2.1. Passive diffusion

The term 'passive diffusion' implies that a substance passes through a membrane as a result of random thermal motion. Net flux occurs when there is an activity gradient for the substance across the membrane. The transported molecule does not interact specifically with any membrane component. Passive diffusion may occur through aqueous pores, or, for lipophilic substances, through lipid domains.

7.1.2.2. Facilitated diffusion (mediated diffusion)

Here the transported solute combines with a specific 'carrier' in the membrane. The 'carrier', whether loaded or not, shuttles between the inner and outer membrane surfaces, or rotates within the membrane, releasing or binding molecules on each side. Facilitated diffusion does not require metabolic energy and does not produce an activity gradient. Thermal energy is thought adequate to provide for 'carrier' movement.

Facilitated diffusion accounts for the transport of monosaccharides in most animal cells [4]. (However, intestinal epithelium exhibits both facilitated glucose diffusion and active transport.) The most extensive work has been performed on erythrocytes and has been reviewed by Lieb and Stein [5]. These authors also present an important model which rests somewhat on the 'induced fit' theory of enzyme regulation.

They propose that transport occurs by 'internal' substrate transfer across a tetrameric protein spanning the membrane. The protein is an H_2L_2 tetramer made up of two HL dimers, anti-symmetrically arranged, as in hemoglobin, that is,

HL
LH

L-protomers bear low-affinity substrate binding sites; H-protomers carry high affinity sites. Two energetically equivalent orientations exist: (a) *one* in which the binding sites of one HL dimer orient toward the outer and inner membrane surfaces, and those of the other dimer face each other in the membrane core; (b) the other in which the orientation of the protomers is reversed. Reorientation occurs only when a substrate molecule is associated with a protomer.

Transport proceeds as follows:

(a) a sugar molecule binds to a protomer;
(b) this produces reorientation of the four protomers bringing the sugar molecule into the membrane core;
(c) the sugar molecule is distributed between the two inwardly oriented binding sites according to their *affinities*;
(d) a second reorientation occurs when the substrate then transfers from a low to a high affinity site;
(e) this reorientation brings the substrate across the membrane.

The model deals well with all important features of facilitated diffusion, including its lack of direction. Also, together with the known broad specificity of membrane 'carriers' for various sugars, the model accounts for the phenomenon of 'counter transport', that is, coupled mediated diffusion of two different sugars.

The details of the model appear highly mechanistic and the 'induced fit' aspect does not seem essential. However, in its emphasis on macromolecular restructuring as an essential mechanism in transport, the proposal, together with that of Changeux and associates (Chapter 1), adds a new dimension to the integration of membrane structure and function.

7.1.2.3. Active transport

Here, a substance is accumulated against an electrochemical gradient. The process requires a specific membrane translocation mechanism driven by metabolic energy. The classic model for active transport suggests that the permeating molecule binds to a specific membrane carrier and that this, or the carrier substrate complex, then undergoes an energy-requiring alteration. The carrier-substrate complex, formed on the outer membrane surface, crosses to the inside surface where it is modified to lower its affinity for the substrate. The carrier is then free to return to the outside surface to repeat a new transfer cycle. Clearly, the tetramer model of Lieb and Stein [5] can also be applied to active transport by including energy coupling. This modified model is consistent with the fact that many active transport systems will carry out 'mediated diffusion' in cells that are deprived of energy sources.

7.1.3. Sugar transport by neoplastic cells

7.1.3.1. Transformation morphology and enhanced sugar transport

7.1.3.1.1. Murine sarcoma viruses

As detailed in the excellent review by Hatanaka [6], recent evidence indicates that neoplastic conversion of cells by certain oncogenic viruses produces significant alterations in the kinetics of membrane sugar transport. This matter has been most intensively studied in the case of neoplastic transformation of mammalian cells by murine sarcoma viruses. The reason is that these RNA viruses produce rapid, complete, oncogenic conversion. However, other recent work indicates that membrane transport changes also upon neoplastic transformation by DNA viruses.

Conversion by murine sarcoma virus [7] produces easily recognizable changes in cell morphology and concomitant enhancement of sugar uptake [8–11]. Significant augmentation in sugar utilization is first seen 15 hr after infection; by this time, morphological abnormalities occur in about 15% of the exposed cells. The sugar uptake by cells of infected cultures then increases with time and coincides with the rising proportion of structurally altered cells. In contrast, transport in control cultures either remains unchanged or declines with time.

Further correlations between the morphologically transformed state and sugar transport come from studies of morphologic 'revertants' [12,13] and experiments on temperature-sensitive virus mutants [14]. Thus, several clones of 3T3 cells, which have reverted to normal growth but contain sarcoma virus, show sugar uptake identical to untransformed cells, that is, much lower rates than those found in converted cell lines [13]. Moreover, transport of 2-deoxyglucose by cells infected with a temperature-sensitive virus mutant depends on temperature exactly as the culture morphology does. Enhanced transport occurs at the permissive temperature (36 °C), but not at the non-permissive one (41.5 °C). Further, when one transfers infected cells from the non-permissive to the permissive temperature, 2-deoxyglucose uptake increases from a rate typical for uninfected cells to a rate equal to that of cells infected by the non-mutant virus.

The reverse occurs when one shifts infected cells from the permissive to the non-permissive temperature.

Similar results have been obtained with ts 68, another temperature-sensitive sarcoma virus mutant [15,16]. Cells infected with this virus exhibit normal cell division, growth patterns and sugar transport at non-permissive temperatures (41 °C). At permissive temperatures (37 °C), however, one observes both enhanced transport and 'transformed' morphology. Inhibitors of DNA or RNA synthesis have no effect on the development of the transformed state in infected cells following a temperature shift from 41 °C to 37 °C; however, puromycin and cycloheximide completely inhibit both the morphological transformation and the increase of sugar uptake that would be produced by a shift from 41 °C to 37 °C.

Another case of temperature-sensitive sugar transport in murine sarcoma-transformed cells has been reported [17]; the converted rat kidney fibroblast line NRV (MSV-16) exhibits the transformed state at the *higher* temperature (39 °C), but not at 33 °C. Concordantly, cells grown at 39 °C exhibit enhanced transport of 2-deoxyglucose, whereas those grown at 33 °C show the same sugar uptake as non-transformed NRK cells. Clearly, enhancement of sugar transport and maintenance of the morphologically transformed state here depend on a cold-sensitive virus function.

All in all, the transformed state induced by murine sarcoma virus – and its temperature dependence – parallel enhanced sugar transport in such cells.

7.1.3.1.2. Papova virus SV40

SV40-transformed mouse fibroblasts (Balb/c 3T3 SV) can be readily distinguished morphologically from parental Balb/c 3T3 cells and certain revertants that still carry the SV40 genome [18]. The transformed cells are smaller, more triangular, and show different culture characteristics. When one cultivates Balb/c 3T3 cells or the revertant clones on conventional surfaces, they grow as monolayers. When these cells become confluent, replication ceases. In contrast, SV40 transformants, which continue to replicate after confluence, produce multilayered plaques and nodules.

These morphologic features correlate well with the transport of 2-deoxyglucose [19]. Normal Balb/c 3T3 cells show a 10-fold drop in sugar uptake at confluency, whereas the SV40 transformants exhibit similar uptake rates over a wide range of cell densities. A concanavalin A-selected revertant gives a 5-fold decrement in 2-deoxyglucose uptake with confluency. We will later relate these changes to the underlying alterations in transport kinetics.

7.1.3.2. Uptake of sugars by normal and transformed cells

7.1.3.2.1. Metabolizable natural sugars

Glucose and mannose. The uptake of natural sugars depends on metabolism as well as transport, and metabolism varies with cell cycle, cell density and other variables. Transport studies with metabolizable sugars, therefore, require short pulses of labelled sugar. Using this approach, Hatanaka et al. [8] demonstrate that mouse embryo fibroblasts transformed by murine sarcoma virus exhibit enhanced uptake of glucose, mannose and galactose [20]. This functional change emerges together with the morphologic and growth characteristics of neoplastic conversion. The transport change is not observed in cells infected by murine leukemia virus. Hatanaka et al. [8] observed no change in the uptake of other transported substances (phosphate, leucine, uridine, thymidine). However, altered amino acid transport has now been detected in comparisons between established untransformed cell lines (BHK21 clone 12, Balb/3T3) and 'transformants' thereof, induced by polyoma or SV40 virus [20], as well as several transformed hamster cell lines [21]. These cells also exhibit enhanced uptake of glucose mannose and galactose.

Galactose. The uptake of *galactose* has been investigated in some depth by Kalckar et al. [22] for untransformed and polyoma-transformed baby hamster kidney fibroblasts (Table 7.1). They report that both cell types grow as well on galactose, a slow catabolite, as on glucose, a rapid catabolite. The rate of uptake of galactose is markedly enhanced in the transformed cells, and this enhancement is as evident in cells propagated on galactose as in those cultivated in glucose. However, cells grown in glucose-containing media consume practically all of the sugar, while cells grown on galactose do not. The galactose taken up by the cells accumulates initially as galactose, galactose-1-phosphate, and UDP-galactose, but after 24 hr in galactose-media, the product accumulated is primarily galactitol.

TABLE 7.1

Galactose uptake and accumulation of UDP-galactose by normal and polyoma-transformed fibroblasts[1]

Cell type	Pre-cultivated in	Accumulation of labelled galactose compounds (nmoles/mg cell protein/hr)	Ratio of hexose-labelled UDP-gal/UDP-glc
BHK	Glc	8.3	2.0
BHK-py	Glc	18.3	3.7
BHK	Gal	10.0	1.6
BHK-py	Gal	40.0	4.5
NIL	Glc	11.8	1.5
NIL-py	Glc	27.5	4.0
NIL	Gal	5.8	2.8
NIL-py	Gal	24.2	16.7

[1] From [22].

Although galactose entry into the transformed cells is enhanced, the metabolic pathway becomes restricted before the stage of glucose-1-phosphate because of limitation by the enzyme UDP-galactose-4-epimerase. This leads to the increased UDP-Gal/UDP-Glc ratios observed. Thus, glucose uptake is enhanced up to 30-fold by 24 hr hexose 'starvation', while galactose entry increases only about 5-fold under similar conditions. The enhancement is blocked by inhibitors of protein synthesis.

It appears that the glucose and galactose carrier systems are not identical because one observes asymmetric 'repression' when glucose and galactose are compared as 'repressors' in chick fibroblasts [23]. Also, the enhanced uptake of galactose by polyoma-transformed hamster cells differs from the increased galactose accumulation produced by hexose starvation [24].

Kalckar and Ullrey [24] have compared the enhanced accumulation of galactose by hamster NIL fibroblasts with that found in polyoma-transformed NIL cells. The untransformed cells accumulate both galactose-1-phosphate and UDP-hexose. Enhanced galactose uptake appears near confluence and the sugar then accumulates as UDP-glucuronic acid at the expense of UDP-hexose.

In polyoma-transformed NIL cells (or very sparse cultures of normal cells), galactose uptake is high even without hexose deprivation. Most of the galactose taken up appears as UDP-hexose and the polyoma-induced augmentation shows none of the metabolic anomalies seen after hexose starvation. This metabolic feature of galactose transport is further demonstrated by the fact that when glucose or galactose in the medium is replaced by 3-*O*-methylglucose, which is also carried by facilitated diffusion but are not phosphorylated, the latter sugar does not alter the high UDP-glucuronate/UDP-hexose ratios typical of hexose starvation.

Glucosamine. Concurrent with the enhanced uptake of glucose following infection of chick fibroblasts, one observes a 20-fold increase in glucosamine uptake [9]. The effect becomes apparent within 16 hr after infection and closely parallels the emergence of morphologically abnormal cells.

7.1.3.2.2. Non-reactive natural sugars

Unlike glucose, galactose, glucosamine and mannose whose uptakes are enhanced after oncogenic transformation, fructose, ribose, deoxyribose, fucose, sucrose, glucose-1-phosphate and glucose-6-phosphate are not accumulated more avidly by transformed cells than by control cells [6]. This suggests that the enhanced sugar uptake by neoplastic cells involves altered transport, rather than changed metabolism.

7.1.3.2.3. Transport of non-metabolizable sugars

The experiments with natural sugars indicate that enhanced sugar transport in transformed cells cannot derive simply from the effects of cell sociology (growth, contact, density) on sugar metabolism, or from a lack of growth regulation. Direct support for this contention comes from the experimentation with L-glucose and the glucose analogues, 2-deoxy-D-glucose and 3-*O*-methyl glucose. 2-Deoxy-D-glucose is phosphorylated by hexokinase to 2-deoxy-D-glucose-6-phosphate, but this compound is metabolized very slowly. 3-*O*-Methyl glucose is not phosphorylated.

L-*Glucose.* Uptake of L-glucose generally occurs very slowly by *passive diffusion*, rather than mediated diffusion; that is, normal cells lack a transport system for this sugar. For this reason, L-glucose is often used as a control in sugar transport studies [25]. L-Glucose is non-toxic but does not support cell growth. When mixed with D-glucose, L-glucose does not interfere with glucose transport or cell growth [26].

Graff et al. [27] have isolated a sarcoma virus-transformed line which can accumulate L-glucose by *facilitated diffusion.* This system operates at 37 °C, but not at 20 °C. The transport of L-glucose found in the transformed cells is not due to D-glucose contamination. No evidence of phosphorylation or induction of further metabolism was found. The amount of L-glucose that can be accumulated within the neoplastic cells at 37 °C corresponds to the limiting levels observed with 3-0-methylglucose [27]. Graff et al. suggest that this new transport for L-glucose may be a temperature-dependent modification of a normal facilitated diffusion system.

*2-Deoxy-*D*-glucose.* 2-Deoxyglucose is phosphorylated by hexokinase-6-phosphate. Some of phosphorylated product is oxidized to gluconic acid phosphate and 10 % is incorporated into the glycolipid of both normal and SV40-transformed hamster fibroblasts [28]. However, the bulk of the transported sugar remains as 2-deoxyglucose-6-phosphate. When one infects mouse embryo fibroblasts with murine sarcoma virus [8], the increase in 2-deoxy-D-glucose transport parallels the morphological signs of transformation (just as in glucose transport). Transport of 2-deoxy-D-glucose is also greater in polyoma-transformed hamster fibroblasts and SV40-transformed mouse fibroblasts, than in the parental cells [20]. Enhanced 2-deoxyglucose transport also occurs in non-permissive, abortive transformations produced by DNA tumor viruses [29].

Romano and Colby [30,30a] claim that the enhancement of 2-deoxyglucose uptake in SV40-transformed 3T3 cells derives from increased intracellular phosphorylation rather than an augmented rate of mediated diffusion. This is in contradiction with the data in [20] and [22]. Moreover, these authors do not demonstrate enhanced hexokinase or glucokinase activity. Furthermore, after infection of mouse fibroblasts with sarcoma virus, there is no increase in hexokinase, even though 2-deoxyglucose uptake rises dramatically [31]. Finally, the enhancement of 3-*O*-methylglucose transport described below cannot be due to elevated activity of intracellular kinases because this sugar is not phosphorylated.

This matter has recently been very carefully evaluated by Kletzien and Perdue [16]. They first developed tissue culture techniques that allowed them to culture chick embryo fibroblasts, both rapidly and slowly under confluent and non-confluent conditions. Measuring 2-deoxy-D-glucose uptake and phosphorylation under all conditions, they found that rapidly replicating cells invariably take up the sugar more avidly, although the phosphorylation rates are identical in all cases. Moreover, phosphorylation is much slower in intact cells than in homogenates thereof, and both uptake and phosphorylation are inhibited by cytochalasin B, a transport inhibitor.

Addition of fetal calf serum to serum-deprived cultures produces an increase in the uptake of 2-deoxy-D-glucose [32]. This is biphasic, with an initial increase in uptake occurring within 10 min and a second increase within 1 hr. The first phase is not inhibited by cycloheximide or actinomycin D, whereas the second is blocked by cycloheximide but not actinomycin D. Again, the differences in uptake are not accompanied by any change in phosphorylation. There is no change in the K_M of transport, and the increase of uptake is due to an augmented V_{max} only; the data accordingly suggest that the cells can both *activate* quiescent receptors and synthesize new ones.

Kletzien and Perdue [33] have extended these studies to chick embryo fibroblasts infected with the temperature-sensitive mutant of Rous sarcoma virus, ts 68. As noted before, infected cells grown at 41 °C show a normal morphology but at 37 °C the transformed characteristic obtains. The uptake by cells propagated at 37 °C is two to four times that of cells cultured at 41 °C, but their phosphorylation rates are identical; also, the difference between the permissive and non-permissive transport rates is abolished by cytochalasin B and cycloheximide.

3-O-Methylglucose. This sugar is transported into cells by 'facilitated diffusion'. It cannot be phosphorylated. As is characteristic for such transport, efflux and influx occur at similar rates [34,35], and efflux can be accelerated by other sugars (through counter-transport).

3-*O*-Methylglucose enters sarcoma-transformed fibroblasts much more rapidly than normal cells [36,34]. The sarcoma-transformed cells saturate within 15 sec, whereas transport in normal cells continues linearly for 4 min. This difference has produced some confusion as to changes of 3-*O*-methylglucose transport upon transformation. The transport of 3-*O*-methylglucose shows a different response to cytochalasin B than normal cells.

7.1.3.3. Kinetics of sugar transport

The specific mechanisms underlying the transport changes described cannot be very easily defined. This is because the kinetic analysis of transport on whole cells (or even membrane fragments) is complicated by some inaccessible variables. Ideally, a change in V_{max} would indicate an alteration in the *number* and/or the *turnover rates* of the 'carrier' sites involved in transport. In contrast, changes in K_M would imply that neoplastic alteration *qualitatively* modified the carriers.

However, measured K_Ms depend on indirect factors, such as unstirred layers, nonspecific steric hindrance, as well as more specific processes for example metabolism, feedback inhibition, substrate pools and multiple transport processes for a single substrate. Such factors readily distort kinetic analyses. Indeed, the kinetic patterns shown by intact cells often differ drastically from those found in the transport systems that are isolated therefrom. Also, even with a purified enzyme system. the linearity of Lineweaver–Burke plots can depend on the concentration ranges of reagents, allostericity, as well as positive and negative cooperativity of enzymes. All these processes must be considered in kinetic analyses of membrane transport.

Studies on cultured normal and *SV40*-transformed fibroblasts [8, 12] reveal no significant changes in the K_M for glucose transport upon neoplastic conversion (Table 7.2), despite the greater sugar uptake by the transformed cells. Concordantly, Isselbacher [20] finds that normal and polyoma-transformed hamster fibroblasts exhibit identical K_Ms (1.2×10^{-3} M) for 2-deoxyglucose uptake.

In contrast to the situation observed following SV40 transformation, Hatanaka and associates [12] find that oncogenic conversion of mouse fibroblasts by *murine sarcoma virus* leads to a drastic reduction in the K_M for glucose; under these conditions, no change is seen in SV40-transformed cells (Table 7.2).

However, Weber [37] reports no significant change in the apparent K_M for

TABLE 7.2

K_M for glucose transport in normal and neoplastically converted mouse fibroblasts

Cell type	K_M (M $\times 10^4$)
Normal	
Balb/3T3-clone 31	25.1
SV40-transformed	
SV-A31 clone T2	42.0
SV-A31 clone M1	36.5
SV-A31 clone S4	30.0
Mouse-sarcoma transformed	
MSV A31 clone 1	3.7
MSV A31 clone 24	2.4
MSV A31 clone 261	2.6

From Hatanaka et al. [6].

2-deoxyglucose transport upon Rous sarcoma virus transformation (0.46–0.80 × 10^{-3} M, when 0.019–1.0 mM of the substrate was used; 1.0–4.7 × 10^{-3} M, with 0.019–10.0 mM of substrate). Bose and Zlotnick [38], using 0.25–20 mM of 2-deoxyglucose for measurements of their transport on mouse sarcoma-transformed cells, also find no change in K_M. Conceivably, the discrepancy relates to the high substrate concentrations used in [37] and [38]. Thus, Hino and Yamamoto [39] working with 0.1–1.0 mM 2-deoxyglucose in their transport studies on normal and sarcoma-transformed mouse fibroblasts obtain K_M values of 5.6 × 10^{-4} M in converted cells compared to 43.4 × 10^{-4} M in nontransformed mouse cells. Also, using the same substrate range, Hatanaka et al. [31] find a K_M of 5 × 10^{-4} for mouse cells infected with mouse sarcoma viruses, compared to 13.3 × 10^{-4} M for normal cells. Finally, May et al. [17], working with 0.05–1 mM sugar, also find a significant reduction in the K_Ms for 2-deoxyglucose transport by NRK cells (Table 7.3). The last studies are particularly important, because they show a direct correlation between the transformed state and reduction of K_M in a temperature-sensitive mutant of murine sarcoma virus. At temperatures at which the virus does not induce morphologic transformation, the K_M for deoxyglucose is the same in infected as in uninfected cells. At the permissive temperature, however, K_M drops about 30-fold.

TABLE 7.3

K_Ms for 2-deoxyglucose transport in normal NRK cells, cells transformed by wild type murine sarcoma virus and cells transformed by a temperature-sensitive strain of murine sarcoma virus[1]

Cell type[2]	K_M (M × 10^4)
NRK	19.2
NRK-MSV	2.8
NRK-MSV 1b (33 °C)	22.0
NRK-MSV 1b (39 °C)	0.7

[1] From May et al. [17].
[2] MSV = wild type murine sarcoma virus, MSV 1b = temperaturesensitive mutant, which induces transformed state at 39 °C, but not at 33 °C.

However, substrate concentration cannot be the sole factor, because Kletzien and Perdue in their careful studies on the transport of 2-deoxy-D-glucose found no changes of K_M with enhanced transport, but marked alterations of V_{max} (Table 7.4), although they worked at substrate concentrations of 0.2–3 mM. This result holds for the transport enhancement associated with rapid cell growth, Rous sarcoma virus transformation and serum replenishment.

TABLE 7.4

Kinetic parameters for the transport of 2-deoxy-D-glucose by chick embryo fibroblasts

Cells	Transport parameter	
	K_M^3 (mM)	V_{max}^3 (nmoles/mg protein/min)
Chick embryo fibroblasts[1]		
Rapidly growing, non-confluent	2.0	33.2
Slowly growing, non-confluent	1.9	19.3
Rapidly growing, confluent	2.1	27.0
Slowly growing, confluent	2.1	16.7
ts-68 Infected[2]		
Transformed (37 °C)	2.0	40.0
Non-transformed (41 °C)	2.0	17.0

[1] From Kletzien and Perdue [16].
[2] From Kletzien and Perdue [33].
[3] 3-*O*-Methylglucose is transported with a K_M of 3.5–5 mM in all cases, but the V_{max} is 35–38 nmoles/mg protein/min in rapidly growing cells compared with 18–20 nmoles/mg protein/min in slowly replicating cells.

Bose and Zlotnick [38] report the activation energies of the 2-deoxy-glucose uptake system, using low substrate concentrations at three different temperatures. They find values of 3.5 kcal/mol for Balb 3T3 mouse fibroblasts and 7.4 kcal/mol for *sarcoma virus*-transformed Balb 3T3 cells. These data also suggest that transformation produces a qualitative change in the ‘carrier’ for 2-deoxyglucose. However, the activation energies determined from Arrhenius plots are not reliable unless one obtains true V_{max} values at each temperature and at high substrate concentrations. At low substrate concentrations analysis is difficult, because K_M and V_{max} vary independently with temperature. Indeed, Kletzien and Perdue [33] obtain similar activation energies for 2-deoxy-D-glucose transport by chick embryo fibroblasts infected with the ts 68 mutant of Rous sarcoma virus at permissive and non-permissive temperatures.

The balance of available data thus suggests that transformation with PAPOVA viruses *does not* produce a *qualitative* change in the membrane carriers for glucose or 2-deoxyglucose. The same appears to be the case for chick fibroblasts transformed by Rous sarcoma virus. In contrast, neoplastic conversion by murine sarcoma viruses appears to produce such a change, as suggested by reduced K_M. In any event, *all the data point to enhanced sugar uptake following viral neoplastic conversion**.

* Uptake of 3-*O*-methylglucose into sarcoma transformed cells is extremely rapid, preventing useful kinetic analyses.

7.1.3.4. Repression and derepression of sugar uptake

An examination of glucose transport in avian cell cultures [23] shows that the rate of uptake of glucose (or 2-deoxyglucose) by chick fibroblasts depends strongly on the glucose concentration in the growth medium. High glucose during growth seems to 'repress' glucose transport, whereas low concentrations produce 'derepression' of this transport system. Since the heavy consumption of glucose by tumor cells reduces sugar concentrations rapidly, the 'derepression' phenomenon, if it operates in such cultures, might well be involved in the enhanced transport of sugars found in tumor cells. However, Kalckar et al. [22] conclude that the enhancement of galactose entry observed in polyoma-transformed hamster fibroblasts is largely independent of the 'repression–derepression' phenomenon. This conclusion rests on the fact that transformed cultures, grown in a medium with galactose and consuming only a minor fraction of the galactose of the growth medium, still show full enhancement of galactose entry.

7.1.3.5. Effect of cytochalasin B

An important difference between sugar transport in normal and sarcoma-transformed fibroblasts is revealed by cytochalasin B.

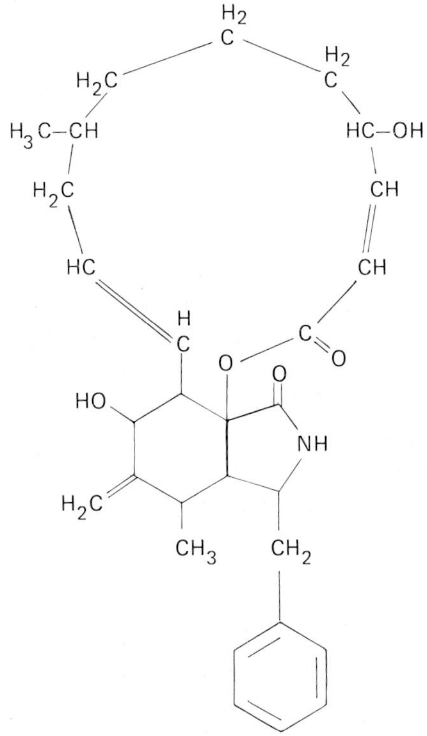

Cytochalasin B

The binding of this drug to cells has been studied by Lin et al. [40]. Hela cells bear about 10^6 high-affinity ($K \cong 10^7$ M) binding sites per cell as well as a class of low-affinity sites ($K \cong 10^5$ M). Human erythrocytes and SV40-transformed mouse fibroblasts also bear 10^5–10^6 high-affinity sites ($K \cong 10^7$ M) per cell plus low-affinity sites. Binding is rapid (< 2 min) and quickly reversible.

Cytochalasin B inhibits the uptake of 2-deoxy-D-glucose, D-glucose and D-galactose by chick embryo fibroblasts [41]. The inhibition is competitive for facilitated diffusion by the sugars (apparent equilibrium constant $\sim 1.45 \times 10^7$ M); phosphorylation is not inhibited. The drug competitively inhibits the transport of 2-deoxy-D-glucose, glucose and glucosamine by cultured Novikoff hepatoma cells [42,43]. Transport of uridine and thymidine into Novikoff cells is also competitively inhibited. However cytochalasin B does not affect intracellular phosphorylation of the nucleosides and has no direct action on RNA or DNA synthesis. Transport of choline is not impaired.

In addition, cytochalasin B depresses the initial uptake rates of 2-deoxy-D-glucose and 3-*O*-methylglucose in normal and sarcoma virus-transformed Balb/c-3T3 cells [34]. Without the inhibitor, normal cells become saturated in about 15 min and transformed cells within less than one minute. In the presence of cytochalasin B, untransformed cells accumulate 3-*O*-methylglucose to a small extent but the sugar enters the *transformed* cells at about the rate seen in untreated, untransformed cells. The cytochalasin B effect occurs over a 10^4-fold substrate concentration range.

The exact mechanisms of cytochalasin B action are not known. Several experimenters [45–47] argue that this drug acts on plasma membrane receptors and the binding data of Lin et al. [40] are not inconsistent with this view. However, cytochalasin B can also react with actin or related substances [48] and recent evidence indicates that cytochalasin B interferes with the contractivity of cytoplasmic alpha-filaments [49]. These 70 Å diameter fibers perhaps attach to some intramembranous protein particles (Chapter 1). Significantly, normal Balb/c 3T3 cells, their SV40 transformants and SV40 revertants differ in the distribution of alpha-filaments [50]. The normal and revertant cells show a much greater abundance of these microfibers than the transformed fibroblasts. This correlates well with the enhanced sugar uptake of SV40-transformed Balb/c 3T3 cells compared with the parental cells and the revertants (Table 7.5) and could implicate impaired membrane-cytoplasmic relationships in the defective sugar transport found in neoplastic cells. However, Lin et al. [40] point out that their high-affinity sites may be the ones involved in hexose transport, whereas the low-affinity loci could be the ones involved in mobility phenomena. This is because higher concentrations of cytochalasin B are required to interfere with cytokinesis, cell locomotion, etc., than are necessary for transport impairment. Clearly, the issue cannot be resolved without identification of the two classes of binding sites. Also, at least in early *Xenopus embryos*, cytochalasin B can only enter the dividing cells

TABLE 7.5

Effect of insulin on Balb/c 3T3 cells SVT2 cells and SV40 reverted cells[1]

Cells	Additive	2-Deoxyglucose uptake (cpm/mg protein × 10^{-3})	% Stimulation
Balb/c 3T3 (subconfluent)	None	28.5	—
	Insulin (10 μg/ml)	34.5	22.0
Balb/c 3T3 (confluent)	None	6.0	—
	Insulin (10 μg/ml)	9.9	65.0
SVT2 (confluent)	None	93.5	—
	Insulin (10 μg/ml)	97.5	4.8
Revertant (confluent)	None	23.4	—
	Insulin (10 μg/ml)	31.8	36.0

[1] From Bradley and Culp [51]. Insulin treatment for 10 min. Definite stimulation was also observed at insulin levels of 1 μg/ml but not at 0.1 μg/ml.

shortly after cleavage [52]. This appears to be due to the increased permeability of the 'new' membrane deposited during cell division and not to an effect of the drug on the membranes.

7.1.3.6. Effects of concanavalin A (Con A)

We have pointed out earlier that the kinetic analysis of membrane transport might be complicated by non-specific, steric effects. Such are invoked by Inbar et al. [21] on the basis of their studies on carbohydrate (and amino acid) transport by secondary culture fibroblasts of golden hamster, rat and SWR mouse embryos and neoplastic variants thereof (polyoma- and chemically-transformed hamster fibroblasts, SV40-transformed, SWR, and hamster fibroblasts and polyoma-transformed rat fibroblasts). Inbar et al. [21] studied the transport properties of the diverse cells after binding equivalent amounts (37–41 × 10^6 molecules/cell) of Con A (Table 7.6). They find that Con A inhibits transport of D-glucose and D-galactose in both normal and neoplastic cells. D-Glucose transport is always impaired more than uptake of D-galactose, but transformed cells appear more sensitive to Con A than parental cells, with respect to both sugars. Concanavalin A inhibition can be specifically reversed by α-methyl-D-glucopyranoside.

The authors observe similar uptake rates for fucose in normal and transformed cells; this result fits the general pattern [6]. However, they fail to observe enhanced uptake of 3-*O*-methylglucose in transformed cells, and this failure is contrary to observations of others. This may be a technical matter: the authors first data point is after 15 min incubation at 37 °C, at which time all cells tested are saturated. As pointed out, the difference between normal and transformed cells can only be detected during a very early time interval because transport

TABLE 7.6

Effect of concanavalin A on transport of D-glucose and D-galactose by normal and transformed fibroblasts[1]

Sugar	Sugar transport[2]								
	Normal			Polyoma			SV40		
	Control	ConA	% Inhibition	Control	ConA	% Inhibition	Control	ConA	% Inhibition
D-Glucose	933	823	33	1,164	559	48	1,102	610	45
D-Galactose	641	552	14	896	635	29	815	601	26

[1] From Inbar et al. [21].

[2] Cpm/mg cell protein/min. Cells treated with Con A for 60 min prior to transport measurements. The number of Con A molecules bound per cell were 3.8×10^7, 3.7×10^7 and 4.0×10^7 for normal cells, polyoma and SV40 transformants, respectively.

occurs at a constant rate for only 4 min in normal cells and less than 15 sec (37 °C) in converted cells.

Inbar et al. [21] also find no effect of Con A on the transport of 3-*O*-methylglucose. Again, their test system could not detect such an effect. We note, however, that kidney bean phytagglutinin strongly accelerates the 'facilitated diffusion' of 3-*O*-methylglucose into lymphocytes within 10 min [35]. This is due to an increase of V_{max} for both the translocation of 3-*O*-methylglucose and linked counter-transport.

Inbar et al. [21] argue that their studies on sugar transport and related experiments on amino acid uptake (p. 315) show that concanavalin A exerts a different steric interference with sugar (or amino acid) transport in normal and neoplastic cells. The data are taken to indicate that the topologic relationships of concanavalin A receptors and transport sites differ significantly in normal and neoplastic cells. Specifically, the concanavalin A sites are thought to lie *closer* to the transport sites for D-glucose and D-galactose in tumor cells than in normal cells. Their argument assumes that concanavalin A binding changes the number of effective transport sites per cell. However, it is also conceivable that the lectin inhibits the shuttling *rate* of the carrier, that is, its rate of oscillation between the two membrane surfaces. In this regard, we note that Inbar et al. [21] (Table 7.6) show that the rate of concanavalin A-inhibited galactose transport in transformed cells is between 109 % and 115 % of the inhibited rate in control cells. The rate of inhibited glucose transport of transformed cells, however, is variably less than that of concanavalin A-treated normal cells (85–98 %). The possibility that the difference between separate clones of the normal or converted cells might amount to ± 15 % cannot be excluded.

Edelman et al. [53] suggest a hypothesis unifying the membrane effects of Con A, cytochalasin B and colchicin. Their discussion focuses on lymphocytes and does not deal with transport. We treat this hypothesis in Chapter 10, but mention it briefly here because of its direct relevance.

Edelman et al. [53] propose that various membrane receptors are topologically linked. They agree that concanavalin A, through its cross-linking function, impairs membrane mobility and that colchicin binding, in contrast, releases Con A-induced membrane immobilization. Cytochalasin B does not affect Con A-induced receptor immobilization – a fact which leads the authors to exclude participation of cytoplasmic microtubules. They accordingly suggest some other membrane framework linking various receptors.

The authors' implication of colchicine as a modifier of membrane mobility is supported directly by the studies of Wunderlich et al. [54] on the alveolar membrane of *Tetrahymena pyriformis*. These authors' studies using freeze-fracture electron microscopy show that colchicin, by a direct membrane action not involving microtubules, *inhibits* temperature-induced lateral and vertical mobility of the intramembranous particles of the alveolar membrane.

7.1.3.7. *Insulin effects*

Considerable evidence (cf. p. 317) shows that insulin enhances membrane transport of certain amino acids in diverse cells. Bradley and Culp [51] further demonstrate differential insulin effects on 2-deoxyglucose transport by Balb/c 3T3 cells, an SV40 transformant thereof, SVT2 and an SV40 revertant (Table 7.5). Specifically, insulin rapidly stimulates 2-deoxyglucose transport in contact-inhibited normal or revertant cells (1.65-fold and 1.35-fold). Smaller effects are found with non-confluent normal or revertant cells. SV40 transformants, whether confluent or not, appear almost immune to insulin action. Similar to its effect on amino acid transport (p. 317), insulin increases V_{max}; that is, the hormone alters the number and/or intramembranous turnover of 'carrier' sites.

As noted elsewhere (p. 321), light trypsin treatment can stimulate the action of insulin on amino acid transport. A similar effect has been observed in confluent chick embryo fibroblasts whose 2-deoxyglucose transport is enhanced within 15 min after trypsin addition [55]. Trypsin (10 μg/ml) also stimulates uptake of 2-deoxyglucose by confluent 3T3 cells, but β-galactosidase has no effect on uptake [56].

7.1.3.8. *Effect of serum*

Cunningham and Pardee [56] report 2- to 4-fold increases of uridine uptake by confluent 3T3 cells within 10–15 min after addition of fresh serum. They found no stimulation of amino acid or 3-*O*-methylglucose uptakes, but Sefton and Rubin [55] did stimulate the uptake of 2-deoxyglucose of confluent chick embryo fibroblasts by serum renewal.

These observations have been extended by Bradley and Culp to Balb/c 3T3 cells, SV40 transformants thereof, as well as an SV40 revertant [51]. They find that increasing concentrations of fresh fetal calf serum stimulate 2-deoxyglucose uptake by confluent Balb/c 3T3 cells and SV40 revertants to levels seen in sparse cultures. SV40 transformants or cells in sparse culture show *no* transport enhancement. In 3T3 cells, the serum effect exhibits two phases. The first, lasting about 1 hr, is characterized by an increase in V_{max} from 125 nmoles/mg protein/hr to 400 nmoles/mg protein/hr. The second step is characterized by a drop in K_M from 2.5 mM to 0.9 mM. In contrast, the transport alteration in the revertants can be attributed solely to a gradual increase in V_{max}. The authors accordingly conclude that the transport properties of the revertant clones are not truly identical to those of the parental cells.

7.1.3.9. *Sugar transport and cell physiology*

The data presented in this section demonstrate clearly that neoplastic transformation of fibroblasts by oncogenic viruses produces enhances *in vitro* sugar transport. Before assessing the biologic significance of this phenomenon, one must recall that enhanced sugar transport has not yet been established for neoplastic *epithelial* cells.

As far as *fibroblasts* are concerned, how can one be certain that transport enhancement represents a membrane-specific change rather than a consequence of altered growth patterns, cell sociology, etc.?

As pointed out in (e.g. [14,38]), fibroblasts in sparsely seeded monolayers vigorously take up glucose and 2-deoxyglucose; however, transport rates drop off markedly when the monolayers reach confluence. Thus, Bose and Zlotnick [38] demonstrate that in Balb/3T3 fibroblasts, which are highly contact inhibited, one of the earliest, measureable aspects of density-dependent growth inhibition is a decrease in 2-deoxyglucose transport. Indeed, the transport rate decreases even during the exponential growth phase, but as the cultures approach saturation density, the rate drops to about one tenth of maximum. However, cells infected with sarcoma viruses show no density-dependent inhibition of sugar uptake, and no correlation with growth rate. Thus, cells infected with murine sarcoma virus initially replicate more slowly than control cells, yet show enhanced sugar transport. In fact, the transport rate is greater than in rapidly-replicating cells infected with leukemia viruses. Finally, as demonstrated by Bader [57] (who blocked cell division by using vinblastine immediately following infection with Rous sarcoma virus) virus-induced enhancement of 2-deoxyglucose uptake occurs even without cell replication.

Enhanced glucose transport in neoplastic fibroblasts also does not relate to the sizes of intracellular *pools* of glucose or its metabolites. Long-term incubation with radioactive glucose shows that the glucose pools in normal and sarcoma-transformed cells do not differ significantly in magnitude [58].

7.1.3.10. *The effects of sugars on the neoplastic transformation by murine sarcoma viruses*

Hatanaka [26] shows that the exposure of NIH swiss mouse embryo fibroblasts *in vitro* to D-mannose or 2-deoxy-D-glucose markedly reduces the number of transformed colonies produced by murine sarcoma virus infection. Such infected colonies are also less able to produce tumors *in vivo*. Since fewer cells are transformed, the enhancement of sugar transport typical of sarcoma virus transformation is also depressed. The full basis for these observations remains to be established.

7.1.4. Amino acid transport

7.1.4.1. Introduction

In contrast to the considerable effort directed at sugar transport in neoplastic cells, relatively little work has been conducted in the area of amino acid transport. This is surprising since the pioneering work on transport in animal cells was performed on Ehrlich ascites carcinoma [59].

The plasma membranes of most animal cells bear at least four distinct transport systems for neutral amino acids (Table 7.7), each of which translocates a class of closely related substances. For Ehrlich ascites carcinoma, Christensen [59] classifies these as follows:

(a) The A-system, which requires Na^+ and metabolic energy.
(b) The ASC system, which required Na^+ but not metabolic energy.
(c) The L-system, which is Na^+ independent.

TABLE 7.7

Overview of the reactivities of some neutral amino acids with various transport systems[1]

Aminoacid	System			
	Gly	A	ASC	L
Gly	3	2	0	±
Pro	?	3	±	0
Ala, Ser	0	3	3	1
Thr	0	3	3	2
α-aminoisobutyrate (AIB)	0	2	3	2
GluNH2	0	2	2	2
Met	0	4	0	4
His, Tyr, Trp	0	2	0	2
Val	0	1	0	2
Leu, Ileu, Phe	0	1	0	4

[1] From Christensen [59]. Data for Ehrlich ascites carcinoma. Reactivities rated from 0–4 (maximal).

Additional systems exist for dibasic and diacidic amino acids. The discrimination among the various systems cannot be considered absolute. For example, α-aminoisobutyrate is transported by the A-, ASC- and L-systems, alanine moves as well through the A-system as via the ASC route, and methionine is translocated equally efficiently by the A- and L-systems. Tissue differences must also be considered. For example, Na^+ is a requirement for transport of cycloleucine in many cells, but Na^+ is not a requirement for transport in thymocytes.

Clearly, amino acids can be transported by multiple routes. All presumably require a carrier system and some require that this system be coupled to metabolic energy production. In the latter case, transport can proceed against a concentration gradient. Several amino acid transport systems are also coupled to Na^+ transport. Finally, in certain cases, amino acid transport can proceed by facilitated diffusion, that is, without energy expenditure.

7.1.4.2. Amino acid transport in cultured tumor cells

Working with unrelated cell lines, Eagle et al. [60], found that all seemed to concentrate amino acids to about the same degree. Similarly, Hare [61], working with normal and polyoma-transformed hamster fibroblasts grown in monolayers on plastic substrates, could not detect any difference between these cells in their incorporation of L-phenylalanine over a 5-min incubation period when substrate concentrations ranging from 0–1.0 mM were used.

However, Foster and Pardee [62] obtain different results with normal and neoplastic 3T3 cells grown on glass coverslips; they report that α-aminoisobutyrate and cycloleucine, both non-metabolizable, accumulate twice as rapidly in polyoma-transformed mouse 3T3 fibroblasts as in control cells*. Glutamine and arginine do not concentrate preferentially in the transformed line. The K_Ms in normal and transformed cells are similar. Foster and Pardee conclude that enhanced amino acid transport reflects density dependence rather than neoplastic membrane changes, that is, amino acid transport declines upon confluence and transformed cells do not reach confluence. As already pointed out, however, similar studies of hexose transport show that transport enhancement is not an 'all or none' phenomenon, that it varies with the cell line and the transforming viruses used, as well as the substrate concentration investigated and that confluence or cell sociology are not determining in enhanced sugar transport. Also, in his recent review [59], Christensen points out that glutamine and arginine are not transported by the same transport system as cycloleucine and AIB.

Recently, Isselbacher [20] showed that polyoma-transformed baby hamster kidney cells (BHK-21) and SV40-transformed Balb/3T3 cells (grown in the glass coverslip system described by Foster and Pardee [62]) as well as murine sarcoma-transformed rat hepatocytes, exhibit 2.5–3.5 times normal uptake rates (Table 7.8). Also, in contrast to the work of Foster and Pardee [62], Isselbacher reports enhanced uptake of arginine and glutamic acid (as well as of 2 deoxy-D-glucose). His kinetic studies fail to reveal any differences in K_M (but see difficulties discussed on p. 301), and he ascribes observed transport changes to enhancement of V_{max}.

An important point, brought out by the K_M values in Table 7.8 is that they lie

---

* α-Aminoisobutyrate is handled by the A-, ASC- and L-systems and requires an Na^+ gradient for maximum efficiency. Cycloleucine ttansport requires Na^+ in some, but not all, cells.

TABLE 7.8

Kinetic parameters for the uptake of cycloleucine and α-aminoisobutyric acid by normal and virally transformed fibroblasts[1]

Cell line	Substrate			
	Cycloleucine		α-Aminoisobutyric acid	
	$(V_{max})^2$	$(K_M)^3$	$(V_{max})^2$	$(K_M)^3$
Balb/3T3				
Normal	5.5	0.8	6.5	1.5
SV40 transformed	15.0	1.0	17.0	1.3
BHK 21, clone 13				
Normal	13.0	1.0	11.0	1.4
Polyoma transformed	32.0	0.9	30.0	0.9

1 From Isselbacher [20].
2 V_{max} in 10^{-9} moles/mg protein/min.
3 K_M in mM.

far above the plasma concentrations of the essential amino acids (Table 7.9). The concentrations experienced by cells in tissues must be lower still, because these cells depend on diffusion of the amino acids from the capillaries, that is, tissue cells in general operate very far from saturation of their amino acid transport.

TABLE 7.9

Plasma concentrations of essential amino acids

Amino acid	Concentration (*mM*)
Lys	0.153
Trp	0.048
His	0.074
Phe	0.053
Leu	0.11
Ileu	0.063
Thr	0.129
Met	0.022
Val	0.21
Arg	0.074

7.1.4.3. Amino acid transport in Morris hepatomas

Amino acid transport has been extensively studied in rat liver and various Morris hepatomas. Initial studies suggested a positive correlation between the control of the transport carriers for α-aminoisobutyrate and cycloleucine and 'short half-life' enzymes such as tyrosine aminotransferase [63, 64]. However, the patterns of induction of amino acid transport by diverse agents, for example, epinephrin, glucagon, dibutyryl-cAMP, vary widely from one hepatoma to the next. The data support the theory [65] that the maintenance of 'short half-life' enzymes depends on the influence of amino acid transport on intracellular amino acid pools [66]. It is suggested that highly differentiated hepatomas resemble mature livers, whereas other tumors are more akin to fetal livers. The latter group exhibits high uptake of α-aminoisobutyrate, both before and after glucagon. The tumors with high uptake of α-aminoisobutyrate show levels of cyclic AMP that are unresponsive to glucagon stimulation [67].

7.1.4.4. Effects on concanavalin A

Exposure of various cells to concanavalin A generally *increases* the V_{max} for α-aminoisobutyrate transport. However, Isselbacher [20] shows that this lectin, as well as wheat germ agglutinin (WGA), actually *inhibits* transport of cycloleucine and α-aminoisobutyrate, but differentially in normal and transformed fibroblasts (Table 7.10). In fact, lectin inhibition is always greater in transformed cells. The lectin effects can be blocked by specific carbohydrate antagonists.

If the effects of concanavalin A can be attributed to steric interference (which is by no means certain), one could argue that lectin receptors lie in closer proximity to amino acid transport sites in transformed cells. Indeed, this is the contention of Inbar et al. [21]. These authors employed a number of SV40 and polyoma-transformed fibroblasts from various rodent species, together with the parental cell types. They observed that transformed cells accumulate leucine, cycloleucine and α-aminoisobutyrate more avidly than normal cells (Table 7.10). This is in accord with Isselbacher's results [20]. The transport of L-arginine, L-glutamate, and L-glutamine are similar in normal and transformed cells.

They then designed experiments so that a similar number of Con A molecules ($37–41 \times 10^6$) were bound to both normal and transformed cells. They find that Con A binding inhibits the transport of amino acid differentially in normal and transformed cells (Table 7.11). This holds for the metabolizable amino acids, L-leucine, L-glutamate and L-glutamine, as well as for the non-metabolizable amino acids, cycloleucine and α-aminoisobutyrate. The lectin effects are blocked by the specific inhibitor, α-methyl-D-glucopyranoside. The authors argue for nonspecific steric inhibition by concanavalin A and suggest that amino acid transport sites lie nearer to Con A receptors in tumor cells than in normal fibroblasts. However, the data can be otherwise interpreted.

It is certainly remarkable that in all the cell lines tested, whether normal or transformed, concanavalin A inhibits amino acid transport, because it *stimulates* the uptake of α-aminoisobutyrate in thymocytes (p. 320) and adipocytes. In

TABLE 7.10

Effect of concanavalin A and WGA on α-aminoisobutyrate uptake by normal and neoplastic fibroblasts[1]

Conditions	Uptake[2] (% of control)			
	BHK21	BHK21-Py[2]	Balb/3T3	Balb/3T3-SV
Control	100	270	100	250
Concanavalin A (100 μg/ml)	26	98	73	100
WGA (100 μg/ml)	63	73	83	140

[1] From Isselbacher [20].
[2] Uptake of normal BHK21 of Balb/3T3 cells set at 100 %. Clearly the transformed cells, even when lectin-inhibited accumulate α-aminoisobutyrate more avidly than control cells WGA inhibits cycloleucine transport equivalently.

addition, the lectin protects against impairment of amino acid transport of thymocytes after ionizing radiation (p. 320). Also, a detailed analysis of the data (Table 7.11) shows that the amino acid transport of the normal hamster cells is generally Con A-insensitive, *whereas all amino acid transport systems tested, in three types of converted cells* are inhibited by concanavalin A. This is equally true for amino acids whose uptake is enhanced by transformation (L-leucine, cycloleucine, α-aminoisobutyrate), as for amino acids whose transport is not enhanced (L-arginine, L-glutamine and L-glutamate).

TABLE 7.11

Effect of concanavalin A on amino acid transport by normal and transformed hamster fibroblasts[1]

Amino acid	Transport[2]					
	Normal		Polyoma		SV40	
	Control	Con A	Control	Con A	Control	Con A
L-Leucine	100	98	174	91	159	94
L-Arginine	100	97	103	47	112	61
L-Glutamate	100	98	106	61	104	70
L-Glutamine	100	95	101	46	104	45
Cycloleucine	100	95	116	49	112	49
α-Aminoisobutyrate	100	99	139	82	135	73

[1] From [21].
[2] Transport presented giving the values for normal hamster cells as 100. Concanavalin A inhibition computed as the rate obtained without the specific inhibitor (α-methyl-D-glucopyranoside) relative to the rate obtained in the presence of 500 μg concanavalin A/ml. Concanavalin A exposure = 60 min. Transport measured for 5 min.

Wherever tested, the effect of concanavalin A has been to modify V_{max}. Also, enhanced amino acid transport in cells transformed by oncogenic DNA viruses appears to be due to an increase in V_{max}. One is thus led to suspect that the concanavalin A inhibition of amino acid transport in neoplastic cells is due to a decrease in V_{max}. Steric hindrance is possible, but unlikely in view of the many transport systems afflicted. A more likely possibility is a more general membrane modification which reduces the shuttling rate of diverse carriers nonspecifically and thereby decreases V_{max}.

7.1.4.5. Effects of hormones

Considerable evidence demonstrates that insulin stimulates the uptake of α-aminoisobutyric acid in lymphoid cells and adipocytes by increasing V_{max}, but there is little information concerning its action on amino acid transport by tumor cells. However, Risser and Gelehrter [68] report an interesting interaction between dexamethasone and insulin in terms of amino acid transport by cultured hepatoma (HTC) cells.

Dexamethasone depresses the accumulation of α-aminoisobutyrate. The effect is 50% maximal within one hour and reaches 75% within 2 hr. The depressed transport can be traced to a decreased V_{max} of influx. The efflux rate of the amino acid remains unchanged.

Insulin slightly augments AIB transport by hepatoma cells that are not treated with dexamethasone. However, in cells whose uptake has been depressed by dexamethasone pretreatment, insulin drastically augments AIB transport. The effect appears within 30 min and becomes maximal in 2 hr. The effect of insulin is due to a 4-fold increase in V_{max} and a 2-fold drop in K_M. Both hormonal effects require protein synthesis.

The data suggest that the hormones control the protein composition (organization) of the plasma membrane. This view fits the observations of Ballard and Tomkins [69] that dexamethasone can modify some physical and immunologic properties of HTC cells.

7.1.5. Possible physiologic consequences of transport alterations in neoplasia

Since enhanced transport appears to be a real property of virally transformed neoplastic cells, what might be the biologic consequences? We will return to this matter in Chapter 11, particularly in relation to the *in vivo* behavior of malignant tumors, that is, their capacity to invade and metastasize. Here we will merely point out an important consequence for cell population dynamics.

In a tissue domain with a limited supply of essential metabolites, the growth and/or survival of different cell types in this domain will depend significantly upon the various cells' capacities to accumulate the critical substrates. Assume that a cell clone arises that has a distinct advantage over its parental cells in terms of essential transport. Barring other regulatory mechanisms, the new cell clone will eventually eliminate the original cell population. This is a fundamental biologic law.

However, we do not know whether the enhanced *sugar* transport in virally transformed fibroblasts actually confers a critical selective advantage.

We consider this possibility questionable since normal serum glucose levels lie near 5 mM, well *above* the K_Ms for glucose transport in both normal and transformed cells.

The enhanced V_{max} for amino acid transport in neoplastic cells might, in contrast, be important, since the serum concentrations of essential amino acids (0.02–0.2 mM) lie far below the transport K_M ($\sim$1 mM).

In vitro tests of this issue should not be difficult. Certainly, most tumor cell lines exhibit a clear competitive growth advantage *in vitro* over the cells of parental lines. However, this *in vitro* superiority has yet to be clearly linked to transport. Moreover, at present, one can only point out possible *in vivo* consequences of enhanced *in vitro* transport.

One must also consider the possibility that the extravascular movement of small and large molecules in tumor tissues may differ significantly from that of normal tissues because of different diffusion coefficients in the two types of tissue. The diffusion coefficient of tumor tissues can be appreciably lower than those of normal tissue because degradation of connective tissue ground substance by tumor enzymes, altered organization of the connective tissue in the vicinity of tumor cells, etc. As has been documented in a number of instances, convective transport may assume a significant role under such circumstances. For this reason, Swabb et al. [70] have evaluated the potential relative contributions of diffusive and convective extravascular transport in normal and neoplastic tissues. Their data indicate that small molecules, such as oxygen, carbon dioxide, glucose, lactic acid and amino acids are transported primarily by diffusion in both normal and neoplastic tissues. Larger molecules such as certain chemotherapeutic agents would also be transported primarily by diffusion. On the other hand, macromolecules, such as antibodies, complement or protein-bound drugs, would migrate, not only by diffusion, but also by convection. The authors point out that if antigen antibody reactions are involved in the host defense against the tumor, these may be rate-limited by extravascular transport. The authors also point out that large particles, such as cancer viruses would be essentially immobile in the interstitial space of neoplastic and normal tissues and that the spread of such particles has to occur by other mechanisms, such as mobile cells.

7.1.6. Membrane transport and tumor therapy

7.1.6.1. Chemotherapy

The transport characteristics of tumor cells may bear significantly on tumor chemotherapy. This is because the drug resistance of some neoplastic cells derives from poor penetration of the drug through the plasma membrane.

This phenomenon has been documented for several chemotherapeutic agents, for example nitrogen mustard (HN2). This compound is transported actively by a carrier for choline, a structurally similar, natural metabolite [71]. In an HN2-resistant strain of L51784 lymphoblasts, drug resistance is closely linked to impaired transport [72]. The same appears true in the case of Yoshida sarcoma. Thus, Inaba and Sakurai [73] show that the HN2-sensitivity of several strains of Yoshida sarcoma varies in the same direction as the permeation of the drug. More extensive studies [74] indicate that the binding of HN2 to DNA in resistant tumor lines can be restored to the levels found in sensitive lines by increasing membrane permeability. However, HN2-binding to DNA is not dependent on membrane permeability in sensitive lines. The resistance that certain lines of Ehrlich ascites tumor exhibit to HN2 can also be attributed to transport restrictions [75].

A similar phenomenon has been reported for methotrexate. This drug is accumulated by active, carrier-mediated transport [76]. To exert its action, the drug must enter the cell and combine with and inhibit dihydrofolate reductase. However, in some resistant tumor cells the drug is not transported into the cells [77,78]. In the L1210 murine leukemia, methotrexate uptake can be enhanced by a variety of agents (e.g. corticosteroids, *Vinca* alkaloids) with a marked increase of chemotherapeutic efficiency [79].

However, the accumulation of chlorambucil by drug resistant strains of Yoshida sarcoma is due to increased metabolism of the drug rather than to decreased transport [80]. Indeed, this compound does not enter by transport but permeates passively.

Nucleoside transport occurs by a facilitated diffusion process of broad specificity and can be inhibited by certain S-substituted derivatives of 6-thioinosine and 6-thioguanosine. Such transport inhibitors (at $\sim 10^{-6}$ M concentrations) also interfere markedly with the uptake and action of antiproliferative nucleoside derivatives in the case of L51784 lymphoblasts [81]. However, the transport inhibitors did not in themselves interfere with cell proliferation at $<10^{-5}$ M, unless the cells were made dependent upon exogenous thymidine by use of methotrexate.

7.1.6.2. Radiotherapy

The use of ionizing radiation constitutes a major tool in the treatment of neoplasia. As summarized in several reviews [82, 83], there is now considerable evidence that high energy radiation can severely impair membrane function. The data on intracellular membranes will not be treated in this volume.

Information on radiation effects on the plasma membranes of normal and neoplastic cells is generally not easily evaluated because the radiation doses employed far exceed levels used in radiotherapy [82, 83]. This is not so in the case of lymphocytes: low radiation doses cause interphase killing, apparently by impairment of membrane transport. We treat this membrane phenomenon here because lymphocytes are the mediators of the immune response against tumor membrane antigens.

That therapeutic radiation indeed alters the immune status of treated patients has been well documented [84–87]. The effects of radiation are to reduce the number of thymus-dependent lymphocytes in the circulatory pool.

The mechanisms involved have been somewhat clarified, at least in animal models. It was shown very early that interphase cell death of thymocytes occurs even after 100 R. The K^+ flux of thymocytes is also highly radiosusceptible, but only at temperatures near the physiological. At 0 °C, 50 % K^+ loss occurs after exposure to 30 kR, this result is no different from the loss in erythrocytes. However, at 37 °C, 50 % K^+ loss is observed already after 30–50 R (a factor of approx. 1000) and the cells become permeable to eosin [88–90].

Recent studies [91] show that the *in vitro* exposure of rat thymocytes to 0.05–10 krads γ-irradiation impairs their capacity to accumulate the non-metabolizable amino acid, α-aminoisobutyrate (AIB). The damage to the AIB transport system increases with time after irradiation and with irradiation dose up to 3–4 krads.

The dose-response curves, which exhibit a plateau between 3 and 8 krads, parallel those observed for potassium transport [92]. Indeed Chapman and Sturrock show that impairment of active transport is the major cause of K^+ leakage from thymocytes after irradiation in the dose range, 0.2–4.0 krads, but the AIB efflux measurements indicate that irradiation doses below 5 krads afflict primarily active transport rather than membrane permeability.

The parallelism in the dose-response effects observed in AIB and potassium transport suggest that a single radiation lesion affects both potassium and AIB transport, and Archer's studies of the effect of radiation on AIB and potassium transport in Ehrlich ascites carcinoma cells suggest the same. The suggestion derives further indirect support from the fact that two amino acids, cycloleucine and *b*-2-aminobicyclo (2.2.1) heptane carboxylic acid, whose uptakes are not linked to alkali cation transport [93] are accumulated normally, even after high radiation doses [94].

Examination of the saturable component characterizing AIB influx in rat thymocytes shows that radiation alters only the maximum rate of AIB uptake,

V_{max} (2.1 ± 0.05 to 0.71 ± 0.03 nmoles AIB/10^7 thymocytes/10 min) without significantly affecting the apparent binding constant K_M (0.48 ± 0.03 mM vs. 0.5 ± 0.04 mM). These data suggest that radiation does not qualitatively alter the AIB binding site. The V_{max} depends on the number of AIB carrier sites, as well as the rates at which these turn over. A decrease in V_{max} following radiation implies a loss in the number of carrier sites and/or a slowing down in the rate of substrate turnover.

One cannot exclude the possibility that increasing irradiation doses preferentially eliminate increasing proportions of the AIB carrier and that this process damages the cells beyond repair. However, this explanation does not deal with the parallel effects of irradiation on AIB and Na^+–K^+ transport.

Alternatively, radiation inactivation of the Na^+ pump might lead to Na^+–K^+ gradients unfavorable to inward transport of AIB [95]. This could lower the V_{max} for AIB uptake without affecting K_M (since Na^+ is not involved in the binding of AIB to the cell membrane). This explanation also fits the observation that radiation does not impair uptake of amino acids, known not to require Na^+ for transport. AIB transport does vary with experimentally induced Na^+ gradients, induced by replacing Na^+ with choline. However, ouabain, which is a specific inhibitor of the (Na^+/K^+) ATPase, has been shown to inhibit AIB uptake only slightly – evidence that indicates the increased AIB uptake is not directly related to the (Na^+/K^+) ATPase. Also, impairment of AIB transport by irradiation occurs at physiologic levels of extracellular (Na^+). The cation gradient hypothesis, thus, does not adequately explain the radiation effects or suggest the existence of lesions in membrane components that are essential to both cation and AIB transport.

The AIB transport system can be protected against radiation damage with insulin (1.0×10^{-6}–1×10^{-8} M). Moreover, concanavalin A, which affects only the V_{max} of AIB transport in rat lymphocytes and not binding affinity, also fully protects AIB transport in rat thymocytes against radiation damage. Indeed, in neither the case of insulin or of concanavalin A does irradiation, even at a dose level of 5 krads, depress AIB transport from the stimulated level induced by these agents. The observations indicate that the radioprotective action of insulin and concanavalin A involves a component which affects the V_{max} of AIB transport; they do not alter the nature of the AIB binding site.

A radioprotective phenomenon, possibly related to the concanavalin A effect, has been reported for kidney bean phytohemagglutinin (PHA) [96]. This lectin, which increases AIB uptake in lymphocytes by affecting mainly V_{max}, protects human peripheral lymphocytes against the lethal effects of *in vitro* radiation. Furthermore, in most cases, lymphocytes from leukemic patients gain a moderate degree of radioprotection from PHA, but the radioprotection is less than that observed with normal cells, and there is a decrease in the number of blast-like cells normally induced by PHA as well.

Available data indicate that the lectins, insulin, and radiation affect the same component involved in AIB transport in lymphocytes. In this connection, considerable evidence suggests that –SH groups constitute major sites for the radiation damage to Na^+–K^+ transport and possibly glucose transport in erythrocytes [82,83]. Moreover, Patt et al. [97] found that lymphocytes were protected against irradiation when incubated with cysteine at 37°C prior to, and for 30–60 min after, irradiation. Similarly, the time courses with insulin and concanavalin A parallel the effects observed for cysteine radioprotection. Ueno [98] shows that treatment of rat thymocytes with *N*-ethylmaleimide (a known –SH inhibitor) induces interphase death – a result suggesting that the –SH groups are essential for the survival of the cells. Also, the fact that radiation levels between 0–4 krads (37°C) reduce non-protein –SH (0.84–0.65 μmoles/ml cells) as well as protein–SH (3.8–3.5 μmoles/ml cells) in thymocytes [99] suggests that radiation-induced loss of sulfhydryl from these cells results from secondary chain reactions, possibly mediated in part by long-lived peroxides. All in all, considerable evidence from diverse sources implicates a membrane –SH protein in the radiosensitivity of thymocyte membrane transport – a radiosensitivity that leads to the interphase death of these cells.

7.2. *Unusual permeability of tumor plasma membranes to macromolecules*

7.2.1. *Protein uptake*

Bush [100] has reviewed extensive early evidence indicating that diverse tumors (for example, Jensen rat sarcoma and Walker 256 rat carcinosarcoma) show an unusual capacity to take up various plasma proteins, in particular serum albumin. The cumulated data indicate that the tumors take up the intact proteins, or at least macromolecular fragments, thereof. These proteins are then broken down within these tumor cells and the amino acids converted into tumor proteins. Because of the prominence of this phenomenon in many experimental tumors it has been proposed that tumors act as 'nitrogen' traps, growing at the expense of other tissues. However, one cannot readily generalize this suggestion [101].

7.2.2. *'Enzyme leakage'*

7.2.2.1. *'Leakage' of enzymes of the glycolytic cycle*

In 1954, Warburg et al. [102] first reported that tumor cells exhibit a tendency to lose aldolase into the extracellular medium. More recent data [103, 104] indicate that Ehrlich ascites carcinoma cells release unusual amounts of glycolytic enzymes during brief incubations. The cells' glycolytic capacity does not appear impaired by this process. Enzymes of the pentose phosphate shunt are similarly released.

The proportions of enzymes released vary according to several environmental variables, for example, pH and presence of glucose. The process is most reasonably attributed as differential leakage rather than cell lysis. An unusual release of hexokinase has also been reported for cultured liposarcoma cells [105].

A considerable amount of work has focused on the glycolytic enzyme, lactate dehydrogenase (LDH). For example, Hill [106] showed already in 1956 that LDH is frequently elevated in the sera of patients with cancer. Other workers [107,108] report that LDH is commonly increased in the plasma of tumor-bearing mice and that it diminished after tumor removal. This increase can be 4–10-fold over the level found in normal animals. It usually begins within 6 hr of transplantation and cannot be easily ascribed to the enzyme content of the transplanted tumors. Transplantation of normal tissues does not increase the LDH level in the recipient animals. Worblewski [108] suggests that the phenomenon is due to the release of the enzyme from the tumor tissues.

The release of LDH from tumor cells has been studied in greater detail by Holmberg [109] who compared the release of LDH from L-cells (clone 929) and from Ehrlich–Landschuetz ascites carcinoma. After incubation of these two cell types under identical conditions he finds a 2–3-fold increase of LDH in the incubation medium of the ascites tumors within 2 hr. In contrast, there is no change in the medium of the L-cells. GSS reductase was also preferentially released by the ascites tumor cells. Enzyme release was identical under aerobic and anaerobic conditions. The process could not be attributed to cell lysis: thus after a two-hour period of incubation, only 6% of the Ehrlich ascites carcinoma were ‘nonviable’ by cell counts and by uptake of Lissamine Green, compared with 20% for the L-cells. There was also no impairment in the respiratory capacity of the ascites tumor cells after incubation. Moreover, there was no release of cytochrome *c* reductase. Holmberg ascribed the release of LDH to a membrane defect in the tumor cells.

This work has been confirmed by Kanamaru [110]. After aerobic cultivation of two types of malignant cell (KB and HeLa S_3), much larger amounts of LDH appear in the culture fluid than after culture of normal hamster embryonic cells.

The preferential release of LDH appears rather quickly after infection of chick embryo fibroblast with Rous sarcoma virus (RSV) but not after infection of the same cells with the nontumorigenic Rous-associated virus (RAV) [111]. Bissell et al. [111] examined the LDH activity in both intact cells and in their culture medium at varying time intervals after infection. The data are shown in Table 7.12. There are two important findings: (a) The LDH activity in transformed (RSV) cells is approximately twice that of uninfected cells, whereas that of RAV-infected cells is identical to that of the controls. (b) The LDH activity in the medium of RSV-transformed cells is $\sim$10 times that of uninfected cells or cells infected with RAV.

The cytochrome *c* reductase activities were approximately the same in the

TABLE 7.12

LDH activities in normal chick embryo fibroblasts, cells infected by B-RSV and RAV and the culture media theoref[1,2]

Conditions	LDH	
	Cells (units/mg prot)	Medium (units/culture plate)
4 Days		
Uninfected	2.3	0.05
B-RSV	4.1	0.8
RAV	2.5	0.1
6 Days		
Uninfected	2.8	0.3
B-RSV	6.2	3.0
RAV	3.0	0.5

1 From Bissell et al. [111].
2 LDH = lactic dehydrogenase; B-RSV = Bryan strain of Rous sarcoma virus; RAV = Rous associated virus.

normal and transformed cells and no release of cytochrome reductase into the medium could be detected. The data thus supports previous work that the tumor cells have an unusual permeability to certain cytoplasmic enzymes.

To our knowledge, the release of cytoplasmic enzyme has not been evaluated in other systems comparing the biological properties of normal cells with neoplastic variants (e.g., in liver and 'minimum deviation hepatomas' or fibroblasts and SV40 or polyoma induced variants thereof). Such studies appear highly desirable to clarify both the role of plasma membrane defects in neoplasia and the general pathologic physiology of neoplasia.

No information exists as to the mechanisms involved in the exodus of glycolytic enzymes. The process probably reflects the membrane alteration in the tumor cells. Exocytosis is a possible mechanism but no information exists on this point. Another possibility to be considered is that the glycolytic enzyme can transiently associate with the membrane and actually pass through it, although this is admittedly a highly speculative concept. Nevertheless it is worthy of note that at least one glycolytic enzyme, namely glyceraldehyde-3-phosphate dehydrogenase has a tendency to associate with the internal faces of erythrocyte membranes under hypotonic conditions [112, 113].

7.2.2.2. Lysosomal enzymes

Cathepsins. An abundance of evidence indicates that diverse tumor cells propagated *in vivo* release unusual amounts of lysosomal enzymes into their environment [114–125].

This problem has been examined very elegantly by Sylven and associates [114–122]. These workers use micropuncture techniques, combined with ultra-microanalyses to sample and analyse the interstitial fluid of three tissue zones, namely:

(a) Non-tumorous regions;

(b) The zone at the periphery of the growing tumor, that is, the interface between the tumor and the surrounding normal tissue. This region is generally well vascularized and the interstitial fluid containing material arising from the tumor is continuously drained off to nearby lymph capillaries.

(c) The core of the tumor. The interstitial fluid from this region more strongly reflects the characteristics of the tumor because it tends to be cellularly uniform and is most distant from lymphatic drainage.

The interstitial fluid from normal tissue regions contains rather low levels of lysosomal enzymes, Fig. 7.1. In contrast, the fluid from the tumor core exhibits enormous levels of diverse lysosomal hydrolases (both on a per volume and a per protein basis), including various peptidases. Perhaps the most interesting enzyme activity present is cathepsin-B. This has a trypsin-like action and its optimal activity lies between pH 5 and 7. (One should note, in this regard, that because of the high aerobic glycolysis of many tumors (Chapter 6) the pericellular pH is likely to be below neutrality).

The concentration of the lysosomal enzymes in the peripheral interstitial fluid is considerably less than in the core. This is not surprising since the maximum

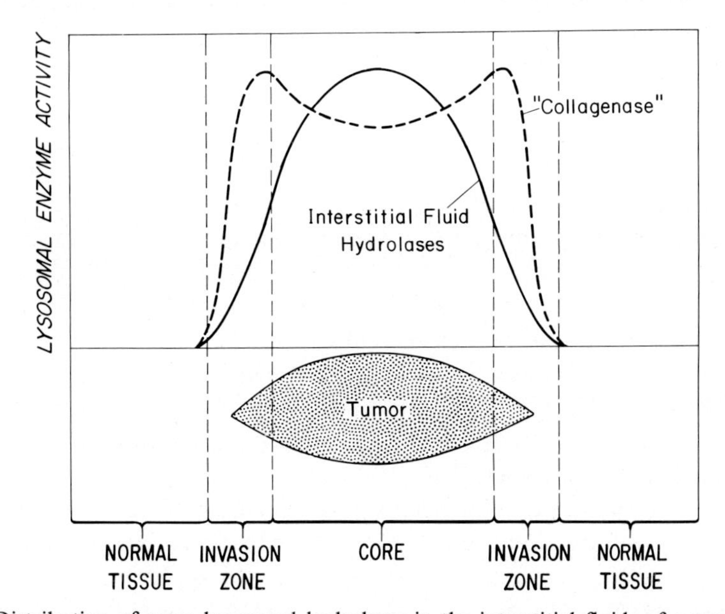

Fig. 7.1. Distribution of some lysosomal hydrolases in the interstitial fluids of a tumor-bearing region.

tumor cell density is in the center of the tumor. Nevertheless the data generally show some increase in lysosomal enzyme activity at the invasive zone.

Several workers have argued that the enzyme activities measured in the interstitial fluid are exclusively due to release of material through the destruction of tissue at the periphery of the tumor [125–127]. However, Sylven and associates argue that the released enzymes have two origins; namely, (a) extrusion from viable tumor cells and (b) leakage from damaged tumor and host cells. They argue that leakage is not the most important factor and, in fact, demonstrate that the central interstitial fluid of highly necrotic tumors contains considerably less lysosomal enzyme (catheptic) activity than non-necrotic regions. Indeed, the high lysosomal enzyme concentration in the core fluid relative to the peripheral fluid is fully consistent with a situation where the tumor cells act as 'source' of the enzymes and the vascularized normal tissue as 'sink'.

Moreover, the high core activities are far more consistent with continuous enzyme production than with autolysis, since lysosomal enzymes, including the peptidases are themselves prone to peptidolysis and consequent inactivation.

Several model experiments support the hypothesis of Sylven and associates: Thus, Holmberg [109], in his studies on the release of glycolytic enzymes from Ehrlich ascites carcinoma during *in vitro* incubation, found that these cells also released considerably more cathepsin than L-cells. He observed no change in cell viability or total cellular enzyme content after the short period of incubation involved. Also, Guerin T8 tumor cells, isolated mechanically from solid tumors and incubated aerobically at 37 °C in physiological media for 2 hr, released up to 15% of their total initial content of cathepsin-B and β-glucuronidase, but maintained a viability of 95% [128]. Both sets of data suggest an active or facilitated secretion of the lysosomal enzymes by the tumor cells.

There is also less direct evidence. Thus, Burger [129, 130] reports a growth stimulatory effect of L1210 cells on confluent mouse fibroblasts, equivalent to that produced by addition of soluble protease or proteases covalently attached to Sephadex B. Normal lymphocytes cannot trigger such a response. However, a membrane fraction (poorly characterized) from the L1210 leukemia cells could initiate the same response. Strangely, the medium from 3-day-old L1210 leukemia cell cultures did not produce growth stimulation. The effect could be blocked by a number of protease inhibitors including the synthetic reagent *n*-α-tosyllysylchloromethylketone (TLCK). The authors attribute the phenomenon to a plasma membrane-associated protease, but their membrane fractions may have been contaminated with lysosomes.

Infection of normal chick embryo fibroblasts, baby hamster kidney fibroblasts and HEp-2 fibroblasts with Newcastle disease virus and Herpes simplex virus, both non-oncogenic, increases their susceptibility to agglutination by concanavalin A and WGA [131]. Significantly, treatment of normal HEp-2 cells with anticellular serum and vitamin A alcohol, both of which produce lysosomal enzyme release and 'sublethal autolysis' [132,133], also enhanced the cells' agglutinability to the levels produced by viral infection. This suggests that the release of lysosomal proteases modifies the cell surfaces in a manner similar to that of exogenously added proteases.

Kazakova et al. [134] report different cathepsin-D activities in rat liver and a transplantable rat sarcoma (M1). The sarcoma cathepsin splits synthetic substrates containing *either* aromatic *or* dicarboxylic side chains. The peptide segments must be greater than 5 residues and optimum activity occurs when a peptide bond bears a dicarboxylic acid residue on both sides.

In contrast, cathepsin-D from rat liver preferentially cleaves peptides with aromatic residues. Also, the synthetic pepsin substrate, acetyl-Phe-Ty, competitively inhibits rat liver cathepsin, but not the sarcoma enzyme. It remains to be determined whether the difference observed is due to differences in cell type or whether it relates to the neoplastic character of M1 sarcoma cells.

Collagenase. Collagenase is another enzyme of probably lysosomal origin [135] which appears to be elevated in certain tumors and also appears to be extruded from viable cells.

Birbeck and Wheatley [136] show by thin-section electron microscopy that invading Ehrlich ascites carcinoma cells break the connective tissue at the invasive zone. They suggest that invading tumors secrete proteolytic enzymes, including collagenases which are active at neutral pH and thereby destroy normal tissue architecture. Additional electron-microscopic evidence for collagen degradation by invading tumors has been presented (137).

Initial biochemical evidence for the participation of collagenase was provided by Strauch et al. [138] who isolated a neutral collagenase from the culture fluid of human HeLa cells. A neutral collagenase has also been isolated from the rat Guerin T8 epithelioma [137] and collagenolytic activity has been found in diverse human tumor cells cultured from biopsy material by Taylor et al. [140]. Specifically, nearly all tumors of epithelial origin showed definite collagenase activity, as did some sarcomatous tumors.

Dresden et al. [141] have demonstrated that cultured cells from a number of human neoplasms have the capacity to produce collagenolytic enzymes. As noted by Taylor et al. [140], cultures from epithelial tumors generally produced large amounts of collagenase. In contrast, neoplastic cells of mesenchymal origin, as well as material non-neoplastic tissues did not produce this enzyme. Certain epithelial neoplasms also lacked collagenase. Tumors of the colon and carcinomas of both squamous and basal cell origin showed very high collagenase activity. Several tumor collagenases were partially purified; they exhibit rather similar properties to the collagenases found in normal skin.

An extensive treatment of the presence of collagenases in tumors and the possible implications of collagenase release has been published by Strauch [142].

This investigator here summarizes his own and others' experience showing that malignant tumors (human and animal) of epithelial origin exhibit high collagenase activity, particularly in their invasive zones. Tumors of mesenchymal origin also generally exhibit strongly enhanced collagenase activity compared with normal tissue. In contrast, benign human tumors (of the breast) and non-invasive malignant breast tumors exhibit only slight enhancement of collagenase activity. Some representative data are given in Table 7.13.

TABLE 7.13

Collagenase activities in some benign and malignant human tumors and in related non-neoplastic tissue[1]

Tissue	Collagenase activity[2]	
	Tumor	Surrounding tissue[3]
Breast-benign		
Tubular adenoma	3.2–7.8	2.3
Intracanaticular fibroadenoma	1.8–3.6	0.7
Glynecomastia	9.9–10.6	3.5
Breast-noninvasive, malignant	5.3–6.6	1.5–7.5
Breast malignant	14–∼45	2.5–5
Colon malignant	$\gtrsim 30$	<5

[1] From Strauch [142].
[2] Cleavage of *p*-phenylazobenzyloxylcarbonyl–Pro–Leu (μg/mg N/hr).
[3] The high range values may include contribution of pathologic tissue.

Strauch also used analyses of serial sections to show that the highest collagenase activity in a malignant tumor often occurs in the invasive zone at the tumor periphery. This contrasts with the findings of Sylen and associates on the central concentration of other lysosomal enzymes (Fig. 7.1).

Collagenase appears to be released from tumor tissue in soluble form [143]. Accordingly, one would expect the highest concentration of this diffusible material in the tumor core. However, there are certain difficulties in evaluating the collagenolytic activity of tissue section material, because many tumor tissues contain low molecular weight substances which inhibit the collagenolytic effect [144]. Also, there is evidence of considerable collagen *synthesis* in the cores, particularly of large neoplasms. This material would compete with the chromophoric substrate used in collagenase assays. The data presented by Strauch [142] may thus only reflect 'excess' collagenase activity and are thus not necessarily in conflict with Sylven's hypothesis.

The significance of the collagenase activity in tumors is rather clear. Collagen is a main constituent of connective tissue and cannot be readily degraded by other proteases. Expression of the malignant, that is, invasive phenotype, may thus require collagenase synthesis. However, the problem of *release*, which concerns us here, is presumably related to the extrusion of other lysosomal enzymes.

Fibrinolysis. In 1925, Fischer [145] compared the culture characteristics of normal and neoplastic tissue explants and found that the malignant cells lysed the culture substrate (clotted plasma) whereas normal cells did not. This demonstration of fibrinolytic activity has been neglected until recently. Now Unkeless and associates [146] have shown that various cultured fibroblasts, transformed by either oncogenic DNA or RNA viruses, exhibit increased fibrinolytic activity compared with the normal cells.

These experimenters detected fibrinolytic activity by coating cultured cells with ^{125}I-labelled fibrin and measuring the release of soluble peptide fragments between 24 and 48 hr after continued culture. They detected no fibrinolysis by primary chick embryo fibroblasts, nor by cells infected with cytocidal RNA or DNA viruses, non-transforming strains of avian leukosis virus, or a temperature-sensitive strain of Rous sarcoma virus grown at a non-permissive temperature (44 °C). In contrast, fibrinolysis was marked in cells transformed with wild-type Rous sarcoma virus or the temperature-sensitive mutant at the permissive temperature (34 °C). In the latter case, fibrinolytic activity appeared *before* morphologic transformation after switching from non-permissive to permissive temperatures. The appearance of fibrinolytic activity could not be related to an increased plasma membrane 'leakiness' for macromolecules, nor to increased levels of intracellular lysosomal enzymes. Appearance of fibrinolytic activity required protein synthesis but was not blocked by inhibitors of RNA- or DNA-synthesis.

Transformation of rodent fibroblasts (hamster, rat, mouse) by oncogenic RNA- or DNA-viruses (murine sarcoma virus; SV40) also induced fibrinolytic activity in the cultures (Table 7.14), in direct correlation with altered culture morphology.

TABLE 7.14

Fibrinolytic activity of normal rodent cells and cells transformed by oncogenic viruses[1]

Cells	^{125}I-solubilized (% of total)
None	1.2
Hamster fibroblasts, normal	1.4
Hamster fibroblasts, SV40 transformed	31.0
Mouse (57 Bl) fibroblasts, normal	1.6
Mouse (C57 Bl) fibroblasts, MSV transformed	9.6

[1] From Ossowski et al. [147].

Unkeless et al. [148,149], Ossowski et al. [147,150], and Quigley et al. [151] have demonstrated that the fibrinolytic phenomenon seen with neoplastic cells depends on the interaction of two components, namely:
(a) the plasminogen present in normal sera and,
(b) a plasminogen activator released by the neoplastic cells.

Concerning the role of (a), the growth of normal and SV40-transformed hamster embryo fibroblasts in a liquid medium is identical whether plasminogen is present or not. However, depletion of plasminogen markedly decreases two characteristic features of the transformed cells, namely their plating efficiencies in soft agar and their tendency to aggregate in soft agar. Plasminogen also decreased the migration rate of SV40 3T3 cells into a culture 'wound' [152]. This effect is shown in Table 7.15. Notably, the SV40 transformants show a much greater migration rate than normal cells and this behavior clearly depends on the availability of plasminogen.

TABLE 7.15

Effects of plasminogen on the migration of 3T3 cells and SV40 transformants thereof into tissue culture 'wounds'[1]

Preincubation[2] in	Postincubation[3] in	Cells migrating into wound (Number/24 hr)[1]	
		3T3	SV40 3T3
Native serum	Native serum	24	147
Plasminogen depleated serum	Depleted serum	7	66
Reconstituted serum	Reconstituted serum	—	170

1 Representative data from Ossowski et al. [150].
2 Before wounding.
3 After wounding.

Rifkin et al. [153] as well as Christman and Acs [154] and Unkeless et al. [149] have begun to characterize the plasminogen activators, (b), released from diverse cell types. They show that the plasminogen activators from neoplastic human, chick and hamster cells are serine proteases, inhibited by diisopropylphosphofluridate. The proteases activate plasminogen by peptide cleavage. The plasminogen activators from human melanomas fall into two categories of mol. wt. = 50,000 and 60,000 respectively.

Importantly, high release of plasminogen activators occurs not only in the case of established cell lines, but also in spontaneous tumors. In an important study, related to enhanced fibrinolysis by malignant cells, Ossowski et al. [150] show that SV40-transformed hamster fibroblasts take up as much as 10-fold the amount of ^{125}I-labelled plasminogen as normal cells. Part of this is due to rapidly-occurring binding (0–4 hr) but the very large effects seen at 24 hr suggest accumulation.

The cell factor isolated from RSV-transformed chick cells is an arginine-specific serine protease with a mol. wt. of ~39,000. Subcellular fractionation shows that the factor is particle-bound with highest total and specific activities in the fragments sedimented at high *g*-forces, Table 7.16. The data are not incompatible with a lysosomal localization but do not prove it. The solubilization of the factor with Triton X-100 is also consistent with a lysosomal localization. However, more sophisticated fractionations are required to pin-point the subcellular localization. Subcellular fractions from normal fibroblast fail to reveal activity with or without Triton X-100.

TABLE 7.16

Intracellular distribution of plasminogen activator of RSV-transformed chick fibroblasts[1]

Fraction	Activity (units)	Specific activity (units/mg)
Homogenate	8400	135
8×10^3 g · min pellet	700	115
60×10^3 g · min pellet	860	100
3×10^5 g · min pellet	4000	620
6×10^6 g · min pellet	2800	350
6×10^6 g · min supernatant	170	1

[1] From Unkeless et al. [149].

Some important morphologic correlates of fibrinolysis by SV40-transformed hamster embryo fibroblasts have been reported by Ossowski et al. [155]. They show that when one co-cultivates normal and transformed cells under conditions allowing plasminogen activation by the SV40 transformants, one rapidly observes morphologic features in the normal cells which are considered characteristic of neoplastic cells. These are rounding up and multilayering. Indeed, exposure of normal cells to plasmin also induces these morphologic alterations. This suggests a response to an enzymatic alteration of the normal cells surfaces.

Significantly, in certain systems fibrinolysis and morphologic change dissociate, that is, active fibrinolysis is a necessary but not sufficient requirement for the morphologic alterations.

Recent studies [156] show that dimethylbenzanthracene-induced rat mammary carcinomas also produce and release the plasminogen activator. In contrast, cells from lactating and non-lactating mammary glands as well as rat mammary fibroadenoma cells, do not. The last represents the only 'benign' tumor tested so far. These data raise the possibility that enhanced release of plasminogen activator might be related to malignancy. This notion receives support from recent studies on tumorigenic and non-tumorigenic cells derived from the malignant clone $B_5 59$

of melanoma B-16 [157]. Clone C_3471, derived from B_559 by continuous growth in the presence of 5-bromodeoxyuridine, has the same plating efficiency as B_559, but is nontumorigenic and releases no plasminogen activator. The conversion from the tumorigenic to the non-tumorigenic state is reversible. Thus after one cell cycle growth in the presence of 5-bromodeoxyuridine, both tumorigenesis and fibrinolysis are reduced to 50% and after two cell cycles both tumorigenicity and fibrinolytic activity dropped to $<30\%$. However, return to normal media for 1–2-cell cycles leads to a recovery of fibrinolytic activity and tumorigenicity.

In evaluating the significance of the high fibrinolysic activity of neoplastic cells in culture it is important to consider possible cell density effects. This has been done for 3T3 fibroblasts [158]. After these cells are subcultured at low density, they 'excrete' virtually no fibrinolysin activator into the culture medium for 2–3 days. Enzyme activity then appears, peaks on day 4, diminishes when the cells become confluent on day 5, and becomes unmeasurable 8 days after replating. No such density dependent modulations are observed with SV40 transformants.

7.2.3. *Relation to cellular levels of lysosomal enzymes*

A number of recent studies indicate that neoplastic cells of diverse origins are actually enriched in lysosomal enzyme activities. Thus, Bosmann and Pike [159] find that polyoma-transformed BHK cells exhibit higher activities than their normal parental cells for every lysosomal enzyme tested except β-D-glucuronidase (Table 7.17). (Also, the activity of fetuin-acetylglycosamine transferase in transformed cells was 2.2 times the activity of the normal cells.)

Bosmann [160] has also observed that the activities of 7 glycosidases and of proteases (on the cell or protein basis) of 3T3 fibroblasts are substantially less

TABLE 7.17

Activities of some lysosomal enzymes in normal and Py-transformed BHK21 fibroblasts[1]

Enzymes	Activity[2]	
	BHK21	Py-BHK
Acid phosphatase	103 ± 3.2	231 ± 1.7
β-D-xylosidase	0	16 ± 2.1
α-D-glucosidase	0	5 ± 0.1
β-D-glucosidase	0	3 ± 0.1
Protease	1403 ± 28	2623 ± 106

[1] From Bosmann and Pike [159].
[2] Activities other than protease are in nmoles/hr^{-1}/mg $proteins^{-7}$. Protease in pronase equivalents.

than those of polyoma transformed variants, SV40-transformed variants and doubly transformed variants.

Bosmann and Hall [161] have examined the enzyme content of a number of invasive human tumors of breast and colon and have compared these with the enzyme activity in normal cells. Their data, given in Table 7.18, show rather clearly that the malignant tumors have anomalously elevated contents of a number of lysosomal enzymes; this is not found in non-malignant neoplasms. These data are perhaps more significant than those previously reported on *in vitro* systems since they actually represent the *in vivo* malignant situation. The authors point out that many of the surface membrane anomalies found in transformed cells could arise from enzymatic modification of normal cell surface components by lysosomal enzymes.

TABLE 7.18

Lysosomal enzymes in normal and neoplastic human tissue[1]

Enzyme	Activity (nmoles/mg of protein per hr)				
	Colon		Breast		
	Normal	Malignant tumors	Normal	Malignant tumors	Fibro-adenoma
β-Galactosidase	108 ± 35	236 ± 41	25 ± 10	78 ± 16	25 ± 8
β-Fucosidase	1 ± 1	1 ± 1	12 ± 7	3 ± 1	4 ± 2
β-*N*-Acetylgalactosaminidase	223 ± 64	317 ± 42	51 ± 15	172 ± 35	89 ± 25
β-*N*-Acetylglucosaminidase	413 ± 98	355 ± 41	518 ± 81	738 ± 91	520 ± 104
α-Mannosidase	42 ± 13	116 ± 23	10 ± 5	40 ± 9	18 ± 4
Acid phosphatase	343 ± 54	340 ± 37	161 ± 34	388 ± 54	238 ± 39
Neuraminidase	103 ± 42	303 ± 41	47 ± 14	218 ± 41	92 ± 27
Protease, pH 3.4	489 ± 68	1399 ± 418	960 ± 123	1861 ± 191	1149 ± 130
Protease, pH 7.4	164 ± 65	211 ± 107	243 ± 93	377 ± 169	144 ± 48

[1] From Bosmann and Hall [161]. Values given are means ± standard errors. The levels of D-galactosidase, α-mannosidase and acid protease are significantly greater ($P < 0.05$) than normal.

Not all tumors contain unusual amounts of lysosomal enzymes. Thus, the studies of Unkeless et al. [146, 148] and Ossowski et al. [147, 150] reveal no increase in the activity of lysosomal proteases of rodent fibroblasts, converted by oncogenic viruses, compared with normal cells. Also, the lysosomal enzymes of rat hepatomas HC, 779A, 779B, 5123A, 7316A and 16 exhibit specific activities between 50% and 150% of those found in normal liver [162]. However, there are also some indications that the lysosomal membranes of the hepatomas differ from those obtained from normal cells.

7.2.4. *Protease activity of neoplastic cells and control of growth* in vitro

Various experimenters (e.g. [163, 164]) have demonstrated that mild protease treatment of density-inhibited, cultured fibroblasts stimulates these cells' proliferative metabolism and allows them to go through one round of cell division. Thereafter, they return to the resting state. This phenomenon has been related to an enhanced lectin induced agglutinability of these cells after protease treatment (see Chapter 10).

Because neoplastically transformed fibroblasts show a similar agglutinability by various lectins as do protease treated normal fibroblasts, and because protease treatment can apparently alter density-dependent growth control *in vitro*, several authors [166–168] have proposed that the surface membranes of neoplastically transformed cells bear proteolytic enzymes, whose action might be important in the release of these cells from contact inhibition of growth. These workers have employed two major strategies to assess possible proteolytic activity on the surface membranes of cells as well as its action in growth control. One of these [165] is to test intact monolayers of normal and transformed fibroblasts for their ability to produce low molecular weight peptide fragments from ^{14}C-labelled *Chlorella* proteins. Thus, Schnebli [165] showed that polyoma-transformed 3T3 cells can degrade the protein 3 to 10 times as fast as normal 3T3 cells, although the final degree of protein breakdown was the same in both cell types.

The second approach has been to evaluate the effects of various protease inhibitors on the growth characteristics of transformed cells. Pursuing this strategy, Schnebli and Burger [166] and Schnebli [168] have found that various protease inhibitors with different modes of action can inhibit the growth of various neoplastically transformed mouse cells, much more readily than the growth of normal cells.

An example is the action of the inhibitor *N*-α-tosyl-L-lysylchloromethane (TLCK) on the growth of normal 3T3 cells and their SV40-transformed variants [166]; TLCK reacts with the active site of trypsin. It is found that TLCK (50 μg/ml) markedly inhibits the growth of the neoplastically transformed cells and reduces their saturation density to approximately that of untransformed cells. The growth of normal 3T3 cells was not affected by TLCK, but the growth of polyoma transformed 3T3 cells was inhibited in a similar fashion to that of the SV40 transformants.

In addition to TLCK, a number of other protease inhibitors with different specificities and modes of action have been tested on the growth of normal and transformed 3T3 cells. These include N-α-tosyl-L-phenylalanylchloromethane (TPCK), an active site titrant for chymotrypsin; N-α-tosyl-L-arginylmethylester (TAME), a substrate and competitive inhibitor for trypsin; Trasylol, a protease inhibitor from pancreas; Leupeptin, a protease inhibitor from *Actimomycetes*;

and ovomucoid which forms strong macromolecular complexes with trypsin.

In an attempt to determine whether the growth-inhibiton action of protease inhibitors might be due to the action of these reagents at the cell surface or due to their penetration into the cell interior, Talmadge et al. [167] have tested the interaction of the protease inhibitor ovomucoid, bound covalently to inert particles (polyacrylamide), with polyoma-transformed 3T3 cells. In these experiments, the beads, which have a spherical diameter of 37 nm or less, were allowed to settle on the cultures at the bottom of a Petri dish. The data demonstrate that the insolubilized ovomucoid has retained the ability to interact with the transformed cells and can produce substantial but not complete inhibition of growth. There was little or no effect on the growth characteristics of normal 3T3 cells. A 57 % inhibition of the growth of polyoma transformed cells was obtained with immobilized ovomucoid corresponding to a maximal equivalent concentration of 50 μg of ovomucoid per ml of medium.

Inhibition became apparent after 3 to 4 days at the earliest. If the beads were washed off the plates after the fourth or fifth day, the cells recover their normal growth properties, but take up to 3 days to regain a density of the untreated polyoma-transformed 3T3 cells. Only ovomucoid beads were effective; beads coupled with equivalent amounts of hemoglobin, ovalbumin, bovine serum albumin and fetuim were ineffective.

The data are interpreted to show that growth inhibition by ovomucoid does not involve uptake of this material, but that it reflects a critical event 'in the plasma membrane'. However, the authors admit that an alternative explanation could be that the beads interact with a component that has been 'excreted' from the cells.

The authors, as well as Burger [168], propose that there are a series of proteases located in plasma membranes. They argue that these proteases can be activated in a cascade-like fashion analogous to the chain reactions mechanisms known to exist in the activation of complement and in the blood clotting process. They suggest that the membrane proteases are normally inactive. The first protease (P1) can be activated by external initiators such as insulin, serum factors, etc.; it in turn then activates a second protease (P2). This chain reaction is proposed to continue to a critical step where the membrane is altered in such a fashion as to be unusually agglutinable by lections. At this point, growth control, an intercellular event, is also impaired, as is interaction with neighboring cells. In our view, the authors' experiments give no evidence for such a mechanism. Indeed, other data presented in this section reveal an entirely different process.

Since TAME is a trypsin substrate as well as a competitive inhibitor for other substrates, hydrolysis of tritiated TAME can be used to measure trypsin-like protease activity. Schnebli [169] has used this approach to evaluate the possible involvement of proteases in the differential growth characteristic of 3T3 cells and their SV40-transformed variants. He studied the hydrolysis of TAME by intact cells, as well as by cells ruptured by repeated freezing and thawing. He observed a rapid phase of hydrolysis lasting less than 20 sec and another phase taking at least 200 times as long. 3T3 cells and their SV40- or polyoma-transformed variants did not differ significantly with respect to the rapid and slow phases. The freeze thawing experiments combined with centrifugation studies showed appearance of a soluble TAME hydrolase after this treatment, but it should be noted that this kind of procedure disrupts lysosomes. It was found that TLCK inhibits TAME hydrolysis at concentrations that are also effective in limiting growth of transformed cells. TPCK was also inhibitory but ovomucoid did not inhibit TAME hydrolysis.

Schnebli's studies [169] indicate the protease inhibitors do restore several 'criteria' for 'contact inhibition' in transformed cells; that is, they reduce the saturation density, they reduce lectin-induced agglutinability and they produce the 'normal' flat morphology. However, thymidine incorporation studies, as well as cytophotometric analyses of DNA content, indicate that the protease inhibitors do not arrest cells in the G_1 phase of the cell cycle; that is, it appears that contact inhibition of replication is not really restored by the inhibitors. The data also raise questions as to whether the inhibitor's sensitive activity plays a significant role in escape from 'contact inhibition' as speculated by Burger and associates [168].

Also, Chou et al. [170,171] have studied the effects of TPCK and TLCK on the growth of SV40-transformed mouse 3T3 fibroblasts. They observed that TPCK causes a non-selective dose dependent and rapidly reversible inhibition of the growth of the SV40-transformed 3T3 cells. Moreover, the inhibitor causes a dose-dependent and reversible blockage of both protein and DNA synthesis. The authors' data show that selective growth inhibition of transformed cells by TPCK and TLCK is not a general property of all SV40-converted 3T3 cells and can therefore not be an invariable consequence of neoplastic transformation by SV40 virus. Moreover, the data indicate that a proteolytic activity which can be inhibited by the mentioned blockers cannot be solely responsible for the loss of density-dependent inhibition in transformed fibroblast. However, the authors do not exclude the possibility that other proteolytic enzymes may play a role in 'impaired contact inhibition' in neoplastic cells.

It is of interest that tumorigenesis in mouse skin by dimethylbenzanthracene can be inhibited by synthetic inhibitors of proteases [172,173]. The promotion of tumorigenesis in mouse skin by phorbol esters is also suppressed by protease inhibitors. Moreover, the effect of protease inhibitors is not limited to mouse cells, because growth inhibition of hamster tumor cells has also been observed [174].

7.2.5. Release of lysosomal enzymes

How might the release of lysosomal enzymes come about? One possibility is that the primary lysosomal membrane fuses with the plasma membrane and thus leads to extrusion of the lysosomal contents. This hypothesis has been extensively discussed by Lucy [175]. Lucy argues that extrusion of lysosomal material into the extracellular space comes about by adhesion between the cytoplasmic surfaces of lysosomes and plasma membrane, fusion of the two cell types and discharge of the lysosomal contents (Fig. 7.2).

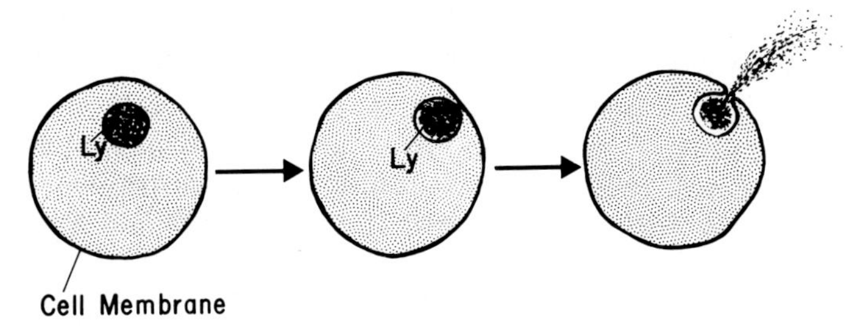

Fig. 7.2. Possible mechanism for the release of lysosomal enzymes into the extracellular space.

Lucy focuses his attention on membrane phosphatides and argues that a primary requirement for fusion of two membranes is that both should have a rather high proportion of their phospholipid molecules in a micellar rather than a bilayer configuration. He argues that agents such as lysolecithin, vitamin A, etc. which induce membrane fusion, do so by producing a bilayer → micellar equilibrium. However, current information on the structure of biomembranes indicates that such micelle-bilayer transitions are much less likely than order-disorder transitions in a lipid phase. This is discussed in Chapters 3 and 5. Such changes in lipid state could come about by compositional alterations or through the action of lipid-perturbing agent. Another factor to be considered here is that the primary action of hydrophobic perturbants, such as lysolecithin may not be on the membrane lipids but on the hydrophobic interphases between the membrane lipids and proteins.

In any case, it is possible that lysolecithin may play an important role in the extrusion of lysosomal enzymes. Most cells contain appreciable proportions of phospholipase A1 and phospholipase A2 which can degrade lecithin to produce lysolecithin and a saturated (A1) or an unsaturated (A2) fatty acid. Such phospholipases have been demonstrated in the lysosomes of various cells. Similarly phospholipases A1 and A2 have been localized in the plasma membranes of some normal cells, including lymphocytes. Moreover, the plasma membranes of many cells are able to reacylate the lysosophosphatides produced by phospholipase action. Thus, the possible role of phospholipases and also acyl transferases in reversible fusion of lysosomal and plasma membranes is a rather attractive one, but there is no information concerning these mechanisms in tumor cells.

The possible mechanisms involved in membrane fusion have been much elucidated by Papahadjopoulos et al. [178]. These workers have examined the capacity of unilamellar phospholipid vesicles of varying composition to induce fusion of cultured BHK21, 3T3 and L929 cells. Their data can be summarized as follows:

(a) Neutral vesicles (phosphatidylcholine) do not induce fusion.
(b) Negatively-charged vesicles (e.g., phosphatidylserine/phosphatidylcholine 1 : 9; dipalmitoylphosphatidylglycerol/dipalmitoylphosphatidylcholine 1 : 9) are highly effective in causing fusion, provided the temperature is above the melting temperature of the fatty acids.
(c) *Lysolecithin* (3–10 moles %) increases the fusion capacity of phosphatidylserine–phosphatidylcholine vesicles but it reduces cell viability.
(d) The presence of cholesterol in phosphatidylglycerol–phosphatidylcholine vesicles, in a 1 : 1 M ratio cholesterol/phospholipid, markedly reduces fusion. However the sterol has no effect on phosphatidylserine–phosphatidylcholine vesicles.

These data deal with plasma membranes and cannot be readily applied to the concept of fusion between *lysosomal* and plasma membranes. Also, the studies suffer in that the authors did not consider the capacity of cholesterol-free vesicles to *extract* cholesterol from plasma membranes (Chapter 5), which generally have a cholesterol/phospholipid molar ratio near 1.0. Indeed it is conceivable that cholesterol *extraction* induces and enhances mobility in lipid membrane domains, thereby allowing plasma membrane fusion. In this connection we point out that what reliable information is available indicates that lysosomes have a much lower cholesterol/phospholipid ratio than plasma membranes, namely 0.27 [179]. Lysosomal membranes are thus intermediate in cholesterol content between the plasma membrane and mitochondria (liver); the last show a cholesterol/phospholipid ratio of 0.055 ± 0.006. The lysosomal membranes also contain approximately 1% of lysolecithin and approximately 3% of lysophosphatidylethanolamine. Unfortunately, no comparable data exists for lysosomes from tumor cells and the information available is neither consistent with the Lucy hypothesis nor adequate for the formulation of any general model, particularly one dealing with neoplastic cells.

The release of lysosomal peptidases from tumor cells can be expected to lead to one or more of the following consequences:

(a) Destruction of normal tissue architecture, allowing the penetration and permeation of tumor cells.
(b) Utilization of normal tissue material as nutrients for tumor growth.
(c) Modification of the surface characteristics of the cells, both normal and neoplastic, interfering with normal surface recognition phenomena.
(d) Alteration of the antigenic properties of the tumor cells. (This could be antigenic deletion, because limited proteolysis can remove mouse histocompatibility antigens (H2) as well as human HLA determinants from cell surfaces. Interference with the immune response against tumor cells might be another consequence, since proteolysis can cleave peptide fragments bearing

the immunological specificity of tumor antigens from cells transformed by SV40 virus.)

(e) Release of cells from growth control. This could be by mechanisms analogous to those observed by Burger and Noonan [180] who show that protease treatment of confluent cultures of non-malignant mouse 3T3 fibroblasts, in amount too small to produce detachment, releases these cells from contact inhibition of growth. Protease treatment might also destroy or alter the characteristics of cell surface receptors for extracellular regulators, such as protein hormones.

7.3. Coda

Anomalous membrane permeability and/or membrane transport could have profound biological consequences and might contribute significantly to malignant behavior.

First, abnormal release of lysosomal proteases, for example, collagenase, might facilitate invasion by destroying normal tissue architecture.

Second, the release of proteases and glycosidases might alter or destroy the cell surface components determining tumor cell individuality. Cell recognition processes would be impaired thereby. Moreover, elimination or dilution of neo-antigens which would otherwise cause immunologic rejection of the tumor cells, would allow the cells to escape the host's immunological defenses.

Third, an exceptional capacity by transformed cells to accumulate essential metabolites, for example, essential amino acids, could confer a critical competitive advantage on to the tumor cells.

Fourth, the effectiveness of many chemotherapeutic agents depends upon their transport into the tumor cell. Modulation or deletion of the relevant transport system can allow the tumor to escape from the action of the agent.

References

[1] Shrivastava, G.C. and Quastel, J.H., *Nature*, *196*, 876 (1962).
[2] Wallach, D.F.H., *Proc. Natl. Acad. Sci. U.S.*, *61*, 868 (1968).
[3] Holley, R.W., *Proc. Natl. Acad. Sci. U.S.*, *69*, 2840 (1972).
[4] Kotyk, A. and Javacck, K. in *Cell Membrane Transport*, Plenum Press, New York, 1970.
[5] Lieb, W.R. and Stein, W.D.S., *Biochim. Biophys. Acta*, *265*, 187 (1972).
[6] Hatanaka, M.H., *Biochim. Biophys. Acta*, *355*, 77 (1974).
[7] Hanafusa, H., *Proc. Natl. Acad. Sci. U.S.*, *63*, 318 (1969).
[8] Hatanaka, M., Huebner, R.J. and Gilden, R.V., *J. Natl. Cancer Inst.*, *43*, 1091 (1969).
[9] Hatanaka, M. and Hanafusa, H., *Virology*, *41*, 647 (1970).
[10] Hatanaka, M. and Gilden, R.V., *J. Nat. Cancer Inst.*, *45*, 87 (1970).
[11] Hatanaka, M., Gilden, R.V. and Kelloff, G., *Virology*, *43*, 734 (1971).
[12] Hatanaka, M., Todaro, G.J. and Gilden, R.V., *Int. J. Cancer*, *5*, 224 (1970).

[13] Stephenson, J.R., Reynolds, R.K. and Aaronson, S.A., *J. Virology*, *11*, 218 (1973).
[14] Martin, G.S., Venuta, S., Weber, M. and Rubin, H., *Proc. Natl. Acad. Sci. U.S.*, *68*, 2739 (1971).
[15] Kawai, S. and Hanafusa, H., *Virology*, *46*, 470 (1971).
[16] Kletzien, R.F. and Perdue, J.F., *J. Biol. Chem.*, *249*, 3366 (1974).
[17] May, J.T., Somers, K.D. and Kit, S., *Int. J. Cancer*, *11*, 377 (1973).
[18] Culp, L.A., Grimes, W.J. and Black, P.H., *J. Cell Biol.*, *50*, 682 (1971).
[19] Schultz, A.R. and Culp, L.A., *Exp. Cell Res.*, *81*, 95 (1973).
[20] Isselbacher, K.J., *Proc. Natl. Acad. Sci. U.S.*, *69*, 585 (1972).
[21] Inbar, M., Ben-Bassat, H. and Sachs, L., *J. Memb. Biol.*, *6*, 195 (1971).
[22] Kalckar, H.M., Ullrey, D., Kijomoto, S. and Hakomori, S., *Proc. Natl. Acad. Sci. U.S.*, *70*, 839 (1973).
[23] Martineau, R., Kohlbacher, M., Shaw, S.N. and Amos, H., *69*, 3407 (1972).
[24] Kalckar, H.M. and Ullrey, D., *Proc. Natl. Acad. Sci. U.S.*, *70*, 2502 (1973).
[25] Carter, J.R., Jr. and Martin, D.B., *Proc. Natl. Acad. Sci. U.S.*, *64*, 1343 (1969).
[26] Hatanaka, M., *Proc. Natl. Acad. Sci. U.S.*, *70*, 1364 (1973).
[27] Graff, J.C., Hanson, D.G. and Hatanaka, M., *Int. J. Cancer*, in press.
[28] Steiner, S. and Steiner, M.R., *Biochim. Biophys. Acta*, *206*, 403 (1973).
[29] Stoker, M.G.P., *Proc. Royal Soc. Lond. (Biol.)*, *181*, 1 (1972).
[30] Romano, A.H. and Colby, C., *Science*, *179*, 1238 (1973).
[30a] Romano, A.H. and Colby, C., *J. Cell. Physiol.*, *85*, 15 (1975).
[31] Hatanaka, M., Augl, C. and Gilden, R.V., *J. Biol. Chem.*, *245*, 714 (1970).
[32] Kletzien, R.F. and Perdue, J.F., *J. Biol. Chem.*, *249*, 3383 (1974).
[33] Kletzien, R.F. and Perdue, J.F., *J. Biol. Chem.*, *249*, 3375 (1974).
[34] Graff, J.C., Hanson, D.J. and Hatanaka, M., *Int. J. Cancer*, *12*, 602 (1973).
[35] Peters, J.H. and Hausen, P., *Eur. J. Biochem.*, *19*, 509 (1971).
[36] Venuta, S. and Rubin, H., *Proc. Natl. Acad. Sci. U.S.*, *70*, 653 (1973).
[37] Weber, M.J., *J. Biol. Chem.*, *248*, 2978 (1973).
[38] Bose, S.K. and Zlotnick, B.J., *Proc. Natl. Acad. Sci. U.S.*, *70*, 2374 (1973).
[39] Hino, S. and Yamamoto, T., *GANN*, *62*, 539 (1971).
[40] Lin, S., Santi, D.V. and Spudich, J.A., *J. Biol. Chem.*, *249*, 2268 (1974).
[41] Kletzien, R.F., Perdue, J.F. and Springer, E., *J. Biol. Chem.*, *247*, 2964 (1972).
[42] Estensen, R.D. and Plagemann, P.G.W., *Proc. Natl. Acad. Sci. U.S.*, *69*, 1430 (1972).
[43] Plagemann, P.G.W. and Erbe, J., *Cancer Res.*, *33*, 482 (1973).
[44] Plagemann, P.G.W. and Estensen, R.D., *J. Cell Biol.*, *55*, 179 (1972).
[45] Blueminck, J.G., *Cytobiologie*, *3*, 176 (1971).
[46] Estensen, R.D., Rosenberg, M. and Sheridan, J.D., *Science*, *173*, 356 (1971).
[47] Krishan, A., *J. Cell Biol.*, *54*, 657 (1972).
[48] Spudich, J.A. and Lin, S., *Proc. Natl. Acad. Sci. U.S.*, *69*, 442 (1972).
[49] Allison, A.C., *Chem. Phys. Lipids*, *8*, 374 (1972).
[50] McNutt, N.S., Culp, L.A. and Black, P.H., *J. Cell Biol.*, *56*, 412 (1973).
[51] Bradley, W.E.C. and Culp, L.A., *Exp. Cell Res.*, *84*, 335 (1974).
[52] DeLaat, S.W., Luchtel, D. and Blueminck, J.G., *Developmental Biol.*, *31*, 163 (1973).
[53] Edelman, G.M., Yahara, I. and Wang, J.L., *Proc. Natl. Acad. Sci. U.S.*, *70*, 1442 (1973).
[54] Wunderlich, F., Muller, R. and Speth, V., *Science*, *182*, 1136 (1973).
[55] Sefton, B.M. and Rubin, H., *Proc. Natl. Acad. Sci. U.S.*, *68*, 3154 (1971).
[56] Cunningham, D.D. and Pardee, A.B., *Proc. Natl. Acad. Sci. U.S.*, *64*, 1049 (1969).
[57] Bader, J.P., *Science*, *180*, 1069 (1973).
[58] Hatanaka, M., *GANN Monograph*, *10*, 45 (1971).
[59] Christensen, H.N., *Fed. Proc.*, *32*, 19 (1973).
[60] Eagle, H., Piez, K.A. and Levy, M., *J. Biol. Chem.*, *236*, 2039 (1961).
[61] Hare, J.D., *Cancer Res.*, *27*, 2357 (1967).
[62] Foster, D. and Pardee, A., *J. Biol. Chem.*, *244*, 2675 (1969).
[63] Baril, E.F., Potter, V.R. and Morris, H.P., *Cancer Res.*, *29*, 2101 (1969).
[64] Reynolds, R.D., Scott, D.F., Potter, V.R. and Morris, H.P., *Cancer Res.*, *31*, 1580 (1971).
[65] Scott, D.F., Reynolds, R.D., Pitot, H.C. and Potter, V.R., *Life Sci., Part 2*, *9*, 1133 (1970).

[66] Scott, D.F., Butcher, F.R., Potter, V.R. and Morris, H.P., *Cancer Res.*, *32*, 2127 (1972).
[67] Butcher, F.R., Scott, D.F., Potter, V.R. and Morris, H.P., *Cancer Res.*, *32*, 2135 (1972).
[68] Risser, W.L. and Glehrter, T.D., *J. Biol. Chem.*, *248*, 1248 (1974).
[69] Ballard, P.L. and Tomkins, G.M., *J. Cell Biol.*, *47*, 222 (1970).
[70] Swabb, E.A., Wei, J. and Gullino, P.M., *Cancer Res.*, *34*, 2814 (1974).
[71] Goldenberg, G.J., Vanstone, C., Israels, L.G. and Bihler, I., *Cancer Res.*, *30*, 2285 (1970).
[72] Goldenberg, G.J., Lyons, R.M., Lepp, J.A. and Vanstone, C.L., *Cancer Res.*, *31*, 1616 (1971).
[73] Inaba, M. and Sakurai, Y., *Int. J. Cancer*, *7*, 430 (1971).
[74] Inaba, M., Moriwaki, A. and Sakurai, Y., *Int. J. Cancer*, *10*, 411 (1972).
[75] Wolpert, M.K. and Ruddon, R.W.A., *Cancer Res.*, *29*, 873 (1969).
[76] Goldman, I.D., Lichtenstein, N.S. and Oliverio, V.T., *J. Biol. Chem.*, *243*, 5007 (1968).
[77] Fischer, G.A., *Biochem. Pharmacol.*, *11*, 1233 (1962).
[78] Sirotnak, F.M. and Donsbach, R.C., *Cancer Res.*, *33*, 1290 (1973).
[79] Zager, R.F., Frisby, S.A. and Oliverio, V.T., *Cancer Res.*, *33*, 1670 (1973).
[80] Hill, B.T., *Biochem. Pharmacol.*, *21*, 495 (1972).
[81] Warnick, C.T., Muzik, H. and Paterson, A.R.P., *Cancer Res.*, *32*, 2017 (1972).
[82] Myers, D.K., *Advan. Biol. Med. Phys.*, *13*, 219 (1970).
[83] Wallach, D.F.H. in *Biomembranes*, L. Manson, ed., Plenum Press, New York, Vol. 5, 1974, p. 213.
[84] Millard, R.E., *J. Clin. Path.*, *18*, 782 (1965).
[85] Thomas, J.W., Coy, P., Lewis, H.S. and Yuen, A., *Cancer*, *27*, 1046 (1971).
[86] Stjernswärd, J., Jondal, M., Vanky, F., Wigzell, H. and Sealy, R., *Lancet*, *1*, 1352 (1972).
[87] Chee, C.A., Ilbery, P.L.T. and Rickinson, A.B., *Br. J. Radiol.*, *47*, 37 (1974).
[88] Myers, D.K. and Skov, K., *Canad. J. Biochem.*, *44*, 839 (1966).
[89] Myers, D.K. and Sutherland, R.M., *Canad. J. Biochem.*, *40*, 413 (1962).
[90] Myers, D.K., DeWolfe, D.E., Araki, K. and Arkinstall, W.W., *Can. J. Biochem.*, *41*, 1181 (1963).
[91] Kwock, L. and Wallach, D.F.H., *Biochim. Biophys. Acta*, *352*, 135 (1974).
[92] Chapman, I.V. and Sturrock, M.G., *Internat. J. Radiat. Biol.*, *22*, 1 (1972).
[93] Christensen, H.N., Handlogten, M.E., Lam, I., Tager, H.S. and Zand, R., *J. Biol. Chem.*, *244*, 1510 (1969).
[94] van den Berg, K.J. and Betel, I., *FEBS Letters*, *29*, 149 (1973).
[95] van den Berg, K.J. and Betel, I., *Exp. Cell Res.*, *76*, 63 (1973).
[96] Schreck, R. and Stefani, S., *J. Natl. Cancer Inst.*, *32*, 507 (1964).
[97] Patt, H.M., Blackford, M.E. and Straube, R.L., *Proc. Soc. Exp. Biol. and Med.*, *80*, 92 (1952).
[98] Ueno, A.M., *Internat. J. Radiat. Biol.*, *21*, 43 (1972).
[99] Myers, D.K., *Internat. J. Radiat. Biol.*, *22*, 203 (1972).
[100] Bush, H. in *An Introduction to the Biochemistry of the Cancer Cell*, Academic Press, New York, 1962, p. 356.
[101] Costa, G., *Progr. Exp. Tumor Res.*, *3*, 321 (1963).
[102] Warburg, O, Gawehn, K. and Lange, G., N. *Naturforsch*, *9B*, 109 (1954).
[103] Bosch, L., *Biochim. Biophys. Acta*, *30*, 444 (1958).
[104] Wu, R., *Cancer Res.*, *19*, 1217 (1959).
[105] McNair-Scott, D.B., Sanford, K.K. and Westfall, B.B., *Proc. of the American Association for Cancer Research*, *3*, 41 (1959).
[106] Hill, G.R., *Cancer Res.*, *16*, 460 (1956).
[107] Hsieh, K., Suntzeff, V. and Cowdry, E.V., *Proc. Soc. Exp. Biol. Med.*, *89*, 627 (1955).
[108] Worblewski, F. *Ann. N.Y. Acad. Sci.*, *75*, 322 (1958).
[109] Holmberg, B., *Cancer Res.*, *21*, 1386 (1961).
[110] Kanamaru, R., *Tohoku J. Exp. Med.*, *97*, 317 (1969).
[111] Bissell, M.J., Rubin, H. and Hatie, C., *Exp. Cell Res.*, *68*, 404 (1971).
[112] Kant, J.A. and Steck, T.L., *J. Biol. Chem.*, *248*, 8457 (1973).
[113] Maretzki, D., Groth, J., Tsamaloukas, A.G., Grundell, M., Kruger, S. and Rapoport, S., *FEBS Letters*, *39*, 83 (1974).
[114] Malmgren, H., Sylven, B. and Revesz, L., *Br. J. Cancer*, *9*, 473 (1955).
[115] Sylven, B., *Acta Un. Int. Canc.*, *14*, 61 (1958).

[116] Sylven, B. in *Biological Interactions in Normal and Neoplastic Growth*, M.J. Brennan and W.L. Simpson, eds., Little, Brown, Boston, Ma., 1962.
[117] Sylven, B. and Bois, I., *Cancer Res.*, *20*, 831 (1960).
[118] Sylven, B. and Malmgren, H., *Acta Radiol. Suppl.*, *154*, 1 (1955).
[119] Sylven, B., Ottoson, R. and Revesz, L., *Br. J. Cancer*, *13*, 551 (1959).
[120] Sylven, B. and Bois-Svensson, I., *Cancer Res.*, *25*, 458 (1965).
[121] Sylven, B., *Eur. J. Cancer*, *4*, 463 (1968).
[122] Sylven, B. in *Endogenous Factors Influencing Host-Tumor Balance*, R.W. Wissler, T.L. Dao and S. Wood, Jr., eds., University Chicago Press, 1967, p. 267.
[123] Koono, M., Ushijima, K. and Hayashi, H., *Int. J. Cancer*, *13*, 105 (1974).
[124] Koono, M., Katsuya, H. and Hayashi, H., *Int. J. Cancer*, *13*, 334 (1974).
[125] Benz, G.N., Lehmann, F.E., Helv. Physiol, Pharmacol. Acta, *17*, 380 (1959).
[126] Goldberg, D.M., McAllister, R.A. and Roy, A.D., *Br. J. Cancer*, *23*, 735 (1969).
[127] Mohr, H.J., *Frankfurt Z. Pathol.*, *77*, 107 (1967).
[128] Poole, A.R. in *Lysosomes in Biology and Pathology*, J.T. Dingle, ed., North Holland, Amsterdam, Vol. 3, 1973, p. 304.
[129] Burger, M.M. in *Biomembranes*, L.A. Manson, ed., Plenum Press, New York, 1971, p. 247.
[130] Burger, M.M., *Fed. Proc.*, *32*, 91 (1973).
[131] Poste, G., *Exp. Cell Res.*, *73*, 319 (1972).
[132] Poste, G., *Exp. Cell Res.*, *67*, 11 (1971).
[133] Poste, G., *Exp. Cell Res.*, *65*, 359 (1971).
[134] Kazakova, O.V., Orekhovich, V.N., Porchot, L. and Schuck, J.M., *J. Biol. Chem.*, *247*, 4224 (1972).
[135] Lazarus, G.S. in *The Lysosomes*, J.T. Dingle, ed., North Holland, Amsterdam, Vol. 3, 1973, p. 228.
[136] Birbeck, M.S.C. and Wheatley, D., N. *Cancer Res.*, *25*, 490 (1965).
[137] Hashimoto, K., Yamanishi, Y. and Dabbous, M.K., *Cancer Res.*, *32*, 2561 (1972).
[138] Strauch, L., Vencelj, H. and Hannig, K., *Hoppe-Seyler's Z. Physiol. Chem.*, *349*, 171 (1968).
[139] Robertson, D.M. and Williams, D.C., *Nature*, *221*, 259 (1969).
[140] Taylor, C.A., Levy, B.N. and Simpson, J.W., *Nature*, *228*, 366 (1970).
[141] Dresden, M.H., Heilman, S.A. and Schmidt, J.D., *Cancer Res.*, *32*, 993 (1972).
[142] Strauch, L. in *Tissue Interactions in Carcinogenesis*, D. Tarin, ed., Academic Press, London and New York, 1972, p. 399.
[143] Lang, R., 1971, quoted by Strauch, 1972.
[144] Robertson, D.M., 1970, cited by Strauch, 1972.
[145] Fischer, A., *Arch. Entwicklungsmech. Org. (Wilhelm Roux)*, *104*, 210 (1925).
[146] Unkeless, J.C., Tobia, A., Ossowski, L., Quigley, J.P., Rifkin, D.P. and Reich, E., *J. Exp. Med.*, *137*, 85 (1973).
[147] Ossowski, L., Unkeless, J.C., Tobia, A., Quigley, J.P., Rifkin, D.P. and Reich, E., *J. Exp. Med.*, *137*, 112 (1973).
[148] Rifkin, D.P., personal communication.
[149] Unkeless, J.C., Danø, K., Kellerman, G.M. and Reich, E., *J. Biol. Chem.*, *249*, 4295 (1974).
[150] Ossowski, L., Quigley, J.P., Kellerman, G.M. and Reich, E., *J. Exp. Med.*, *138*, 1056 (1973).
[151] Quigley, J.P., Ossowski, L. and Reich, E., *J. Biol. Chem.*, *249*, 4306 (1974).
[152] Burk, R.R., *Proc. Natl. Acad. Sci. U.S.*, *70*, 369 (1973).
[153] Rifkin, D., Loeb, J.N., Moore, G. and Reich, E., *J. Exp. Med.*, *139*, 1317 (1974).
[154] Christman, J.K. and Acs, G., *Biochim. Biophys. Acta*, in press.
[155] Ossowski, L., Quigley, J.P. and Reich, E., *J. Biol. Chem.*, *249*, 4312 (1974).
[156] Reich, E. in *Control of Proliferation in Animal Cells*, B. Clarkson and R. Baserga, eds., Cold Spring Harbor Laboratory, 1974, p. 351.
[157] Christman, J.K., Acs, G., Silagi, S., Newcomb, E.W. and Silverstein, S., *J. Cell Biol.*, *63*, 61a (1974).
[158] Chou, I.-N., Black, P.H. and Roblin, R., *J. Cell Biol.*, *63*, 61a (1974).
[159] Bosmann, H.B and Pike, G.Z., *Life Sciences*, *9*, *pt. II*, 1433 (1970).
[160] Bosmann, H.B., *Exp. Cell Res.*, *54*, 217 (1969).
[161] Bosmann, H.B. and Hall, T.C., *Proc. Natl. Acad. Sci. U.S.*, *71*, 1833 (1974).

[162] Wattiaux, R., Wattiaux-De Coninck, S., van Dijck, J.M., Dupal, M.F. and Morris, H.P., *Eur. J. Cancer*, *6*, 503 (1970).
[163] Burger, M.M., *Nature*, *227*, 170 (1970).
[164] Sefton, B.M. and Rubin, H., *Nature*, *277*, 843 (1970).
[165] Schnebli, H.P., *Schweiz, Med. Wochenschr.*, *102*, 1194 (1972).
[166] Schnebli, H.P. and Burger, M.M., *Proc. Natl. Acad. Sci. U.S.*, *69*, 3825 (1972).
[167] Talmadge, K.W., Noonan, K.D. and Burger, M.M. in *Control of Proliferation in Animal Cells*, B. Clarkson and R. Baserga, eds., Cold Spring Harbor Laboratory, 1974, p. 313.
[168] Burger, M.M. in *The Neurosciences: Third Study Program*, F.O. Schmitt and F.G. Worden, eds., MIT University Press, Cambridge, Mass., 1973, p. 773.
[169] Schnebli, H.P. in *Control of Proliferation in Animal Cells*, B. Clarkson and R. Baserga, eds., Cold Spring Harbor Laboratory, 1974, p. 327.
[170] Chou, I.-N., Black, G.H. and Roblin, R. in *Control of Proliferation in Animal Cells*, B. Clarkson and R. Baserga, eds., Cold Spring Harbor Laboratory, 1974, p. 339.
[171] Chou, I.-N., Black, G.H. and Roblin, R., *Proc. Natl. Acad. Sci. U.S.*, *71*, 1748 (1974).
[172] Troll, W., Klassen, A. and Janoff, A., *Science*, *169*, 1211 (1970).
[173] Hozumi, M., Ogawa, M., Sugimura, T., Takeuchi, T. and Umeuzawa, H., *Cancer Res.*, *32*, 1725 (1972).
[174] Goetz, I.E., Weinstein, C. and Roberts, E., *Cancer Res.*, *32*, 2469 (1972).
[175] Lucy, J.A. in *Lysosomes in Biology and Pathology*, J.T. Dingle and H.B. Fell, eds., North Holland, Amsterdam, Vol. 2, 1969, p. 313.
[176] Lucy, J.A., *Nature*, *227*, 815 (1970).
[177] Stoffel, W. and Trabert, U., *Hoppe-Seylers, Z. Physiol. Chem.*, *350*, 836 (1969).
[178] Papahadjopoulos, D., Poste, G. and Schaeffer, B.E., *Biochim. Biophys. Acta*, *323*, 23 (1973).
[179] Henning, R. and Heidrich, H.-G., *Biochim. Biophys. Acta*, *345*, 326 (1974).
[180] Burger, M.M. and Noonan, K.D., *Nature*, *228*, 512 (1970).

8

Cell contact and cell recognition

Thirty years ago Coman [1] first proposed the important hypothesis that the *malignancy* of cancer cells – their capacity to invade and metastasize – derives from an impairment of the adhesive properties of these cells. Coman summarized early experimentation in this area in 1953 [2]. Since then, the field has expanded enormously, as attested by a large number of reviews and monographs (e.g. [3–7].

The mechanisms which are important in cell contact phenomena continue to remain elusive. We will, therefore, not exhaustively review the vast literature in the field, but will instead critically assess several prominent aspects of its present status and will evaluate the major working hypotheses in the area.

8.1. Physical mechanisms of attraction and repulsion between membranes

8.1.1. Introduction

The interactions between cells and cell organelles will include some of the following mechanisms:

Attractions

(a) Covalent chemical bonds between apposed surfaces, for example, amide bonds, ester linkages, S–S bonds.
(b) Hydrogen bonding.
(c) Specific non-covalent interactions, (e.g., 'hydrophobic bonding').
(d) Ion-pair and ion-triplet formation, for example, $-NH_3^+ \ldots\ {}^-OOC-$ and $COO^- \ldots Ca^{2+} \ldots {}^-OOC$.

(e) Non-specific hydrophobic bonding.
(f) Attractions due to charge fluctuations.
(g) Electrostatic attractions between surfaces of opposite charge.
(h) Attractions due to 'charge mosaics' on surfaces of like or opposite overall charge.
(i) Electrostatic attractions between surfaces of like charge.
(j) Electrostatic attraction due to image forces.
(k) Surface tension or surface energy.
(l) Van der Waals' forces.

Repulsive interaction
(a) Charge repulsion between surfaces of like charge.
(b) Van der Waal's forces of repulsion.

Not included in this list are steric effects (e.g., solvated layers, intervening macromolecules) which might prevent close contact.

Items (a) and (b) are highly specific. There are indications that desmosomes might represent an example of (a). In these structures protein filaments appear to span the gap between apposed cell surfaces [8].

Item (b) *per se* appears improbable, since cell surfaces and intercellular spaces are highly hydrated and H_2O competes for H-bond donors.

Item (c) is a highly specific mechanism. It is involved in the cross-linking of cells by divalent antibodies or lectins. It is also invoked in the theories of specific cell recognition treated later on. Hydrophobic 'forces' can lead to very large interactions provided the contacts are appropriate. Thus, an evaluation of the surface hydrophobicity of hemoglobin [9] from the hydrophobicities of its amino acid residues and the sequence positions of these residues, indicates that the free energy of dimer formation in hemoglobin is in the order of 43 kcal. If intercellular contact and intercellular recognition took place through the formation of dimers with monomers contributed by each cell surface, as suggested by Najjar [10], one could expect a very large adhesive contribution from hydrophobic interactions. At the moment, however, there is no basic information whether this kind of mechanism exists.

Ion-pair or ion-triplet formation (d), may be involved in some very tight membrane associations (e.g., tight junctions [8]) and may provide one basis for the role of Ca^{2+} in cell adhesion.

Charge fluctuation attractions (f), might come into play when charged groups at or near the membrane are only partially ionized, that is, when the surface pH is near to the pK of the ionizable group. Then the actual location of the charge varies from instant to instant as individual ionizable residues become protonated or deprotonated. The charge fluctuation on one surface sets up induced fluctuations on the apposing surface, producing a net attraction. The attraction is maximal when half of the ionizable groups are ionized. The attraction has a range similar to ordinary electrostatic interactions in electrolyte solutions and is attenuated by the screening effect of ordinary electrolytes between the surfaces. At close range the effect could be geometrically rather specific when the ionizable groups are suitably arranged on the surfaces. Of the various ionogenic residues at cell surfaces (Table 8.1), histidines appear the major candidates for this type of interaction.

TABLE 8.1

Some approximate ionization constants of acidic and basic residues possibly exposed at cell surfaces

Residue	pK_a
Proteins[1]	
Glu γ-COO^-	4.2
Asp β-COO^-	3.6
C-terminal COO^-	3.0–3.2
Lys $-NH_3^+$	9.4–10.6
Arg (Guan)	11.6–12.6
N-terminal NH_3^+	6.5–8.5
His (Im)	5.6–7.0
Tyr –OH	9.4–10.1
Phospholipids[2]	
Phosphatidylcholine	
Phosphate	<2
$N^+(CH_3)_3$	13.9
Sphingomyelin	
phosphate	<2
$N^+(CH_3)_3$	13.9
Phosphatidylethanolamine	
phosphate	<2
NH_3^+	9.1
Phosphatidylserine	
phosphate	<2
COO^-	4.6
NH_3^+	10.3
Phosphatidic acid	
phosphate	3.4;8.6
Carbohydrate[3]	
Sialic acid COO^-	2.6
RNA	
phosphate	<2

[1] From Cohn and Edsall [11].
[2] From [12–15].
[3] From [16].

Items (g) and (h) represent electrostatic attractions with ranges typical of Coulombic interactions. Attractions between oppositely charged surfaces appear highly improbable. On the other hand, attractions due to charge mosaics could provide specificity of interaction, but only at short range. At distances, ~10–14 Å from a charged surface, the distribution of anions and cations will be come random and any discreteness of charge patterns on the cell surfaces will be lost.

The electrostatic attractions between surfaces of like charge (i) arise only when the two surfaces possess different potentials of the same sign. For two surfaces of identical potential, the electrostatic interaction is always repulsive. These attractions would be significant only at very small intermembrane distances.

Electrostatic attractions due to 'image forces' (j), occur between low-dielectric particles (e.g., oil droplets) dissolved in a high dielectric solvent containing dissociated salts (e.g., an aqueous salt solution). In general anions tend to concentrate at the interfaces between the particles and solvent, whereas cations tend to be desorbed. As the low-dielectric particles approach, the desorbed ions between them will have parts of their fields in the low-dielectric regions of the particles. They will thus tend to migrate out of the inter-particle region, depleting this of ions. This then leads to a net attraction between the particles. The effect is not likely to be significant since membrane surfaces bear an appreciable high-dielectric shell (only the apolar membrane cores are of really low dielectric). Moreover, the image effect would be largely screened out at physiological ionic strengths.

Item (k) has been invoked by Weiss [4] as participating in cellular attractions. It is argued that the reduction of interfacial free energy upon cell contact favors adhesion. However, it is now known that the interfacial tension between membranes and aqueous buffers are low, and (k) is not likely to play an important role in intermembrane attractions.

Item (l) occupies a central position in the theories of colloidal stability, discussed below.

Of the *repulsive interactions* item (a) constitutes a principal factor in theories of colloidal stability, which, in fact deals with the balance of Van der Waals' attractions and electrical repulsion. Van der Waals' forces of repulsion (b), operate at small distances only and are small compared to other repulsive mechanisms.

It has been suggested (e.g. [17, 18]) that a major mechanism in the interaction between cells is the balance between electrostatic repulsion and certain attraction forces (particularly Van der Waals–London's forces). This argument represents an extension of the colloid stability theory [19, 20]. We will examine the applicability of this theory to membrane interactions after first exploring the nature of the major attractive and repulsive interactions that are likely to be involved.

8.1.2. *Energies and forces of attraction*

The most important physical mechanisms of attraction are Van der Waals' interactions. Several phenomena are involved, but from a quantitative point of view, the most significant Van der Waals' interactions are those arising from charge fluctuations – the London–Van der Waals' forces. Charge fluctuations themselves may arise from Brownian motion, from shifts in electron position with respect to different quantum levels or from proton migrations around basic sites. London–Van der Waals' dispersion forces may thus occur between molecules having no permanent dipole moments.

According to the London theory, the force between two molecules is C/R^7, where R = intermolecular distance, C is a constant depending on several molecular electric properties, particularly polarizability. The term polarizability refers to the extent to which an electrical field distorts a molecule and changes the position of its electrons relative to the nuclei of its constituent atoms. The polarizabilities of individual molecules are unknown and London's formula does not permit accurate calculations of intermolecular forces.

Polarizability effects are electromagnetically transmitted at the speed of light. The radiation emitted by molecules has a wavelength of the order of 10^{-5} cm; therefore, when molecules are separated by 10^{-5} cm or less, considerable phase-shift may occur in the radiation exchanged between them. For this reason, at distances of $\simeq 10^{-5}$ cm, the attractive forces may vary as the inverse 8th power of R.

The inverse 7th power rule thus applies when phase shifts are negligible, that is, at separations of 8 Å or less. The inverse 8th power rule applies when phase shifts cannot be ignored, that is. at distances of > 1000 Å.

The energy of London–Van der Waals' attraction V_A, is given by

$$V_A = \frac{-Ax^{-2}}{48\pi} \tag{8.1}$$

where A is the interatomic London–Hamaker attractive force constant and x is the distance. The attractive force, F_A, is defined as – the change of energy with respect to distance, i.e.

$$F_A = \frac{-dV_A}{d(2x)} = \frac{-Ax^{-3}}{48\pi} \tag{8.2}$$

These equations are the same for constant charge and constant potential models.

One of the major problems in evaluating the colloidal interactions between membranes is that the value of the London–Hamaker constant has not been established unambiguously for cellular interactions. Thus, Albers and Overbeek [21] found a value of 4×10^{-15} erg for an emulsion of water in oil. Wilkins et al. [22] compute values between 10^{-14} and 10^{-15} erg from leukocyte flocculation kinetics. Ottewill and Wilkins [23] report a value of 6×10^{-15} ergs for an arachidic acid colloid. In contrast, Srivastava and Haydon [24] report a value of 1.6×10^{-13} erg for paraffin and water while diverse values for polystyrene in water range from 5×10^{-13} erg [25] to 7×10^{-14} erg [26]. Derjaguin et al. [27] report the high value of 2.5×10^{-12} erg for platinum wires in electrolytes. Theoretical studies [28] show that adsorbed, oriented surface layers reduce the effective value of the London–Hamaker attractive force constant. Thus, if A' is the value of the constant between two particles, and A'' is that between solvent molecules, the resultant constant for interparticle interaction, A, is reduced according to the equation.

$$A = (\sqrt{A'} - \sqrt{A''})^2 \qquad (8.3)$$

Interaction between particles and adsorbed surface material introduces another term, A''', which can further reduce A by a factor of up to 50% in the case of spherical particles. In view of the above uncertainties in the magnitude of London–Hamaker constant, various authors have assigned it rather arbitrary values in attempts to calculate net interaction forces and energies between cells. However, Haydon and Taylor [29] have estimated A using 'black films' of glycerol mono-oleate in distilled water and obtained a value of 5.6×10^{-14} erg. This fits computations based on the infrared reflectivity of water [30] which yield a range of $3.4–6.8 \times 10^{-14}$ erg. The value of 5.6×10^{-14} erg thus appears to be a reasonable one for biomembranes.

8.1.3. Electrostatic repulsions

8.1.3.1. The sources of surface charge

Fixed charges. The surfaces of all cells examined bear ionogenic groups. In animal cells these groups produce a net negative charge at physiological pH. Potentially charged residues known to exist at cell surfaces, and also at the surfaces of subcellular organelles, are listed in Table 8.1. Probably the largest contribution on mass basis is due to ionizable residues of membrane proteins.

Some idea of the proportion of ionogenic amino acids in membranes can be gained from amino acid analyses, but such data do not inform about the location of charged amino acids at a given membrane surface. However, thermodynamic principles dictate that polar and charged amino acid residues would occupy only two types of location, namely (a) the external or internal membrane surface and (b) possible polar channels penetrating through membranes. The proportion of potentially ionogenic amino acids in plasma membranes lies near 34–36%, of which 52–71% are potentially anionogenic (Asp, Glu, Tyr). However, Asp and Glu may be largely amidated, reducing the proportion of potentially charged amino acids to 21–27% of which 23–48% are anionogenic.

At physiological pH, the protein carboxyls due to Glu (28–36% of potentially ionogenic groups, excluding amidation), Asp (24–25%) and C-terminal residues will be predominately deprotonated, that is, they will bear a negative charge. However, some carboxyl residues in globular proteins may exhibit anomalously high p*K*'s. (For example, there is substantial evidence for carboxylate residues in hemoglobins, which change from a p*K* of 5.25 in the deoxygenated form of the protein to a p*K* of 5.75 upon oxygenation.)

The Lys and Arg residues of membrane proteins (15–20% and 6–7% respectively, of potentially charged residues) can be expected to be fully protonated at physiological pH; that is, they will bear a positive charge. On the other hand, histidine residues (6–7%) will be only partially protonated near neutrality and the actual degree of protonation depends very much on the local pH at the membrane surface. Moreover, the association constants of histidines vary over a wide range, depending on protein structure. (This is illustrated in the case of hemoglobins where the pKs of certain residues, probably histidines, vary from 7.93 in the oxygenated state to 6.68 in the deoxygenated form.) However, the proportion of histidines is very small. Tyrosine – OH groups (~10% of chargeable residues) most probably bear no net charges at neutral pH. They titrate only at alkaline pH. N-terminal amino groups will also not be fully protonated. The proportion of these groups is also small.

As far as the lipids are concerned, except for phosphatidic acid, the phosphate residues can be expected to be negatively charged at all but very acid pH's and the choline residues of lecithin and sphingomyelin can be anticipated to bear a positive charge over virtually the entire pH range. Accordingly, the head groups of phosphatidylcholine and sphingomyelin behave as dipolar ions and contribute no *net* charge. On the other hand, the amino group of phosphatidylethanolamine is a relatively weak base and is thus not completely protonated at neutral pH; phosphatidylserine can contribute negative charge due to its carboxyl group, and because its amino group is a relatively weak base. Phosphatidic acid is negatively charged at physiological pH; its net charge increases from -1 at pH 7, to -2 at pH >8. Diphosphatidylglycerol, a prominent component of mitochondrial and nuclear membranes, contributes two negative phosphate charges per molecule. Phosphatidylionositol also contributes negative charges.

There is some evidence [31] that the surfaces of some cells bear a negative charge due to RNA phosphate residues which are exposed at the external surfaces.

A very important surface-ionogenic group is the carboxyl residue of sialic acid, whether present in sialoglycoproteins (Chapter 2) or acid sphingoglycolipids (Chapter 4). The contribution of sialic acid to the charge characteristics of tumor cell surfaces has been a particularly active area of investigation and will be discussed in detail further on.

Adsorbed charges. The ionogenic groups discussed above constitute *fixed* charges, that is, they are tightly associated with the membrane structure. Another source of surface charge derives from the unequal adsorption of anions and cations on apolar surfaces. Anions tend to be absorbed preferentially, probably because they are usually less hydrated than cations. It is for this reason that droplets of triglycerides suspended in neutral salt solutions, for example, isotonic sodium chloride, exhibit a net negative charge. However, this phenomenon is probably not very prominent in cellular membranes. This is because the lipid elements which are exposed to the aqueous phase are either polar and/or charged. Moreover, it is highly probable that the portions of membrane proteins which contact the aqueous phase are also predominantly polar in character; the apolar residues will be buried either in the cores of globular proteins or in contact with apolar regions of the membrane phosphatides. That these expectations are probably correct has been demonstrated by Seaman and Cook [32]. These authors aldehyde-fixed erythrocytes, to eliminate positive charges due to lysyl ε-amino groups, arginyl side chains and other amino residues, and then converted ionized carboxyl groups to carboxyl methyl esters to eliminate the negative charges contributed by these residues. Such modified erythrocytes are isoelectric in sodium chloride solutions between pH 6 and 8, indicating *no* preferential absorption of chloride over sodium ion. These observations and similar studies on *A. aerogenes* [33] indicate that the charge characteristics of cells arise predominantly from ionogenic residues fixed to the membrane surfaces.

8.1.4. Estimation of surface charge

8.1.4.1. Cell electrophoresis

Various methods have been used to detect or demonstrate the existance of negative charges on cell surfaces, but only one, namely *cell electrophoresis*, allows any degree of quantitation, or permits the evaluation of isoelectric points.

In cell electrophoresis, cells suspended in a buffer are placed in an electrical potential gradient. Because of the potential gradient, a direct current passes through the suspension and the cells move in the potential gradient, depending on their surface charge. The velocity of migration is generally measured with respect to the cathode and is conventionally expressed in terms of μm/sec/V/cm.

The interface between a migrating charged cell and the bulk phase is not that of the membrane proper, but lies somewhere within a *diffuse ionic double layer* (Fig. 8.1).

In the simplest case, the innermost part of this double layer can be considered to be the interface between the charged membrane and its environment, whereas the outer boundary of the double layer consists of a gradual merging of the layer with the bulk phase (e.g. [34–38]).

The effective thickness of the diffuse double layer is given by $1/\kappa$, the Debye–Hückel parameter; this has the dimension of length. The Debye–Hückel reciprocal

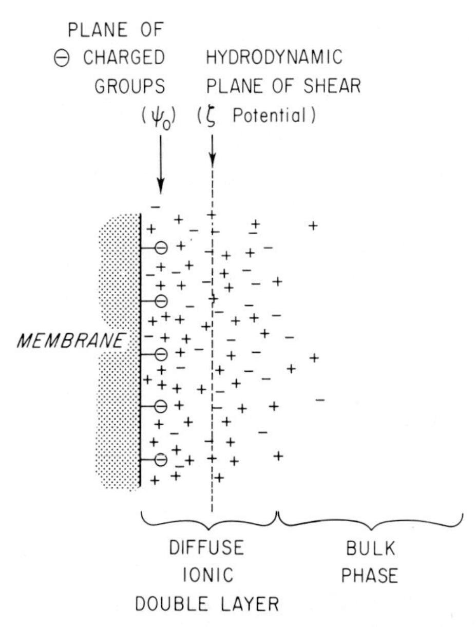

Fig. 8.1. Schematic of the diffuse ionic double layer showing electrophoretic plane of shear.

length, κ is defined as

$$\kappa = \left(\frac{8\pi\varepsilon^2}{DkT}\right)^{\frac{1}{2}} \cdot N^{\frac{1}{2}} \tag{8.4}$$

where N = the number of ions/cm^3, ε = the electronic charge (4.774×10^{-10} esu), D = the bulk dielectric constant of water, k = Boltzmann's constant and T = in °K. kT, the thermal energy unit = 4.14×10^{-14} erg (gm · cm^2 · sec^{-2}) at 25 °C.

For surfaces where the surface potential, ψ_0 (mV), is less than ~25 mV, $1/\kappa$ is that distance, x, where the potential, ψ_x equals $1/e$ of ψ_0, the potential at the surface proper, i.e.

$$\psi_x = \psi_0 \exp.(-\kappa x) \tag{8.5}$$

When a cell migrates in a potential gradient, it carries part of the diffuse electrically double layer with it; the 'plane' between that portion of the double layer moving with the cell and that of the bulk phase is known as the *hydrodynamic plane of shear*. The location of this shear plane clearly depends upon $1/\kappa$, that is, upon ionic strength. Electrophoretic mobility measurements thus really reflect the potential at the shear plane, a potential which comprises contribution from both the surface potential proper of the cell and from the ionic double layer between the bound charges and hydrodynamic plane of shear. However, measurements of cellular electrophoretic mobility allow one to make some estimate of the potential at the cell surface and also of the average net surface charge density.

An extensive body of practical and theoretical literature deals with factors involved in evaluations of surface potential and surface charge density. We will not treat this here, since it has been adequately reviewed (see, for example [39]), and will instead present some basic principles and point out some major areas of uncertainty.

The electrophoretic mobility U/X of animal cells is related to ζ, the zeta potential, by Helmholtz–Smoluchowski equation

$$\frac{U}{X} = \frac{\zeta D}{4\pi\eta} \tag{8.6}$$

where U = the electrophoretic velocity ($\text{cm} \cdot \text{sec}^{-1}$), X = the strength of the applied field ($\text{V} \cdot \text{cm}^{-1}$), D = the dielectric constant of the suspending fluid, and η = the viscosity ($\text{gm} \cdot \text{cm}^{-1} \cdot \text{sec}^{-1}$).

Surface charge density can be estimated from zeta potentials in the following way:

For spherical cells with radius, a, and zeta potentials <25 mV, the surface charge density, σ, is given by

$$\sigma = \frac{D\zeta}{4\pi}\left(\frac{1 + \kappa a}{a}\right). \tag{8.7}$$

In cases where $\kappa a > 100$ this approximates to

$$\sigma = \frac{\zeta D}{4\pi\eta} \cdot \kappa$$

Then U/X becomes

$$\frac{U}{X} = \sigma/\eta\kappa. \tag{8.8}$$

For spherical cells with zeta potentials >25 mV, in the presence of polyvalent electrolytes, where $\kappa a > 100$.

$$\sigma = \left[\frac{\eta D\,\kappa T}{2000\pi}\right]^{\frac{1}{2}} - [\Sigma\, c_i \exp(-z_i \varepsilon\zeta/kT) - 1]^{\frac{1}{2}} \tag{8.9}$$

Where N = *Avogadro's* number, k = Boltzmann's constant, T = in °K, c_i = molar concentration of ions, z_i = ion valency and ε = the electronic charge.

The applications of the Helmholtz–Smoluchowski relationship and attempts to derive values of surface charge density from electrophoretic measurements are severely complicated by several uncertainties, including the following:

First: The conductivity of the cell surface is generally equated with that of the bulk phase. This is probably reasonable because membrane surfaces are polar and are not likely to exhibit preferential anion adsorption. Moreover, for anion absorption to produce a significant increase in surface-conductivity, the environmental concentration of univalent electrolyte would have to be less than 10^{-3} M. It is also assumed that there are no significant ionic 'pockets' in the surface membrane. This again, appears to be reasonable.

Second: The dielectric constant and viscosity of the diffuse double layer are assumed to be identical to those of the bulk phase. This is not totally justified, since, as discussed by Drost–Hansen [40] water is highly structured near the membrane surface. Drost–Hansen points out that the properties of water show anomalities which may extend over 100 nm from the cell surface. Accordingly, the dielectric constant may vary from approximately 10 near a membrane surface interphase to 30 further out.

Third: The electric field of the double layer and the applied electric field are taken to be additive, despite the distortion of the applied field by the particle. This assumption probably does not create major difficulties, since potentials are generally additive.

Fourth: The product of the particle radius or curvature, a (in cm) and the Debye–Hückel constant, κ (in cm^{-1}) should give a large dimensionless number; that is, κa should be greater than 100. In physiological buffers, $1/\kappa$ approximates 10 Å, that is 10^{-7} cm, and radii of curvature down to 10^{-5} cm will give values of $\kappa a > 100$. But what is important, are the radii of curvature of the *charge-bearing sites*; these are generally not known. Certainly, some cells, for example, erythrocytes, appear to have very smooth surfaces down to less than 10^{-7} cm. However, other cells, particularly lymphocytes may bear numerous protruding microvilli, whose tips have radii of curvature often less than 10^{-5} cm. The molecular geometry of some cell surfaces may therefore not permit the direct use of the Helmholtz–Smoluchowski equation. Possibly, cell surfaces could be considered as consisting of many charge-bearing patches with low radii of curvature.

Henry [41] has proposed a corrected formula for cases where $100 > \kappa a > 0.1$, i.e.

$$\frac{U}{X} = \frac{\zeta D f(\kappa a)}{6\pi\eta} \tag{8.10}$$

In the limiting case where $\kappa a \rightarrow \infty$ (i.e. when the double layer is very thin compared with particle radius), $f(\kappa a) = 3/2$ and Henry's equation reduces to that of Helmholtz–Smoluchowski. When $\kappa a \rightarrow 0, f(\kappa a) = 1$ and the Helmholtz–Smoluchowski equation yields values 2/3 of those computed by Henry's equation. Clearly the effect of radius of curvature depends on ionic strength. At physiological levels, where $1/\kappa < a$, the hydrodynamic shear plane will be essentially smooth and, as pointed out originally by Glaeser [42], the gross radius of curvature will then be the most relevant quantity in electrophoretic measurements.

8.1.4.2. Electrophoretic identification of potentially ionogenetic cell surface groups

8.1.3.2.1. Amino groups

The presence of amino groups at the electrophoretic shear plane has been inferred from the effects of lower aldehydes on the electrophoretic mobilities and pH-mobility profiles of treated cells [43]. A preferable approach is that of Mehrishi [44], who used citraconic anhydride or 2,3-dimethylmaleic anhydride to reversibly block surface amino groups. The reaction for 2,3-dimethylmaleic anhydride is as follows:

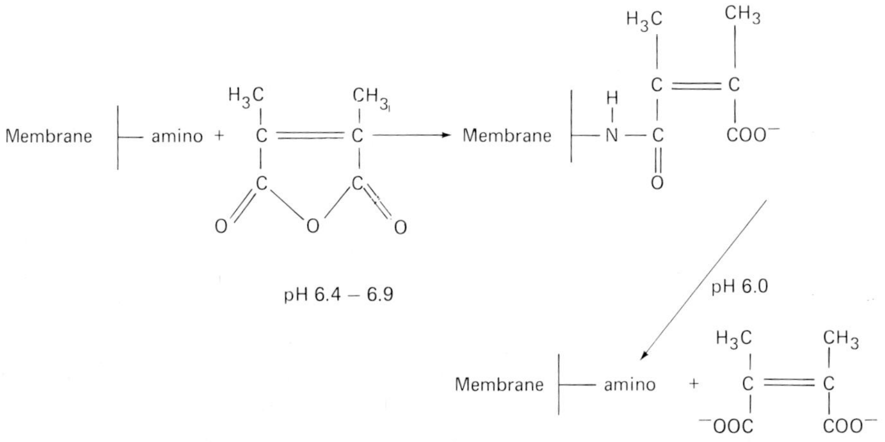

Membrane ⊢ amino + 2,3-dimethylmaleic anhydride (H_3C–C=C–CH_3, cyclic –C(=O)–O–C(=O)–) → Membrane ⊢ N(H)–C(=O)–C(H_3C)=C(CH_3)–COO^-

pH 6.4 – 6.9

pH 6.0

Membrane ⊢ amino + ^-OOC–C(H_3C)=C(CH_3)–COO^-

that is, a positively-charged group is replaced by a negatively-charged one, giving an increased electrophoretic mobility (Table 8.2). The adduct is dissociable by brief exposure to pH 5–6.

SH Groups might also react, but the adducts should be stable at pH 5–6. Since experiments on various cell types show complete reversibility of enhanced electrophoretic mobility, Mehrishi [44] argues that the contribution of reacted SH groups (if any) is negligible.

In recent years, a whole series of reagents have been developed for purposes of radiolabelling amino residues located at the external surfaces of cells. Most of these will convert positively charged amino groups on cell surfaces to negatively charged adducts (cf. review [45]). Examples are sulfonic acid derivatives such as:

diazonium sulfanilate

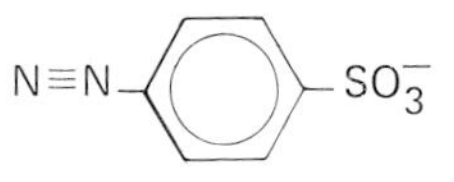

$N{\equiv}N$–C_6H_4–SO_3^-

4–isocyano–stilbene–2, 2′–disulfonate

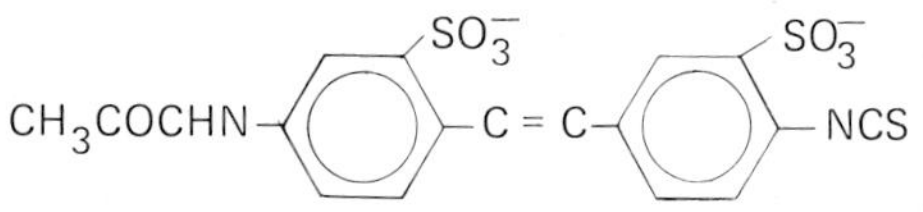

CH_3COCHN–$C_6H_3(SO_3^-)$–C = C–$C_6H_3(SO_3^-)$–NCS

trinitrobenzene sulfonate

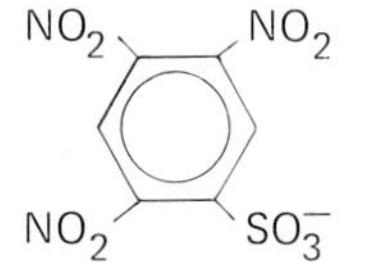

NO_2 NO_2

NO_2 SO_3^- and

formylmethionyl sulfone methyl phosphate

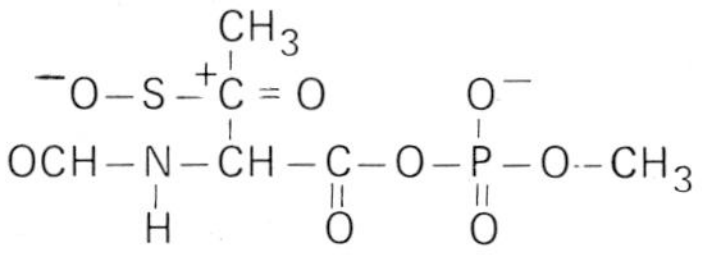

CH_3

$^-O-S-^+C=O$ O^-

$OCH-N-CH-C-O-P-O-CH_3$

H O O

Also pyridoxal phosphate

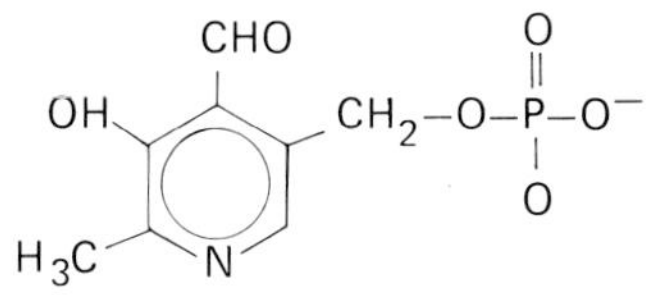

CHO O

OH $CH_2-O-P-O^-$

O

H_3C N

forms Schiff bases with free protein amino groups, generating two negative charges.

Although all of these reagents are used as either fluorescent or radioactive labels, and might thus allow quantification of reacting groups, we know of no study correlating labelling with alterations in cellular electrophoretic properties.

TABLE 8.2

Modification of positively charged surface groups of Ehrlich ascites carcinoma cells and mouse peripheral lymphocytes[1]

Cells[2]	Electrophoretic mobility ($\mu m \cdot sec^{-1} \cdot V \cdot cm^{-1}$)	Zeta potential (mV)	Number of positive charges/cell
Ehrlich ascites carcinoma			
untreated	1.05	13.9	
+ CA	1.46	18.8	1.3×10^7
+ DMA	1.42	18.2	1.1×10^7
+ NEM	1.42	18.2	1.1×10^7
+ DMA; pH 6[3]	1.08	13.9	
Lymphocytes			
untreated	1.05	13.5	
+ CA	1.28	16.4	9.5×10^5
+ DMA	1.28	16.4	9.5×10^5
+ DMA; pH 6[3]	1.08	13.9	

[1] From [44].
[2] CA = citraconic anhydride; DMA = 2,3-dimethyl maleic anhydride; NEM = *N*-ethyl maleimide.
[3] The complex of DMA with positive groups is reversed at pH 6.

8.1.4.2.2. Carboxyl groups

General. A general reaction employed for membrane carboxyl groups has been treatment with diazomethane (e.g. [33]), i.e.

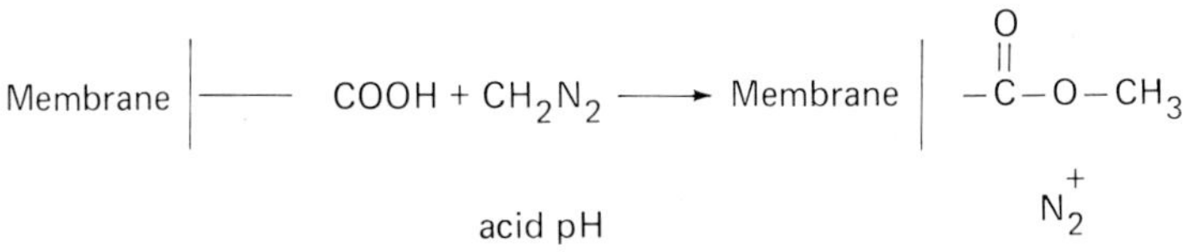

$$\text{Membrane}\Big|\!-\!COOH + CH_2N_2 \xrightarrow[\text{acid pH}]{} \text{Membrane}\Big|\!-\!\overset{\displaystyle O}{\overset{\|}{C}}\!-\!O\!-\!CH_3 + N_2$$

However, this method requires drastic denaturing conditions.

Sialic acid. One of the dominant ionogenic groups at cell surfaces is *N*-acetyl neuraminic acid.

[Structural formula of N-acetylneuraminic acid in α-glycosidic linkage: ring carbons bearing H, H; OH, H; C–N(H)–C(=O)–CH$_3$; H–C bearing side chain HCOH–HCOH–HOCH$_2$; ring O; anomeric carbon with α–O– and carboxylate C(=O)O$^-$]

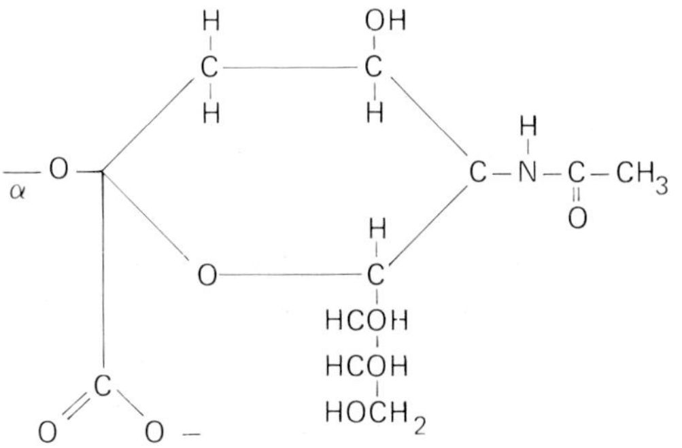

This is bound in α-glycosidic linkage to other sugars of membrane glycoproteins or glycolipids (Chapters 2 and 4).

Much of the neuraminic acid of intact cells can be cleaved off by neuraminidase without impairment of viability. The resulting loss of ⊖ charge reduces electrophoretic mobility.

Phosphate. Certain cells bear RNAase susceptible phosphate linkages on their surfaces. RNAase treatment thus reduces electrophoretic mobility.

Tyrosines. Much recent experimentation indicates that tyrosine OH– groups are exposed at the external surfaces of plasma membranes (cf. reviews in [45,46]). The evidence rests on the procedure of Phillips and Morrison [47] which employs lactoperoxidase, in the presence of H_2O_2 and $Na^{125}I$, to iodinate tyrosyl residues. Since the enzyme cannot permeate the plasma membrane, only externally located residues are labelled. The procedure has been successfully applied to numerous cell types, but we know of no systematic attempt to evaluate the surface density of tyrosyl residues. In any event, such groups are extremely unlikely to contribute charges because of their low proportion and high p*K*.

Histidines. At pH 7.2, *N*-ethylmaleimide should combine with surface histidines, but Mehrishi (e.g. [48]) finds no effect of this reagent on the electrophoretic mobility of lymphocytes and certain tumor cells. This is not unexpected, since histidines constitute only 6–7% of available charged residues and are in any case expected to be only partly protonated at physiological bulk pH.

–SH groups. Surface SH groups do not bear a charge, but can be reversibly coupled to yield a ⊖ charge and increased negative electrophoretic mobility [49]. The procedure uses a 6,6′-dithionicotinic acid (DTDNA) as follows.

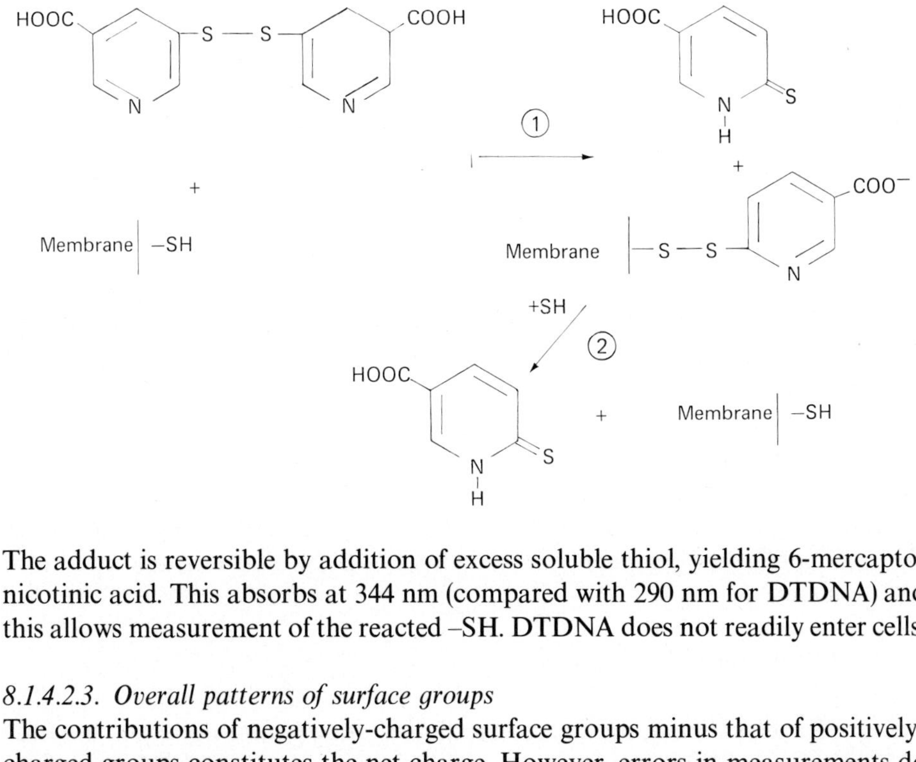

The adduct is reversible by addition of excess soluble thiol, yielding 6-mercapto-nicotinic acid. This absorbs at 344 nm (compared with 290 nm for DTDNA) and this allows measurement of the reacted –SH. DTDNA does not readily enter cells.

8.1.4.2.3. Overall patterns of surface groups

The contributions of negatively-charged surface groups minus that of positively-charged groups constitutes the net charge. However, errors in measurements do not allow a precision >10%. This could mean that small, but anomalously migrating subpopulations of a cell suspension may escape detection. Also, there are few data concerning the distribution of ionogenic groups, except for the case of sialic acid.

However, Mehrishi [48, 50] has taken available data and has computed surface distribution of certain groups for either cubic or hexagonal lattices. Both packing patterns yield similar values (Table 8.3), but neglect the probability of clustering. Mehrishi's [48, 50] computation of the numbers and mean distributions of diverse surface groups on lymphocytes and Ehrlich ascites carcinoma cells are interesting. These cell types showed no significant differences in the densities of anionogenic sites, other than RNA phosphates, but differ markedly in terms of positively charged groups and –SH residues.

Although this type of 'balance sheet' is highly desirable and, in fact, essential for any meaningful evaluation of possible differences in the surface properties of normal and neoplastic cells, it is admittedly based on inadequate information. The calculations need to include evaluations of charge distribution, surface topology and effective radii of curvature (e.g., role of microvilli). Techniques to achieve these refinements are available but have not been employed.

Another problem derives from the fact that not all ionogenic groups on a cell surface actually contribute to electrophoretic mobility. This has been well illustrated for the case of neuraminic acid, which contributes importantly to the electrophoretic mobility of many cells. However, only few cell types (erythrocytes, Ehrlich ascites tumor cells) show a linear relationship between the amount of neuraminic acid removed by neuraminidase and anodic electrophoretic mobility [51, 52]. In other cell types, for example, MC1M sarcoma cells [52] only some of the neuraminic acid accessible to neuraminidase contributes to the electrokinetic potential. The same is true for normal and neoplastic hepatocytes [53, 54].

Finally, correlation of the change of electrophoretic mobility produced by neuraminidase action, with the amount of charge removed per cell (by chemical assay) requires some dubious assumptions. Indeed, even in the case of erythrocytes [55] and Ehrlich ascites carcinoma [56], less change of charge is detected electrophoretically after neuraminidase action than expected from chemical assays of released neuraminic acid. Available information thus indicates that the neuraminic acid residues on a given cell type occupy various positions relative to the hydrodynamic shear plan and that different cell types vary in this respect [52]. 'Activation' of other ionogenic groups by removal of neuraminic acid must also be considered.

8.1.4.3. pH-Mobility relationships

Figure 8.2 schematically shows the variation of anodic electrophoretic mobility with pH that is typical for most cells other than erythrocytes [48]. Most cells show no change in electrophoretic mobility between pH 6.5 and 9.5. The increase in anodic mobility above pH 9.5 can be attributed to the titration of amino groups. Except in erythrocytes, there is a continuous decrease in electrophoretic mobility with pH below 6.5. This reflects titration of carboxyl groups predominantly, although at pH ~6–7, histidines and phosphates may also contribute. Neuraminic acid (pK ~2.6) titrates at very low pH. Nucleated cells are generally

TABLE 8.3

Cell surface groups detected electrokinetically[1]

Group or charge	Lymphocyte			Ehrlich ascites carcinoma		
	Number per cell ($\times 10^5$)	Area per group/charge $(Å)^2$	Average inter-group/distance (Å)	Number per cell ($\times 10^5$)	Area per group/charge $(Å)^2$	Average inter-group/distance (Å)
Groups contributing charge	102.9			800		
Positive charges	9.5	11,900	109	117	7,700	97.6
Negative charges	93.4			688		
Sialic acid	29.2	3,880	62	235	3,830	62
Phosphate-RNAase susceptible	8.7	13,000	114	119	7,550	87
Unidentified anionogenic	55.5	2,050	45	391	2,300	48
–SH groups	19.8	5,760	76	369	2,440	49.4

[1] From Mehrishi [48].

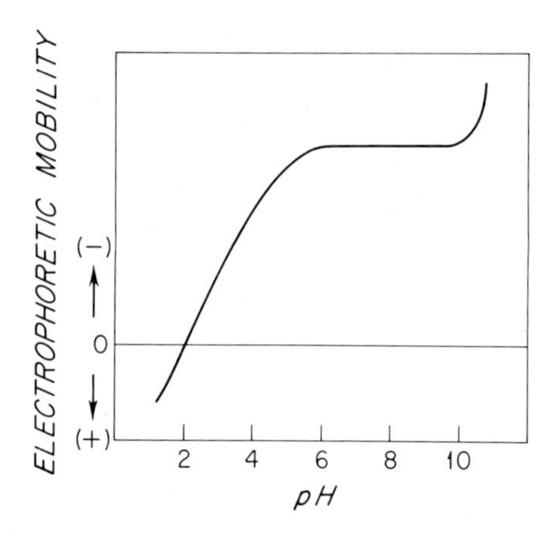

Fig. 8.2. Variation of electrophoretic mobility typical of most animal cells.

'isoelectric' between pH 2 and 4. Notably, erythrocytes do not show any change in anodic mobility between pH 3 and 10, although they are known to bear protein carboxyls, protein amino groups and tyrosyl residues on their surfaces. Clearly, diverse cells can vary in the degree to which various ionogenic groups contribute to electrophoretic mobility.

8.1.4.4. Surface pH

The pH in the vicinity of a membrane surface may deviate markedly from that of the bulk phase. This is because the presence of fixed negatively charged groups on a membrane induces a relative deficiency of mobile anions in the vicinity of the charged groups and an excess of mobile cations including protons; that is, the pH in the immediate vicinity of the membrane surface will lie below that of the bulk phase.

The possible consequences of this phenomenon can be appreciated from a comparison of the behavior of enzymes in free solution with that of the same enzymes coupled to supports which are charged by virtue of fixed ionizable groups. Goldstein et al. [57] have developed a quantitative treatment for this kind of situation. They show that if the supporting polyelectrolyte is positively charged, the hydrogen ion concentration in the vicinity of the coupled enzyme will be lower than in the bulk phase. In contrast, if the polyelectrolyte is negatively charged – the case relevant to the cell surface – the local hydrogen ion concentration will be higher. The pH activity curve of enzymes attached to charged supports is therefore displaced to higher bulk pH values when the support is negatively charged.

The quantitative treatment of these effects is as follows: suppose that the support membrane bears a negative charge, which creates a negative potential, ψ.

Then the ratio of the concentrations of hydrogen ions in the support $[H^+]'$, is related to that in the bulk phase $[H^+]$, by the Boltzmann factor

$$\frac{[H^+]'}{[H^+]} = e^{-\varepsilon\psi/kT} \tag{8.11}$$

where ε is the charge on the proton and e the base of the natural logarithm.

Since for free solution

$$pH = -\log_{10}[H^+] \tag{8.12}$$

and for the support

$$pH' = -\log_{10}[H^+]' \tag{8.13}$$

it follows that

$$\Delta pH \equiv pH' - pH = \frac{\varepsilon\psi}{2.303\,kT} \tag{8.14}$$

The factor 2.303 is for the conversion from natural to common logarithms.

As a consequence of these effects, large differences between bulk and surface pH can occur. A clue to the magnitude of these effects is given by experiments where enzymes are used as 'molecular pH meters' (Fig. 8.3). Thus, when trypsin is covalently coupled to a negatively charged support, the bulk pH for maximal tryptic activity shifts by as much as 2.5 pH units towards more alkaline values from the pH optimum of the free enzyme. This is true only at low ionic strength; as ionic strength is increased, the pH activity curve shifts towards less alkaline pH values and gradually approaches the value observed for trypsin in solution.

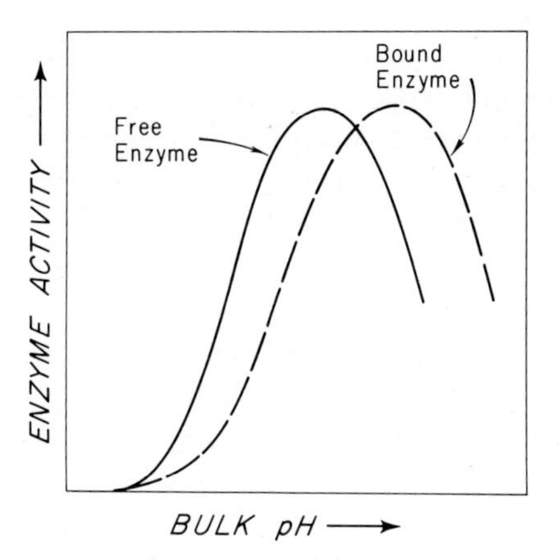

Fig. 8.3. Different pH-activity profiles of free trypsin and trypsin bound to a negative charged substrate. After [57].

This indicates that the effects involved are indeed electrostatic (Table 8.4).

In analogous experiments, Goldman et al. [58] have shown that the local pH of papain coupled to colloid ion membranes (which bear a negative charge) is also considerably lower than that of the bulk phase.

TABLE 8.4

Effect of ionic strength on bulk pH for maximal activity of trypsin bound to a polyanionic carrier[1]

Ionic strength	Bulk pH for maximal enzyme activity
0.006	10.4
0.010	10.0
0.035	9.6
0.200	9.3
1.000	8.4

[1] From [57].

There is rather little experimental information about the local pH at the surfaces of biomembranes. This is to a large extent due to the ambiguities involved in estimating ψ from ζ, insufficient information about surface-located enzymes, which can also be studied in solution, and lack of experience in coupling soluble enzymes to cell surfaces. However, McLaren [59] has compared the pH vs. activity curve of yeast invertase attached to cells with the pH activity curve of the enzyme free in solution. His data suggest that, when bound to the negatively charged surfaces of the yeast cells, the invertase functions at a pH lower than that of the bulk phase. Of course, there is difficulty in properly localizing an enzyme with respect to the electrophoretic shear-plane of cells, but one can anticipate that the low-pH region might extend 10 Å from the membrane surface proper. This topic clearly requires further study, using, for example, erythrocyte acetylcholinesterase and surface glycosyltransferases (p. 382).

Another complication involving pH derives from the fact that the metabolism of most cells yields some non-volatile acids. These diffuse from the cells to the draining capillaries and lymph channels. Their concentration will, therefore, be higher at the cell surface than in the bulk phase and can cause a locally depressed pH. This effect can be appreciable ($\sim$1.5 pH units) in cells, such as many tumor cells, with high aerobic lactic acid production [60].

8.1.4.5. Effects of heavy metals and polycations

Addition of various heavy metals reverses the negative electrophoretic mobilities of cells in the order Th > UO_2 > La > Cu, Ni > Ca, Sr, Ba, Mn, and this has been ascribed to phosphate complexing. However, as noted, the negative surface charge of most cells arises primarily from carboxyl groups especially those *N*-acetylneuraminate. Polycations, such as polyornithine and polylysine also alter electrophoretic mobility. Thus, Merishi [61] has studied the interaction of some basic polypeptides with Ehrlich ascites carcinoma cells. Both polylysine (mol. wt. 90,000) and polyornithine (mol. wt. 175,000) adsorb strongly to the cells and, depending on the degree of absorption, can change electrophoretic mobility to zero or even to positive. This effect can be reversed with a polyanion such as heparin. Protamine sulfate (which is 65% in arginine) also reduccs electrophoretic mobility, but is far less efficient than the polylysines. The maximum reduction of electrophoretic mobility is about 27%. Merishi argues that the structural properties of the polyornithine and polylysine make these much more effective in neutralizing and/or reversing the charges on the tumor cell membranes.

Because of their effects on electrostatic repulsions, various polyvalent inorganic cations (Table 8.5), as well as polyornithine and polylysine, can induce cell agglutination. The agglutination effect of polycations depends as expected on ionic strength; in addition, the effects of heavy metals increase with pH. Polycarboxylic acids block polycation agglutination of erythrocytes, as do plasma proteins. The former compete for membrane carboxyls, the latter appear to act in a more complex manner. Polyornithine and polylysine (mol. wt. $\sim 10^5$) produce erythrocyte agglutination at levels of $\sim 10^{-8}$ M.

TABLE 8.5

Cation induced agglutination of human erythocytes[1,2]

Cation	M of cation required for erythocyte agglutination in 0.15 M NaCl
Ca^{2+}, Cu^{2+}, Th^{4+}, Ti^{3+}	0.1–1.0
Al^{3+}	0.1–1.0
Fe^{3+}, Cr^{3+}, Sn^{4+}, Zn^{2+}	1.0–5.0
Rb^{2+}	5.0–10.0
Ag^{+}, HAu^{4+}	10.0–50.0
Cd^{2+}	50.0–100.0
Ni^{2+}	50.0–100.0
Mn^{2+}, Co^{2+}	100.0–500.0
Hg^{2+}	100.0–500.0

[1] Data from Jandl and Simmons [62].
[2] Silver, gold, cadmium and mercury produced strong hemolysis under conditions of minimal agglutination.

To summarize, measurements of cellular electrophoretic mobility, when performed under strictly controlled conditions of metabolism and of environmental pH, temperature, viscosity, ionic strength, ionic, and macromolecular composition allow certain comparisons. They do not, however, allow unambiguous conclusions, particularly regarding surface charge and surface potential. Moreover, comparisons of differing cell types, for example, normal vs. neoplastic, by electrophoretic mobility measurements, must be interpreted most cautiously. However, experimental manipulations such as variation of pH, enzymatic removal of specific ionogenic groups and chemical modifications of ionogenic groups can prove very useful in defining the charge characteristics of cell surfaces.

8.1.4.6. Energies and forces of electrostatic repulsion

The original theory on the stability of colloidal dispersions [19,20] does not really apply to biomembranes, because it treats cases where the ions responsible for the surface charge are in thermodynamic equilibrium between the surface and the bulk solution. Then, interaction between closely approaching particles occurs at a *constant electrostatic surface* potential and at variable surface charge densities. However, the surfaces of cell membranes bear changes *fixed* to membrane protein, lipid, and carbohydrate. Indeed, as shown on p. 351, absorbed ionic charges appear to play a very minor part in determining surface potential of cells under physiologic conditions. Gingell [63] has therefore developed the theory for the case of *constant surface charge* (Fig. 8.4).

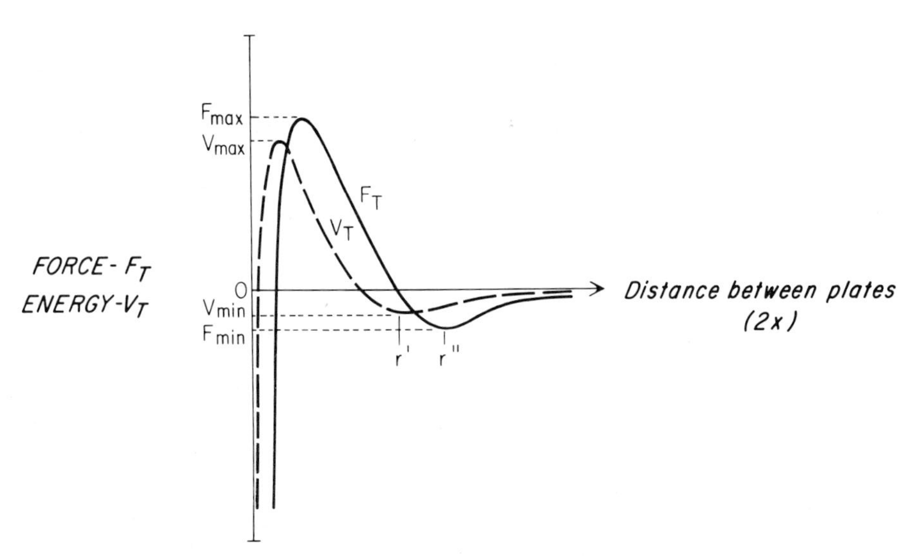

Fig. 8.4. Forces and energies of attraction between parallel plates with fixed charge densities. After Gingell [63].

The repulsive force is defined as the negative change of energy with distance, that is, $-dV_T/d(2x)$ where V_T = total repulsive interaction energy and $2x$ = separation distance. Conversely $V_T = -2\int_{\infty}^{r/2} F_T dx$, that is, the work done moving from ∞ to $2x = r$, where F_T and V_T are total force and total energy at $2x = r$. $-dV_T/d(2x)$ is zero (i.e., force is zero) at maximal and minimal value of V_T.

For approaching parallel, planar ionic double layers one can express the repulsive force in terms of the potential, ψ_d, mid-way between the plates, without specifying whether the surface potential or charge is constant during interaction. The equation valid for both cases is [63]

$$F_R = 2nkT\left(\cosh\frac{\varepsilon\psi_d}{\kappa T} - 1\right) \tag{8.15}$$

where n = number of ions/cm^3, and κ = Debye–Hückel reciprocal length. ψ_0, the surface potential, is a known function of x when $\psi_0 < 25$ mv [64], namely

$$\psi_0 = 4\pi\sigma/\kappa D(\theta \tanh \kappa\theta t + \tanh \kappa x) \tag{8.16}$$

where $\theta = (1 - \alpha)^{\frac{1}{2}}$.

$1 - \alpha$ = fractional volume inside the surface layer available to counter ions [65];

σ = surface charge density, with correct sign;

t = thickness of the layer.

A particular solution of the differential equation in the region between the charged planes of the approaching double layer shows that

$$\psi = \psi_0 \cosh \kappa(x - 1) \operatorname{sech} \kappa x. \tag{8.17}$$

where ψ is potential at l and l is any region between the charged planes of the approaching double layers. Setting $l = x$, where x is the mid-point between the charged planes, yields $\psi_d = \psi_0 \operatorname{sech} \kappa x$.

From equations 8.15–8.17 the repulsive force for low potentials therefore becomes

$$F_R = 2nkT\left\{\cosh\left[\frac{e}{kT}\cdot\frac{4\pi\sigma \operatorname{sech} \kappa x}{\kappa D(\theta \tanh \kappa\theta t + \tanh \kappa x)}\right] - 1\right\} \tag{8.18}$$

The repulsive energy at separation ξ is then defined as

$$V_R = \int_{2x=\infty}^{2x=\xi} F_R d(2x) = -2\int_{x=\infty}^{x=\xi/2} F_R dx \tag{8.19}$$

The expression for interaction at *variable surface charge* [20] are as follows for cases where the interactions are weak and $\kappa x \simeq 1$. The repulsive energy, V_R, is

$$V_R = \frac{64nkT\gamma^2}{\kappa}\exp(-2\kappa x) \tag{8.20}$$

where

$$\gamma = \frac{\exp(z\varepsilon\psi_0/2kT) - 1}{\exp(z\varepsilon\psi_0/2kT) + 1} \tag{8.11}$$

(ψ = surface potential [esu], ε = the electronic charge and z = the valency of symmetrical electrolyte).

The repulsive *force*, F_R, is

$$F_R = 64nkT\gamma^2 \exp(-2\kappa x). \tag{8.22}$$

8.1.5. The theory of colloid stability: balance between physical attractions and repulsions between membranes

8.1.5.1. Introduction

Tissue differentiation, regeneration and repair involve extensive movements of cells relative to each other and relative to non-cellular substrates. During such cell migrations many established cell connections are dissolved and numerous new encounters occur between cells. Some of these are transient and some lead to the development of specific functional linkages.

For a cell to detach from a neighbor requires that (a) specific connections are severed and (b) that non-specific repulsions outweigh non-specific attractions.

Analogously, establishment of new, specific connections requires that the distances between the participating membranes remain within appropriate limits for a time sufficient for the elaboration and assembly of the molecules required for the specialized contact. This condition can only be met when the non-specific forces of attraction and repulsion are in appropriate balance.

These considerations also apply to 'free floating' and migrating cells (erythrocytes, leukocytes and macrophages) whose function requires that they operate as individuals, but some of which, particularly lymphocytes, must make transient, highly specific contacts with other cells.

The theory of colloid stability is particularly appropriate to malignant tumors, whose cells are prone to detach from each other and to form only loose, transient relationships.

The colloidal theory also applies to the stability of intracellular organelles. For example, it is known that the mitochondria in numerous animal cells are rather mobile, but rarely aggregate; but what are the forces which stabilize these structures and what determines the separation of inner and outer mitochondrial membranes? Similarly, how can one account for the ordered packing of the lamellar membranes of the Golgi apparatus? How do the lysosomes which are also small mobile particles, ordinarily remain unaggregated? Also, during mitosis, when the nuclear envelope breaks into numerous small membrane vesicles, how do these remain segregated and how do they then reaggregate or rearrange into an ordered nuclear membrane?

Finally, the colloidal theory is directly relevant to the interaction of viruses with membranes. In the case of enveloped viruses, this is again a membrane-membrane interaction. This is not so in the case of, for example, small DNA tumor viruses, but colloidal interactions must still be involved. We know that in many cases virus entry requires a specific membrane receptor but colloidal interactions between the virus and the cell curface must precede attainment of the close contact involved in the reaction with specific receptors.

8.1.5.2. Collisions of small particles

Small particles, such as cells, or subcellular organelles, undergo continuous random thermal motion. Other, more directed, energy-requiring forms of motion can occur, but are superimposed on thermal motion. Particles undergoing thermal motion tend to collide in a random fashion. Whether they do collide, and whether collisions are elastic (not leading to flocculation) or inelastic (leading to aggregation) depends upon the balance between the forces of attraction and repulsion between the particles.

For disperse systems, the particles should collide randomly at a rate, dn/dt

$$dn/dt = 4\pi DRn^2 \tag{8.23}$$

where dn/dt is the number of collisons/sec, D = the diffusion coefficient in $cm^2 \cdot sec^{-1}$, R = the collision radius in cm (approximately double the particle radius) and n = the particle number/cm^3.

In the case when every collision is inelastic and removes a particle from the system, one can obtain useful expressions for the aggregation rate. Thus, integration of the above rate equation gives

$$1/n - 1/n_0 = 4\pi DRt \tag{8.24}$$

where n = the concentration of nonaggregated particles at time = t, and n_0, a constant for the system, is the particle concentration at $t = 0$.

If $t_{\frac{1}{2}}$ is the time required to reduce the particle concentration to $\frac{1}{2}$ of n_0, that is, $n = n_0/2$ at $t = t_{\frac{1}{2}}$:

$$t_{\frac{1}{2}} = 1/(4\pi DRn_0) \text{ or } t_{\frac{1}{2}} = 3\eta/4kTn_0 \tag{8.25}$$

where η is the viscosity ($gm \cdot cm^{-1} \cdot sec^{-1}$), k = Boltzmann's constant (1.38×10^{-16} erg (°C^{-1}) and T is in degrees Kelvin. For water $\eta = 0.01$ and at 25°C $t_{\frac{1}{2}} = 2 \times 10^{11}/n_0$ sec. According to this, cells suspended in aqueous buffers at concentrations of $10^8/cm^3$ should show a $t_{\frac{1}{2}}$ of <1 hr. This is not the case. In fact, erythrocyte suspensions are stable for prolonged periods of time, even at concentrations of 10^{10} cells/cm^3.

Suspensions of cells, or subcellular organelles, can be stable for long periods of time when repulsive interactions outweight attractive interactions. Then the net interaction energy, V, is repulsive and the collisions are elastic. Such stabilized systems can be described by the approximate equation

$$-dn/dt = 4\pi DRn^2 \exp(-V'/NkT) \quad (8.26)$$

where N = Avogadro's number (6.02×10^{23}). Combining and rearranging equations 8.23, 8.25 and 8.26:

$$t_{\frac{1}{2}} = (3\eta/4RkTn_0) \exp(V'/NkT) \quad (8.27)$$

Depending upon concentration, V' values of $15NkT$ to $25NkT$ are required to maintain stability for $\sim 10^6$ sec.

One must also consider the likelihood that once cells or organelles aggregate, they may again separate. One should therefore introduce a term for the redispersion e.g.

$$-dn/dt = 4\pi DRn^2 \exp(-V'/NkT) - k'[A] \quad (8.28)$$

where $[A]$ = concentration of aggregates and k' is the rate constant for redispersion.

8.1.5.3. Evaluation of the interaction of repulsive and attractive interactions

The model most appropriate to the biological situation is the fixed-charge hypothesis of Gingell [63]. Here the total energy of interaction, $V_T = V_A - V_R$, at a separation, ξ, is defined as

$$V_T = V_A - V_R = -2 \int_{x=\infty}^{x=\xi/2} (F_R + F_A)\, dx = -2 \int_{x=\infty}^{x=\xi/2} F_T\, dx \quad (8.29)$$

Now imagine two planar, parallel, ionic double layers associated with membrane surfaces, separated by a few hundred Å with 0.15 M sodium chloride as electrolyte solution (Fig. 8.4). A small mutual attraction between these two surfaces exists and this attraction increases until the separation distance nears 100 Å. With further approximation, the attractive force decreases abruptly and becomes 0 at about 90 Å separation. At this *so-called* 'secondary minimum' of potential energy, the attractive and repulsive forces are in equilibrium. This phenomenon has been invoked to explain the interaction between the cell surfaces separated by gaps of 100–200 Å, such as are commonly seen in the electron microscope [18]. If the plates are pushed closer together, a progressive increase in resistive force results. This force reaches a maximum at about 20 Å separation and then falls off rapidly to 0, to reach the so-called 'primary minimum'. The 'primary minimum' represents a very unstable position; a slight perturbation which results in closer apposition pushes the surfaces into a region of strong attraction, leading to molecular contact and very large cohesive energies. Perturbation in the opposite direction will cause large repulsive effects and will accelerate separation of the surfaces.

Since force is defined as minus the change of energy with distance, the variation of total force with plate separation produces a curve very similar to that of the energy variation but displaced to higher distances. As shown in Fig. 8.4, the force is zero at maximal and minimal values of V_T. Both force and energy asymptotically approach negative infinity as distance, $2x$, approaches zero and asymptotically approach zero as $2x$ approaches plus infinity. The primary minimum in potential energy occurs at $2x = 0$. The secondary minimum in potential energy, V_{min}, corresponds to the value of $2x = r'$. The force curve also shows a secondary minimum which occurs at $2x = r''$.

The shapes of the force and energy curves depend upon a number of critical variables. Some of these affect primarily V_R, others V_A and still others influence both.

8.1.5.3.1. Debye–Hückel parameter, $1/\kappa$

The value of $1/\kappa$ depends upon electrolyte concentration (and composition). This is likely to be stable in the physiological situation. Consequently, variations in $1/\kappa$ are not likely to be important in altered cellular interactions in neoplasia.

Curtis [5] has argued divalent ions, such as Ca^{2+}, may play an unusual role. However, the change in ionic strength due to addition of physiological levels of calcium ion (2 mM) to physiologic saline (0.145 M) introduces only negligible variations into force and energy calculations [63]. Moreover, Gingell's [63] computations show that the secondary minimum of potential energy, as well as the force required to escape from this position, are only slightly affected by a change in surface potential as large as 8 mV, which is considerably greater than the reduction in the negative zeta potential due to calcium binding in physiological media.

These considerations derive from studies on cells, for example, erythrocytes, lymphocytes and ascites tumor cells, which do not ordinarily aggregate. They do not deal with the specific binding of calcium to membranes, but this is rather low at physiological ionic strengths. Thus, intact erythrocytes bind $\sim 4 \times 10^{-12}$ moles of calcium at low ionic strength [66]. At physiological ionic strength, the value is much less than 10^{-12} moles. Mikkelsen and Wallach [67] used the rare earth cation Tb^{3+} as a calcium probe and obtained only 3×10^5 sites/cell for erythrocytes in physiologic media. These binding sites are protein in nature and involve tyrosyl OH groups as chelating partners. Similar specific binding can be demonstrated on lymphocyte membranes [68], but the number of sites is again very small. Whatever the role of calcium in cell adhesion, it can probably not be explained simply on the basis of colloidal theory.

8.1.5.3.2. Dielectric constant

Forces between membranes depend upon the dielectric properties of the medium between the membranes as well as those of the membranes themselves. The electrostatic repulsive force, V_R, varies in a complicated way with dielectric constant of the medium, but it is essentially independent of the nature of the membrane. In contrast, attractive London–Van der Waals' forces vary with the dielectric properties of both the membrane and the medium.

The membranes of normal and neoplastic cells do not differ significantly in overall composition and it is unlikely therefore that the anomalous contact relationships of tumor cells are related to differences in membrane dielectric constant. However, the interstitial fluid of tumors commonly contains anomalously high protein levels and one must enquire as to the possible significance of this phenomenon on the physics of cellular adhesion.

For the flat plate model [63] one can define the minimal force required to bring the membranes into molecular contact as F_{max}. Similarly, one can define the minimal perpendicular force required to separate plates in the 'secondary minimum' positon, as F_{min}. Assuming a reasonable value of the London–Hamaker constant, that is, $\sim 10^{-15}$ erg, F_{max} will vary essentially linearly with the dielectric constant of the bulk medium. In contrast, F_{min} will vary as the inverse square of the dielectric constant. The maximum energy of repulsion, V_{max}, varies linearly with dielectric constant (in the dielectric range of 50–200), whereas V_{min}, the attraction energy at the secondary minimum, changes as a function of somewhat less than the inverse square of dielectric constant. Finally, with a surface potential of -20 mV and a London–Hamaker constant of 10^{-15} erg, the hypothetical intercellular gap distances corresponding to $V_T = V_{min}$ will vary almost linearly with bulk dielectric constant (between 50–200) at 37 °C (~ 10 Å for each increment of 15 in dielectric constant; [63]).

Gingell [63] suggests that variations of intercellular protein might be highly significant because the static dielectric constant of serum is ~ 180 [69] compared to ~ 74 for physiologic buffers at 37 °C. Using this fact, he argues that the force required to bring the surfaces into molecular contact, F_{max}, in serum would be only about twice that in saline, while the force required to separate membranes at the secondary minimum, F_{min}, would be approximately one sixth the value in aqueous buffers. However, this contention is not justified since the dielectric constant relevant to colloidal theory is not the DC value, but the high frequency AC value; this cannot be greater than the static value of pure water.

The possible role of medium dielectric constant in cellular adhesion has been clarified by Jones [70]. Jones finds that at frequencies ≤ 1 MHz, the dielectric constants of solutions of dextran, Ficoll and D-sorbitol do not vary with concentration (86–87, between 1 and 5% w/v) and are only slightly in excess of that of water (78.4). These data are consistent with those of Brooks and Seaman [71] who also found that the dielectric constants of dextran solutions are indistinguishable from those of water. However, solutions of glycine and diglycine produce substantial increases of dielectric constant (89.31 and 95.01, respectively at 0.1 M concentration, 500 MHz and 25 °C). Jones has added various proportions of glycine and diglycine to tissue cultures media and measured their effect on the reaggregation of embryonic chick neural-retina or limb-bud cells. The limb-bud cells show greater adhesive interaction energies than the neural retina cells, but, significantly, both cell types exhibit a regular increase in cell aggregation as the medium dielectric constant is raised by addition of glycine or diglycine. Jones interprets his data to indicate that the dielectric properties of the medium affect both the repulsive and the attractive interactions between membranes. Unfortunately, he does not exclude other actions of glycine and diglycine.

Many studies of this type are required to provide a meaningful evaluation of the application of colloidal theory to cellular interactions. However, present information suggests that alterations in bulk dielectric constants do not provide a basis for the differential behavior of normal and neoplastic cells.

8.1.5.3.3. Surface potential ψ_0

Alterations in surface potential could critically change the physical interactions between cells. ψ_0 Clearly depends on surface charge density and this may vary considerably between cells and between different physiological states of the same cell. Changes in the conformation of membrane proteins may alter the p*K*'s of exposed ionogenic groups and thus lead to an alteration in charge, density and surface potential. The role of sialic acids is also critical, because of the existence of enzymes which can add or cleave these residues from cell surface components. Generally speaking, one can expect that the higher the charge densities on cell surface areas involved in intercellular contact, the greater the repulsive force between surfaces. Very likely the ionogenic groups fixed on cell surfaces are not in a random topologic distribution, but in zones of high charge density and therefore high surface potential. This possibility is of special interest in the case of cells which contact other cells through microvilli (e.g., lymphocytes).

8.1.5.3.4. Interaction potential and mobility of charges in the membrane plane

When two membrane surfaces bearing fixed constant surface charges approach an increase in the absolute value of the membrane surface potentials should occur, unless the charge-bearing groups are laterally mobile in the plane of the membranes as is suggested by the fluid mosaic model. This possibility has been considered briefly by Gingell [72] with particular reference to membrane glycoproteins. He points to evidence that the major glycoproteins of erythrocyte membranes, which bear about 1 sialic acid residue per 17,000 Å^2, can aggregate under certain conditions [73]. (Actually, this aggregation phenomenon occurs only in isolated membranes after extraction of some stabilizing proteins and it appears that, in eukaryotic cells, lateral mobility of proteins in the plasma membrane is not a prominent phenomenon under physiologic conditions.) However, there is evidence that the proteins of some intracellular membranes, for example nuclear membranes, can move within the membrane plane and that such movement can be induced by changing the state of membrane lipids from an unordered one to a crystalline one (Chapter 1). Lateral mobility of charged membrane proteins may thus not be very significant as far as tumor cell *plasma membranes* are concerned. On the other hand, it could very markedly alter the electrostatic terms of overall interaction in subcellular organelle.

Electrostatic effects may be much more important in membrane lipids. That this may be the case has been shown by Träuble and Eibl [74]. These authors have studied the effects of pH, as well as monovalent and divalent cations on the gel → liquid-crystalline transitions of lecithin, phosphatidylethanolamine, phosphatidylserine and phosphatidic acid in artificial bilayers. As mentioned previously, lecithin would behave independently of pH over almost the entire pH range. Phosphatidylethanolamine would be pH-dependent because the ethanolamine group is a relatively weak base. Phosphatidylserine has a negatively charged carboxyl group plus a weak amino group base and phosphatidic acid has two ionizable protons, one with a pK of 3.4 and the other with a pK of ~8.6.

For a uniform distribution of fixed charge, the theory of the diffuse ionic double layer predicts that gel → liquid-crystalline transition in bilayers of charged lipids should be accompanied by a decrease in electrostatic free energy, primarily as a result of bilayer expansion. The theory also predicts that increasing charge density should produce a decrease in the transition temperature. Träuble and Eibl [74] systematically investigated this in phosphatidic acid bilayers. They found that a change in the bulk pH from 7 to 9 increases the charge per polar group from one to two elementary charges. This is consistent with the established second pK of the exposed phosphate group. In agreement with theory, this change of charge lowers the transition temperature by about 20 °C. The important point of the experiment is not that the transition temperature changes; in higher animals, the biologic temperature is kept within close limits. What is significant is the fact that in the pH regions between 7 and 9, very small changes in pH suffice to induce a phase transition at a *constant* temperature. This effect is extremely pronounced in the case of phosphatidic acid and somewhat less in phosphatidylserine.

A plot of transition temperature versus pH, shows a sharp deflection at pH < 2 for lecithin (as the phosphate group is titrated). In the case of phosphatidylethanolamine an inflection is observed at approximately pH 2 and another one at above pH 8; the latter is due to the titration of the ethanolamine group. In contrast, phosphatidylserine shows an inflection between pH 2 and 5, presumably due to the carboxyl residue and a reverse inflection between pH 6 and 9 which can be attributed to the serine amino residue. The most pronounced effect is seen in the case of phosphatidic acid where there is a huge change in transition temperature between pH 6 and pH 8.

The effect of ionic strength is important. Monovalent solutions of salts, such as sodium chloride, potassium chloride, etc., lower the transition temperature at a given pH. This effect can be rather dramatic at the pH for maximal change of the transition temperature. For example, at approximately pH 8.5, a phosphatidic acid bilayer in water is in an essentially crystalline state, whereas in 0.5 molar sodium chloride it is in a largely liquid state. The monovalent cations, therefore, tend to fluidize the bilayer at a given temperature and pH.

Divalent cations, such as magnesium and calcium behave in an opposite fashion. They tend to increase the transition temperature by charge neutralization. Divalent cations can thus induce a liquid-crystalline → gel transition at a constant temperature.

The *isothermal* transitions of lipid state described by Träuble and Eibl [74] appear of major relevance to the physical intractions between cells, whether normal or neoplastic and, in general, to membrane anomalies in neoplasia. This is because:

(a) Tumors are commonly abnormal in the distribution of ionizable phosphatides among diverse cellular membranes (Chapter 3).
(b) Tumors typically produce large amounts of lactic acid and are thus expected to exist at a lower localized pH than normal cells (Chapter 6).
(c) There is evidence that neoplastic cells handle divalent cations differently from normal cells (Chapter 6).
(d) Tumors commonly deviate in cholesterol metabolism and membrane cholesterol distribution (Chapter 5) and cholesterol strongly modifies the cooperativity of lipid state transitions.

In membranes where lateral translation of ionogenic groups within the plane of the membrane is permitted, one can expect such lateral displacement to occur upon close approach of the charge-bearing surfaces. This would produce a localized decrease in electrostatic repulsion and allow close approximation of restricted membrane domains. It is even possible that membrane fusion might occur in such restricted areas. One can speculate that such a mechanism might be involved in the reformation of the nuclear envelope after mitosis. The notion might also be extended to the lysosomal membranes and their fusion behavior (Chapter 7). Lysosomal membranes contain appreciable proportions of phosphatides which will contribute negative charge at neutral pH (phosphatidylinositol 9.7%, phosphatidylserine 3%, phosphatidylethanolamine 26%, diphosphatidylglycerol 4%; [75]).

8.1.6. Possible physical basis for the altered cellular interaction in neoplasia electrophoretic data

Many investigators have attempted to explain the asocial character of neoplasia according to colloid theory. Increase in V_R has been invoked particularly frequently, because of the often enhanced electrophoretic mobilities associated with neoplastic conversion.

Thus, hamster kidney fibroblast cells have a lower electrophoretic mobility than cells from stilbesterol-induced kidney tumors [76], and the anodic mobility of hepatoma cells is greater than that of normal hepatocytes [77]. Also, the anodic mobility of mouse MC1M sarcoma sublines increases as they progress from the solid to the ascites form, a change associated with greater invasiveness [78]. Forrester et al. [79,80] report that the anodic mobility of hamster fibroblasts usually, but not always, increases after conversion with polyoma virus (Table 8.6); the same appears true for mouse spleen cells in Friend virus disease [81]. Fuhrmann et al. [82] and Sakai [83] report a positive correlation between anodic electrophoretic mobility and *in vivo* malignancy for several strains of rat ascites hepatoma.

Experiments with normal and neoplastic human tumor cells show that, in general, mesenchymal tumor cells exhibit higher electrophoretic mobilities than carcinoma cells or normal epithelial cells [84]. Hepatoma cells commonly exhibit higher anodic electrophoretic mobilities than normal hepatocytes.

Certain leukemic mouse cells exhibit an abnormally low *anodic* mobility compared to normal lymphocytes [43]. The increased charge contribution of

TABLE 8.6

Electrophoretic characteristics of normal and polyoma-converted hamster kidney fibroblasts[1]

Clone	Comment	Electrophoretic mobility (μm/sec/V/cm)
C13	Control-untransformed	-1.02 ± 0.06
C13	Neuraminidase-treated	-0.64 ± 0.05
N	Py-transformed	-1.30 ± 0.06
N	Neuraminidase-treated	-0.65 ± 0.06
P	Py-transformed	-1.26 ± 0.06
Q	Py-transformed	-1.29 ± 0.05
S	Py-transformed	-1.25 ± 0.05
V	Py-transformed	-1.27 ± 0.06
X	Py-transformed	-1.23 ± 0.05
Y	Py-transformed	-1.28 ± 0.09
J	Py-transformed	-0.97 ± 0.06
M	Py-transformed	-1.02 ± 0.05

[1] From [80].

TABLE 8.7

Electrokinetic behavior of normal mouse lymphnode cells and mouse lymphoblastic leukemic cells under carious conditions[1]

Cell	Electrophoretic mobility ($\mu m \cdot sec^{-1} \cdot V \cdot cm^{-1}$)	Estimated isoelectric point
Normal		
untreated	−1.27	<2.2
aldehyde-fixed	−1.28	
neuraminidase-treated	−0.94	
Leukemic		
untreated	−1.04	3.3
aldehyde-fixed	−1.28	
neuraminidase-treated	−0.53	
aldehyde-fixed plus neuraminidase	−1.01	

[1] From [43].

sialic acid is outweighed by a concomitant rise in surface *cationic* groups (Table 8.7). These are detected by the increased electrophoretic mobility of the leukemia cells after aldehyde fixation and pH vs mobility curves indicated a $pK \sim 10.5$. However, the use of aldehydes as reagents introduces some ambiguities. Also, Mehrishi [44] has demonstrated ⊕-charged groups also on the surfaces of normal lymphocytes. In this he uses reagents such as citraconic aldehyde, 2,3-dimethyl maleic anhydride, all of which react near neutrality, primarily with lysines and arginines, generating a negative charge instead of a positive one. As shown in Table 8.2, all reagents markedly increase the negative mobility of the lymphocytes, as well as of Ehrlich ascites carcinoma cells. Treatment at pH ~6 of cells-coupled with 2,3-dimethyl maleic anhydride leads to rapid regeneration of the amino groups.

On the whole, one can discern no consistent pattern of electrophoretic difference between normal leukocytes and leukemic obes. Thus, Fuhrmann and Ruhenstroth-Bauer [85] report significant increases in electrophoretic mobility of cells from human lymphocytic and myeloid leukemia, but Thomson and Mehrishi [86] could observe no such difference. They carried out pH-electrophoretic mobility curves on human peripheral lymphocytes and chronic leukemic lymphocytes. Both cell types show titration curves suggesting carboxylic acid groups of varying pH's, including that of sialic acid near pH 3. However, the data indicate that groups are titrating all the way down from pH 7 (probably including histidines). The basic groups titrate at approximately 10.5 in both normal and neoplastic cells.

Similarly, Patinkin et al. [87] could not detect any significant differences in the electrophoretic mobilities of normal lymphoid cells or macrophages of C58,

AKR and C57/B1 mice, and corresponding leukemic cells. However, each cell type exhibited its own characteristic mobility, with spleen and lymph node cells moving more rapidly toward the anode than thymocytes and macrophages. There were no significant differences in pH–mobility relationships in response to calcium ion or to neuraminidase. Lymph node cells show pronounced heterogeneity in electrokinetic parameters.

Schubert et al. [88] have compared bone marrow cells of patients with chronic myeloid leukemia, chronic monocytic leukemia, myeloblast leukemia, plasma cell myeloma and chronic lymphocytic leukemia, using free-flow, preparative electrophoresis. Only in the case of the myeloma did the electrophoretic profiles differ from those of normal bone marrow cells.

In studies of heterogeneous cell populations such as exist in the bone marrow, lymph nodes, blood and spleen, one must take account of the generally higher electrophoretic mobility of T-type lymphocytes compared with B-type cells [89–91]. Interestingly, the proportion of lymphoid cells with high electrophoretic mobility increases during the immune response of C57/Bl mice to injected BP8 ascites tumor cells (Table 8.8; [92]).

Experimentation with embryonic cells [93] and cells from regenerating liver [94,95] suggest a possible correlation of electrophoretic mobility and growth rate [96]. Thus, RPMI 41 cells exhibit a 26% higher anodic electrophoretic mobility during mitosis than during interphase. Also, at high growth rates, RNA contributes more to the surface charge density, while the contribution of sialic acid is about the same at high and low growth rates.

TABLE 8.8

Electrophoretic mobilities of C57/Bl mouse lymphoid cells and macrophages before and after transplantation of BP8 ascites tumor cells[1]

Cell[2]	Electrophoretic mobility (μm/sec/V/cm)	
	Day 0	Day 10
Lymphocyte		
fast	-2.27 ± 0.13	-2.52 ± 0.05
slow	-1.12 ± 0.01	-1.26 ± 0.04
Thymocyte		
fast	-1.79 ± 0.01	-2.60 ± 0.07
slow	-1.16 ± 0.01	-1.27 ± 0.03
Macrophage	-1.11 ± 0.02	-1.04 ± 0.04

[1] From [92].
[2] Population analyses show an increased proportion of 'fast' cells after 10 days.

There is no correlation between the electrophoretic mobility of various strains of TA 3 mouse ascites tumor cells and malignancy *in vivo* [97]. Similarly, Patinkin et al. [87] find no difference between the anodic electrophoretic mobilities of normal chick embryo fibroblasts and cells transformed by Rous sarcoma virus. The two cell types also exhibit identical pH–mobility curves and no significant differences in reduction of electrophoretic mobility after neuraminidase treatment.

8.1.7. Summation

It is a fact that cells can neither make nor break specific contacts unless the balance of the physical attractions and repulsions between the cell surfaces permit such processes.

In considering the possible contribution of physical forces to the altered interactions between normal and neoplastic cells, one can state without hesitation that the major attractive forces, namely Van der Waals' forces of attraction, are unlikely to differ significantly between cell types. On the other hand, charge fluctuation attractions, charge mosaic attractions, and, of course, more specific attractions could vary significantly with alterations of cell surface components. Such alterations could involve either the type of charge groups present, their degree of ionization at the membrane surface, the numbers of such groups, their distribution, or their mobilities. The same considerations apply to the most important repulsive interactions, namely repulsions between surfaces of like charge. Finally, significant imbalance between attractive and repulsive forces could interfere with the establishment of specific intercellular contacts.

It is necessary, therefore, to define the ionic characteristics of normal cell surfaces and to compare these with the surface properties of neoplastic cells of the same type. The only technique used for this purpose to date, simple cell electrophoresis, is not adequate for this task and the data obtained do not allow meaningful conclusions. The field requires extensive technical development and the use of surface modification methods may represent an important step in this direction.

8.2. Concepts of specific intracellular adhesion

8.2.1. Introduction

The hypotheses invoked to explain specific interactions between cell surfaces fall into three categories: In the *first*, surface specificity is encoded in macromolecular surface-located substances, which act as ligands for equally specific macromolecules. Reactions of both the 'antigen–antibody' type and the 'enzyme–substrate' type have been invoked. The *second* class of hypothesis argues that the cognitive process in specific cellular interactions involves the topological matching of macromolecular subunits as in the case of the protomers of an oligomeric protein. The *third* type of hypothesis suggests that specific cellular interactions depend on specific surface topologies, that is, 'surface coding'. All but the 'surface coding' concept would be effective only over very short distances and demand that specific recognition proceeds via 'close junctions'. Also, all but the 'surface coding' hypotheses effectively 'require' one gene for each specific contact.

8.2.2. Receptor–ligand hypotheses

8.2.2.1. The Ehrlich–Tyler–Weiss hypothesis

This concept was originally proposed by Ehrlich [98] in 1909 and reintroduced by Tyler [99] and Weiss [100]; it is essentially an antigen-antibody theory and suggests that cell surfaces bear antigen-like and antibody-like components. Specific intercellular adhesion will occur through a typical antibody–antigen type of reaction, when the surfaces of apposed cells bear specific complementary receptors in appropriate alignment (Fig. 8.5). The Ehrlich–Tyler–Weiss concept requires that cells become very closely apposed before specific adhesion can proceed. Such close apposition is found in 'tight junctions' and 'nexuses'.

In favor of this proposal is the fact that all animal cells bear surface components which are known to act as antigens under appropriate conditions, that is, they can react with specific antibodies. Also, cell surfaces bear diverse carbohydrate residues which can act as receptors for specific plant agglutinins, again in an

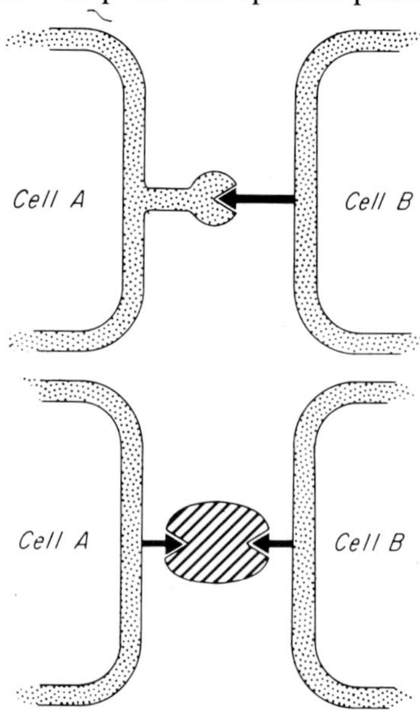

Fig. 8.5. Concepts of specific adhesion. *Top*: Ehrlich–Tyler–Weiss hypothesis. *Bottom*: specific adhesion factor.

antigen–antibody type of reaction. Moreover, it is now well recognized that there exist classes of antibodies which have unusually high affinities for cell surfaces [10] and that certain types of lymphocytes bear tightly-bound surface immunoglobulins. Indeed, one of the major hypotheses of 'immune surveillance' proposes that specific cell surface immunoglobulins constitute the receptors for immune recognition.

One objection to the Ehrlich–Tyler–Weiss concept is as follows: During various stages of differentiation as well as during tissue regeneration, cells adhering to each other must dissociate, migrate to other locations and then again adhere to each other, or to other cells. This requires that the 'antigens' or 'antibodies' involved in recognition or adhesion have to be altered in reactivity to allow dissociation, and then be reactivated at appropriate sites on the cell surfaces to foster reassociation. This is improbable but not impossible; indeed, modulation of surface-antigen activity is now a well recognized phenomenon as is the masking of antigenic expression by various non-antigenic groups.

A far more serious objection applies also to all other surface recognition other than the 'surface coding' hypothesis. This is that there is not enough genctic information to account for all the specific contacts known to occur. We will return to these hypotheses later.

8.2.2.2. Specific adhesion factors

Many investigators have proposed that specific cellular adhesion is mediated by soluble macromolecules (Fig. 8.5). Species-specific aggregation of sponge cells was noted already in 1907 by Wilson [101] and this observation has been repeatedly confirmed. Thus, Moscona [102] as well as Humphreys [103] have shown that a water-soluble protein released by sponge cells, specifically stimulates the aggregation of the cells of the same species. The chemical composition of the aggregation factor [104] indicates a glycoprotein, consisting of 50% amino acids and the rest carbohydrate (glucosamine, fucose, mannose, galactose and glucose). The ratios of the sugars in material isolated from two different species are distinctly different. It appears that the aggregation factors are secreted after intracellular synthesis.

Tissue-specific reaggregation has also been described for embryonic cells of higher forms [105–108]. Such tissue-specific reaggregation, particularly in the case of trypsin-dissociated cells apparently requires synthesis of some protein components; it is inhibited by puromycin and the reaggregation-capacity of the cells is also reduced at low temperature.

Several investigators have examined the liberation from embryonic chick neural retina cells of a macromolecular agent that specifically enhances the these cells [109]. This is a controversial topic and recent studies by Roth and associates (p. 385) tend to favor the hypothesis that specific reaggregation of neural retina cells involves specific interactions between the carbohydrates and glycosyltransferases on apposed cell surfaces.

Adhesion induced by soluble aggregation factors is hypothecated to involve multi-valent glycoproteins with binding sites specific for receptors on the surface of the cells producing the aggregation factor. The interaction would thus resemble the agglutination of cells by plant-agglutinins. However, the possible participation of a concanavalin-A-like substance in this kind of mechanism has been examined and excluded (See Chapter 10.) The possible participation of bivalent immunoglobulins in specific intercellular interactions via cell surface determinants such as histocompatibility antigens has been neither proven nor excluded.

8.2.2.3. The enzyme substrate hypothesis

8.2.2.3.1. Theory

Roseman [110] has presented a very intriguing hypothesis of specific intracellular adhesion. He argues that cell recognition and specific cell adhesion are mediated by interactions between complex surface-carbohydrates and specific glycosyltransferases for which these carbohydrates serve as substrates (Fig. 8.6). Glycosyltransferases comprise several categories of enzymes, each of which catalyzes the following reactions (Chapters 2, 4):

Sugar-nucleotide + oligosaccharide-acceptor → Sugar oligosaccharide-acceptor + nucleotide.

Different glycosyltransferases in specific sequence can thus elongate the oligosaccharide chains of complex carbohydrates by addition of monosaccharide units. All of the glycosyl transferases within one family utilize the same sugar-nucleotide as a glycose donor (e.g., cytidine monophosphate–CMP-sialic acid). Each of the enzymes is specific for a specific sugar or its analogs.

The Roseman hypothesis rests on two experimental observations: (a) cell surfaces bear complex carbohydrates attached to glycoproteins and/or glycolipids and (b) plasma membranes appear to carry glycosyltransferases on their external surfaces. Roseman suggests that a glycosyltransferase on one cell surface can serve as specific ligand for the carbohydrate on the surface of an appropriate, apposed cell; in this it would function as an enzyme only if the sugar nucleotide required for the glycose transfer were secreted into the extracellular space.

This model incorporates several important features: *First*, the interaction between the glycosyltransferase on one surface and the complex carbohydrate on the other would be as that between an enzyme and a substrate; that is, it should exhibit a rather substantial binding constant. Binding would be much stronger than any interactions mediated by hydrogen bonds, image forces, mosaic attractions, etc., and would not require the geometric constraints demanded by such interactions.

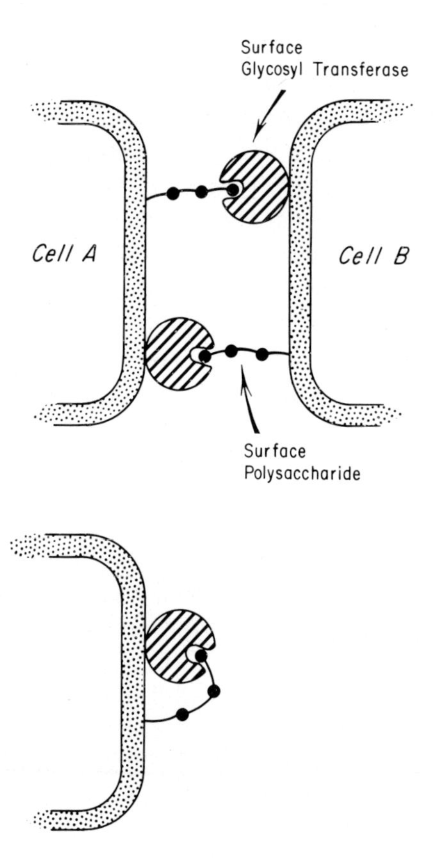

Fig. 8.6. The enzyme-substrate hypothesis of cell recognition and adhesion of Roseman [110]. *Top*: '*Trans*-glycosylation' normal adhesion. *Bottom*: '*Cis*-glycosylation' impaired adhesion.

Second, the binding constant of an enzyme–substrate interaction depends on a wide variety of external factors and adhesion mediated by such a process would be more subject to regulatory mechanisms than an antigen-antibody type of reaction. Sugar nucleotides would constitute one class of regulatory agent. This is because all of the glycosyltransferases catalyze *bimolecular* reactions in which one of the substrates is a sugar nucleotide. Without the sugar nucleotide, the complex between the glycosyltransferase and the complex sugar will not dissociate. However, sugar nucleotides are synthesized in the cytoplasm. Since the level of one substrate in a bimolecular reaction regulates the association of the enzyme with the other substrate, the rates of synthesis and 'secretion' of the *sugar nucleotides* could act to regulate the adhesive process in a highly selective fashion. Divalent cations might constitute another class of regulatory substance, since most of the glycosyltransferases require divalent cations for activity. Many of the enzymes require Mn^{2+}; some of them are activated by Mg^{2+}. Also, a variety of cations, for example, Ca^{2+}, are inhibitory. The listed suggestions are based on the known properties of the glycosyltransferases; other modifying agents could be invoked, for example, allosteric effectors.

Roseman has extended the concept of intercellular adhesion via glycosyltransferases and complex carbohydrates by suggesting that contacting cells can specifically modify their mutual adhesive properties. He argues that, if the glycosyltransferase on one cell surface reacts with the sugar on another cell surface, the availability of an appropriate sugar nucleotide will lead to glycosylation of the carbohydrate acceptor. The original substrate would thus be converted into a product which would bind poorly to the enzyme; a decrease in intercellular adhesion would result. Roseman points out the possibility that the next enzyme of a multi-glycosyltransferase system might also be located on the surface leading to the formation of a new enzyme-substrate and reestablishment of intercellular adhesion. Accordingly, *intercellular* glycosylation (*trans*glycosylation) could result in either a loss or gain of adhesiveness depending on the nature of the surface glycolsyltransferases and the surface density of each enzyme species.

Roseman has extended his hypothesis to explain the apparent impaired adhesiveness of tumor cells. He argues as follows:

Normal cells bear all of the enzymes of a multi-glycosyltransferase complex together with the corresponding substrates and products. In contrast, the tumor cells lack a glycosyltransferase somewhere along the sequence. Tumor cells would thus be deficient in their capacity to form all of the enzyme–acceptor complexes required for normal cell–cell associations. The extent of the defect would depend on which enzyme in a series is deficient. For example, deletion of the first transferase in a series would eliminate participation also of the second and third enzyme in the series. In contrast, deletion of the second transferase in the series would still allow adhesive interaction via first enzyme in the series.

We shall now treat the experimental data which bear on the Roseman hypothesis.

8.2.2.3.2. *The participation of specific carbohydrates in processes of cell recognition*

Normal cells. Many investigators have long speculated about the role of cell surface carbohydrates in specific cell interactions but solid experimental evidence has only begun to develop since the publication of Oppenheimer et al. [111]. These authors found that single cells obtained from 'embryoid bodies' of an ascites form of mouse teratoma (402 AX strain 129) could aggregate in a complex tissue culture medium but not in glucose-containing balanced salt solutions

In these studies the rate of reaggregation of trypsin-dissociated teratoma cells was compared with that of trypsin dissociated neutral retina cells, limb-buds and other embryonic tissues in various media. This was done by use of a Coulter counter. It was found that the cell derived from the various embryonic tissues generally reaggregated very rapidly in Hank's balanced salt solution. In contrast, the dissociated teratoma cells did not reaggregate for at least 60–90 min in this medium. Indeed, about half the preparation showed only slight aggregation after 2–3 hr of incubation. However, dissociated teratoma cells aggregated very rapidly in *supplemented* tissue culture media.

A thorough search revealed that only one component of the tissue culture medium, namely L-glutamine, was responsible for the rapid aggregation of the cells in the complete media. Moreover, the activity of L-glutamine could be replaced by two hexosamines, D-glucosamine and D-mannosamine. This stimulation of aggregation by L-glutamine and its replacements, usually involved a time lag. Also, the response could be completely blocked by inhibitors of electron transport or glycolysis, and by uncouplers of oxidative phosphorylation. These inhibitory effects were not due to cell death. Two metabolic antagonists of L-glutamine blocked the action of this amino acid but did not inhibit the effects of D-glucosamine or D-mannosamine. Aggregation occurred at 37 °C but did not occur at an appreciable rate at 25 °C or 5 °C.

The authors explored the following possibilities: (a) L-glutamine may be required for cell survival, (b) aggregation requires the interaction of cell surface components with L-glutamine and (c) 'adhesive substances' are formed by a metabolic reaction utilizing L-glutamine. The authors' evidence strongly indicates the operation of metabolic processes. Specifically, (a) the cells remain viable in the absence of L-glutamine during the periods of the experiments; that is, the amino acid is not required for survival of the cells under the conditions used, (b) aggregation requires elevated temperatures, while simple binding should proceed also at 25 °C, (c) L-glutamine could be replaced by two hexosamines which, although toxic *in vitro* above certain levels, enter the same metabolic pathway as L-glutamine, (d) metabolic inhibitors block the action of both L-glutamine and hexosamines, and (e) specific antagonists of L-glutamine inhibit the action of the amino acid but not of the hexosamines. The authors suggested possible involvement of the metabolic pathway schematized in Fig. 8.7 and propose that intercellular adhesion might occur by the formation of complexes between cell surface carbohydrates and respective cell surface glycosyltransferases.

To test this hypothesis, Roth et al. [112] developed an assay for specific intercellular adhesion. The method involves the collection of ^{32}P-labelled cells by aggregates of the same (isotypic aggregates) or different (heterotypic aggregates) types of tissues, and determination of the numbers of cells collected, by liquid sintillation counting. They find that an aggregate of a given cell type preferentially attracts labelled cells of the same cell type rather than cells of another cell type.

These studies were extended in depth to the behavior of neural retina cells. The collection of labelled neural retina cells by neural retina aggregates was studied as a function of time, cell concentration, aggregate diameter, temperature and aggregate number. The effects of various enzymes were also evaluated in an attempt to determine whether specific changes in the cell surfaces of the component cells interfere with the collection of the labelled neural retina cells.

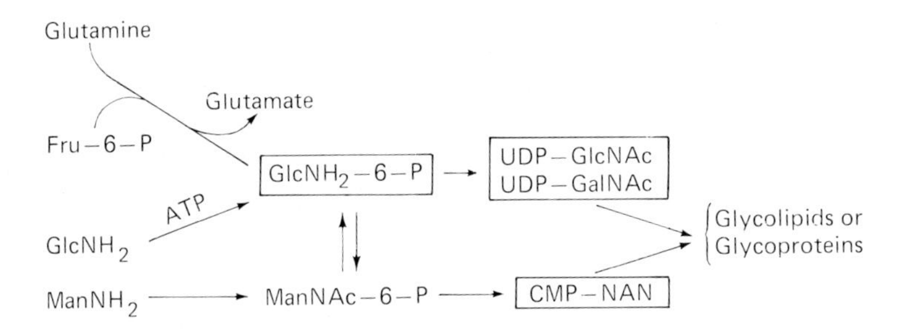

Fig. 8.7. Possible partial common pathways for L-glutamine, D-glucosamine and D-mannosamine involved in the synthesis of the sugar moieties of glycoproteins and glycolipids. In presence of D-glucose, the key compound, D-glucosamine 6-phosphate ($GlcNH_2$-6-P), is synthesized by a transamidase from L-glutamine and D-fructose-6-P. *Abbreviations*: $ManNH_2$-D-mannosamine; UDP-GlcNAc and UDP-GalNAc, the uridine diphospho derivatives of *N*-acetyl-D-glucosamine and *N*-acetyl-D-galactosamine, respectively; CMP-NAN, cytidine-5'-monophospho-*N*-acetylneuraminic acid (CMP-sialic acid). After [111].

It was shown that blockers of oxidative phosphorylation and respiration interfered with aggregation. Low temperatures also hindered aggregation but inhibitors of protein synthesis had no effect for periods up to 80 minutes. Exposure of the collecting aggregates to trypsin and collagenase had no effect on aggregation. On the other hand, treatment of the collecting aggregates with β-galactosidase made these much less specific in their adhesive properties. Whereas untreated aggregates collect almost exclusively neural retina cells, the β-galactosidase-treated aggregates collect other cell types also. The authors suggest that multiple cell surface sites are responsible for specific adhesion but that not all of them are sensitive to β-galactosidase. They propose that treatment with β-galactosidase results in a cell surface that is 'mixed' with respect to its adhesive properties. Terminal β-galactosyl residues on the surface of the neural retina cells could still be at least partially responsible for the adhesive selectivity of these cells.

In a subsequent study, Roth et al. [113] studied the glycosyl transferase activities in chick-embryo neural retina cells. In view of earlier experiments demonstrating that removal of terminal β-galactoside residues alters the adhesive recognition of chick embryo neural retina cells, Roth et al. [112, 113] focussed on the ability of these cells to transfer ^{14}C from [^{14}C]uridine diphosphate galactose (UDP-galactose) to endogenous acceptors of high molecular weight, as well as exogenous acceptors. A comparison of the galactosyl transferase activities toward endogenous and exogenous acceptors in intact cells and cell sonicates is given in Table 8.9.

Importantly, in the absence of detergents, the whole cell preparations incorporated 75% as much ^{14}C-labelled galactose into endogenous acceptor as the homogenates. When sialidase-treated, ovine submaxillary mucin was added as acceptor the whole cell preparations were 92% as active as the homogenates. However, in the presence of Triton X-100 the sonicates show a 2-fold increase in

TABLE 8.9

Galactosyltransferase activities in intact cells and cell sonicates[1]

Enzyme source	Endogenous[2] activity (no acceptors)	Exogenous[2,3] activity (plus acceptor)
Intact cells	100	100
Sonicated cells	125	109
Sonicated cells plus Triton X-100 (1 %)	256	715

[1] From Roth et al. [113].
[2] Activity of intact cells set at 100.
[3] Exogenous acceptor – Sialidase-treated, ovine submaxillary mucin.

activity towards endogenous acceptors and more than 6-fold increase in activity towards exogenous acceptors. Other experimentation showed no significant, soluble galactosyl transferase in the supernatant fluids of cell sonicates.

The authors argue that the galactosyl transferases catalyzing the observed reactions are at least in part localized on the external surfaces of the neural retina cell plasma membranes. Their arguments are as follows:

(a) There was no evidence for significant uptake of sugar-nucleotide. Moreover, unlabelled galactose, galactose-1-phosphate and UDP-glucose did not interfere with the transfer of labelled galactose from UDP-galactose to endogenous acceptors (but see p. 391).

(b) The cells remained essentially intact during the reaction, that is, one could not assume leakage of galactosyltransferase activity. In addition, there was insufficient galactosyltransferase activity in the cell supernatant to account for the incorporation of the galactose into the cell pellets.

(c) The intact cells could transfer galactose to acceptors with molecular weight of 10^6, and the product of this reaction was then found in the extracellular fluid.

These data tend to exclude the possibility that the galactosyltransferase activity is located in the soluble cytosol. The evidence also tends to exclude the possibility that labelled sugar nucleotides are hydrolysed externally and the resulting label (galactose or galactose-1-phosphate) is then utilized after transfer to the cell membrane. The authors' data also exclude the possibility that the enzyme activity involved is 'excreted' into the medium. However, the results of the detergent treatment are compatible with a lysosomal localization of at least some of the glycosyltransferase activity. Whatever the source of the galactosyltransferase, the results suggest that a transferase-like acceptor complex may play a role in cellular recognition (but see p. 391).

The data of Roth et al. [112, 113] fit those of Hagopian and Eylar [114], who have purified a polypeptidyl:*N*-acetylgalactosaminyltransferase and a 'glycoprotein' galactosyltransferase from HeLa cells. Both transferases appear to be membrane-associated but 80–95% of the activities are masked in the isolated membranes. Disruption of the membranous structures with non-ionic detergents or phospholipase A liberated the transferases. The authors' fractionation procedure does not allow distinction of the subcellular organelle bearing the transferases, although the detergent effects suggest lysosomal involvement. In another study, the same authors [115] have purified a polypeptide:*N*-acetylgalactosami-aminyltransferase from bovine submaxillary glands. The enzyme is highly concentrated in a membranous pellet sedimented by a centrifugation at 480,000 $g \cdot$min. It can be extracted therefrom by release with Triton X-100. This study also suggests a lysosomal (Golgi?) localization of the transferase and one must seriously consider the possibility that these enzymes have more than one subcellular locus.

Jamieson et al. [116] and Bosman [117] have studied the glycosyltransferase activities of human platelets. Bosman finds that platelets contain collagen: glucosyltransferase, collagen: galactosyltransferase, polypeptide: *N*-acetyl galactoseaminyltransferase and glycoprotein: glucosyltransferase, but they do not show any fetuin: *N*-acetyl-glucoseaminyltransferase activity. Moreover, the platelets show no endogenous acceptor activity and 89–94% of the enzyme activities were localized in a plasma membrane fraction isolated according to the method of Barber and Jamieson [118]. This membrane isolate is a rather pure one and, according to its developers, contains less than 3% contamination by lysosomal enzymes. Bosman's data appear fairly convincing that human platelets contain glycosyltransferases in their plasma membranes. Their role is not totally clear, but Bosman suggests that they may actually act as mediators of collagen-platelet adhesion or platelet–platelet adhesion, rather than serving an enzymatic function.

Jamieson et al. [116] have independently demonstrated the presence of a collagen: glucosyltransferase on platelet plasma membranes and show that inhibitors of this enzyme activity also interfere with platelet adhesion to collagen. They argue that the enzyme functions in mediating platelet–collagen adhesion.

In a directly relevant study, Pricer and Ashwell [119] show that isolated liver bile-front plasma membranes can bind various glycoproteins which have been desialyated. On the other hand, this binding could not occur if the sialic acid of the membranes was removed by sialidase. The authors suggest the presence of an intrinsic, membrane-associated sialyltransferase.

Recently the Roseman group [120] has developed an ingenious technique to evaluate the role of cell surface carbohydrates in cell adhesion. They devised a method to couple various sugars covalently to beads of cross-linked dextran (Sephadex), 10–40 μm in diameter. They then measured the adhesion of SV40-transformed 3T3 fibroblasts to the beads, using mixtures of 10^{-5} cells and 5×10^{-4} beads, incubating for 2 hr at 37°C. They found that these fibroblasts adhered readily to beads derivatized with D-galactose, but not to similar beads derivatized with D-glucose or *N*-acetyl-D-glucosamine.

Interestingly, cells that adhered to galactose-beads also became more adhesive towards cells in suspension, leading to the formation of large aggregates containing both cells and galactose beads. The possible participation of non-specific factors (serum proteins) in the growth medium were eliminated as follows:

(a) One could not interfere with the interaction of the beads and the cells by pretreatment with serum.
(b) Cells growing for 24 generations in immunoglobulin-free serum, in place of calf serum gave the same results as cells grown normally. This should exclude adsorbed immunoglobulins as a participating factors.
(c) The adherence of the cells to the galactose beads occurred only at 37 °C and not 4 °C; possible antibody-involving reactions should also occur at 4 °C.

The data reported seem to indicate that the SV40/3T3 cells can specifically distinguish β-galactosyl- from β-glucosyl- and *N*-acetyl-β-glycosaminyl-groups; that is, they adhere specifically to beads bearing covalently linked β-galactosyl groups. This is consistent with the Roseman concept of the participation of surface carbohydrates in cell adhesion.

One of the most intriguing findings, however, is the nucleation phenomenon, which implies that the carbohydrate mediated binding produces a *cooperative effect, increasing adhesive affinity of the cells for each other.*

8.2.2.3.3. *Tumor cells*

Bosman et al. [121] describe changes in four membrane 'glycoprotein' glycosyltransferases in mouse 3T3 fibroblast transformed by SV40 or Py viruses compared with the untransformed variants. They studied the following systems: (a) polypeptide: *N*-acetyl-galactosaminyltransferase; (b) glycoprotein: galactosyltransferase; (c) glycoprotein: glucosyltransferases (1 enzyme, the fetuin: glucosyltransferase, transfers fucose onto *N*-acetyl-galactosamine in a receptor prepared from fetuin; the other, PSM: fucosyltransferase, transfers fucose onto galactose in a receptor prepared from porcine submaxillary glycoprotein); these enzymes are involved in the synthesis of membrane glycoproteins.

Each of the four transferases studied was more than twice as active in the transformed cell lines than in the parental fibroblasts. The most pronounced differences were found with the fibroblasts doubly transformed by SV-40 and Py. In these cells, the *N*-acetyl-galactosaminyltransferase and galactosyl transferase activities were 8- to 16-fold higher than in the non-transformed cells. The authors exclude the possibility that the effects observed were due to the influence of lysosomal glycosidases. They argue that their data demonstrate enhanced levels of several enzymes involved in the synthesis of the carbohydrate moiety of membrane glycoproteins. They also show that the activity without added receptors, that is, using endogenous receptors representing partially synthesized glycoprotein products, is much greater in the transformed cells; indeed, such activities are often only barely detectable in untransformed cells. This is taken to indicate a much higher level of endogenous substrate in the transformed cells.

Roth and White [122] have studied the capacity of Balb/c 3T3 and Balb/c 3T12 cells to transfer galactose from UDP-Gal to cellular or exogenously added acceptors.

They find that both cell types can carry out this transfer, but in the case of the 3T3 line, cells from sparse cultures incorporate (transfer) much more galactose than cells from confluent cultures, although the galactosyltransferases activities are identical in both cases. The authors suggest that this is due to an increased concentration of receptors in the cells. In contrast, 3T3 cells show equivalent transfer activity at low and high cell densities. The authors also present the following evidence suggesting that the galactosyltransferase activity is located on the plasma membrane: (a) they find no inhibition of the endogenous reaction by unlabelled UDP-Glu, Gal, or Gal-1-P, indicating that entry of the UDP-Gal into the cell is not an essential step (but see p. 391). (b) Whole cells catalyze the reaction but the 500 *g* cell supernatant does not. This should exclude the possibility of enzyme 'secretion'. (c) Radio-autographs show a 'peripheral labelling pattern'. This, in itself, is not convincing, but it certainly suggests that the enzyme activity leads to a cell-associated rather than a secreted product.

The authors suggest that they are dealing with a contact-dependent catalysis phenomenon. They suggest that in the 3T3 cell, *trans*-glycosylation occurs linking the cells together. In contrast, in the 'malignant' 3T12 cells, a *cis*-glycosylation phenomenon may occur (Fig. 8.6), that is, enzyme and substrates are close enough on the cell surface allowing cross-glycosylation. The authors argue that the defect in the 3T12 cell is architectural rather than compositional.

Podolsky et al. [123] show that normal hamster BHK fibroblasts as well as their polyoma transformants showed appreciable capacity to transfer galactose from UDP-Gal to fetuin. Importantly, exposure of cells to Con A significantly impaired galactosyltransferase, but did not affect other glycosyl transferases.

The galactosyltransferase activity of galactosyltransferase-positive rat erythrocyte ghosts could be selectively purified by adsorption onto immobilized Con A and elution with α-methyl-D-mannopyranoside. Human erythrocytes, which do not normally agglutinate in the presence of Con A, adsorb the enzyme purified from galactosyltransferase-positive rat erythrocytes and then become Con A-aggultinable. Importantly, other glycoproteins extracted from rat erythrocytes lacked this capacity.

The fact that BHK cells do not agglutinate in the presence of Con A, whereas polyoma transformed BHK cells do so, and the suggestion that both cells bear galactosyltransferase, shows the presence of this surface enzyme is not in itself sufficient to confer Con A agglutinability. However, the data can be taken to indicate that agglutinability occurs preferentially with enzyme-bearing cells. Two possibilities must be considered: (a) the presence of the enzyme is essential for Con A agglutination; (b) agglutinable and non-agglutinable cells bear qualitatively different enzymes. It is possible that the galactosyl transferase is merely a protein with Con A receptors which adsorbs to membranes. Alternatively, its substrate affinity may be essential in agglutination as suggested by Roseman [110].

LaMont et al. [124] compare the activities of cell surface galactosyl transferase and *N*-acetylglucosaminyltransferase of normal rat intestinal crypt cells and colonic epithelial cells with those of corresponding adenocarcinoma cells (induced by dimethylhydrazine). They used exogenous macromolecular acceptors in these studies, but did not specifically determine whether their nucleotide sugars were excluded from the various cell types. Moreover, no assays for possible extracellular glycosidase activity were carried out.

No consistent difference could be observed between normal and neoplastic small-intestine cells, but the carcinomatous colonic cells showed substantially lower glycosyltransferase activities than normal cells. The significance of these observations is obscure, particularly since one cannot be certain that the authors actually measured *surface* transferase activity.

Deppert et al. [125] dispute the notion that galactosyltransferases play a general role in cell recognition and/or adhesion. They show that normal, intact BHK fibroblasts bear (or secrete) α-glycosidases (Chapter 7) which hydrolyze UDP-Gal first to Gal-1-P and then to Gal. They further show that the experiments of Roth et al. [112,113] do not adequately exclude intracellular utilization of Gal, because they did *not saturate the Gal transport system* with competing non-radioactive Gal. They also show that the medium used in [112] damages BHK cells. Deppert et al. [125] finally show that, *provided adequate controls are employed, no galactosyltransferase activity can be detected on the surfaces of BHK cells.* In view of these data and the demonstrated tendency of cells to 'leak' glycosidases (Chapter 7), the Roseman hypothesis must be viewed with some reserve.

8.2.2.4. *Multimer interaction hypothesis*

Both the 'antigen–antibody' hypothesis and the 'enzyme–substrate' concept envisage that intercellular recognition involves the interaction of a macromoleculebearing a receptor site with a small molecule specific for this receptor. However, one should consider the possibility that recognition depends on the macromolecular topology of cell surfaces. Thus, Roseman [110] has suggested a mechanism of intercellular adhesion based on the formation of hydrogen bonds between specific glycose units on adjacent cell surfaces. He points out that although hydrogen bonding in aqueous solutions is generally not favorable because of the competition of water molecules for the hydrogen bonds, hydrogen bonds between glycose units can form extremely stable structures. For example, cellulose fibers are held together by hydrogen bonds between parallel chains of polysaccharide molecules [126].

Najjar [10] has raised an even more intriguing possibility. He emphasises the extremely high specificities of interaction between the protomers of oligomeric proteins, for example, the α- and β-chains of hemoglobin. He suggests that specific intercellular adhesion could come about through such interactions. He suggests that the protomers of the oligomeric 'adhesive' protein are distributed on the surfaces of apposing cells. Specific interaction occurs when a protomer on one surface combines with the complementary subunit on the other surface (Fig. 8.8).

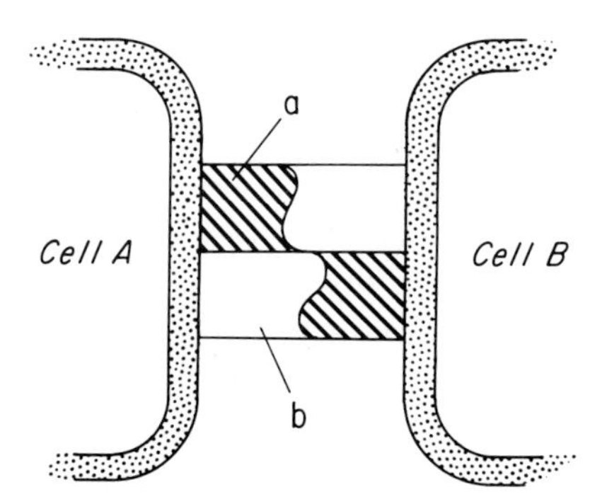

Fig. 8.8. The multimer interaction hypothesis of specific cell recognition [10]. The clear and hatched sectional represent two subunits of a tetramer.

Both concepts have weaknesses in addition to the ones previously mentioned. A particularly serious difficulty with the hydrogen-bonding concept is that specific interactions via hydrogen bonds require an extremely high degree of directional specificity, of an order, indeed, which is rarely achieved in a single macromolecule. A similar objection applies to the protein-subunit hypothesis.

8.2.2.5. *Membrane coding lattices*

Changeux [127] points out a major objection to all of the preceding specific cell-recognition or cell-adhesion theories. This is that the *number of specific cell contacts known to exist in a mature vertebrate organism is much too large to allow simple coding of each individual contact by a separate protein entity.*

Two numerical examples will serve to illustrate this point: A typical diploid mammalian cell contains $\sim 10^{-11}$ gm DNA. This is adequate to code for $\sim 10^7$ polypeptide chains with mol. wt. $\sim$50,000. However, estimates of the number,of separate immunoglobulin chains in an adult mammal lie in the vicinity of 10^6. This would mean that 10% of the genetic information in the fertilized ovum is utilized for immunologic processes. This is a staggeringly high proportion, but quite small if one considers the organization of the nervous system. Thus, the human cerebrum contains $\sim 5 \times 10^9$ neurons, each of which makes 10^4–10^5 contacts with other neurons, that is, a total of $\sim 10^{14}$ contacts! If 10% of the human genome were allocated to cerebral organization and a specific contact required a protein molecule of 50,000 mol. wt., only 10^{-8} of neuronal contacts could be specific.

Changeux [127] therefore argues that cell recognition phenomena require gene-conservation processes. He proposes the following:

(1) The affinity of cell contacts and the stability of cell–cell associations depends upon interactions between the apposed cell surfaces.
(2) The membrane surfaces are made up of various classes of subunits as discussed elsewhere (Chapters 1, 11) and a given combination of subunits within the plane of the membrane makes up a 'membrane coding unit'.
(3) The relative affinity of two apposed membrane surfaces depends on the characteristics and proportions of membrane subunits.

Changeux argues that specific interactions between cells depending on fixed or controlled positions of subunits in the membrane plane cannot account for cellular recognition, since it demands the subunits to be non-equivalent and requires some mechanism for the positioning of the subunits in a 'recognition code'. This is not a 'gene saving process'. Changeux, accordingly, proposes a 'statistical code' based on the concept of structural homology between membrane subunits (Chapters 1, 11), an essential feature for the formation of ordered membrane lattices. Such a 'statistical code' can yield an almost infinite number of different 'coding units'.

A very important theoretical study bearing on this subject has been published by Oosawa and associates [128]. The authors consider the distribution of globular protein molecules, all of the same kind, fixed in a two-dimensional hexagonal lattice. (Fig. 8.9.) Each molecule is allowed to rotate about an axis perpendicular to the plane of lattice and the interaction energy between two neighboring molecules in the lattice is assumed to be a function only of the relative orientation of the molecules. In such a system, if the orientation of all molecules in the lattice is known, the total interaction energy of all pairs of molecules can be obtained.

The authors obtained numerical solutions for two dimensional hexagonal lattices composed of 50×50 molecules. They allowed only six directions of orientation about the axis of rotation, corresponding to the six interacting sites at hexagonal positions on the surface of each molecule (Fig. 8.9). The six orientations of the molecules relative to the lattice are given the numbers 1, 2, 3, 4, 5, and 6; the pairs 1 and 4, 2 and 5, and 3 and 6 correspond to an antiparallel orientation. The orientational distribution of molecules in the lattice can then be expressed by the distribution of these numbers.

In the model, the authors consider Van der Waals interactions and Coulombic interactions at the areas of contact. When two protein molecules are apposed at two sites of their surfaces, the total interaction energy is given by the properties

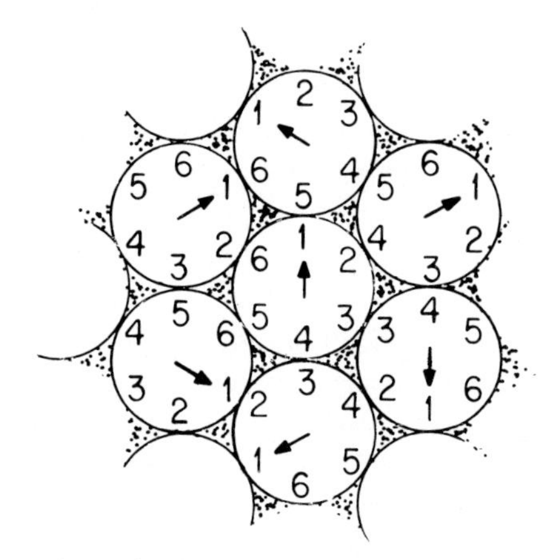

Fig. 8.9. Six unique orientations of a globular protein in a two dimensional hexagonal lattice, after Oosawa et al. [128]. The proteins are identical and can rotate only about an axis perpendicular to the lattice plane. The plane of the lattice is parallel to the plane of the paper. The six orientations are designated by 1–6.

of these sites. The authors assign certain numbers to the 6 sites of each molecule and assume that the interaction energy due to the contact of the two molecules equals the product of the numbers assigned to the contacting sites. The 6 numbers given to 6 interaction sites of each molecule are listed as *a, b, c, d, e,* and *f* (Fig. 8.9). Thus, if an *a* site on one molecule contacts a *b* site on another molecule, the interaction energy would be $a \times b$. The interaction energy of the central molecule with 6 surrounding molecules is given by the sum of 6 products of the two numbers, one on the sites of the central molecule and the other on the sites of the surrounding molecules. For Van der Waals' interactions, the energy is assumed to be approximately proportional to the product of the polarizabilities of the sites; for Coulombic interactions the energies are proportional to the products of the charges of the sites.

The authors used a computer technique to determine the distribution of the protein orientations, which give a minimum energy for the whole system. Each of the six interaction sites of the 2500 identical molecules is assigned a reasonable Van der Waals' and/or Coulombic interaction energy and the orientations of the molecules (1, 2, 3, 4, 5 or 6) distributed randomly in the lattice. The computer then iteratively rotates the subunits in 60°C steps about their axes until the energy of the lattice reaches a minimum. At various stages of the computation the orientation of the subunits are printed out as their number-equivalents, providing a map of the subunit orientations in the lattice.

An example of this type of computation is given in Fig. 8.10. In this case only Coulombic interactions were considered and the numerical values for the interaction parameters of the six sides were:

$$a = 7; \quad b = 90; \quad c = 9; \quad d = 80; \quad e = 5; \quad f = -63.$$

The computation begins with a random distribution of subunit orientations. With progressive iteration a pattern emerges (Fig. 8.10a). This contains two types of domain: one (solid) contains only odd numbers, the other contains only even numbers. With the progress of computation the number of domains changes and small domains are eliminated or fuse into large ones (Fig. 8.10b–c). Concomitantly, the total extent of the domain boundaries decreases, but the energy minimum is attained with rather complex orientation pattern (Fig. 8.10d). Assignment of different sets of reasonable interaction parameters yields separate patterns with diverse degrees of order.

Oosawa and associates ([128]; personal communication) have also examined situations where one (or more) interaction parameters are altered by the hypothetical combination of the subunits with a ligand. As anticipated, such interactions alter the orientation patterns. Moreover, depending upon the change in interaction parameters, gradual, transitional, 'two-state' and metastable patterns can be obtained. Finally, Oosawa and associates have evaluated the *time* dimension in pattern formation. This is again illustrated in Fig. 8.10, which shows that distinct orientational patterns occur at various stages of iteration (time).

We consider the model of Oosawa and associates unique in that it encompasses all currently known requirements for a comprehensive cell recognition hypothesis. The critical features are as follows:

(a) *Coding capacity:* genetic information is encoded in DNA through the specific spatial array of four fundamental coding units. In the Oosawa model, topologic information is analogously encoded in the spatial orientation of the lattice subunits.

(b) *Gene conservation:* a virtually unlimited number of surface codes can be generated utilizing a small amount of genetic information.

(c) *Temporal dependence:* the model can account for time-dependent variations in surface specificity. Such temporal modulations are critical in tissue differentiation and repair (e.g. [129, 130]).

(d) *Environmental responsiveness:* surface coding can be altered in a specific fashion upon binding of specific ligands.

(e) *Versatility of interaction modes:* the model is compatible with all the specific interaction modes proposed before, for example, charge mosaics, charge fluctuations, ionic bridges, macromolecular bridges, etc.

(f) *Relevance to neoplastic change:* the model might account for the altered contact behavior of neoplastic cells in at least the following ways: (1) altered lattice subunits; (2) lattice impurities; (3) lattice disruption; (4) altered interaction parameters; (5) altered ligand binding.

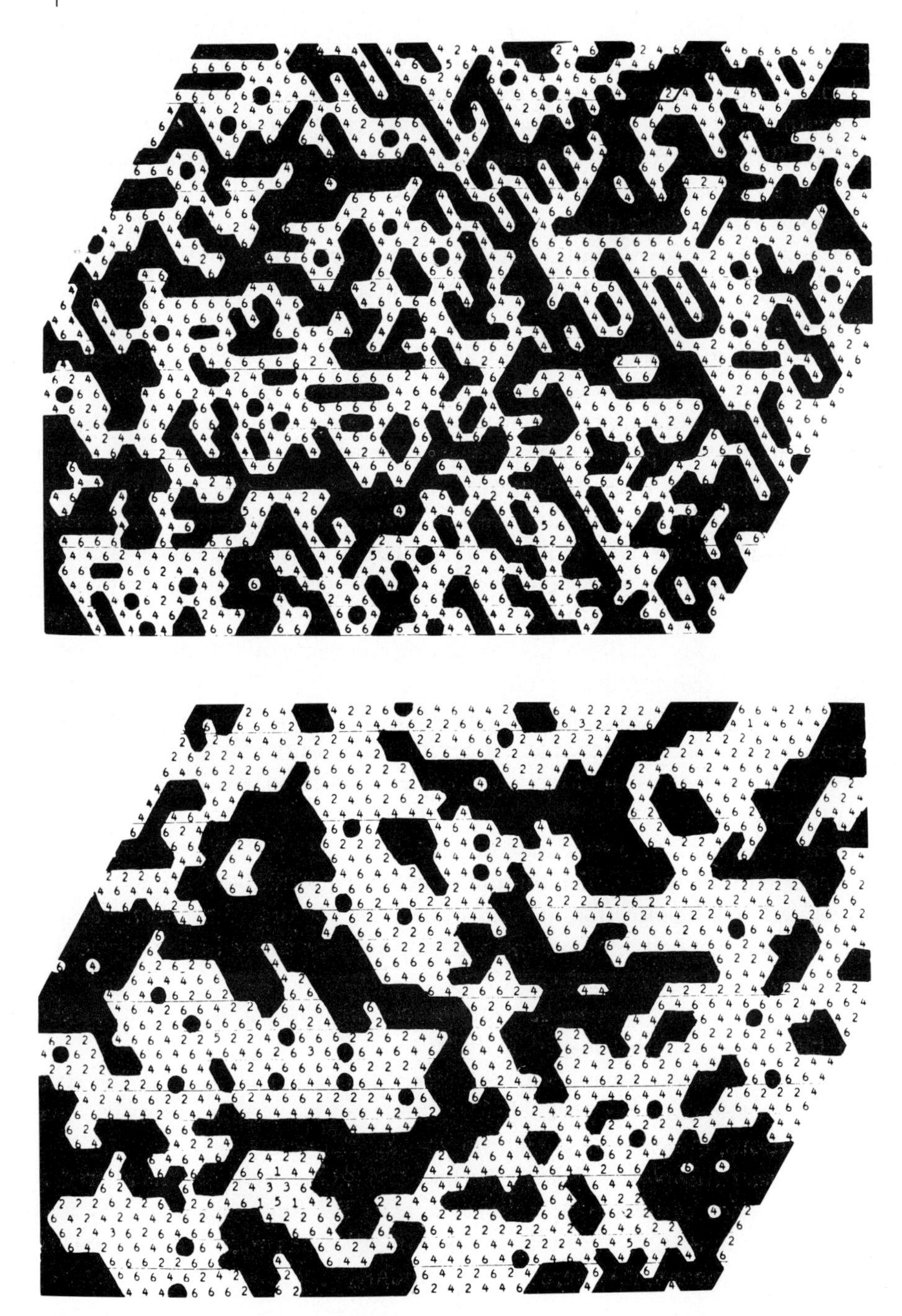

Fig. 8.10. Development of stable orientational pattern in a two-dimensional protein lattice during progression from a random distribution to one of minimal free energy. *Top-left*: Beginning of iteration. *Bottom-right*: End of iteration. From Oosawa et al. [128] by Courtesy of the Authors and Academic Press Inc. (London).

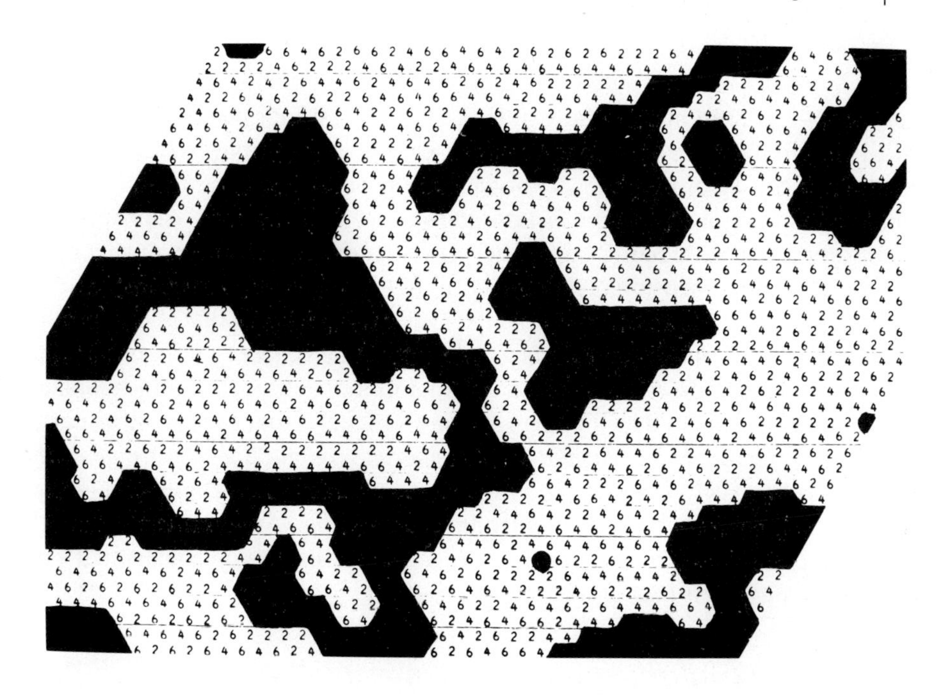

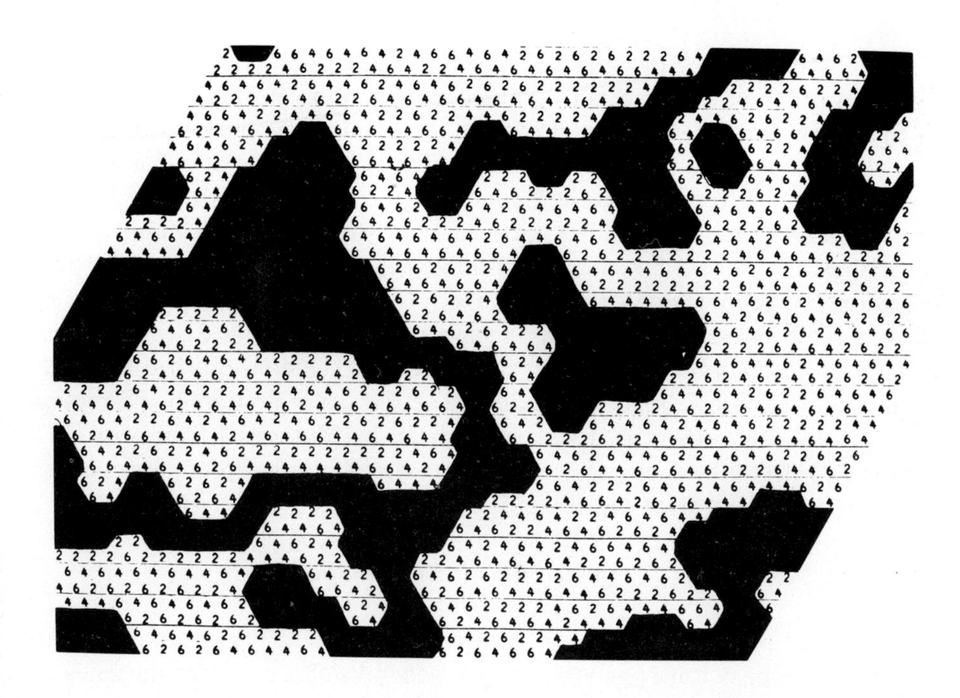

8.2.2.6. *Coda*

Of all proposals presented to date, only the Changeux–Oosawa models meet the critical requirement for gene conservation in cell recognition processes. At the same time, it can accommodate all known aspects of cell recognition and does not exclude concomitant operation of interaction mechanisms such as proposed in the hypotheses of Roseman, Najjar and others. Indeed, such mechanisms might add an additional dimension of specificity to that provided by the coding patterns. A very important aspect of the Changeux–Oosawa concept is that it can provide an explanation not only for the surface recognition defects in tumors but also for the multiplicity and diversity of the other membrane defects found in neoplasia.

An essential feature of the Changeux-Oosawa model is that specific information is encoded in the *topologic distribution of membrane codons. Specific, stable coding patterns can therefore occur in a 'fluid membrane' only if this contains non-fluid, biologically ordered domains.* As a corollary, *dispersion of the lattice* so that the subunits become independent entities free to diffuse in a two-dimensional lipid continuum *would lead to a disruption of cell recognition processes.*

8.3. *Contact inhibition of movement*

8.3.1. *Introduction*

Contact inhibition of movement was initially observed by Abercrombie and associates [131] in their cinematographic observations of fibroblast migration in cell culture. Since their discovery of substantial differences in behavior between normal fibroblasts and sarcomatous cells [132,133], the topic of contact inhibition has become a very important if controversial one in cancer research. However, an unambiguous correlation between anomalous contact behavior *in vitro* and malignancy *in vivo* has not been established. Moreover, the term 'contact inhibition' has become applied to a number of different cell-culture phenomena which are not necessarily related in mechanism. These include a set of culture-phenomena commonly termed 'contact inhibition of cell division', because of the assumption that intercellular contact could lead to an inhibition of cell replication [134–136]. However, as discussed in detail by Martz and Steinberg [137] there is no reliable evidence that cell contact as such can depress mitotic rate. We will therefore not treat this topic here.

8.3.2. *The diversity of contact inhibition phenomenon*

In their excellent analytical review, Martz and Steinberg [138] point out that six separate tissue culture phenomenon are generally classed under the category of contact inhibition of movement. These are:

(a) Inhibition of surface 'ruffling'.
(b) Inhibition of locomotory speed.
(c) Inhibition of nearest neighbor exchange.
(d) Inhibition of cellular overlapping.
(e) Inhibition of directional mobility.
(f) Inhibition of colony expansion.

(a) *Contact inhibition of surface 'ruffling'.* The surfaces of motile fibroblasts exhibit a 'ruffled' appearance [139–142]. The ruffles represent local expansions of the thin plasma membrane, which is otherwise flattened against the culture substrate. Ruffles are not static structures but continuously move inward from the cell periphery. The direction of the movement of the ruffles does not necessarily correspond to the locomotory direction of the whole cells. There is a consistent correlation between ruffling and locomotion in the sense that whenever there is cellular contact, the ruffling activity of the cell periphery becomes inhibited. However, there is no evidence unequivocally indicating that ruffling is an essential part of fibroblast locomotion or that inhibition of ruffling is necessarily related to contact inhibition.

(b) *Inhibition of locomotory speed.* In their initial studies of contact inhibition, Abercrombie and associates [131–133] used time lapse cinematography to measure locomotory speeds of fibroblasts by following the displacement of the cellular nuclei. They found that the average cellular locomotory rate varied inversely with the number of contacts between cells. Free cells (no contact with other cells) exhibited speeds about twice those of cells making contact with five or more other cells. A careful statistical analysis showed that this relation between cell locomotion and cell collision did not depend on local cell density or on diverse culture variables. These observations have been confirmed in more recent investigations on various established cell lines.

(c) *Inhibition of nearest neighbor exchange.* It is implicit in Abercrombie's original concept of contact inhibition that, once culture confluence is attained by all cells in contact with other cells, all cells should be immobilized. However, Garrod and Steinberg [143] in their study of the relationship of contact inhibition of movement to tissue-specific cell sorting, show very extensive alteration of nuclear nearest-neighbor relationships in confluent monolayers of embryonic chick hepatocytes. Similar movements of cells relative to each other also occur in confluent monolayers of mouse 3T3 fibroblasts [144].

(d) *Inhibition of cellular overlapping.* The concept of contact inhibition of movement implies that inhibited cells will not migrate over, under, or upon each other in a monolayer. Contact-inhibited cultures should therefore show relatively little nuclear overlap. In contrast, cells, lacking contact-inhibition should form multilayers which should exhibit numerous nuclear overlaps.

Abercrombie and Heaysman [131] have proposed that contact inhibition could be defined in terms of the nuclear overlap ratio, that is, ([density of nuclear overlap observed]/[density of nuclear overlap expected with nuclei randomly distributed]). At high degrees of contact inhibition, the nuclear overlap ratio should be substantially below 1.0. Very low nuclear overlap ratios (0.02) have been observed for cells such as the 3T3 mouse fibroblasts [138] propagated on conventional culture surfaces. However, the same cell line when grown on a gas-permeable surface (FEP Teflon) exhibits nuclear overlap ratios >1.0 [145]. Wide variations in nuclear overlap ratio have also been observed in primary cultures of various cells under different culture conditions [146–148].

(e) *Inhibition of motional direction.* It is a general characteristic of non-neoplastic, cultured fibroblasts to assume rather elongated spindle-shapes and for the cells to align each other with their long axes essentially parallel [149]. Elsdale [149] has proposed that this alignment arises from contact inhibition of overlapping. He argues that movements which deviate very markedly from a direction parallel to the long axis of the neighboring cells would be inhibited, whereas those which occur essentially parallel to the axes of the neighboring cells would be permitted. This alignment phenomenon is not dependent on 'guidance' effects such as possible alignment of collagen fibers, since it proceeds also in the presence of collagenase [150].

(f) *Inhibition of colony expansion.* As shown by Abercrombie and associates [132, 151] when small pieces of tissue are placed into culture, cells migrate outward therefrom in a radial fashion. This process is called colony expansion. Normally the outgrowth from one explant will inhibit the outgrowth from a nearby explant. This is defined as contact inhibition of colony expansion, and is evaluated as follows:

Two explants are placed close to each other and outgrowth allowed to proceed. The distances from centers of the colonies to their peripheries are then measured at various times to obtain information about the net outward movement of the cells arising in the two colonies. The ratio between the outgrowth distance *along* the axis connecting the colonies to the radius perpendicular to that axis–the between/side radio–is determined for each colony. When the between/side ratio is significantly below 1.0, inhibition of colony expansion has occurred.

8.3.3. What is 'contact'?

In their review on contact inhibition of motion, Martz and Steinberg [138] point out that there is no quantitative evidence that the ruffling behavior of cell surfaces, the tendency of fibroblasts to align themselves parallel to their cellular axes and the rate of nearest-neighbor interchange are in any way influenced by contact. On the other hand, they argue that contact inhibition of locomotory speed, contact inhibition of cellular overlapping and contact inhibition of colony expansion are true contact phenomena because of evidence indicating no 'action at a distance'.

However, present information in this area does not really define the term 'contact'. The microscopic observations used in evaluating contact inhibition phenomena have a resolving pattern of not more than 0.2 μ and we know of no experimentation which unambiguously distinguishes between a 'merging of cellular microenvironments' and specific *molecular* contact between cells.

At issue here is the fact that the transfer of nutrients to cells in culture and the removal of metabolites from the cells occurs by diffusion. A rigorous theoretical analysis by Jensen et al. [60] indicates that conventional culture procedures cause an abnormal accumulation of waste products and an abnormal depletion of nutrients in the immediate environment of the cultured cells. These effects are highly density-dependent. One of the most critical variables is that of pH, and Jensen et al. [60] show that the pH in the immediate environment of the cell surface is likely to be considerably below that of the bulk phase. The biological relevance of this phenomenon has been demonstrated by Jensen et al. [152]. These authors have compared the growth characteristics of various 'contact inhibited' fibroblast lines on two different types of culture surfaces. The first type of substrate was of the conventional category, namely, either glass or plastic; these surfaces are impermeable to all gases and nutrients. The second type of surface tested consisted of very thin films of ion-etched gas-permeable plastics. Both types of substrate allowed equivalent cell adhesion but, because of the gas permeability of the plastic films, the pO_2 in the pericellular environment could be very closely controlled and maintained at high levels. Similarly, by use of bicarbonate buffers, and a controlled carbon dioxide atmosphere on the other side of the film, the pericellular pH could be precisely regulated. Jensen et al. [152] observed that cells which are highly contact inhibited on glass, or impermeable plastics, lose contact inhibition of cellular overlapping when propagated on the gas-permeable films; indeed, up to five superimposed cell layers could be produced without medium change. Subcultures of such multilayered cells onto conventional surfaces again produce contact inhibited growth, whereas subcultures on the gas-permeable surfaces again produce multilayer overlapping growth. The degree of overlapping could be clearly related to the gas permeability of the membranes in the following ways:

(a) When the gas-permeable films were cemented onto an impermeable substrate, contact inhibited growth resulted.

(b) The degree of contact inhibition of cellular overlap could be modified by alterations in gas phase pO_2 and pCO_2.

These data add to the suspicion that, even in cases where 'action at a distance' appears to have been eliminated, 'contact' does not necessarily imply more than the merging of cellular microenvironments.

8.3.4. *Contact inhibition and malignancy*

The possible relevance of *in vitro* contact phenomena to neoplastic processes was first suggested by the studies of Abercrombie and associates [132, 151] which demonstrated that, while normal fibroblasts could induce contact inhibition of cell overlapping and contact inhibition of colony expansion (of normal cells), they could not thus inhibit sarcomatous fibroblasts. These authors found that outgrowths from explants of neonatal mouse muscle and chick embryo heart were mutually inhibitory and that these outgrowths were inhibited by outgrowths from mouse sarcomas 37 or 180. In contrast, expansion of the tumor colonies was not inhibited by contact with outgrowths from the normal tissues. It was also observed that, while the nuclear overlap ratios of the normal cells were well below 1.0, the nuclei of the sarcoma cells overlapped those of normal cells to a degree expected from a random nuclear distribution, indicating that the overlapping movements of the neoplastic cells were not inhibited by contact with normal cells.

These fundamental observations have been widely and often quite uncritically extended. Indeed, it is a common practice to 'diagnose' neoplastic transformation *in vitro* on the basis of 'altered contact inhibition'. Thus, primary cultures of normal fibroblasts, as well as many established fibroblast cultures, form cellular monolayers in which the cells are spindle shaped and align with their long axes parallel, but oncogenic transformation (e.g., by oncogenic viruses) tends to change the culture characteristics towards the formation of foci of randomly oriented, 'piled up' cells. These foci can then be isolated and subcultured leading to the establishment of transformed lines.

Although this phenomenon may be highly reliable within well defined situations (cf. [153]), it is not universal. Thus, as already noted in 1963 by Defendi and associates [154] one can isolate variants of neoplastically transformed fibroblasts (e.g., polyoma transformed BHK 21) which retain the normal culture characteristics of the untransformed parental lines but produce lethal tumors when injected into isogeneic animals. Also as pointed out by Abercrombie [155–157] nuclear overlap of tumor cells among themselves is not consistently high and it is actually rather low for some tumors. 'Loss of contact inhibition' defined in a rigorous sense therefore does not necessarily involve increased nuclear overlap of tumor cells upon tumor cells, altered cellular alignment, or 'piling up'. Its relation to density dependent inhibition of cell division is also unclear.

The original observation that outgrowths from normal cells do not reduce the between/side ratios of outgrowths from tumor explants has been consistently observed in a few tumor systems, but has not been tested very often. Thus,

Abercrombie [155] has confirmed his results with mouse sarcoma 37 and 80 using several methyl-cholanthrene-induced mouse sarcomas (both primary and established lines). In all cases, the outgrowths from the tumor explants were not inhibited by the outgrowths from the normal explants. Similarly, Barski and Belehradek [158] have found that mouse sarcomas M1 and M6 are not inhibited by outgrowth from normal mouse heart fibroblasts. Interestingly, leukocytes, which are not malignant but invasive *in vivo*, are also able to move over normal heart fibroblasts [159]. This finding is in accord with a possible correlation between invasiveness and lacking contact inhibition of overlapping. On the other hand, Barski and Belehradek [158] observed that, although both sarcomas M1 and sarcoma M6 are invasive *in vivo* and are not inhibited by normal fibroblasts, they cannot move over *endothelial* cells in culture. Abercrombie [157] has reported that sarcoma 180 cells can also not move over sheets of endothelial cells in culture.

In a recent study on contact inhibition of movement involving virally transformed cells, Guelstein et al. [160] measured cell to cell collisions by time lapse microcinematography. Normal mouse embryo fibroblast-like cells were compared with cells, primarily fibroblasts, transformed by Moloney mouse sarcoma virus and neoplastic CIM-strain fibroblasts. It was observed that the culture morphology of the CIM cells and the MSV-transformants is typical of transformed fibroblasts: there is considerable multilayering and impaired cellular alignment. However, a critical analysis of cell–cell collisions gave no evidence for impaired contact inhibition of movement. The neoplastic cells showed no abnormal cellular overlapping nor any significant change in collision behavior. However, analysis of mixed cultures of normal cells and neoplastic CIM cells revealed deficient inhibition of the neoplastic cells by the normal cells. Concordant observations have been reported for L-cell fibroblasts [161]. Present information thus indicates that the anomalous culture morphologies of transformed cells cannot be unequivocally explained by altered contact inhibition of movement.

8.3.5. *Mechanisms of contact inhibition*

Martz and Steinberg [138] have reviewed possible mechanisms of contact inhibition of overlapping. They argue that contact inhibition of overlapping and its impairment by neoplastic cells is best interpreted in terms of a 'differential adhesion hypothesis', and that contact inhibition represents a special case of the general tendency of cellular systems to seek steady state positions determined by their adhesive interactions. The adhesive interactions involved are those between the cell and the culture substrate on the one hand, and between cell and cell. Three major cases are considered:

(1) The adhesive interaction between cells is greater than between cells and substrate: then multilayering is favored and the nuclear overlap ratio will be greater than 1.0.

(2) The strength of cell to cell adhesion is weaker than that of cell to substrate: it is then energetically favorable for the cells to remain associated with the substrate. This situation will induce monolayers and contact inhibition of overlapping.

(3) The strength of adhesion between cells and substrate are approximately equal: the cells will then move over and off each other as freely as across the culture substrate. The nuclear overlap ratio will approach 1.0 and contact inhibition of overlapping will not be observed.

We agree with the general logic of Martz and Steinberg [138] but feel that we lack the information necessary to extend this reasoning to the molecular level. We also ask whether the relative adhesivities relate to molecular contact or whether they depend upon the microenvironments of the cultured cells, which could vary drastically between normal and neoplastic cells. Thus, sarcoma cells typically exhibit high aerobic glycolysis and therefore tend to have a much lower pericellular pH than normal cells [60]. The altered contact behavior of the tumor cells may thus be a phenomenon secondary to an anomalous energy metabolism. Another element to be considered is the commonly altered transport behavior of neoplastic cells (Chapter 7). It is essential, therefore, to establish the possible metabolic correlates of anomalous contact behavior of transformed cells and to design studies which can unambiguously test the significance of such correlations. The development of novel culture techniques which allow manipulations of the pericellular environment in a biologically meaningful fashion [152, 162] may elucidate this very important problem.

References

[1] Coman, D.R., *Cancer Res.*, *4*, 625 (1941).
[2] Coman, D.R., *Cancer Res.*, *13*, 397 (1953).
[3] Huxley, J., *Biological Aspects of Cancer*, Harcourt Brace, 1958.
[4] Weiss, L. in *The Cell Periphery, Metastasis and other Contact Phenomena*, North-Holland, Amsterdam, 1967.
[5] Curtis, A.S.G. in *The Cell Surface: Its Molecular Role in Morphogenesis*, Logos Press, London, Academic Press, London, 1967.
[6] Tarin, D. in *Tissue Interactions in Carcinogenesis*, Academic Press, London, New York, 1972.
[7] Curtis, A.S.G., *Progress in Biophysis and Molecular Biology*, *27*, 315 (1973).
[8] McNutt, N.S. and Weinstein, R., *Prog. Biophys. Mol. Biol.*, *26*, 45 (1973).
[9] Chothia, C., *Nature*, *248*, 338 (1974).
[10] Najjar, V.N. in *Biological Membranes*, D. Chapman and D.F.H. Wallach, Academic Press, London, Vol. 3, p. , 1975.
[11] Cohn, E.J. and Edsall, J.T., *Proteins, Aminoacids and Peptides*, Reinhold, New York, 1943, Ch. 4.
[12] Abramson, M.B., Katzman, R. and Gregor, H.P., *J. Biol. Chem.*, *239*, 70 (1964).
[13] Abramson, M.B., Katzman, R., Wilson, C.E. and Gregor, H.P., *J. Biol. Chem.*, *239*, 4066 (1964).
[14] Garvin, J.E. and Karnovsky, M.L., *J. Biol. Chem.*, *221*, 211 (1956).
[15] Wallach, D.F.H. and Garvin, J.E., *J. Am. Chem. Soc.*, *80*, 2157 (1958).

[16] Gottschalk, A. in *The Chemistry of the Sialic Acids and Related Substances*, Cambridge University Press, 1960.
[17] Bangham, A.D. and Pethica, B.A., *Proc. Roy. Phys. Soc. (Edinburgh)*, *28*, 73 (1960).
[18] Curtis, A.S.G., *Am. Nat.*, *94*, 37 (1960).
[19] Derjaguin, B.V. and Landau, L., *Acta Physicochim., URSS*, *14*, 633 (1941).
[20] Verwey, E.J.W. and Overbeeck, J.Th.G., *The Theory of the Stability of Hydrophobic Colloids*, Elsevier, Amsterdam, 1948.
[21] Albers, W. and Overbeeck, J.Th.G., *J. Colloid. Sci.*, *15*, 489 (1960).
[22] Wilkins, D.J., Ottewill, R.H. and Bangham, A.D., *J. Theor. Biol.*, *2*, 176 (1962).
[23] Ottewil, R.H. and Wilkins, D.J., *Trans. Faraday Soc.*, *58*, 608 (1962).
[24] Srivastova, S.N. and Haydon, D.A., *Trans. Faraday Soc.*, *60*, 971 (1964).
[25] Schenkel, J.H. and Kitchner, J., *Trans. Faraday Soc.*, *36*, 161 (1960).
[26] Watillon, A. and Joseph-Petit, A.M., *Discuss. Faraday Soc.*, *42*, 143 (1966).
[27] Derjaquin, B.V., Voropaneva, T.N., Kaanov, B.N. and Titiyepskaya, A.S., *J. Colloid Sci.*, *19*, 113 (1964).
[28] Vold, M.J., *J. Colloid Sci.*, *16*, 1 (1961).
[29] Haydon, D.A. and Taylor, J.L., *Nature*, *217*, 739 (1969).
[30] Gingell, D. and Parsegian, V.A., *J. Theor. Biol.*, *36*, 41 (1972).
[31] Mayhew, E. and Weiss, L., *Exp. Cell Res.*, *50*, 441 (1968).
[32] Seaman, F. and Cook, V.F. in *Cell Electrophoresis*, E.J. Ambrose, ed., Little Brown, Boston, 1965, p. 48.
[33] Gittens, G.J. and James, A.M., *Biochim. Biophys. Acta*, *66*, 237 (1963).
[34] Gouy, G., *J. Phys.*, *9*, 457 (1910).
[35] Gouy, G., *Ann. Phys.*, *7*, 129 (1927).
[36] Chapman, D.L., *Phil. Mag.*, *25*, 475 (1913).
[37] Debye, P. and Hückel, E., *Physikal. Z.*, *24*, 185 (1923).
[38] Debye, P. and Hückel, E., *Physikal. Z.*, *24*, 97 (1924).
[39] Brinton, C.C. and Lauffer, M.A. in *Electrophoresis*, N. Beer, ed., Academic Press, New York, 1959, p. 427.
[40] Drost-Hansen, W. in *Chemistry of the Cell Interphase*, B. Harry Darrow Brown, ed., Academic Press, New York, 1971, Ch. 6.
[41] Henry, D.C., *Proc. Roy. Soc. A*, *133*, 106 (1931).
[42] Glaeser, R.M. in *The Electric Charge and Surface Properties of Intact Cells*, Office of Technical Services, Washington, 1963.
[43] Cook, J.M. and Jacobson, W.J., *Biochemistry*, *107*, 549 (1968).
[44] Mehrishi, J.N., *Eur. J. Cancer*, *6*, 127 (1970).
[45] Wallach, D.F.H. and Winzler, R., *Evolving Strategies and Tactic in Biomembrane Research*, Springer Verlag, New York, 1974, Chs. 1,2.
[46] Wallach, D.F.H. and Lin, P.-S., *Biochim. Biophys. Acta*, *300*, 211 (1973).
[47] Phillips, D.R. and Morrison, M., *Biochemistry 10*, 1766 (1971).
[48] Mehrishi, J.N., *Prog. Biophys. Mol. Biol.*, *25*, 1 (1973).
[49] Mehrishi, J.N. and Grassetti, D.R., *Nature*, *224*, 563 (1969).
[50] Mehrishi, J.N., *Internat. Arch. Allergy Apply. Immunol.*, *42*, 69 (1972).
[51] Cook, G.M.W., Heard, D.H. and Seaman, G.V.F., *Nature*, *191*, 44 (1961).
[52] Wallach, D.F.H. and Perez-Esandi, M., *Biochim. Biophys. Acta.*, *83*, 363 (1964).
[53] Ruhenstroth-Bauer, G. and Fuhrmann, G.F., *Z. Naturforsch*, *16b*, 252 (1961).
[54] Fuhrmann, G.F., Granzer, E., Kuebler, W., Rueff, F. and Ruhenstroth-Bauer, G., *Z. Naturforsch*, *17b*, 610 (1962).
[55] Eylar, E.H., Madoff, M.A., Brody, O.V. and Oncley, J.L., *J. Biol. Chem.*, *237*, 1992 (1962).
[56] Wallach, D.F.H. and Eylar, E.H., *Biochim. Biophys. Acta*, *52*, 594 (1961).
[57] Goldstein, L., Levin, Y. and Katchalski, E., *Biochemistry*, *3*, 1913 (1964).
[58] Goldman, R., Kedem, O., Silman, I.H., Caplan, S.R. and Katchalski, B., *Biochemistry*, *7*, 486 (1968).
[59] McLaren, A.D., *Science*, *125*, 697 (1957).
[60] Jensen, M., Wallach, D.F.H. and Sherwood, P., *J. Theor. Biol.*, in press.
[61] Merishi, J.N., *Eur. J. of Cancer*, *5*, 427 (1969).

[62] Jandl, S. and Simmons, R.L., *Brit. J. Hematol.*, *3*, 19 (1953).
[63] Gingell, D., *J. Theor. Biol.*, *30*, 121 (1971).
[64] Gingell, D., *J. Theor. Biol.*, *17*, 451 (1967).
[65] Haydon, D.A. and Taylor, F.H., *Proc. Roy. Soc. Acta*, *252*, 225 (1960).
[66] Long, C. and Mouat, B., *Biochem. J.*, *123*, 829 (1971).
[67] Mikkelsen, R.B. and Wallach, D.F.H., *Biochim. Biophys. Acta*, *363*, 211 (1974).
[68] Mikkelsen, R.B. and Wallach, D.F.H., *Biochim. Biophys. Acta*, in press.
[69] Pollack, W., Hager, H.J., Reckel, R., Torem, T.A. and Singer, H.O., *Transfusion*, *5*, 158 (1965).
[70] Jones, G.E., *J. Memb. Biol.*, *16*, 297 (1974).
[71] Brooks, D.E. and Seaman, G.V.F., *J. of Colloid Interphase Science*, *43*, 670 (1973).
[72] Gingell, D., *J. Theor. Biol.*, *38*, 677 (1973).
[73] Pinto da Silva, P., *J. Cell. Biol.*, *53*, 777 (1972).
[74] Träuble, H. and Eibl, H., *Proc. Natl. Acad. Sci. U.S.*, *71*, 214 (1974).
[75] Henning, R. and Heidrich, H.-G., *Biochim. Biophys. Acta*, *345*, 326 (1974).
[76] Ambrose, E.J., James, A.M. and Lowick, J.H.B., *Nature*, *117*, 576 (1956).
[77] Lovick, J.H.B., Purdom, L., James, A.M. and Ambrose, E.E., *J. Roy. Micr. Soc.*, *80*, 47 (1961).
[78] Purdom, L., Ambrose, C.J. and Klein, G.G., *Nature*, *181*, 1865 (1958).
[79] Forrester, J.A., Ambrose, E.J. and Macpherson, J.A., *Nature*, *196*, 1068 (1962).
[80] Forrester, J.A., Ambrose, E.J. and Stoker, M.G., *Nature*, *201*, 945 (1964).
[81] Forrester, J.A. and Salamann, M.H., *Nature*, *215*, 279 (1967).
[82] Fuhrmann, G.F. in *Cell Electrophoresis*, E.J. Ambrose, ed., Little Brown, Boston, 1965, p. 92.
[83] Sakai, S.I., *Nagoya Med. J.*, *13*, 51 (1967).
[84] Vassar, P.S., Seaman, G.V.F. and Brooks, D.E., *Proc. VIIth Canad. Cancer Conf., Ontario*, Pergamon, Oxford, 1967.
[85] Fuhrmann, G.F. and Ruhenstroth-Bauer, R.B.G. in *Cell Electrophoresis*, E.J. Ambrose, ed., Churchill, London, 1965.
[86] Thomson, A. and Mehrishi, J.N., *Eur. J. of Cancer*, *5*, 195 (1969).
[87] Patinkin, D., Schlesinger, M. and Doljanski, F., *Cancer Res.*, *30*, 489 (1970).
[88] Schubert, J.C.F., Walther, F., Holzberg, E., Paschev, G. and Zeiller, K., *Klin Wochenschr.*, *51*, 327 (1973).
[89] Zeiller, K. and Hannig, K., *Hoppe-Seyler's Z. Physiol. Chem.*, *352*, 1162 (1971).
[90] Zeiller, K., Hannig, K. and Pascher, G., *Hoppe-Seyler's Z. Physiol. Chem.*, *352*, 1168 (1971).
[91] Zeiller, K., Holzberg, E., Pascher, G. and Hannig, K., *Hoppe-Seyler's Z. Physiol. Chem.*, *353*, 105 (1972).
[92] Hartveit, F., Cater, D.B. and Mehrishi, J.N., *Brit. J. Exp. Path.*, *49*, 634 (1968).
[93] Heard, D.H., Seaman, G.V.F. and Simon-Reuss, I., *Nature*, *190*, 1004 (1961).
[94] Ben-Or, S., Eisenberg, S. and Doljanski, F., *Nature*, *188*, 1200 (1960).
[95] Eisenberg, S., Ben-Or, S. and Doljanski, F., *Exptl. Cell Res.*, *26*, 451 (1962).
[96] Mayhew, E., *J. Genl. Physiol.*, *49*, 717 (1966).
[97] Weiss, L.A. and Houschka, T.S., *Int. J. Cancer*, *6*, 270 (1970).
[98] Ehrlich, P., *Gesammelte Arbeiten, Band II*, Springer Verlag, Heidelberg, 1957, p. 188.
[99] Tyler, A., *Growth Supplement*, *10*, 7 (1946).
[100] Weiss, P., Yale *J. of Biol. and Med.*, *90*, 235 (1947).
[101] Wilson, H.V., *J. of Exptl. Zoology*, *5*, 245 (1907).
[102] Moscona, A.A., *Proc. Natl. Acad. Sci. U.S.*, *49*, 742 (1963).
[103] Humphreys, T., *J. Exptl. Zoology*, *160*, 235 (1965).
[104] Margoliash, E., Schenk, J.R., Hargie, M.P., Burokas, S., Richter, W.R., Barlow, G.H. and Moscona, A.A., *Biochem. Biophys. Res. Commun.*, *20*, 383 (1965).
[105] Moscona, A.A., *Exptl. Cell Res.*, *22*, 455 (1961).
[106] Moscona, A.A. in *Biological Interactions in Normal and Neoplastic Growth*, M.J. Brennan and W.L. Simpson, eds., Little Brown, Boston, 1962, p. 113.
[107] Steinberg, M.S., *Science*, *141*, 401 (1963).
[108] Curtis, A.S.G. and Greaves, M.F., *J. Embryology and Exp. Morph.*, *13*, 309 (1965).
[109] Roth, S., *Developmental Biol.*, *18*, 602 (1968).

[110] Roseman, S., *Chemistry and Physics of Lipids*, *5*, 270 (1970).
[111] Oppenheimer, S.B., Edidin, M., Orr, C.W. and Roseman, S., *Proc. Natl. Acad. Sci. U.S.*, *63*, 1395 (1969).
[112] Roth, S., McGuire, E.J. and Roseman, S., *J. Cell. Biol.*, *51*, 525 (1971).
[113] Roth, S., McGuire, E.J. and Roseman, S., *J. Cell. Biol.*, *51*, 536 (1971).
[114] Hagopian, A. and Eylar, E.H., *Arch. Biochem. Biophys.*, *129*, 447 (1969).
[115] Hagopian, A. and Eylar, E.H., *Arch. Biochem. Biophys.*, *129*, 515 (1969).
[116] Jamieson, G.A., Urban, C.L. and Barber, A.J., *Nature*, *234*, 5 (1971).
[117] Bosmann, H.B., *Biochem. Biophys. Res. Commun.*, *43*, 1118 (1971).
[118] Barber, A.J. and Jamieson, G.A., *J. Biol. Chem.*, *245*, 6357 (1970).
[119] Pricer, W.E., Jr. and Ashwell, G., *J. Biol. Chem.*, *246*, 4825 (1971).
[120] Chipowsky, S., Lee, Y.C. and Roseman, S., *Proc. Natl. Acad. Sci. U.S.*, *70*, 2309 (1973).
[121] Bosman, H.B., Hagopian, A. and Eylar, E.H., *J. Cell. Physiol.*, *72*, 81 (1968).
[122] Roth, S. and White, V., *Proc. Natl. Acad. Sci. U.S.*, *69*, 485 (1972).
[123] Podolsky, D.K., Weiser, M.M., LaMont, J.T. and Isselbacher, K.J., *Proc. Natl. Acad. Sci. U.S.*, *71*, 904 (1974).
[124] LaMont, J.T., Weiser, M.M. and Isselbacher, K.J., *Cancer Res.*, *34*, 3225 (1974).
[125] Deppert, D.W., Werchau, H. and Walter, G., *Proc. Natl. Acad. Sci. U.S.*, *71*, 3068 (1974).
[126] Percival, E.G.V. and Percival, E. in *Structural Carbohydrate Chemistry*, J. Garnet Miller, London, 1962.
[127] Changeux, J-P. in *The 11th Nobel Symposium; Symmetry and Function of Biological Systems at the Macromolecular Level*, A. Engström and B. Strandberg, eds., Wiley Interscience, John Wiley, New York, 1969, p. 235.
[128] Oosawa, F., Maruyama, M. and Fujima, S., *J. Theor. Biol.*, *36*, 203 (1972).
[129] DeLong, G.R. and Sidman, R.L., *Developmental Biol.*, *22*, 584 (1970).
[130] Gottlieb, D.I., Merrell, R. and Glaser, L., *Proc. Natl. Acad. Sci. U.S.*, *71*, 1800 (1974).
[131] Abercrombie, M. and Heaysman, J.E.M., *Exptl Cell Res.*, *6*, 293 (1954).
[132] Abercrombie, M. and Heaysman, J.E.M., *Nature*, *174*, 697 (1954).
[133] Abercrombie, M., Heaysman, J.E.M. and Karthauser, H.M., *Exptl Cell Res.*, *276* (1957).
[134] Golde, A., *Virology*, *16*, 9 (1972).
[135] Abercrombie, M., *Symposia for Quantitative Biol.*, Cold Spring Harbor, *27*, 427 (1962).
[136] Eagle, H., *Science*, *148*, 42 (1965).
[137] Martz, E. and Steinberg, M.S., *J. Cell. Physiol.*, *79*, 189 (1972).
[138] Martz, E. and Steinberg, M.S., *J. Cell. Physiol.*, *81*, 25 (1973).
[139] Abercrombie, M., Heaysman, J.E. and Pegrum, S.M., *Exptl Cell Res.*, *59*, 393 (1970).
[140] Abercrombie, M., Heaysman, J.E. and Pegrum, S.M., *Exptl Cell Res.*, *60*, 437 (1970).
[141] Abercrombie, M., Heaysman, J.E. and Pegrum, S.M., *Exptl Cell Res.*, *62*, 389 (1970).
[142] Abercrombie, M., Heaysman, J.E. and Pegrum, S.M., *Exptl Cell Res.*, *67*, 359 (1971).
[143] Garrod, D.R. and Steinberg, M.S., *Nature New Biol.*, *244*, 568 (1973).
[144] Martz, E. and Steinberg, M.S., *J. Cell Science*, *15*, 201 (1974).
[145] Jensen, M., Wallach, D.F.H. and Lin, P.-S., *J. Cell Biol.*, *63*, 155 (1974).
[146] Curtis, A., *J. Nat. Cancer Inst.*, *26*, 253 (1961).
[147] Curtis, A. and Varde, M., *J. Nat. Cancer Inst.*, *33*, 15 (1964).
[148] Abercrombie, M., LaMont, D.M. and Stephenson, E.M., *Proc. Roy. Soc.*, *170*, 349 (1968).
[149] Elsdale, T.R. in *Homeostatic Regulators*, G.E. Wolstenholme and J. Knight, eds., Churchill, London, 1969, p 291.
[150] Elsdale, T.R., *Exptl Cell Res.*, *51*, 439 (1968).
[151] Abercrombie, M., Heaysman, J.E.M. and Karthauser, H.M., *Exptl. Cell Res.*, *13*, 276 (1957).
[152] Jensen, M., Wallach, D.F.H. and Lin, P.-S., *Exptl. Cell Res.*, *84*, 271 (1974).
[153] Sanford, K.K., Westfall, B.B. and Jackson, J.L., *J. Natl. Cancer Inst.*, *44*, 611 (1970).
[154] Defendi, V., Lehman, J., Kraemer, P., *Virology*, *19*, 592 (1963).
[155] Abercrombie, M., *Canadian Cancer Conf.*, *4*, 101 (1960).
[156] Abercrombie, M. in *Mechanisms of Invasion in Cancer*, P. Denoix, ed., Springer-Verlag, Berlin, 1967, p. 140.
[157] Abercrombie, M., *Nat. Cancer Inst. Monograph*, *26*, 249 (1967).
[158] Barski, G. and Belehradek, J.Jr., *Exptl. Cell Res.*, *37*, 464 (1965).

[159] Oldfield, F.E., *Exptl. Cell Res.*, *30*, 125 (1963).
[160] Guelstein, V.I., Ivanova, O.Y., Margolis, L.B., Vasiliev, J.U.M. and Gelfand, I.M., *Proc. Natl. Acad. Sci. U.S.*, *70*, 2011 (1973).
[161] Domnina, L.V., Ivanova, O.Y., Margolis, L.B., Olshevkaja, L.B., Rovensky, Y.A., Vasiliev, J.M. and Gelfand, I.M., *Proc. Natl. Acad. Sci. U.S.*, *69*, 248 (1972).
[162] Munder, P., Modolell, M. and Wallach, D.F.H., *FEBS Letters*, *15*, 191 (1971).

9

Cell coupling

9.1. *Introduction*

In 1925, Schmidtman [1] observed that micro-injection of minute amounts of pH indicator into individual cells of a variety of tissues frequently led to the rapid spread of dye from the punctured cells to its neighbors. This suggested that small molecules could pass directly between contacting cells. This discovery was soon forgotten, even after the demonstration by Weidman in heart Purkinje fibers [2] and Furshpan and Potter [3] on crayfish axons that these impulse conducting cells were linked directly by ion-permeable pathways. Both groups considered these linkages synaptic in nature, but in 1963 Loewenstein and Kanno [4] demonstrated non-synaptic, ionic coupling between the salivary gland cells of *Chironomus* larvae and the phenomenon has, since then, been documented for numerous cell systems. The mechanisms involved in intercellular coupling have been reviewed from various points of view [5–8]. The membrane ultra-structure at mammalian intercellular junctions has also been reviewed [9].

Intercellular coupling is a membrane phenomenon. We deal with it here, because intercellular communications are often, but not always defective in tumors. Some aspects of the phenomena involved are treated by Sheridan and Johnson [10].

9.2. *Methods for detecting intercellular coupling*

9.2.1. *Electrical measurements*

One of the major approaches employed to detect intercellular coupling is electric. Here each of two (or more) cells is impaled with a fine microelectrode filled with electrolyte solution and connected electronically to allow measurements of potential and to permit injection of ions (current) (Fig. 9.1). When cells are in cytoplasmic continuity, or are linked by low resistance junctions, introduction of a current pulse into one cell causes a burst of net ion flow to the second cell which induces a change of potential in the second cell. However, the potential change

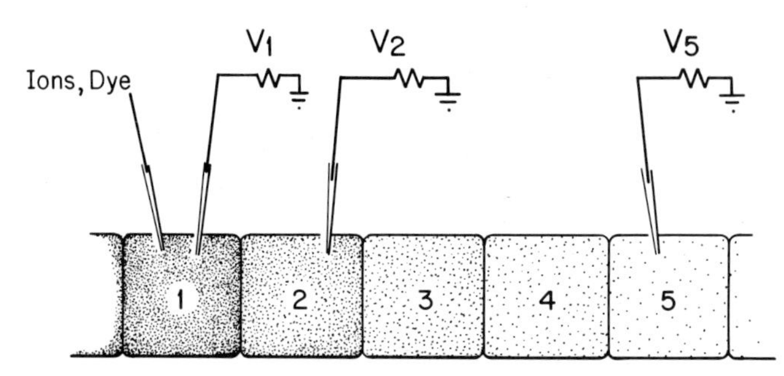

Fig. 9.1. Measurement of cell-coupling using micropuncture. A set of microelectrodes is used to measure the potentials, (V_1, V_2 ... V_5) in a chain of cells. Density of stippling represents concentration of dye injected into cell 1 of a coupled chain.

in the second cell varies in a complex way with the intensity of the current pulse, as well as the junctional and nonjunctional resistances of the cells. It is preferable, therefore, to impale the first cell with a second microelectrode, and use this to record the potential V_1 (Fig. 9.1) in that cell and compare this potential with that of the second cell, V_2 (or other cells). Then the 'coupling' depends solely on the relative junctional and non-junctional resistances of the cells. The degree of coupling is often expressed as the ratio V_2/V_1, the 'coupling coefficient'.

Quantitative evaluation of the coupling coefficient can be complicated. Thus, if the permeability of a junctional area increases relative to nonjunctional membranes, coupling will 'improve'. On the other hand, an increase in the permeability of *nonjunctional* membranes, could appear as 'impaired' coupling. Geometric factors can also be very important [6]. Thus, a junction with given permeability and area can produce greater coupling between two small cells than two large cells. Also, if a small cell is linked to a large one, coupling will be asymmetric; a larger coupling coefficient is obtained when the current is injected into the large cell than when it is injected into the small cell. The situation becomes even more complex when the cells under examination are coupled to other cells by similar junctions; such additional junctions simply drain current away from the test cells. Sheridan [6] suggests that this is the reason for the lower coupling coefficient found for liver cells *in vivo* [11] compared to liver cells *in vitro* [12]; epithelial cells *in vitro* tend to have fewer coupled neighbors than the same cells *in vivo*.

In order to measure *junctional resistance*, one must use four microelectrodes, two in each test cell. Then one can measure the potential response due to current injection into either cell and, knowing the current supplied, compute junctional resistances. For two linked cells, the resistances range between $1 \times 10^5\ \Omega$ and $5 \times 10^5\ \Omega$. For multicoupled systems, estimates range from $3 \times 10^4\ \Omega - > 10^6$ [7].

What are the areas of intercellular junctions? Sheridan [6] has addressed this question micromorphologically. He assumes, as a limit, that the nexus junctions (p. 418) represent the ion permeable paths between cells and computes the ratio of nexus area/cell volume for hepatocytes, brown fat cells and BHK fibroblasts. He obtains ratios of 0.006, 0.005 and 0.001 for the three cell types listed; that is, liver and brown fat cells exhibit similar junctional area/volume ratios while fibroblasts are much lower.

In another approach, Sheridan [6] compared coupling coefficients, junctional areas and junctional area/volume ratios for normal liver, *in vivo*. Ranking the cells according to coupling coefficient: Normal liver, *in vitro* (0.9) = Novikoff cells (0.9) > normal liver, *in vivo* (0.7) ≫ H35 cells (0.3). The order on the basis of junctional area per cell is: normal liver *in vitro* (1.26 μm^2) > Novikoff cells (0.6 μm^2) > H35 cells (0.3 μm^2). (Normal liver *in vivo*, calculated for cells contacting neighbors on all sides gives a value of 20 μm^2). A ranking according to junctional area/volume ratios gives 0.006 for normal liver *in vivo*; 0.001 for normal liver *in vitro*, 0.0008 for H35 hepatoma cells and 0.0008 for Novikoff cells. This last ranking appears to be physiologically, the most meaningful.

Sheridan [7] has computed specific resistances for hepatoma cells combining micromorphologic data with electrophysiologic measurements. The lowest junctional resistance found was $2.5 \times 10^7\ \Omega$ and the largest nexus area was 4 μm^2, giving a specific resistance of 60 Ω/cm^2.

Electrical measurements when feasible are valuable particularly because they provide some quantitation of coupling. However, they are often difficult to execute *in vivo*. Precise interpretation can also be difficult; for example, one cannot distinguish electrically between two large apposed surfaces of moderate resistivity and very small channels of extremely low resistivity. Finally electrical coupling only indicates the presence of 'passageways' for very small particles, that is, ions. It does not inform about the 'sieving' and 'gating' properties of these pathways.

9.2.2. *Tracer techniques*

A *second* approach is to inject a radioactive or fluorescent molecule into one cell and measure its appearance in vicinal cells (Fig. 9.1). The use of radioactive tracers requires either direct assay or radioautography. In contrast the spread of fluorescent dyes can be monitored directly by fluorescence microscopy. Such studies show, for example, that K^+ passes freely between myocardial cells [13], as do Na^+, K^+, Cl^-, I^- and $SO_4^=$ between septate axons of Crayfish [14].

Investigations on the movement of dyes between cells give information about larger molecules. Thus, fluorescein (mol. wt. 330, diameter ~ 10 Å) passes readily between the cells of dipteran salivary glands [15], between cultured neoplastic hepatocytes [16] between various electrically coupled, cultured fibroblasts [17]

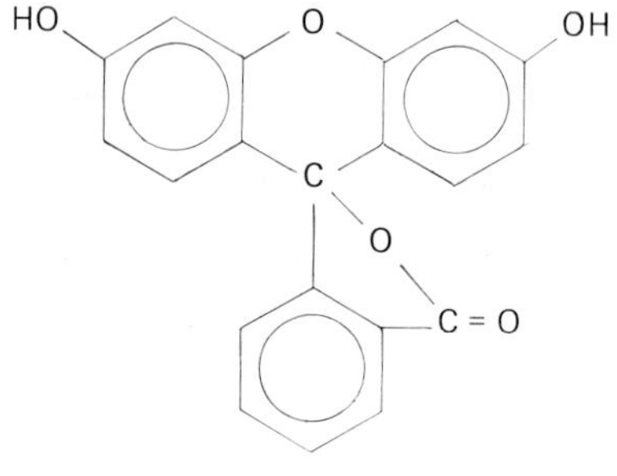

Fluorescein

and through the junctions of the crayfish septate axons [14]. Procion yellow (mol. wt. 625) also moves freely between contacting hepatoma cells as do dansyl-L-glutamate and dansyl-L-aspartate [16]. Procion yellow also passes readily between salivary gland cells [15] and through the junctions of the crayfish septate axon [14].

Interestingly, fluorescein does not cross between coupled cells of *Fundulus embryos* [18] or of *Xenopus blastulae* [19]. This indicates that electrical coupling can occur between cells which cannot interchange molecules the size of common metabolic intermediates.

A related technique for the detection and quantitation of intercellular junctions has been developed for 'free-floating' cells such as lymphocytes, ascites-tumor cells, etc., which do not lend themselves to micro-injection techniques [20]. In this approach cells suspect of junction-forming capacity are incubated with a non-fluorescent fluorescein ester, such as fluorescein dipropionate:

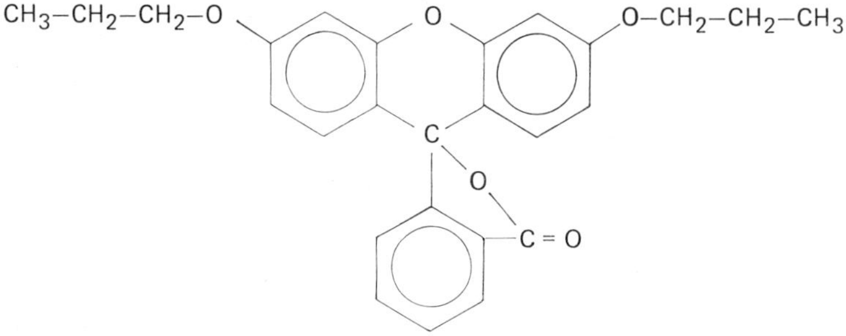

These lipophilic substances easily diffuse into most cells where they are rapidly cleaved into free (fluorescent) fluorescein by non-specific, intracellular esterases [21]. The efflux rate of free fluorescein is much slower than the combined rates of ester-influx and hydrolysis. The dye thus accumulates in the cytoplasm, allowing the detection of labelled cells by their intense fluorescence; this sequence is termed 'fluorochromasia'. High intracellular accumulations of dye can be achieved in this fashion, so high in fact that the normal diffusion of fluorescein from cells, or the loss of dye from damaged cells, does not contribute significant background interference during short term experiments. Thus, if collision between cells produces pathways for intercellular transfer of molecules the size of fluorescein, these will result in the transfer of dye from pre-labelled cells to unlabelled ones. Such transfer can be detected by fluorescence microscopy [20] and quantified using a 'fluorocytograph' technique [22].

The microscopic data demonstrated that mouse lymphocytes, sensitized against mouse mastocytoma cells, rapidly form transient, fluorescein-permeable junctions with the tumor cells. Moreover, a single lymphocyte can form such junctions repeatedly with different target cells. Using the flow-fluorocytograph, Sellin et al. [22] show that sensitized lymphoid cells tend to form fluorescein-permeable junctions very readily with unsensitized lymphocytes. In contrast, unsensitized lymphocytes are resistant to junction formation with each other or with different cells. Hülser and Peters [23] have confirmed this observation using electrophyssiological measurements. These authors used microelectrodes to measure intercellular ion flow between lymphocytes aggregated by a non-mitogenic agglutinin. No coupling was observed between unstimulated cells, but ionic communications appeared within seconds after exposure of the cells to mitogenic lectins. Lymphocytes, agglutinated after stimulation with mitogenic lectins, have also been shown to exhibit extensive intercellular electrical communication [24,25]. Introduction of Ca^{2+} into the coupled cells produces decoupling within two to four minutes after injection.

Lymphocytes appear to occupy a unique role in terms of intercellular coupling. Virtually all other normal cell types with proliferative potential are usually coupled, but in lymphocytes coupling capacity appears only as a concomitant of immunologic differentiation. Interestingly, a number of neoplastic lymphocytes carried in culture have been observed to form fluorescein-permeable junctions quite indiscriminately with unrelated cells (Fig. 9.2), but the generality of this finding remains to be established [26].

9.2.3. *Metabolic coupling*

The above three approaches directly measure the passage of substances between cells in contact. Another criterion of intercellular communication has been termed 'metabolic coupling'. This phenomenon was discovered by Subak–Sharpe et al. [27,28] who isolated a variant of polyoma transformed baby hamster fibroblasts (BHK 21) lacking the enzyme IMP, GMP: pyrophosphoribosyltransferase; therefore these mutant cells cannot incorporate radioactively labelled, exogenous hypoxanthine into nucleic acid, when *grown alone*. However, radioautographic studies show that they can incorporate the label when *grown together* with non-defective BHK 21 sublines, under conditions of close contact. In these experiments

Fig. 9.2. Dye transfer between fluorescein-labelled VERO fibroblasts (background) and 2 contacting leukemic lymphocytes. Courtesy Dr. P.-S. Lin, Magnification 1900 ×.

a few wild-type 'donor' cells were scattered onto the defective cells (ratio 1:300). The donors could then be identified by their intense radioautographic image. A gradient of labelling could often be observed between non-defective 'donor' cells, defective cells in primary contact with the donor cells and defective cells in 'secondary' contact with 'donors', that is, via other defective cells. Radioautographic studies show that this transfer occurs also between the defective cell type and cells of a different kind and from different species [29]. It appears that the material transferred from the normal to the deficient cells constitutes nucleotides, or nucleotide derivatives, which bypass the enzymatic step lacking in the mutant [30].

Similar metabolic cooperation has been observed between normal, cultured fibroblasts and cells from Lesh–Nyhan syndrome, that is, deficiency of IMP, GMP: pyrophosphoribosyltransferase [31,32]. Bürk et al. [33] demonstrate that metabolic cooperation can be bi-directional. They co-cultivated cells from a polyoma transformed BHK line deficient in IMP, GMP: pyrophosphoribosyltransferase, with cells of another line deficient in AMP pyrophosphoribosyltransferase; this last cell line cannot incorporate exogenously-added [^{3}H]adenine into its nucleic acid. Under conditions of contact, the enzymatic defects in both cell types were bypassed, that is, adenine is converted to adenylate in one cell population, hypoxanthine to inosinate in the other and the products are exchanged freely between the cells of the two populations.

Gilula et al. [34] have correlated metabolic cooperation with ionic coupling and morphological junctions in variants of cultured Chinese hamster fibroblasts. Cells of the permanent DON line can incorporate exogenously added [^{3}H]hypoxanthine into their nucleic acids. This requires the participation of IMP,GMP:pyrophosphoribosyltransferase, which can be demonstrated in these cells by autoradiographic and enzymatic techniques. Two other cell lines DA and A9, cannot incorporate the exogenously added purine due to a lack of the transferase. The enzyme defect in the DA and A9 cells can be bypassed by co-cultivating DON cells with DA cells. Moreover, the DON cells readily couple ionically with DA cells. In contrast, co-cultivation of DON cells with A9 cells does not alleviate the metabolic defect and does not lead to electrical coupling between cells. Electron microscopic observations, using both thin section and freeze-fracture approaches, show gap junctions (p. 418) only between DON and DON or DON and DA, but not between DON and A9 or A9 and A9. The data thus indicate that A9 cells are a variant which lack IMP,GMP:pyrophosphoribosyltransferase as well as the capacity to couple ionically and morphologically to each other or to different cells. This coupling defect appears responsible for the lack of 'metabolic cooperation' between A9 cells and DON cells.

The role of intercellular junctions in metabolic cooperation has been further explored by Azarnia et al. [35]. Using an approach somewhat similar to that of Gilula et al. [34] these workers isolated three strains of cell, which are incapable of forming junctions with other cells as, tested electrically and by the passage of injected fluorescein: two are sublines of the Morris H-5123 rat hepatoma and one an epithelial line derived from X-ray transformed embryonic hamster fibroblasts. These cells were preloaded with [^{3}H]hypoxanthine and then cultured in contact with the cells deficient in IMP,GMP:pyrophosphoribosyltransferase. Label transfer was measured by radioautography and intercellular coupling was assayed electrically and by fluorescein transfer in the same preparations. No transfer or coupling was observed. In contrast, co-cultivation of the defective cells with several cell lines capable of junction-formation led to label transfer as well as to coupling. The authors observed that some label transfer can occur via the bulk medium from the cells incapable of junction-formation and the metabolically defective cells. The effect is small, compared with contact-dependent transfer, and is not observed in coupled systems.

In a very extensive study Cox et al. [36] evaluated the ability of various non-deficient cells to bypass enzymatic deficiencies in diverse mutant cells. Their data indicate that the metabolic cooperation between normal cells and cells deficient in IMP,GMP: pyrophosphoribosyltransferase or AMP: pyrophosphoribosyltransferase is due to transfer of nucleotides or nucleotide derivatives. However, glucose-6-phosphate dehydrogenase deficiency cannot be corrected through 'metabolic cooperation'. Clearly neither the enzyme (mol. wt. 240,000), its subunits (mol. wt. 43,000) nor its metabolic products are able to cross between contacting cells in this system.

9.2.4. 'Sieving' or 'gating' in intercellular communications

The electrical measurements tracer techniques and metabolic methods described indicate that small molecules may pass easily from cell to cell, but give no information about exchange of larger substances, including macromolecules. However, Ashkenazi and Gartler [37] have shown that the defect in cultured Lesh–Nyhan cells can be bypassed by incubating these cells with *homogenates* of non-defective cells [38]. Incubation of cells, defective with respect to branched-chain amino acid decarboxylase, with sonicates of normal cells eliminates this deficiency also. The corrective factors appear to be protein in nature, and may be the enzyme lacking in the defective cells. These experiments clearly demonstrate a process different from the previously discussed intercellular transfer between cells; pinocytosis appears a likely possibility. Kolodny [38] has suggested that macromolecular, radioactively-labelled RNA can transfer directly from 'donor' 3T3 fibroblasts to 'recipient' cells, but his experiments do not exclude transfer of nucleotides from donor to recipient and their subsequent incorporation into RNA.

In attempts to evaluate the 'sieving' characteristics of intercellular communications, several investigators have extended the microinjection technique to tracer macromolecules. The results have been less than gratifying. Thus, Kanno and Loewenstein [39] have injected fluorescein-labelled bovine serum albumin into Drosophila salivary gland cells and report appearance of fluorescence in coupled cells (as far as ten cells from the injection site). However, these experiments have not been fully substantiated and the authors concede that the transferred label might represent breakdown products of the injected protein. Reese et al. [40] have reported that a 'microperoxidase' component (mol. wt. 1600) of horseradish peroxidase (HRP) could cross between apposed septate axons of crayfish, whereas HRP (mol. wt. ~ 44,000) could not. However, further experimentation [5], described below, shows that this observation was due to the artefactual creation of large intercellular junctions during fixation. More recently, Loewenstein [4] injected HRP (mol. wt. ~ 44,000) myoglobin (mol. wt. ~ 17,000) and cytochrome *c* (mol. wt. ~ 12,500) into salivary gland cells of *Chironomus* and monitored the distribution of peroxidase by light and electron microscopy using H_2O_2 as substrate and 3,3-diaminobenzidine as oxygen acceptor. Oxidation of this compound ultimately leads to brown, electron-opaque deposits very close to the site of production [42]. These reaction products can reduce OsO_4 leading to osmium deposition and additional opacity. The peroxidase technique is extraordinarily sensitive because a few enzyme molecules can generate large amounts of visible reaction product.

Loewenstein [41] finds that after glutaraldehyde fixation, electron-opaque deposits appear in the injected cell and in cells coupled to it, as shown by fluorescein transfer; no staining is found in cells previously decoupled, or after injection of myoglobin and cytochrome *c*. (These are stated to have peroxidase activity. It has been established, however, that the H_2O_2 reduction by these proteins is non-enzymatic and depends only on their heme content.) Loewenstein [41] argues paradoxically that the large HRP molecule (50 Å diameter) crosses and that the smaller proteins do not. We believe that this interpretation is not correct for the following reasons:

(a) The detection sensitivity for HRP is vastly greater than for the other heme proteins. Besides, the experiments using myoglobin and cytochrome *c* are not documented.

(b) Bennett [5] and associates have demonstrated a major source of artefact. This arises from the use of glutaraldehyde fixation to visualize the peroxidase reaction. Bennett [5] reports that fixation, for example, by glutaraldehyde, decreases junctional permeability as such, but in time (10–30 min) allows the formation of artefactual intercellular channels which allow the passage of large molecules. Thus, horseradish peroxidase can be induced to cross between cells in this fashion. Indeed, the combined use of lanthanum (which can substitute for calcium [43]) and glutaraldehyde very predictably produces large intercellular channels, while lanthanum alone can induce cell fusion.

The experiments of Bennett and associates indicate the critical caution which has to be exerted in experiments designed to estimate the 'sieving' properties of junctions by intercellular passage of labelled macromolecules.

Loewenstein [41] also presents electronmicrographs, purported to demonstrate preferential deposition of oxidized diaminobenzidine in the 'channels' of 'septate' junctions between adjacent salivary gland cells. However, the apparent differences between 'peroxidase-stained' and control junctions fall well within the limits normally observed in thin sections. Besides, as already noted, deposits of oxidized dye do not specifically localize the peroxidase.

In summary, the published information currently available gives no reliable indication that substances larger than Procion yellow (mol. wt. ~ 600) can pass directly between cells. This size, however, is totally adequate for the transfer of low molecular weight, '*intracellular*' messengers, such as cAMP and cGMP (Chapter 6).

9.3. Morphological correlates of intercellular communications

Intercellular communications have long been recognized to require close membrane appositions of the communicating cells. Two types of close appositions have been defined; namely, the 'nexus' or 'gap junction' and the 'tight junction'. There is now considerable indirect evidence compatible with the possibility that nexuses mediate electrotonic coupling and also the transfer of small molecules such as fluorescein and Procion yellow.

The general structural characteristics of intercellular junctions have been extensively reviewed by McNutt and Weinstein [9]. Transmission electron microscopy of stained thin sections perpendicular to the membrane planes of apposed cells show that, in nexus regions, the membranes of the two cells approach to within 20–30 Å. This results in a seven-layered complex, that is, two trilaminar membranes separated by a 20–30 Å zone. The thickness of the complex is ~ 150 Å.

The gaps in nexuses (gap junctions) are defined by their accessibility to extra-

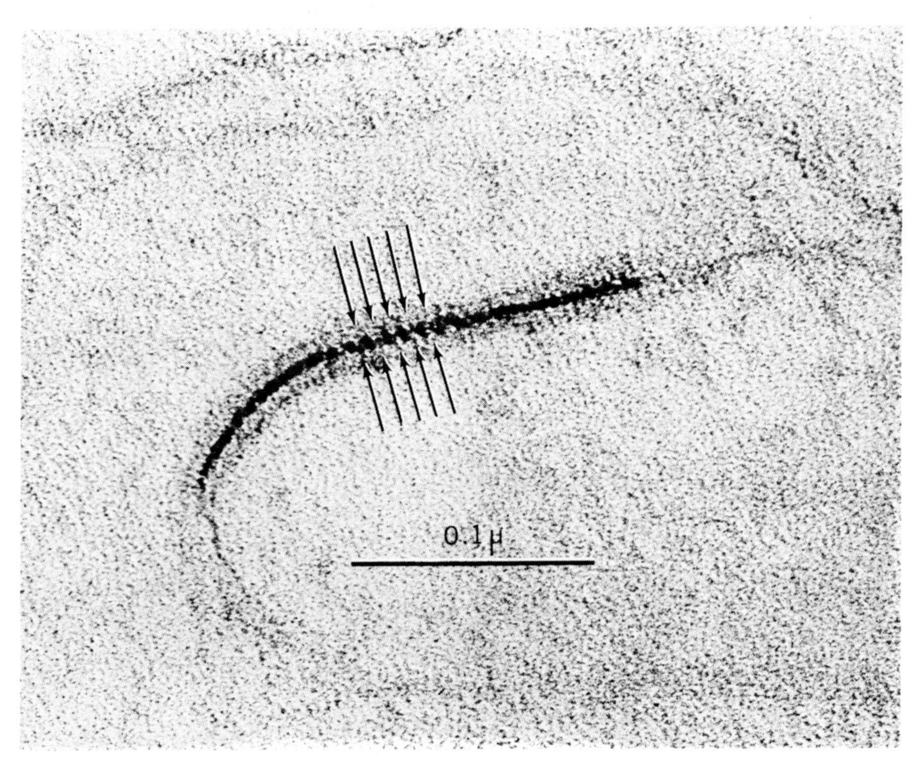

Fig. 9.3. Perpendicularly-sectioned gap-junction or nexus visualized using colloidal lanthanum. Arrows indicate probable points of intercellular communication. From McNutt and Weinstein [9] through the courtesy of the authors and Oxford University Press.

cellular macromolecular tracers, for example, colloidal lanthanum hydroxide (La $(OH)_3$). The gaps can also be identified using the peroxidase technique [44]. After fixation in the presence of colloidal lanthanum, sections normal to the membranes show that the gaps between the apposed cells consist of extracellular channels (electron-opaque), with a periodicity of about 100 Å, Fig. 9.3. Well-defined structures bridge the gaps. These are generally not observed in thick, perpendicular sections of lanthanum-impregnated material but they can be visualized in very thin sections.

Nexuses generally occur as macular patches of contact between cells. However, the size of these patches can vary very markedly. One can probably not recognize a nexus unless it consists of at least 4 subunits. However, sometimes the maculae can be as large as 10 microns.

The structure of nexuses has been very greatly elucidated by the use of freeze-fracture electron microscopy [45,46]. The typical appearance of a nexus by freeze-fracture microscopy is shown in the Figure 9.4. The nexuses appear as hexagonal arrays of intramembranous particles with 90–100 Å depressions. (This is also occasionally observed on non-functional particles.) The nexus probably arises through the apposition of identical structural domains in the contacting membranes. This is schematized in Fig. 9.5.

Most models suggest that the individual particles in a nexus bear channels for intercellular communication. It is not proven, however, that this is really the case. Also, if it is so, do dispersed, single particles allow coupling? If this is so, the absence of recognizable nexuses cannot be used to exclude coupling. Certainly one could not recognize hexagonal packing, one of the major criteria of a nexus with fewer than four associated particles*.

The ultrastructural characteristics of gap junctions suggest that they might connect the cytoplasmic spaces of coupled cells. First, they do constitute a structure bridging the gap between junctional membranes and such a structure is required if channels exist between cells. Second, the junctional areas of cells appear adequate for intercellular communications [6] ranging from 0.05% of the surface area, in cultured fibroblasts, to 1–2% in hepatocytes and brown fat cells. Third, when nexuses are disrupted morphologically with low-calcium buffers [47] or hypertonic solutions [48,49] there is a coincidental loss of electrical coupling. Also, the dimensions of nexus subunits fit their postulated role. Thus the work with tracer molecules to date limits the size of the intercytoplasmic channel to about 10 Å diameter. The entire bridging structure as determined by freeze-cleaving electron microscopy is 70-80 Å and the repeat period of the junction structure is 100 Å in diameter. This makes the diameter of the extracellular

---

* An extensive search for nexuses between contacting lymphocytes under conditions known to produce ionic coupling and dye transfer (p. 412) has not revealed any such structures [46]. However, large regions of closely apposed membranes were observed.

channel, revealed by lanthanum impregnation, about 20 Å. The intercytoplasmic channel in the center of the bridging particle would then have 'walls' about

Fig. 9.4. *En face* view of a nexus in freeze-cleaved and etched myocardium. Nexus face A (NA) is oriented toward the cytoplasm and bears numerous intramembranous particles, designated contact cylinders (CC). These are often in hexagonal array (insert). Nexus face B (NB) is oriented toward the cytoplasm (apposed cell). It is marked with double arrows and contains numerous pits (P). × 170,000 except for insert which has a magnification of 200,000 ×. Shadow direction toward 1 o'clock. From McNutt and Weinstein [9] through the courtesy of the authors and Oxford University Press.

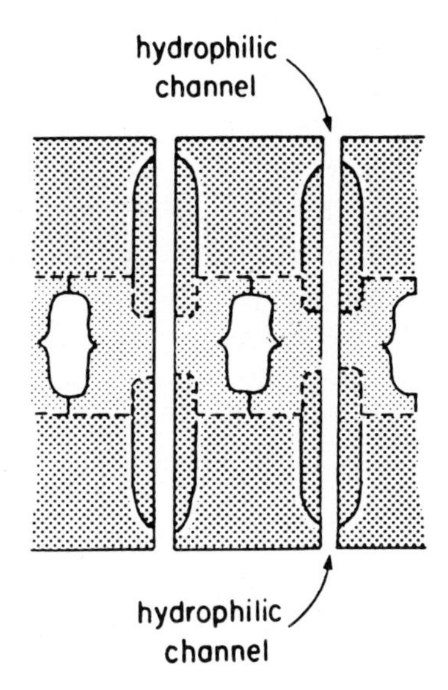

Fig. 9.5. Interpretation of the location of hydrophilic channels in a nexus The hydrophilic channels are envisaged to pass through the centers of apposed intramembranous particles. From McNutt and Weinstein [45] through the courtesy of the authors and of the Rockefeller University Press.

30–35 Å thick. Some authors [5] argue that the computed wall dimension is reasonable for a lipid bilayer membrane, but most such bilayers are much wider. It would appear more probable that the bridging structure is a polymeric protein. This is certainly the most reasonable explanation of the freeze-fracture appearance of nexuses, as well as of the fact that isolated nexuses are degraded by proteolytic enzymes [50].

There has been considerable progress in the isolation and purification of nexus junctions. First, Benedetti and Emmelot [51] have been able to prepare fractions of rat liver bile-fronts containing arrays of hexagonally packed subunits, which can be identified with nexus junctions. More extensive work has been carried out by Goodenough and Revel [52,53]. These authors have developed techniques for the purification of nexuses from mouse liver bile-fronts. These segments of apposed hepatocyte plasma membranes were isolated by conventional techniques and then disrupted by ultrasound in the presence of the detergent Sarkosyl N197. This was followed by ultracentrifugation in sucrose gradients. A small fraction, at a density of approximately 1.16, consists of relatively well purified nexus junctions. These are readily recognized by thin section electron microscopy and by negative staining. The negative staining show the typical hexagonal packing with a periodicity of about 100 Å. Analysis of the nexus fraction by polyacrylamide gel electrophoresis in sodium dodecylsulfate indicates a relatively simple protein composition. Indeed the electrophoretic pattern reveals only two major, non-glycosylated components, with approximate mol. wt. of about 20,000 and 10,000 in a 1 : 3 molar ratio [54]. This fact is remarkable in view of the very complex protein patterns in intact membranes [52]. The two electrophoretic components could be separated into four bands by gradient gel electrophoresis [54].

Semi-quantitative thin-layer chromatographic analyses of the nexus lipids also show a rather simple pattern; phosphatidylcholine predominates and there is rather little phosphatidylethanolamine. This lipid pattern is also considerably simpler than that of the total membrane.

X-Ray diffraction analyses reveal the same periodicity of the hexagonal lattice as is observed electronmicroscopically [52,55,56]. The X-ray data further demonstrate that the nexus proteins penetrate through the thickness of the membrane.

What information proves that nexuses act as the *sole* mediators of intercellular communication? The evidence is largely inductive. Thus, coupled cells when examined electron microscopically reveal nexus junctions. Conversely, cells which are not coupled do not show nexuses. However, how does one recognize a small nexus? Can a single intramembranous particle act as a channel between one cell and another? In the case when coupling is very close, nexuses appear unusually numerous, but careful quantitative correlations remain to be carried out. In some instances the degree of coupling and the incidence of gap junctions appear to be similarly affected by an experimental manipulation [5], but this correlation does not hold in all cases and does not prove what functions the nexus junctions perform. Indeed, glutaraldehyde fixation, which definitely causes decoupling, certainly does not destroy the hexagonal structure of gap junctions. Indeed it is generally required to visualize it.

Although there is an extensive correlation between the incidence of nexuses and coupling, one cannot exclude that coupling may occur by way of the extracellular space. Thus, Bennett [5] has computed that the resistivity per unit area of junctional membranes lies in the vicinity of about 1 Ω/cm^2. This is much smaller than the 60 Ω/cm^2 value of Sheridan [7] but both values lie well below the 10^3–10^6 Ω/cm^2 of non-junctional membranes. If small regions of such 'low-resistance membranes' were apposed but separated by an intercellular space of about 200 Å, one would detect significant electrical coupling [56,57]. All that is required is that the 'low resistance' membrane domains are restricted to the area of apposition. This is not an unreasonable notion, particularly since some regions of apposing membranes can be sealed off from the extracellular space by the so-called 'tight-junctions', which represent bands of intimate contact between two cell surfaces. Without 'tight junctions' electrical measurements cannot readily discriminate how much current leaks out of the cleft and how much flows across the membrane away from the region of apposition. The leakage merely forms part of the overall input resistance of the cell.

Interaction by way of apposed but separated low resistance membranes becomes unlikely only when coupling is very close. However, it could account for many coupling phenomena. Thus, coupling by way of extracellular space has been shown to occur in a number of instances. For example, in *Rana pipiens* [58] and *Xenopus laevis* [59] there is a marked increase in transmembrane potential (inside negative) and membrane resistance during cell division shortly after appearance of the cleavage furrow. This can be attributed to the new membrane inserted during cell division in the regions where the daughter cells appose. DeLaat and Bluemink [60] argue that the new membrane has a high K^+ permeability and compute a specific resistance of 1–2 $K\Omega\,cm^2$ compared with 74 $K\Omega\,cm^2$ for 'old' membrane. They point out that such low specific resistances could easily provide a high coupling ratio. Apparently the high permeability of 'new' membrane formed at cell division is *not limited to ions*, since it is only at this time that cytochalasin B (mol. wt. 474) can also enter the cells [59]. Apposition of 'high permeability' membranes might, therefore, allow intercellular passage not only of ions but of small dyes. This is not an unreasonable supposition since fluorescein, one of the most widely used markers for intercellular communication, diffuses relatively easily through plasma membranes [22,61].

Even the transfer of small nucleoside derivatives between apposed cells might be explained without invoking cytoplasmic continuity between the cells. All that would be required is the location of the appropriate membrane transport machinery in the regions of apposition. Of course, without knowing the specific nature of the material transferred in experiments on metabolic coupling, one cannot go beyond suggesting this possibility.

9.4. *The genesis of coupling*

The plasma membranes of diverse 'free floating' cells including those which have been artificially separated from adhesive tissues, have high resistance characteristics. Clearly, junctional communication is a characteristic of contacting cells. One must therefore ask how communication comes about once contact is established and what mechanisms are involved. O'Lague et al. [62] have demonstrated electrical coupling between dividing and adjacent non-dividing fibroblasts *in vitro*; this finding is significant because dividing cells appear to lose close contact with surrounding interphase cells. This observation has been confirmed for the case of mitotic and interphase epidermal cells *in vitro* [63]. Also as already noted, [20,22,62] under appropriate conditions communications can form between contacting lymphocytes in very short periods of time; they can also be dissolved in equally short periods of time. Coupling thus does not require 'strong', permanent contact.

In this direction Loewenstein and Ito [64,65] have found that when one manipulates embryonic cells into contact while monitoring the electrical conductance between the cells, electrical communications establish between the cells within seconds upón bringing the cells into mechanical contact. Also, when the cells were again separated mechanically or by chemical treatment, the cellular membranes reestablished high resistance provided that sufficient calcium ion was present in the external medium. New, communicating junctions could then be established by pushing the cells into contact at another area. The authors argue that since communications could be formed apparently at almost any spot on the membrane, large parts of the total cell membrane must have the potential to form communicating junctions. This suggestion again raises the alternative that passage of ions and small dyes might proceed via apposed, low-resistance membranes rather than intercellular channels. However, DeHaan and Hirakow [66] find that ionic and morphologic communications can appear very rapidly after contact of cardiac myoblasts. These cells generate spontaneous, rhythmic action potentials, which, in separated cells, occur at different frequencies. When the cells become coupled, their electrical activities become synchronized. This can occur within minutes after contact and seems closely related to the appearance of morphologic junctions, between apposed cells. These can be detected within 2 min after synchronization.

Freeze-fracture electron microscopy shows that non-junctional membrane domains contain uniformly distributed particles, many of which are in the same size range as those forming the closely packed arrays characteristics of nexuses [46,45]. Moreover, nexuses are frequently surrounded by haloes of particle-free membrane [45]. Certainly not all of the distributed particles participate in coupling processes. For example, the membrane of erythrocytes contain numerous intra-membranous particles of a size equivalent to that prevalent in the membranes of cells capable of coupling and nexus formation. Probably only some of the particles are potential elements for intracellular communication and potential precursors of what is defined as a nexus. The process of nexus formation might thus proceed by the lateral diffusion of such elements in a two dimensional fluid membrane and their subsequent aggregation in a nucleation process, where the typical hexagonal arrays may form after aggregation of a few channel bearing units. Clearly, the size and symmetry properties of individual or even aggregated intramembranous particles would only make these 'hemi-channels', which could link up with corresponding elements in opposed cells, thus forming the intercytoplasmic channel of the gap junction.

That this may be the case is suggested by the work of Sheridan and associates (cf. [7]), and others, who have approached this problem both electrophysiologically and micromorphologically.

The experiments of Johnson et al. [67] suggest the following sequence in the formation of nexus junctions and ionic communications between Novikoff hepatoma cells:

(a) When cells contact, planar areas of membrane apposition develop. These

bear 10 nm intramembranous particles (but essentially no others). These particles may interact across the intercellular gap which is initially 10–20 nm wide.
(b) The intercellular gap then narrows to 2–4 nm.
(c) The particles then aggregate into small hexagonal nexuses.
(d) The small aggregates then fuse to form large nexuses.

Somewhere between steps (c) and (d) the cells become ionically coupled and permit intercellular fluorescein transfer.

This concept fits the observations of Yee [68] who monitored the incidence of nexuses between rat hepatocytes at varying intervals after partial hepatectomy. Between 28–35 hr after surgery there is an extensive decrease in junctions. Between 37–40 hr small nexuses reappear and by 44–46 hr, the junctional pattern cannot be distinguished from that of normal liver. A process of junction-maturation has also been observed in the case of hepatoma H35 propagated *in vitro* [69].

The development of electrical coupling and nexus junctions between hepatoma cells can proceed even in cells whose protein synthesis has been blocked by cycloheximide and whose intracellular ATP levels have been depleted by iodoacetate [70]. These studies indicate that the association of nexus proteins into clusters and the alignment of the nexus subunits of apposed cells, are 'self assembly' processes.

9.5. *The role of calcium*

Loewenstein and associates have suggested a critical role for calcium ion in the control of intercellular coupling and have carried out many provocative experiments in this direction. They propose that cells pump divalent ions outward more or less uniformly over their surface in order to maintain a low cell cytoplasmic Ca^{2+} activity [71]. Loewenstein and associates argue that intracellular calcium is critical in the making of the low-membrane resistances in junctional regions. Their hypothesis rests on experiments in which decoupling of cells is achieved by injection of calcium ion into cells [72]. Under such conditions, the permeability of junctional membrane regions falls drastically when intracellular calcium concentration is raised from below 10^{-6} M to above $4–8 \times 10^{-5}$ M. At intracellular concentrations of 10^{-3} M, such as prevail in normal *extracellular* fluids, junctional domains cannot be distinguished in terms of permeability from non-junctional membranes and intercellular communication is blocked. Other alkaline earth cations also depress junctional membrane permeability but with less efficiency than calcium. For example, Mg^{2+} is effective at about 10 times the concentration of Ca^{2+} and Sr^{2+} and Ba^{2+} require 100 and 1000 times greater concentration. The Mg^{2+} finding is paradoxical, since intracellular Mg^{2+} levels are usually well above 10^{-5} M. Indeed, levels of 10^{-4} M are required for the functioning of numerous intracellular enzymes.

There can be no question that intracellular [Ca^{2+}] is regulated at low cytoplasmic levels by active processes, involving both transport *out* of the cell through the plasma membrane and *into* the *mitochondria* (Chapter 6). The role of the mitochondria appears particularly critical, particularly since their calcium binding- and/or pumping-capacity may change upon neoplastic conversion.

9.6. Coupling defects in neoplasia

Defective coupling has been observed in numerous cancerous tissues, for example, malignant hepatocytes in a number of hepatomas [73], neoplastic thyroid cells and malignant gastric cells [74,75]. Defective coupling has also been observed in cell cultures derived from such tissues [11,76–78]. These physiological studies correlate with electronmicroscopic observations which suggest a common deficiency of nexuses between contacting neoplastic cells. Thus, Benedetti and Emmelot [79] report a deficiency of nexuses in 'bile-front' isolates of hepatoma cells, compared with similar preparations from normal liver. Moreover, Martinez-Palomo et al. [80] find that such junctions are much less common in SV40-transformed fibroblasts, compared with the normal variants, while 'tight' junctions and desmosomes are equally common in the two cell types. Also McNutt et al. [45,81] have demonstrated a deficiency of such junctions between the cells of squamous cell cervical carcinomas. They find 5 to 10 junctions per cell in the basal rows of proliferating cervical cells, 100 to 200 junctions per cell in the normally differentiated 'intermediate zone', but *less than one per cell* in invasive carcinomas. Carcinoma *in situ* and severe dysplasias also appear deficient in nexuses.

However, numerous malignant cells do not exhibit junctional defects with respect to the passage of small molecules (11,16,17,33,34,35,63,78 and many others). For example, Johnson and Sheridan [16] studying Novikoff ascites hepatoma Nl Sl 67, which grows in small clumps and chains, discovered tight electrical coupling between cells, ready intercellular passage of fluorescein and numerous nexuses. No tight junctions or desmosomes were observed. Similarly, Flaxman and Cavoto [82] observe comparable electrical coupling between the cells of *in vitro* explants from normal human epidermis and epidermal basal cell carcinoma. Morphologically speaking, Pinto da Silva and Gilula [83] find identical nexus junctions between normal chick embryo fibroblasts and cells transformed by Rous sarcoma virus.

Hülser and Webb [84] have carried out a comprehensive series of experiments to clarify the possible relationships between intercellular coupling and malignancy. They evaluated the *in vitro* coupling properties and *in vivo* tumorigenicity of seven permanent 'fibroblastoid' lines and seven epithelioid lines. Their results are summarized in Table 9.1.

TABLE 9.1

Ionic coupling and tumorigenicity[1,5]

Cell line	Origin	Morphology when forming a monolayer	Ionic coupling	Tumor-igenicity[4]
RE (6 267)	Rat embryo (Sprague–Dawley)	Epithelioid	–	–
ME (7 769)	Mouse embryo (BALB)	Epithelioid	–	–
HE (22 667)	Hamster embryo	Epithelioid	–	NT
HE (21 268)	Hamster embryo	Epithelioid	–	NT
HEBP (131 267)	Hamster embryo benzpyrene-transformed	Epithelioid	–	NT
HeLa	Human, carcinoma of cervix	Epithelioid	–	+
KB	Human epidermoid carcinoma of the nasopharynx	Epithelioid	–	+
RE-F (1 272)	Rat embryo (Sprague–Dawley, selected from RE)	Fibroblastoid	+	–
3T3	Mouse embryo	Fibroblastoid	+	–
BHK-21	Baby hamster kidney	Fibroblastoid	+	+
BHK-21/SV[2]	Baby hamster kidney	Fibroblastoid	+	+
BHK-21/SVf[3]	Baby hamster kidney	Fibroblastoid	+	+
BICR/M1R-K	Transplanted rat mammary tumor (Marshall)	Fibroblastoid	+	+
RN	Transplanted rat neurinoma (BD IX)	Fibroblastoid	+	+

[1] Cells were considered coupled after at least 50 measurements in different dishes from various preparations. In the case of RE and BICR/M1R-K cells several hundred measurements were carried out.
[2] One year persistent infection with Sindbis virus.
[3] One year persistent infection with Sindbis virus eliminated by anti-Sindbis-bovine serum.
[4] NT = Not tested.
[5] From Hülser and Webb [84].

All fibroblastoid lines are coupled and there is no difference in ionic coupling whether the cells are of normal or of malignant origin. In contrast, none of the epithelioid cells are coupled, again regardless of origin. There is no consistent relation between *in vivo* tumorigenicity and the presence or lack of coupling. The electrical communications between fibroblastoid cells was unaffected by variations of serum source, growth in media conditioned by uncoupled cells, or variations of cell cycle. The authors data again confirm the fact that one cannot use ionic coupling as a criterion distinguishing normal from neoplastic cells.

A number of technical problems arise when one attempts to compare normal with neoplastic cells. Among these are the following:

(a) Microelectrode studies always produce mechanical stress, some depolarization and even some Ca^{2+} influx, all of which can produce 'uncoupling'. Some tumor cells may vary from normal in their sensitivity to these effects.
(b) Artificial uncoupling may occur even more readily in dye transfer experiments because these require larger electrode tips.
(c) Tumor cells tend to vary considerably in volume.
(d) Presence of coupling merely indicates that communications exist, not that they are normal.
(e) Coupling coefficients depend on non-junctional resistances.

Additional complications arise when measurements are made *in vitro* [17]. Thus, cultured Sarcoma 180 cells are well coupled when maintained in media which have not been renewed for 4 days or more. However, 1 to 2 days after changing to fresh medium 60% of the cells become decoupled. With time they then again couple. Excluding possible artefacts due to experimental manipulations, these data raise questions about the apparent normal coupling between the cells remaining in communication after the medium change. It could be that the junctions have normal resistance. It could also be that more cells are now coupled in pairs or chains than in sheets [16]. This could giva a *normal coupling coefficient* with *an increase of junctional resistance.* For reasons such as this, Johnson and Sheridan [16] stress the need to measure junctional resistance when comparing cells. Their preliminary data suggest that some hepatoma cells, although coupled, may have impaired junctional permeability.

9.7. Hybridization studies

An important experimental direction has been initiated by Loewenstein and associates, to clarify the relationship between malignancy and junctional deficiency in those cases where it exists. These workers have employed hybridization techniques to determine the genetic determinants of intercellular coupling. In this they have followed the observations of Weiss et al. [85], Klein et al. [86], Bregula et al. [87] and Wiener et al. [88] that fusion of various kinds of neoplastic cells with non-neoplastic cells tend to yield non-neoplastic hybrids.

Loewenstein and associates have published data on two hybrid systems. The *first* [76] is a liver cell system: cells of a neoplastic hepatocyte line were fused by use of Sendai virus with three normal cell lines, two of them fibroblasts and one hepatocyte. The neoplastic hepatocytes do not form junctions between each other, are not contact-inhibited (they grow to very high densities in culture) and injection of 100 cells into a rat produces fatal tumors. In contrast, the three types of normal cells used are electrically coupled, contact inhibited, and do not form tumors. The heterokaryons formed immediately after fusion couple both with each other and with normal parental cells; they also allow the passage of fluorescein between contiguous cells. The appearance of normal coupling is accompanied by alteration of the growth characteristic of the cell and the tested hybrids showed normal contact in addition and did not form tumors when injected into rats. This is also true for early descendants of the primary hybrid. The experiments thus indicate that the junctional defect present in the neoplastic cell is corrected in the hybrid and that the elimination of the junctional defect is due to the normal cell genome in the hybrid.

In the *second* system, the authors made hybrids between human Lesch-Nyhan and mouse cl–1D cells [78]. As noted previously, the Lesch-Nyhan cells are skin fibroblasts which are deficient in IMP, GMP: pyrophosphoribosyltransferase. The mouse cells employed are a neoplastic subline of L-cells and are deficient in thymidine kinase. The mouse cells also are non-coupling, whether this is measured electrically or by fluorescein transfer, and show no micromorphologically detectable nexus junctions.

The initial hybrids are able to utilize exogenously provided hypoxanthine and thymidine. Moreover, they couple electrogenically with each other and allow the passage of fluorescein between each other. Finally, electronmicrographs show the presence of nexus junctions between the hybrid cells. The initial hybrids contain a nearly complete set of parental chromosomes. However, as the hybrid cells lose human chromosomes during repeated passage in culture, some clones appear which have reverted to the non-coupling, junction-deficient trait of the parental mouse cell. Evidently the human cell contributes a factor to the hybrids which eliminates the junction deficiency of the neoplastic mouse cells. The data in essence complement those of Klein et al. [86], Bregula et al. [87] and Wiener et al. [88] suggesting that the malignancy of hybrids between normal and tumor cells behaves as a *recessive* genetic feature and emerges only after loss of some normal chromosomal material from the hybrids.

9.8. Possible roles of junctions between epithelial cells

Evidence already presented indicates that cell coupling can involve the movement of some metabolically important substances between the cells. In cell populations this would tend to equalize or buffer the population; it would certainly tend to support cells which for some reason or other lack some metabolic competence. That this can occur has been demonstrated experimentally but there is little information as to its physiological importance. However, one can conceive that one cell type might be the source of an essential metabolite for another cell type to which it is coupled.

An important hypothesis concerning the role of intercellular coupling in growth control both in normal and neoplastic cells, has been proposed by Loewenstein (e.g. [89]). He argues that control of embryonic differentiation requires some form of close range cellular interaction. This might involve regulating the diffusion of molecules through the extracellular fluid. An alternative possibility, the communication by molecules located on the cell surface, that is, via the plasma membrane, is treated in Chapter 10. However, Loewenstein argues that communication via *intercellular* junctions might play a more important role than the other mechanisms mentioned. He argues that the connected cell ensemble is a system of finite volume and that it accordingly has the capacity of conveying information on cell number simply on the basis of simple properties of chemical concentration gradients.

In Loewenstein's model, sets of cells are connected by intercellular junctions. Each member of the set can both produce and respond to certain regulatory molecules. Regulatory molecules can be released as asynchronous pulses in the linked cells. If the pulse duration is small in relation to the pulse intervals, the rate of diffusion of the regulated substance through the system is short in relation to the pulse intervals, and the bursts are short in relation to the time for diffusion out of the system. The steady state concentration of the regulator molecules inside the system provides information about the number of cells present in the system at a particular time. Then, the average concentration of the regulator molecules is inversely proportional to the volume of the system, or inversely proportional to the number of linked cells. Given an appropriate feedback loop, such information might regulate growth. Loewenstein suggests, for instance, that if cell division proceeds only when the concentration of regulator molecules is above a critical level, multiplication in a growing cell system will cease when the cells reach a number at which the signal molecules become diluted below the critical concentration. In that case, cellular replication would become restrained as a connecting cell population increases by cell division. A similar restraint would occur if the population increased by establishment of junctional connections between formerly uncoupled cells. An example of the later might be cells moving together in tissue culture. Loewenstein argues that this type of regulation may have a very important role in cancer. Specifically, the model as proposed would provide protection against excessive growth by diluting out signal molecules in interconnecting cell systems. The model does not fit very simply with the known behavior of regulatory enzymes, since it implies that the lower the concentration of a product, the less is manufactured.

Socolar [90] reasons somewhat differently. He argues that coupling should buffer each cell of a linked ensemble against fluctuations in cellular concentrations of critical metabolites. This would stabilize cellular processes that depend on these metabolites [17]. Without coupling, some cellular variants in a tissue might deviate in the concentrations of critical metabolites in a fashion deleterious to the whole. One could apply this argument to the possible emergence of variants, which have an unusual capacity for the accumulation of essential metabolites. As noted in Chapter 7, such cells might have a selective advantage over cells with less effective transport mechanisms. Clearly, if a cell with an excessively acquisitive transport mechanism were coupled to cells with less effective means of transport, the potential selective advantage of the former might be eliminated.

In either case, junctional connections permeable to either regulating molecules, or to the essential nutrients, would protect a linked cell assembly against either high signal concentrations or unusual accumulation of essential metabolites in cell variants. Both Loewenstein and Socolar argue that the loss of junctional transmission of essential molecules (be they regulators or nutrients) would allow excessive growth such as is found in neoplasia. Loewenstein continues his argument as follows:

Control of growth requires three elements: (a) regulatory signals; (b) receptors and effectors for these signals; (c) means of transfer of these signals.

He contends that uncontrolled growth, such as found in neoplasia, may involve: (a) anomalies in signal production; (b) anomalies in signal reception and (c) anomalies in signal transmission. (Stoker [91] has reasoned similarly in his discussion of contact inhibition of growth in culture.)

Loewenstein reasons that each of these three anomalies could constitute a sufficient cause of cancer. He argues, therefore, that if intercellular junctions indeed constitute a path for growth regulatory substances, interruption or interference with the transmission of these substances by a genetic or other defect might produce uncontrolled growth. He suggests that a defect in any one of the steps in signal transmission could constitute a sufficient cause for neoplasia; he does not imply that cancer cells should invariably exhibit junctional defects.

9.9. Coda

In assessing the role of intercellular junctions, one must recognize that the techniques employed to date have focused on small molecules. Without knowing the sieving properties of intercellular junctions, it is impossible to assess the presence or absence of coupling defects for much larger molecules, that is, intercellular communications could be blocked for molecules of a certain size, but not for much smaller substances. Also, the passageways might be defective in the sense that they block the migration of molecules of a given charge or configuration, which would otherwise be allowed to cross. Alternatively, the intercellular communications might be partly limited and simply reduce the rate of transmission of important regulating molecules between cells. At the present, these questions appear beyond experimental exploration. However, in cAMP and cGMP, we have excellent candidates for the integration of coupled cells via small molecules. This possibility should now be accessible to test. Finally, one cannot underemphasize the possible role of mitochondria in the regulation of intercellular communications and in possible coupling defects in neoplasia.

References

[1] Schmidtman, M., *Gesamte, Z., Exp. Med.*, *45*, 714 (1925).
[2] Weidman, S., *J. Physiol. (London)*, *118*, 348 (1952).
[3] Furshpan, E.J. and Potter, D.D., *J. Physiol. London*, *145*, 289 (1959).
[4] Loewenstein, W. and Kanno, Y., *J. Cell Biol.*, *22*, 565 (1964).
[5] Bennett, M.V.L., *Fed. Proc.*, *32*, 65 (1973).
[6] Sheridan, J.D., *Am. Zool.*, *13*, 1119 (1973).
[7] Sheridan, J.D. in *Cell Communication*, R. Cox, ed., Wiley Interscience, New York, 1974, in press.
[8] Sheridan, J.D. in *Proc. International Congress for Devel. Biol.*, A. Moscona, ed., Wiley Interscience, New York, 1974, p. 197.
[9] McNutt, N.S. and Weinstein, R., *Progr. Biophys. Mol. Biol.*, *26*, 45 (1973).
[10] Sheridan, J.D. and Johnson, R.G. in *Molecular Pathology*, R. Good and S. Day, eds., Thomas, New York, Ch. 14, 1974.
[11] Borek, C., Higashino, S. and Loewenstein, W., *J. Membrane Biol.*, *1*, 274 (1969).
[12] Penn, R.D., *J. Cell Biol.*, *29*, 171 (1966).
[13] Weidman, S., *J. Physiol. Lond.*, *187*, 323 (1966).
[14] Bennett, M.V.L., Dunham, P.B. and Pappas, G.D., *J. Gen. Physiol.*, *50*, 1094 (1967).
[15] Rose, B., *J. Memb. Biol.*, *5*, 1 (1971).
[16] Johnson, R.G. and Sheridan, J.D., *Science*, *174*, 717 (1971).
[17] Furshpan, E.J. and Potter, D.D. in *Current Topics in Developmental Biology*, New York Academic Press, Vol. 3, 1968, p. 95.
[18] Bennett, M.V.L. and Spira, M.E., *Biol. Bull.*, *141*, 378 (1971).
[19] Slack, C. and Palmer, J.F., *Exptl. Cell Res.*, *55*, 416 (1969).
[20] Sellin, D., Wallach, D.F.H. and Fischer, H., *Eur. J. Immunol.*, *1*, 453 (1971).
[21] Rotman, B. and Papermaster, B.W., *Proc. Natl. Acad. Sci. U.S.*, *55*, 134 (1966).
[22] Sellin, D., Wallach, D.F.H., Weltzien, H.U., Resch, K., Sprenger, E. and Fischer, H., *Eur. J. Immunol.*, *4*, 189 (1974).
[23] Hülser, D.F. and Peters, J.H., *Eur. J. Immunol.*, *1*, 494 (1971).
[24] Oliviera-Castro, G.M., Barcinski, M.A. and Cukierman, S., *J. Immunol.*, *111*, 1616 (1973).
[25] Oliviera-Castro, G.M. and Barcinski, M.A., *Biochim. Biophys. Acta*, *352*, 338 (1974).
[26] Wallach, D.F.H. and Lin, P.-S., unpublished.
[27] Subak-Sharpe, H., Bürk, R.R. and Pitts, J.D., *Heredity*, *21*, 342 (1966).
[28] Subak-Sharpe, H., Bürk, R.R. and Pitts, J.D., *J. Cell Sci.*, *4*, 353 (1969).
[29] Stoker, M.P.G., *J. Cell. Sci.*, *2*, 293 (1967).
[30] Pitts, J.D. in *Ciba Foundation Symposium on Growth Control in Cell Cultures*, G.E.W. Wolstenholme and J. Knight, eds., Churchill-Livingston, London, p. 89, 1971.
[31] Fujimoto, W. and Seegmiller, J.E., *Proc. Natl. Acad. Sci. U.S.*, *65*, 577 (1970).
[32] Cox, R.P., Krauss, M.R., Balis, M.E. and Dancis, J., *Proc. Natl. Acad. Sci. U.S.*, *67*, 1573 (1970).
[33] Bürk, R.R., Pitts, J.D. and Subak-Sharpe, J.H., *Exptl. Cell Res.*, *53*, 297 (1968).
[34] Gilula, N., Reeves, O.R. and Steinbach, A., *Nature*, *235*, 262 (1972).
[35] Azarnia, R., Michalke, W. and Loewenstein, W.R., *J. Memb. Biol.*, *10*, 247 (1972).
[36] Cox, R.P., Krauss, M.R., Balis, M.E. and Dancis, J., *Exptl. Cell Res.*, *74*, 251 (1972).
[37] Ashkenazi, Y.E. and Gartler, S.M., *Exp. Cell Res.*, *64*, 9 (1971).
[38] Kolodny, G.M., *Exptl. Cell Res.*, *65*, 313 (1971).
[39] Kanno, Y. and Loewenstein, W.R., *Nature*, *212*, 629 (1966).

[40] Reese, T.S., Bennett, M.V.L. and Feder, N., *Analt. Rec.*, *169*, 409 (1971).
[41] Loewenstein, W.R., *Fed. Proc.*, *32*, 60 (1973).
[42] Graham, R.C. and Karnovsky, M.J., *J. Histochem. Cytochem.*, *14*, 291 (1966).
[43] Mikkelsen, R. and Wallach, D.F.H., *Biochim. Biophys. Acta*, *363*, 211 (1974).
[44] Brightman, M. and Reese, T.S., *J. Cell Biol.*, *40*, 648 (1969).
[45] McNutt, N.S. and Weinstein, R.S., *J. Cell Biol.*, *47*, 666 (1970).
[46] Weinstein, R.S., unpublished.
[47] Pappas, G.D., Asada, Y. and Bennett, M.V.L., *J. Cell Biol.*, *49*, 173 (1971).
[48] Barr, L.M., Dewey, M. and Berger, W., *J. Gen. Physiol.*, *48*, 797 (1965).
[49] Dreifuss, J.J., Girardier, L. and Forssmann, *Pflügers Arch, Gesamte Physiol., Menschen Tiere*, *292*, 13 (1966).
[50] Goodenough, D.A. and Revel, J.P., *J. Cell Biol.*, *50*, 81 (1971).
[51] Benedetti, E.L. and Emmelot, P., *J. Cell Biol.*, *38*, 15 (1968).
[52] Goodenough, D.A. and Revel, J.P., *J. Cell Biol.*, *45*, 272 (1970).
[53] Goodenough, D.A. and Stoeckenius, W., *J. Cell Biol.*, *54*, 646 (1972).
[54] Gilula, N.B., *J. Cell Biol.*, *63*, 111a (1974).
[55] Goodenough, D.A., Caspar, D.L.D., Makowski, L. and Phillips, W.C., *J. Cell Biol.*, *63*, 115a (1974).
[56] Bennett, M.V.L. and Auerback, A.A., *Anat. Record*, *163*, 152 (1969).
[57] Heppner, D.B. and Plonsey, R., *Biophys. J.*, *10*, 1057 (1970).
[58] Woodward, D.J., *J. Genl. Physiol.*, *52*, 509 (1968).
[59] DeLaat, S.W., Luchtel, D. and Bluemink, J.G., *Develop. Biol.*, *31*, 163 (1973).
[60] DeLaat, S.W. and Bluemink, J.G., *J. Cell Biol.*, *60*, 529 (1974).
[61] Sheridan, J.D., *J. Cell Biol.*, *45*, 91 (1970).
[62] O'Lague, P.H., Dalen, H., Rubin, H., Tobias, C., *Science*, *174*, 467 (1970).
[63] Cavoto, F.V. and Flaxman, B.A., *J. Cell Biol.*, *58*, 223 (1973).
[64] Loewenstein, W.R., *Develop. Biol.*, *15*, 503 (1967).
[65] Ito, S. and Loewenstein, W.R., *Develop. Biol.*, *19*, 228 (1969).
[66] DeHaan, R.L. and Hirakow, R., *Exptl. Cell Res.*, *70*, 214 (1972).
[67] Johnson, R., Hammer, M., Sheridan, J.D. and Revel, J.-P., *Proc. Natl. Acad. Sci. U.S.*, *71*, 4536 (1974).
[68] Yee, A., *J. Cell Biol.*, *55*, 294a (1972).
[69] Porvaznik, M. and Johnson, R., *J. Cell Biol.*, *63*, 273a (1974).
[70] Epstein, M. and Sheridan, J., *J. Cell Biol.*, *63*, 95a (1974).
[71] Loewenstein, W.R., J. *Colloid Interphase Science*, *25*, 34 (1967).
[72] Oliviera-Castro, G.N. and Loewenstein, W.R., *J. Memb. Biol.*, *5*, 51 (1971).
[73] Loewenstein, W.R. and Kanno, Y., *J. Cell Biol.*, *33*, 225 (1967).
[74] Jamakosmanovic, A. and Loewenstein, W.R., *Nature*, *218*, 775 (1968).
[75] Kanno, Y. and Matsui, Y., *Nature*, *218*, 775 (1968).
[76] Azarnia, R. and Loewenstein, W.R., *Nature*, *241*, 455 (1973).
[77] Azarnia, R., Michalke, W. and Loewenstein, W.R., *J. Memb. Biol.*, *10*, 247 (1972).
[78] Azarnia, R., Larsen, W.J. and Loewenstein, W.R., *Proc. Natl. Acad. Sci. U.S.*, *71*, 880 (1974).
[79] Benedetti, E.L. and Emmelot, P., *J. Cell Sci.*, *2*, 499 (1967).
[80] Martinez-Palomo, A., Brailovsky, C. and Bernhard, W., *Cancer Res.*, *29*, 925 (1969).
[81] McNutt, N.S., Hershberg, R.A. and Weinstein, R., *J. Cell Biol.*, *47*, 135 (1970).
[82] Flaxman, B.A. and Cavoto, F.V., *J. Cell Biol.*, *58*, 219 (1973).
[83] Pinto da Silva, P. and Gilula, N.B., *Exptl. Cell Res.*, *71*, 393 (1972).
[84] Hülser, D.F. and Webb, D.J., *Exp. Cell Res.*, *80*, 210 (1973).
[85] Weiss, M.C., Todaro, G.J. and Green, H., *J. Cell. Physiol.*, *71*, 105 (1968).
[86] Klein, G., Bregula, U., Wiener, F. and Harris, H., *J. Cell Sci.*, *8*, 659 (1971).
[87] Bregula, U., Klein, G. and Harris, H., *J. Cell Sci.*, *8*, 673 (1971).
[88] Weiner, F., Klein, G. and Harris, H., *J. Cell Sci.*, *8*, 681 (1971).
[89] Loewenstein, W.R. in *Cellular Membranes and Tumor Cell Behavior*, E.F. Walborg, Jr., ed., Williams and Wilkins, 1975, in press.
[90] Socolar, S., *Exp. Eye Res.*, *15*, 693 (1973).
[91] Stoker, M.P.G., *Current Topics in Developmental Biology*, *2*, 108 (1967).

10

Lectin reactivity

10.1. Introduction

Many animal cells can be linked to each other and/or other cell types by agglutinating proteins, the *lectins*. Numerous lectins have been described and the properties of these substances have recently been reviewed in depth by Sharon and Lis [1,2]. Lectins are very widely distributed in nature. Plants, particularly seeds, constitute the predominant sources, but lectins occur in invertebrates (e.g., snails) and lower vertebrates as well. The primary function of these proteins remains obscure, but they now attract major attention because of their unusual reactivity with certain tumor cells, and their mitogenic action on lymphocytes. The lectins most widely used in the study of plasma membrane alterations in neoplasia are concanavalin A (Con A) from jack bean, wheat germ agglutinin (WGA) and soybean agglutinin (SBA).

In 1963, Aub and associates [3] studied the effect of wheat germ lipase on tumor cell surfaces. They found that wheat germ lectin, a contaminant of their lipase preparation, agglutinated neoplastic lymphocytes much more effectively than it agglutinated normal lymphoid cells. Since then, it has become well established that many tumor cells are unusually susceptible to agglutination by numerous lectins; in fact, the lectin reactivity of neoplastic cells has evolved into an important, if controversial area of tumor biology.

Unfortunately, cancer biologists tend to overlook the complex behavior, under highly defined conditions, of even the few lectins that have been characterized. Indeed, there are few rigorous studies on the reactivity of lectins with any cell type, and studies on tumors seem to raise more questions than they answer.

We will here scrutinize the available information very critically. To do this we must first introduce some general aspects of lectin biology and also look into another important membrane–lectin interaction, that observed in lymphocytes.

10.2. Some general properties of lectins

Some general properties of Con A, WGA, SBA and some other important plant agglutinins are listed in Table 10.1.

As shown in column 5 of this table, the agglutinating activity of a given lectin can often be specifically inhibited by a simple sugar. This implies that sugars, on either mono- or oligo-saccharides, that are attached to membrane glycoproteins and/or glycolipids, form at least a part of the cellular binding sites for the lectins. The facts in accord with this concept are that lectins can precipitate certain polysaccharides, glycoproteins and glycolipids, and that simple sugars can specifically block such precipitation.

The best studied lectins are polyvalent (column 3–4) and their interactions with membranes or other ligands appear to involve cross-linking, analogous to that found in reactions of immunoglobulins with antigens. However, the association constants of the binding sites of lectins for specific sugars lie between 10^3–10^5 L/mole, much below the binding constants for the interactions of antibodies with carbohydrate antigens. (Indeed, the K_A for the reaction of lentil lectin with α-D-mannose is only 2.3×10^2). With such low K_A values, lectins should rapidly dissociate from membranes under the washing conditions usually employed in studies on lymphocytes and tumor cells; this dissociation clearly does not occur and, as shown later, the affinity of some lectins for membranes appears much greater than for free, specific sugars.

Lectins also differ from antibodies in that the combining sites in various molecules of a given lectin are identical and homogeneous. However, as typified by Con A, lectins do not *necessarily* comprise a homogeneous molecular species, that is, various 'isolectins' differ substantially in their protein moieties, yet bear identical combining sites for sugars. Such heterogeneity may not interfere with the interaction of the lectins and small molecules, but could complicate interpretation of the reactivity of cellular membranes.

10.3. Important general properties of some purified lectins

10.3.1. Concanavalin A

Concanavalin A is the most widely studied lectin. It complexes with α-D-mannopyranoside, α-D-glycopyranoside, D-fructofuranoside, their glycosides and

TABLE 10.1

Some properties of certain purified lectins important to tumor biology

Source	Mol.wt.	No. of subunits	No. of binding sites	Sugar specificity	Mitogenicity	Metal requirement
Jack bean[1] Con A	110,000[2]	4	4	α-D-Man	+	Ca^{2+} Mn^{2+}
Soy bean[1] SBA	120,000[3]	4	2	D-Gal NAc	0	Ca^{2+} Mn^{2+}
Wheat germ[1] WGA	17,000[4]	1[4]	4	$(D\text{-GlcNAc})_2$	–	–
Lentil	39,000–48,000	2	2	α-D-Man	+	Ca^{2+} Mn^{2+}
Red kidney[5] bean	140,000	4	?	D-Gal NAc	+	–
Castor bean[6,7]						
I	120,000	4	2	D-Gal and sterically related structures		–
II	65,000	2				–

[1] From Lis and Sharon [2] and Sharon and Lis [1].
[2] At pH 5.6-7.0. Monomer weight 25,500; 4 identical subunits.
[3] Four identical subunits.
[4] Subunit weight. Dimer at neutral pH. WGA may aggregate during agglutination to form intercellular bridges. X-ray data suggest an orthorhombic cell containing 8 molecules. Possibly 2 binding sites per monomer [4].
[5] More than one lectin in this source [5].
[6] From Nicolson et al. [6]. It has subunits of 29,500 and 37,000 mol.wt. The subunits of II have mol.wt. of 29,500 and 34,000. The authors suggest that I and II share the 29,500 subunits, that is; it has an $\alpha_2\beta_2$-structure and II and $\alpha\beta$-structure.
[7] From Podder et al. [7].

sterically related compounds, and with other compounds bearing these saccharides. We comment on the relative binding of Con A to membranes and simple sugars later on. Concanavalin A agglutinates numerous animal cells, as well as microorganisms and viruses, apparently by reacting with specific carbohydrates that are exposed at cell surfaces. Like certain other lectins, Con A is mitogenic to lymphocytes. Concanavalin A can exist in monomeric, dimeric, tetrameric and aggregated form [8,9] depending on pH and ionic conditions. When the pH is dropped below 4.6, the protein dissociates into monomers and undergoes a time-dependent structural change involving release of bound Ca^{2+} and Mn^{2+}. At pH 5.0, the molecule exists as a dimer of 55,000 mol. wt. At pH 7.0, the 110,000 mol. wt. tetramer predominates, but it exists in a non-interacting equilibrium with the dimer [9]. The tetramers and dimers differ substantially in sugar affinity. Above pH 7.5, one observes time-dependent aggregation of the tetramers.

Each Con A monomer has a mol. wt. of 25,000 and bears one sugar binding site. In addition, each monomer binds one ion of Ca^{2+} and one of Mn^{2+} (or of another transition metal) at distinct sites. A transition metal cation appears necessary for Ca^{2+} binding, and both cations are required for sugar binding [10].

The subunits of Con A may consist of one, intact non-glycosylated polypeptide chain, or of chain fragments (mol. wt. 11,000–13,000) which dissociate only in agents such as sodium dodecyl sulfate [9]. The proportion of fragmented chains appears to depend on the purification method. It is unknown whether subunits consisting of fragmented chains bind sugar or metals. However, it appears that fragmented chains do *not* favor tetramer formation and that they tend to be excluded from Con A crystals. Unfortunately, most studies in which Con A has been used to probe surface differences between normal and neoplastic cells have been conducted without considering the various inhomogeneities of Con A.

The structure of Con A has been established at a resolution of 2Å by X-ray crystallography [11]. Each Con A subunit contains 238 amino acids, has a mol. wt. of 25,000 and is globular in shape (dimensions 42 × 40 × 30 Å). The predominant elements comprise two extensive folded sheets of anti-parallel β-structure. Peptide not in the β-conformation is without regular structure. One of the pleated sheets contributes extensively to the association of subunits into dimers and tetramers. The subunits interact to form a pseudotetrahedron. One sugar is bound to each subunit and, in the tetramer, the four sugar binding sites lie in symmetrically equivalent positions. Very close to each sugar binding site lies an extended non-polar region.

The X-ray data indicate that the transition metals lie deep within each subunit, but are in contact with an aqueous channel leading to the bulk solvent. Of the 6 ligands for Mn^{2+} one is histidine and three are either Asp or Glu side chains. The binding site for Ca^{2+} lies 5.3 Å from that for Mn^{2+}, and two of the acidic residues associated with the Mn^{2+} also participate in Ca^{2+} binding. The metal binding sites are not in contact with the sugar binding sites. There is some evidence that the protein undergoes conformational and/or other structural changes upon sugar binding [12,13].

Succinylation or acetylation of tetrameric Con A, converts the lectin to a dimer which has sugar binding properties identical to, and a mitogenicity similar to (but toxicity lower than), the native molecule [14]. Studies using sheep erythrocytes and mouse lymphocytes show binding saturation at 2.5 and 3.1 times the number of bound molecules of native Con A/cell at saturation. The derivatized molecules show much lower agglutination activity and are less able to block 'cap formation' on lymphocytes by IgG receptors (Chapter 1) than the native molecules are. Succinyl-Con A and acetyl-Con A are also unable to induce 'capping' of Con A receptors, a phenomenon observed at low concentrations of Con A. It, therefore, appears that the ability of Con A to agglutinate cells and to influence lymphocyte surface topology depends on its valency.

The affinity constant of Con A for various sugars and saccharide derivatives varies from $2.1 \times 10^4\ M^{-1}$ for α-methyl-D-glucopyranoside to $2.0 \times 10^1\ M^{-1}$ for methyl-6-deoxy-α-glucopyranoside [15]. The affinities vary in a way that implies that lectins associate with sugars through both a specific, primary interaction and via a secondary process, which could involve hydrophobic interactions with non-polar regions near the sugar binding sites. However, the chromophoric hapten sugar, *p*-nitrophenyl-α-D-mannopyranoside, is readily displaced by methyl-α-D-mannopyranoside at 27°C, a temperature at which both sugars have a K_A of $1.5 \times 10^4\ M^{-1}$ [16]; the methyl derivative cannot be expected to complete for an apolar site.

10.3.2. Soybean agglutinin

Extensively purified SBA [2] consists of four, apparently identical subunits with mol. wts. ~30,000. Unlike Con A, SBA is a glycoprotein, containing 4.5% mannose and 1.2% *N*-acetyl-D-glucosamine in four identical carbohydrate chains, one per subunit. SBA (120,000 mol. wt.) bears *two* identical binding sites, which are specific for *N*-acetyl-D-galactosamine. These show rather low affinity constants for the sugar, that is, $\sim 3 \times 10^4$ L/mole.

10.3.3. Wheat germ agglutinin

This lectin can now be obtained in highly purified, crystalline form [17,18]. The mol. wt. is 35,000 at neutral pH and 17,000–18,000 at acid pH or in denaturing solvents [18–20]. The protein contains an unusually high proportion of half cystine residues (17%) and glycines (23%) but lacks carbohydrate. WGA has no metal requirement for sugar binding.

N-Acetyl-D-glucosamine (HlcNAc) specifically inhibits the agglutination of cells by WGA; the dissociation constant is 7.5×10^{-4} M. Oligosaccharides containing GlcNAc are more effective agglutination inhibitors than GlcNAc. In fact, $(GlcNAc)_2$ and $(GlcNAc)_3$ inhibit agglutination much more effectively than GlcNAc does (dissocation constants 4.9 and 1.2×10^{-5} M respectively). The specificity of WGA has been very extensively studied by Allen et al. [20], who suggest that the sugar binding site consists of 3–4 'subsites' with different specificities.

Nagata and Burger [18] report two binding sites for *N*-acetyl-D-glucosamine per 17,000 dalton unit at neutral pH. However, LeVine et al. [21] report only one binding site per subunit. The agglutinating ability of WGA and its feeble dissociation from membranes (unless *N*-acetyl-D-glucosamine is added) both suggest that this lectin also functions as a polymer.

10.3.4. Lentil agglutinin

Highly purified lentil agglutinin [2] consists of dimers of identical polypeptide chains with a mol. wt. of 22,000–25,000. The dimeric agglutinin bears two, low-affinity binding sites for α-D-glucopyranoside or D-mannose (K_A 1×10^2 and 2.3×10^2 L/mole respectively). These also bind both Ca^{2+} and Mn^{2+}. However, the transition metal is apparently not required for sugar or Ca^{2+} binding.

10.4. Binding 1inetics and equilibria of lectins

10.4.1. Wheat germ agglutinin

Cuatrecasas [22] has studied the binding of WGA and Con A, both ^{125}I-labelled, to *adipocytes*. The association of WGA with adipocyte plasma membrane is extremely rapid and Cuatrecasas' data suggest an apparent second order association rate constant of about 10^7 M^{-1} sec^{-1}. Spontaneous breakdown of the fat cell-lectin complex is slow, but addition of *N*-acetyl-D-galactosamine very rapidly dissociates the adipocyte-WGA complex. So does ovomucoid, a glycoprotein containing *N*-acetyl-D-galactosamine. The binding curve shows saturation near $\sim 3 \times 10^8$ molecules/cell and indicates substantial heterogeneity of receptor sites. The affinity of the adipocyte receptors for WGA exceeds that for Con A.

Both WGA and Con A perturb the insulin receptors of adipocyte plasma membranes. Concordantly, both Con A and insulin are found to alter WGA binding to adipocytes. The insulin effect is the same at high and low levels of WGA, but Con A competes more effectively for the low-affinity WGA receptors.

10.4.2. Concanavalin A

Gray and Glew [23] have computed the kinetic and equilibrium constants for Con A binding to *small sugars*, using the chromophoric derivative *p*-nitrophenyl-α-D-mannopyranoside. At pH 6.95 and a temperature of 20° (where Con A is tetrameric), the association rate constant is 8×10^4 M^{-1} sec^{-1} and the dissociation rate constant 4 sec^{-1}. These kinetic constants yield an equilibrium affinity constant of 2×10^5 L/mole which is less than the value of 2×10^4 L/mole for α-methyl-D-mannopyranoside obtained by equilibrium titration [15].

The association constant found for the reaction of Con A with small sugars appears to be much smaller than that observed by Cuatrecasas for the interaction of Con A with the plasma membranes of adipocytes [22]. As pointed out in Chapter 7, binding studies using whole cells or membrane fragments cannot be simply analyzed. Nevertheless, steric considerations alone would lead one to expect that Con A should react much more slowly with a membrane receptor than a free sugar. However, Cuatrecasas [22] observes very rapid binding of ^{125}I-labelled Con A to adipocytes. In fact, the data (initial uptake of $\sim 1.3 \times 10^{-16}$ moles/sec at a binding site concentration of $\sim 1.7 \times 10^{-9}$ M and a lectin concentration of $\sim 10^{-15}$ M) yield an apparent second order association rate constant of $\sim 10^7$ M^{-1}/sec^{-1}, that is, a significantly faster combination rate than found with free sugar. The reaction with adipocytes is also much more rapid than the association reaction of Con A and the glycoprotein lectin from castor bean (3×10^3 M^{-1}/sec^{-1}) [7]. The spontaneous dissociation of the Con A–adipocyte complex is vanishingly slow in comparison with the rates found for specific sugar complexes [23]. However, addition of α-methyl-D-mannopyranoside fosters extremely rapid dissociation of the lectin–adipocyte complex. The binding curve is sigmoidal in character, strongly indicative of heterogeneous binding sites and suggestive of cooperative interaction between receptor sites.

The binding of Con A to quiescent and Con A stimulated human lymphocytes is also complex [24]. Quiescent cells at >20°C bind $1.7\text{–}1.8 \pm 0.2 \times 10^7$ molecules at saturation while stimulated cells bind $3.7 \pm 0.6 \times 10^7$ (Table 10.2). The surface binding density is 1.5×10^5 molecules/μm^2 in both cases. The change caused by stimulation is maximal at 48 hr. Both cell types exhibit similar binding curves with *three* saturation plateaux – at 5×10^{-7} M, 1.1×10^{-6} and 2.6×10^{-6} M.

The data of Krug et al. [24] must be compared with the experimental results of Betel and van den Berg [25] who studied the binding of macromolecular acrylate–Con A complexes to lymphoid cells. Remarkably, they obtain the same number of binding sites per lymphocyte ($\sim 2 \times 10^7$ molecules of monomer/cell) that were observed with native Con A (Table 10.2). Moreover, they obtain

TABLE 10.2

Con A binding by diverse normal tissue cells and by normal and neoplastic cultured cells[1]

	Con A molecules/cell		Con A molecules/cell
Tissue		Culture	
Adipocytes[2,7]	4×10^8	L-cells[5,8]	7.6×10^7
Liver[3]	3×10^7	BHK-21[5]	8.8×10^7
Hepatoma AH13[12]	9.8×10^7	BHK-(polyoma-converted)[6]	6.8×10^7
Kidney[3]	3.4×10^7	3T3[6]	2.2×10^7
Testis[3]	5.1×10^7	3T3 (SV40-converted)[6]	2.2×10^7
Sperm-whole[3]	4.9×10^7		
Sperm-heads[3]	7.9×10^6	3T3 (0 °C)[9]	6.6×10^5
Sperm-tails[3]	2.5×10^6	3T3 (22 °C)[9]	4.1×10^7
Erythocytes[3]	6.8×10^6	CHO (parental)[10]	3.3×10^7
Spleen[3]	1.4×10^6	CHO (clone 15B)	5.5×10^7
Lymphocytes – mouse[3]	1.3×10^7		
Lymphocytes – human[4]		Lymphoblasts, (human)[13]	$2.3–4.5 \times 10^7$
quiescent	1.8×10^7	MOPC 173 murine plasmacytoma[11]	
Con A-stimulated	3.7×10^7	Clone ME_2	1.5×10^8
Thymocytes[3]	1.6×10^7	Clone ME_2C^r	2.5×10^8

[1] Assuming a subunit weight of ~ 25,500.
[2] From Cuatrecasas [22].
[3] From Edelman and Milette [26].
[4] From Krug et al. [24].
[5] From Brown [27,28].
[6] From Ozanne and Sambrook [29].
[7] Surface densities of Con A receptors on adipocytes range between $1.6 \times 10^4/\mu m^2$ for large cells (45 μm radius) and $8.5 \times 10^4/\mu m^2$ for starved cells (17.5 μm radius)[22]. The value for thymocytes (radius = 2.5 μm) computes to $2 \times 10^5/\mu m^2$).
[8] Hunt and Brown [30] have characterized two classes of Con A receptors from L-cells. These have mol. wts. of 100,000 and 50,000. Both appear to bear more than one type of glycopeptide residue. Their Con A reactivity remains to be established.
[9] From Noonan and Burger [31].
[10] From Gottlieb et al. [32]. Clone 15 B exhibits markedly impaired binding for other lectins.
[11] From Guerin et al. [33] ME_2 is maximally agglutinated by 25 μg Con A/ml ME_2C^r shows virtually no agglutination at 300 μg Con A/ml. Data recalculated for subunit weight 25,500. We compute the density of monomer sites to be $1.2 \times 10^5/\mu m^2$ for ME_2 and ME_2C^r respectively. The authors convert radioactivity data to molecules of tetramer. Since they labelled by acetylation and acetylation blocks tetramer formation, there is some doubt as to the exactness of the figures.
[12] From Kano et al. [34].
[13] From De Salle et al. [35]. The number of molecules bound per μm^2 of cell surface ($0.6–0.7 \times 10^5/\mu m^2$ is actually less than the found for normal human lymphocytes) ($\sim 10^5/\mu m^2$).

highly linear Scatchard plots suggesting a single class of binding sites with a high equilibrium constant ($\sim 10^7$ M^{-1}, 37 °C).

Guerin et al. [33] have measured the Con A binding of two clones of cultured MOPC 173 murine plasmacytoma; ME_2 which is readily agglutinated by Con A and ME_2 C^r which is resistant to agglutination. They used highly purified Con A, labelled with [^{3}H]acetic anhydride for the binding studies and found the same number of binding sites on both cell types (Table 10.2). Scatchard plots of binding curves reveal an unexpected homogeneity of binding sites. The authors compute affinity constants (tetramer) of $\sim 10^7$ M^{-1} for both clones. This is considerably greater than that found for Con A hapten sugars.

Use of acetylation as a labelling procedure complicates interpretation of these data because, while binding is measured by label uptake, formation of Con A tetramer is prevented by acetylation [14]. Unfortunately, the degree of acetylation is not provided and, at the temperatures used ($>20\,^\circ$C), Con A is rapidly endocytosed.

In a more recent study [36] the same group compared Con A binding by a series of divergent derivatives of MOPC 173, differing in their susceptibility to the agglutinating and killing effects of Con A. They were unable to find a significant relationship between binding-site density (4.4–$18 \times 10^4/\mu^2$) or binding affinity ($K_A \sim 3$–5.5 L/mole) and agglutinability or Con A toxicity.

In contrast to the binding of Con A (and WGA) to most plasma membranes, the lectin's reaction with the rhodopsin of retinal disc membranes [37] appears straightforward. There is only one class of binding site. However, the stability of Con A complex with these membranes ($K = 2 \times 10^7$) is again unexpectedly large compared with that of Con A and 'hapten sugar'. Nevertheless, the data suggest that α-methyl-D-mannoside acts as a competitive inhibitor.

Cuatrecasas' study on adipocytes [22] has, to our knowledge not been duplicated in comparisons of normal and neoplastic cells. Yet it has the most profound bearing on such comparisons; the interactions of lectins with cell surfaces are clearly far more complicated than with simple sugars.

The rapid binding kinetics could be explained by positive cooperativity between membrane receptors as suggested by Changeux et al. [38], that is, binding increases lectin affinity. Indeed, the lattice hypotheses of Changeux et al. [38] and Wyman [39], described in Chapters 8 and 11, predict very large cooperative effects in membranes. The density of Con A receptors on adipocytes and lymphocytes, $\sim 1/10^3$ Å^2, is very large (Table 10.2), possibly allowing topological linkages between receptors. If this type of interaction participates in Con A binding it could involve cooperative processes extending over large areas of the membrane surface. But how can this type of interaction be reconciled with the demonstrated fact that some Con A receptors can rapidly move on the membrane surface at least in lymphocytes?

One reasonable alternative is that the lectins associate not only to sugar-specific regions but also to other, secondary membrane binding sites. This suggestion receives support from work on the affinity of Con A for diverse hapten sugars [15] and also from the studies of Noonan and Burger [31], who find that α-methyl-D-mannopyranoside dissociates [^{14}C]succinyl–Con A from normal and transformed 3T3 cells much more rapidly and completely than it removes

[^{3}H]acetyl–Con A. Both acetylation and succinylation produce dimerized Con A, but the succinylated form bears new anionic groups. Indeed, Burger [40], reports that *several lectins bind non-specifically to plastic and glass surfaces and that these associations can be partially blocked or reversed by specific hapten sugars.* He suggests that binding of the hapten sugar might alter the structure of the lectin molecules in a way which leads to their dissociation from specific carbohydrate residues as well as from 'non-specific' sites.

Accordingly, one can postulate that interaction with the carbohydrate receptors enhances secondary binding to the membrane, possibly through a structural change in the lectin brought about by occupation of the sugar-specific site. In extending this notion, one can hypothesize that the Con A might not always react specifically with the plasma membrane, or that the binding is specific but does not only involve membrane sugars [40]. In this model, occupation of the specific sugar binding site is envisioned as altering the lectin structure, interfering with its membrane association and causing its release. We consider this an unlikely general proposition. Certainly, the sugar specificity of membrane receptors has been demonstrated for some lectins, for example, PHA [41]. However, for Con A, the possibility cannot be excluded until membrane receptors are properly purified and shown to react uniquely through a specific hapten sugar. Such information is not yet available for Con A or any other lectin.

There are other unexplained phenomena. Thus, how can one reconcile the vanishingly slow dissociation of adipocyte–lectin complexes with their rapid dissociation by specific sugars? Since spontaneous dissociation is very slow, the sugars cannot act to prevent dissociated free lectin from reassociating with the membrane. Since dissociation is restricted to the sugars specific for the lectin, these must react with unoccupied sugar binding sites on lectin molecules already bound to the cell. This is feasible, because Con A is tetravalent at pH ~ 7 and WGA probably forms a polyvalent aggregate under the conditions employed by Cuatrecasas [22].

But why should occupation of a free valence by a small ligand lead to release of the membrane associated valence? Several explanations should be considered:

(a) Occupation of the free valence(s) induces a major change of secondary tertiary or quaternary structure, which reduces the affinity of the lectin for the membrane receptors. If the membrane receptor indeed corresponds to the 'hapten' sugar, one must invoke *negative cooperativity* But this is not observed with the 'hapten' sugars in solution! (However, negative cooperativity between *membrane* components has been proposed, particularly to account for large membrane effects produced by ligands with only partial occupation of receptor sites [42]. Such a model again requires clustering of receptors, as well as negative cooperative interaction between receptor molecules.) We thus return to our earlier suggestion: a structural change due to reaction of specific sugar with unoccupied sites might interfere with 'secondary' lectin–membrane associations, that is, Con A–membrane associations not involving specific sugars are dissociated by a specific structural change, induced by a specific sugar.

(b) The Con A tetramer breaks down into dimers or monomers upon binding of free sugar and releases those monomers (dimers) that are not already membrane associated. This would not represent true dissociation.

(c) Con A is not released *per se*, but together with its membrane receptor. We consider this a possibility to be evaluated seriously in view of the rapid turnover and extrusion of Con A receptors by thymocytes, following Con A binding [43,44]. However, one would again have to hypothesize that substantial changes in lectin structure accompany sugar saturation. In our view, suggestion (a) appears the most likely and is supported by the recent model study of Podder et al. [7]. These authors find that concanavalin A forms an insoluble complex with the 120,000 mol. wt. castor bean lectin. The complex contains one molecule of castor bean lectin per Con A tetramer. It can be dissociated by hapten sugars for Con A. The authors used turbimetric methods to evaluate the kinetics of complex formation and dissociation and obtain rate constants of $3 \times 10^3\ M^{-7}/sec^{-1}$ and $70\ sec^{-1}$ respectively, yielding an affinity constant of 4.5×10^5 L/mole This affinity constant is larger than the K_A for hapten sugar binding despite the low forward rate constant. The data indicates that, following an initial reaction, probably via the carbohydrate moiety of the castor bean glycoprotein, stabilization occurs via *protein–protein* interaction. This stabilization energy is estimated at 3–4 kcal/mol. The data also indicate that not all sugar binding sites of the complexed Con A are occupied and that the association of these modifying secondary effects observed here appear smaller than those noted by Cuatrecasas [22].

10.5. Lectin receptors

10.5.1. Overview

Few studies deal in any depth with the precise chemical nature of membrane lectin receptors. Indeed, virtually all studies of lectin–membrane interactions assume that specific membrane receptors can be identified by the ability of 'hapten' sugars to block or reverse lectin membrane interactions. They also assume that lectin–sugar reactions are equivalent to antibody–antigen reactions. However, this analogy cannot be valid: *First*, in polyvalent lectins, for example, Con A, the receptor-bearing subunits are not covalently linked as in most immunoglobulins. *Second*, lectin-sugar association constants lie several orders of magnitude below the affinity constants for antibody–antigen reactions (including sugar antigens). *Third*, the associations of lectins with membranes appear much stronger than with specific sugars. *Fourth*, the association kinetics of Con A are more rapid with membranes than with simple sugars and the dissociation rate of the membrane–lectin complexes of WGA and Con A are vanishingly small. *Fifth*, the rapid dissociation of stable membrane-lectin complexes by 'hapten' sugars cannot be explained by any one simple mechanism.

The assumption that lectins bind to membranes via simple sugar receptors appears reasonable and appealing. However, at this time, we can find no direct proof of this contention for the lectins used in the study of tumor membranes. Specific receptors for these lectins have not been isolated in pure form and their chemistry and reactivity have not been established. In fact, our information about the specificity of lectin receptors relies almost exclusively on 'hapten' inhibition studies using simple sugars or glycosides. In any case, even a good 'hapten' inhibitor for a solution system need not be identical to a membrane-bound sugar. The latter need not lie in a terminal position and may be modified by neighboring sugar and/or amino acid residues. Also membrane proteins and lipids might provide 'secondary' attachment sites for the lectin.

These considerations also bear on the techniques for purifying receptors, in particular on affinity techniques, in which specific purification depends on elution of the receptor-bearing material from affinity columns (immobilized lectin) by use of the 'hapten' sugar.

An effective receptor preparation should inhibit binding of a specific lectin to a cell and should inhibit lectin-induced agglutination. However, as first shown for PHA by Steck and Wallach [45], agglutination depends on numerous factors unrelated to the lectin, and inhibition of agglutination cannot alone be taken as proof of the presence of lectin receptors.

Moreover, Loor [46] shows that agglutination of lymphocytes by lectins, such as Con A, cannot easily be considered a passive process. He points out that cytochalasin B (Chapter 7) and sodium azide, both of which inhibit the 'capping' process, hinder lymphoid-cell agglutination by low concentrations of Con A and other lectins. He suggests that the agglutination produced by low lectin concentrations may be an active cellular process, possibly involving the participation of cytoplasmic microfilaments, whereas agglutination at high lectin concentrations is due primarily to passive crosslinking.

One must also appreciate that lectin receptors occur not only on plasma membranes, but also on intracellular membranes. For example, Con A and castor bean lectin bind to the nuclei of normal and neoplastic hepatocytes [34]; the apparent association constants are about the same for the nuclei of normal rat liver (1.2×10^6 M^{-1}) and those of ascites hepatoma AH 108A (1.4×10^6 M^{-1}). The liver nuclei bind $\sim 3.8 \times 10^4$ molecules of Con A and 8.6×10^2 molecules castor bean agglutinin per μm^2. Also, Con A, WGA and castor bean lectin strongly agglutinate liver nuclei and mitochondria at concentrations as low as $\sim$ 1 μg/ml and $\sim$ 40 μg/ml respectively [47]. The agglutination of these organelles is blocked by the lectins' hapten sugars.

The treatment of intact and ruptured cells with covalent ferritin–lectin conjugates allows the detection of intracellular binding sites for these lectins by thin-section electron microscopy. Hirano et al. [48] have applied this approach to intact and disrupted P3K mouse plasmacytoma cells. They focus their attention on microsomal vesicles, which derive from disrupted endoplasmic reticulum and plasma membrane.

Significantly, the lectin–ferritin granules were in all cases restricted exclusively to one of the two surfaces of the visualized membrane elements. With ferritin–Con A, about a third of the *ribosome-bearing* vesicles were stained. The stain granules appeared exclusively on the membrane face opposite the ribosomes, that is, on the face corresponding to the cisternal aspect of rough endoplasmic reticulum. In the case of *smooth* membranes, deposition of ferritin–Con A occurred more frequently on the external surfaces of plasma membrane vesicles. The ferritin derivative of castor bean lectin gave similar results with smooth vesicles but did not stain ribosome-bearing membranes. Deposition of ferritin–lectin could be prevented by addition of 'hapten' sugars, but only when these were added in large excess.

Nicolson and Singer [49,50] have used the same ferritin-labelled lectins to determine whether their receptors lie on one or both faces of plasma membranes. They examined S49-1TB2 mouse lymphoma cells, P.3.6.2 myeloma cells, as well as L-cells and HeLa cells. The authors report that lectin-binding sites of plasma membranes lie exclusively on the extracellular surfaces.

These experiments of Hirano et al. [48] and Nicolson and Singer [49,50] provide strong evidence for transverse asymmetry in biomembranes with respect to certain lectin receptors, presumably saccharides. However, the data on plasma membranes are not totally convincing – although much other evidence attests these membranes' asymmetry – because the authors do not provide conclusive proof that the reagents had access to the cell interior. Such proof would have been provided if lectin deposition had been observed on mitochondrial or nuclear outer membranes, which are known to bind Con A in some cells [34,47].

The localization of Con A receptors on one side only of plasma membrane has been employed in the fractionation of plasma membrane fragments. Thus, Wallach et al. [51] have developed an ultracentrifugal technique – 'affinity-density perturbation' – to separate membrane fragments bearing differing densities of Con A receptors (and fragments with different Con A affinities) through the use of ^{125}I-labelled Con A convalently coupled to small phage particles of high density. Using an analogous principle, Zachowski and Paraf [52] selectively isolate 'right-side-out' plasma membrane vesicles, by employing polymerized Con A. Con A-bound to Sepharose has also been employed for the purification of liver 'microsomal membranes' in a centrifugal system [53].

Another instructive illustration of concanavalin A binding to intracellular membranes is provided by the reaction of the lectin with the membranes of retinal rod outer segments [37]. The reaction is with the carbohydrate moiety of rhodopsin. This protein contains covalently linked carbohydrate consisting of three *N*-acetylglucosamine residues and three mannose residues. Rhodopsin also binds WGA. Significantly, the dissociation constant for the reaction of fluorescein-labelled Con A with retinal disc membranes (2×10^{-7} M) is much lower than for the reaction of the lectin with its hapten sugar (5×10^{-5} M).

Finally, one must recall that some lectins penetrate rapidly into cells. This has been demonstrated for Con A and WGA for lymphocytes [54], as well as for normal and transformed hamster fibroblasts [55].

A very important consequence of the intracellular penetration of Con A is apparent in the studies of Noonan and Burger [31]. These authors use [^{3}H]-acetyl-Con A and [^{14}C]succinyl-Con A to demonstrate two components of lectin binding to a variety of cells: in all cases tested, binding is much greater at 22°C than at 0°C. Moreover, Con A bound at 0°C can be more readily eluted by α-methyl-D-mannopyranoside than Con A bound at 22°C. Noonan and Burger argue with reason that the binding at 0°C occurs primarily at the plasma membrane, whereas binding at 22°C also involves intracellular sites.

10.5.2. Isolation of lectin receptors

Allan et al. [56] report on the purification of Con A receptor activity from pig lymphocyte membranes. Glycoproteins solubilized in sodium deoxycholate were selectively purified by affinity adsorption on concanavalin A covalently attached to Sepharose 4B, and by subsequent elution with α-methyl-D-glucopyranoside. SDS-PAGE of the eluted fraction revealed five principal components, of which three were glycoproteins with mobilities slower than IgG; the two others had mobilities corresponding to molecular weights of approximately 27,000 and 33,000, respectively. In subsequent work [57], pig lymphocyte plasma membrane glycoproteins were purified by affinity adsorption on lentil-lectin, immobilized on Sepharose 4B, and then eluted with α-methyl-D-mannopyranoside. This technique yielded twice as much glycoprotein as the previous method. SDS-PAGE of the purified glycoproteins revealed more than *ten* protein bands. All were stained on treatment with periodic acid-Schiff reagent. The purified glycoproteins were found to inhibit lymphocyte transformation by lentil agglutinin. Experiments conducted with an antiserum against the glycoprotein isolate, suggest that the membrane glycoproteins are exposed at the outer surface of the plasma membrane.

Cuatrecasas [22] finds that digestion of fat cells with trypsin produces only minor decreases of Con A binding even under the most vigorous conditions. Pronase is somewhat more effective (~50% decrease in binding). This observation contrasts with experiments on Novikoff ascites tumors [58] in which Con A receptors were released by papain digestion. Neuraminidase treatment does not alter Con A binding. This is in contrast to the action of this enzyme on the binding of WGA. Schmidt-Ullrich et al. [43] have studied the glycoproteins from the membranes of rabbit thymic lymphocytes defined by SDS-PAGE and have found glycoprotein bands of apparent mol. wts. between 230,000 and 55,000. Metabolic studies show that the low molecular weight component responds uniquely to Con A. In these experiments, rabbit thymocytes were labelled *in vitro* with ^{14}C-labelled amino acids either before or during stimulation with Con A. The plasma membranes were then isolated and their proteins fractionated by SDS-PAGE. Normal and Con A-treated cells showed no qualitative and quantitative differences in the SDS-PAGE patterns of their membrane proteins. However, the protein turnover of the plasma membrane was generally augmented by Con A stimulation. The turnover of the glycosylated component with mol. wt. ~55,000 was particularly enhanced. Also, Con A induced excretion of this component into the cultivation medium. Recently, Schmidt-Ullrich et al. [44] have purified this glycoprotein. They solubilized the membranes in 1% Triton, a non-ionic detergent, and purified through affinity chromatography on immobilized Con A, by adding just sufficient α-D-glucopyranoside to elute *high*-affinity material. The glycoprotein was homogeneous on SDS-PAGE with an apparent mol. wt. of ~55,000. Crossed immunoelectrophoresis using anti-rabbit thymocyte serum, indicated a high degree of immunological purity. The protein contained sialic acid but its other sugar components have yet to be characterized.

Using cultured, malignant mouse lymphocytes (L1219), Hourani et al. [59] have succeeded in the relatively large-scale isolation of plasma membrane glycoproteins which inhibit agglutination of intact cells by concanavalin A, WGA and other lectins. The major components showed molecular weights of 84,000, 63,000, 44,000 and 33,000 by SDS-PAGE. Sugar analyses revealed the presence of sialic acid, galactose, galactosamine, glucosamine, mannose and fucose. There are major unexplained differences between this report and the data described by Jansons and associates [60,61].

Con A receptor activity has also been isolated from Novikoff ascites hepatoma [58] and rat erythrocyte ghosts [62]. The receptor from Novikoff tumors is a sialoglycopeptide, released by a papain digestion of intact cells. This material inhibitis agglutination of Novikoff cells by both Con A and WGA. After pronase treatment, material reacting with Con A could be separated from material reacting with WGA. Four sialoglycopeptides with mol. wt. 2000–33,000, reactive against Con A, were isolated. Strangely enough, papain-treated cells could still be agglutinated by both Con A and WGA.

More recently, papain-treatment of AS-30D hepatoma cells has been used to isolate plasma membrane glycopeptides bearing binding sites for concanavalin A and WGA. These peptide fragments have also been partially characterized chemically [63]. The AS-30D cells are readily agglutinated by low concentrations of WGA but only weakly agglutinated by high concentrations of concanavalin A. Treatment of the intact cells with papain, which renders them highly agglutinable by both lectins, at the same time released a glycopeptide fraction from the cell surface. This fraction contains components which inhibit hemagglutination by both WGA and concanavalin A. A sialoglycopeptide was isolated from the total papain lysate; this represented 12% of the mass of this material and possessed the major portion of the WGA receptor activity. Several sialoglycopeptide fractions were isolated with a molecular weight less than 3000, which showed binding activity only for Con A. These appeared rather similar to cell surface sialoglycopeptide fractions isolated from Novikoff ascites carcinoma [58].

The purification from erythrocyte ghosts [62] is very limited, but suggests that these membranes may contain two Con A reactive proteins. However, recently Podolsky et al. [64] have purified a Con A reactive glycoprotein (from rat erythrocyte ghosts) which had galactosyltransferase activity. The protein has an apparent mol. wt. of 54,000 and a pH optimum of 7.4–7.7. It transfers UDP-[H^3]Gal to fetuin with a K_M for UDP-Gal of 4.5×10^{-6} M. Enzyme activity is optimal with 5 mM Mn^{2+}. The enzyme can be precipitated in agar by Con A. After affinity purification on immobilized Con A, purification is 53-fold over starting membranes.

The receptors for castor bean lectin appear to be distinct from those for Con A [32]. Thus, parental chinese hamster ovary cells (CHO) exhibit drastically different saturation binding for diverse lectins than clone 15B derived therefrom (Tables 10.2 and 10.3). Concordantly, crude membrane preparations derived from 15B show 28% less sialic acid, 38% less *N*-acetylglucosamine, 49% less galactose but 53% more mannose than membranes from parental CHO.

The Con A receptor in the influenza virus envelope has been shown to lie on the 'spike' glycoproteins [65]. It appears to be an $\alpha_2\beta_2$-tetramer with two 39,000 dalton subunits and two 49,000 subunits.

A glycoprotein which inhibits cell agglutination by WGA has been isolated from a surface membrane preparation of L1210 mouse lymphocytes [60,61]. Extraction procedures for glycoproteins are followed by affinity chromatography on immobilized WGA with a hapten sugar as eluting agent. The receptor material is a glycoprotein, (with mol. wt. 40,000–60,000) containing $\sim$94 μg neutral sugar per mg/protein and 22 μg/mg sialic acid, as well as unstated amounts of hapten sugar *N*-acetyl-D-glucosamine. Unlike the sialoglycopeptide isolated from Novikoff ascites carcinoma [58], the L1210 glycoprotein does not react with Con A. Its effectiveness in inhibiting WGA-induced agglutination cannot be readily assessed. The preparation does contain *N*-acetyl-D-glucosamine, the hapten sugar, but this saccharide is so ubiquitous in distribution that its presence in a glycoprotein isolate cannot be considered remarkable.

Wheat germ agglutinin binding to adipocytes is virtually completely blocked by treating the cells with neuraminidase. This implicates sialic acid in WGA binding. Tryptic digestion can eliminate up to half the WGA binding. Isolates with different lectin receptor activity have been obtained from L1210 cells by another procedure [59].

TABLE 10.3

Binding of various lectins to parental CHO cells and clone 15B cells[1]

Lectin	No. of lectin molecules bound per cell	
	Parental CHO	Clone 15B
Con A	3.3×10^7	5.5×10^7
WGA	13×10^7	4.9×10^7
Lentil agglutinin	2.1×10^7	0.36×10^7
Castor bean agglutinin	2.6×10^7	0.03×10^7
SBA	0.2×10^7	$<2 \times 10^4$

[1] From Gottlieb et al. [32]. Binding measured using ^{125}I-labelled lectins.

10.6. Interaction of lectins with lymphoid cells

10.6.1. Overview

Several lectins, for example, Con A, PHA, lentil lectin and pokeweed mitogen, produce a blastogenic transformation in lymphoid cells. This mitogenic effect leads to the synthesis of proteins, RNA and DNA prior to cell division, as well as to numerous other metabolic events.

Soluble Con A, PHA and lentil lectin activate only T-lymphocytes, whereas pokeweed mitogen stimulates both T- and B-cells. This differentiation appears to be related to the form in which the lectin is presented. Thus, PHA immobilized on Sepharose beads effectively stimulates B-lymphocytes [66]. B-cells can also be activated by Con A locally immobilized at the bottom of plastic culture dishes [67]. However, T-cells are not activated by the immobilized Con A.

10.6.2. Membrane events

10.6.2.1. Permeability and transport

Within minutes after mitogenic stimulation of lymphocytes, major changes in membrane permeability and transport can be observed. Indeed, ion-permeable junctions form between contacting lymphocytes within *seconds* after PHA stimulation [68]. Also, mitogenic lectins induce increased uptake of a wide range of molecules from the culture medium, including potassium [69,70], calcium [71–73], nucleosides [74], carbohydrates [75], inorganic phosphate [76], and amino acids such as those treated in Chapter 7, for example, α-aminoisobutyric acid [77,78]. The accumulation of α-aminoisobutyrate reflects increased net amino acid influx and not decreased efflux or accelerated exchange diffusion. Kinetic analyses show that Con A and PHA exert their main effect by increasing maximum turnover of the amino acid carrier, with the affinity, (K_M), of the transport site unchanged.

Within an hour after the addition of PHA or Con A to lymphocyte cultures, the uptake of ^{42}K accelerates [75]. The stimulation becomes pronounced within 2 hr, and increases with time. Ouabain inhibits the rate of influx of ^{42}K at all stages of transformation and has no observable effect on the rate of potassium efflux. It is suggested that the increased uptake of potassium seems to be a necessary condition for the biosynthetic and morphological events of lymphocyte stimulation, since lymphocyte activation is inhibited by ouabain [70, 79]. Activation must also require divalent cations, since the process is prevented by the chelating agent EDTA and the effects of EDTA are reversed by Ca^{2+} or Mg^{2+} [73,80]. The increased membrane permeability may play a role in lymphocyte activation by altering the internal environment of the cell and thereby affecting enzyme activities and gene expression in such a way that the cell begins to divide [70].

10.6.2.2. Phospholipid metabolism

Studies on the effect of mitogenic lectins on the incorporation of isotopes into the membrane phospholipids of cultured lymphocytes using [^{32}P]orthophosphate [81,82], [^{14}C]choline [83], and [^{14}C]oleic acid [83,84] show two prominent, early changes in phospholipid metabolism: accelerated incorporation of [^{32}P]phosphate into *phosphatidylinositol* and accelerated turnover of the *fatty acid moieties* of the phospholipids.

PHA-treated lymphocytes incorporate ^{32}P more rapidly into phosphatidylinositol (and selectively into this phosphatide) than unstimulated lymphocytes [81,82]. Incorporation of ^{32}P into other major phospholipids increases also, but begins later. The phosphatidylinositol response warrants special attention both because it is one of the earliest effects of PHA in lymphocyte activation, and because there is a link between phosphate turnover in phosphatidylinositol and numerous active membrane functions. In this context, Fisher and Mueller [82] speculate that immunoglobulins or other proteins, when located at certain sites in the cell, restrict the expression of the genetic information required for the replication of lymphocytes, and that phosphatidylinositol metabolism is activated during their removal. This hypothesis rests upon the apparent similarity between the phosphatidylinositol response induced by PHA in lymphocytes and the one induced by acetylcholine in the pancreas [85]. However, there is little evidence to substantiate the hypothesis. The enhanced ^{32}P incorporation into the phosphatidylinositol of PHA-treated lymphocytes might be due to increased *de novo* biosynthesis of the phosphatide with or without an increased turnover, or to higher ^{32}P specific activities in its phosphorylated precursors, such as ATP or phosphatidic acid.

Another major change in membrane phospholipids is the accelerated turnover of their fatty acid moieties. Resch et al. [83,84] find that after 10 min of incubation, microsomes from PHA-treated lymphocytes incorporate 3 times more [^{14}C]-oleate into lecithins than microsomes from controls. Upon subfractionation of the microsomal fraction, the incorporated [^{14}C]oleate is recovered in plasma membrane lecithins. Resch et al. [83,84] show that the increased oleate incorporation induced by PHA follows activation of phospholipase A within the membrane. This enzyme acts on plasma membrane lipids and splits glycerophosphatides into lysolecithins and free long-chain fatty acids; both can perturb membrane structure. Reacylation of the lysophosphatides by Lands cycle enzymes (Chapter 3) prevents undue lysolecithin accumulation and allows rearrangements of membrane fatty acid composition. This cycle also produces localized membrane lipid turnover.

In this connection it is instructive to examine the insulin-like activities of Con A and WGA on *adipocytes* [86]. Like insulin, both lectins enhance glucose uptake and inhibit epinephrine-stimulated lipolysis in isolated fat cells. These effects occur with $\sim 2.0 \times 10^{-8}$ M Con A and $\sim 0.4 \times 10^{-8}$ M WGA and represent lectin reactions with high affinity binding sites; such sites account for only a fraction of the total lectin capacity. Low WGA levels (1 μg/ml) enhance insulin biding but 16 μg/ml of WGA or 5 μg/ml of Con A diminish insulin affinity competitively. (The same effect is observed with insulin receptors solubilized from liver membranes.) Significantly, both lectins, like insulin, inhibit the basal, as well as the epinephrine-stimulated, adenylate cyclase activity in adipocyte homogenates. This inhibition as well as other lectin effects can be blocked by excess 'hapten' sugar. Cuatrecasas and Tell [86] suggest that inhibition of membrane adenylate cyclase, which appears fundamental to many insulin responses, may also constitute an important mechanism in the action of WGA and Con A.

10.6.2.3. Membrane protein metabolism

Con A stimulation of rabbit thymocytes increases membrane protein metabolism [43]. The turnover of some membrane glycoproteins, possibly active in Con A binding, is particularly enhanced. This topic is discussed in Chapter 2.

10.6.2.4. Receptor redistribution

The presence of lectin receptors on the surfaces of lymphoid cells can be detected by use of fluorescein-labelled lectin. However, the resolution limits of fluorescence lie near 1 μ and far more precise localization of receptors can be achieved by electron microscopy. Here the detection of specific receptors depends on labelling them with specific lectins which have been coupled to (or react with) electron-dense particles, such as ferritin, hemocyanin or bacteriophage [87–89]. The distribution and movement of binding sites under various conditions can then be inferred from the distribution of these electron-dense markers as seen by transmission microscopy.

Fluorescence microscopy and the use of ferritin labelling can reveal topologic events occurring at membrane surfaces. To assess what occurs *within the core of the membrane* requires the use of freeze-cleaving electron microscopy (Chapter 1).

The freeze-cleaving process tends to split a membrane in a plane parallel to its surfaces, exposing the membrane interior. A single fracture within a membrane thus generates two complementary fracture faces, which, in the case of plasma membranes, face the extra-cellular and intracellular spaces respectively. The true inner and outer surfaces of membranes are not revealed by fracturing but can be visualized by etching. The fracture faces of most biomembranes are populated with intramembranous particles 8–10 nm in diameter, which are proteins or protein complexes penetrating into and/or through the membranes (Chapter 1).

In resting lymphocytes, the distribution of lectin receptors is diffuse. However, the addition of lectin (at appropriate levels) induces a polar distribution 'cap formation' of the receptors [88,89] (and then their subsequent pinocytotic engulfment). This redistribution process appears to be analogous to that induced with isoantibody against specific surface antigens and with anti-immunoglobulin reacting with membrane-bound immunoglobulin.

At concentrations below 5 μg/ml, Con A can induce cap formation, but at higher concentrations of this lectin or with succinyl-Con A no caps are observed [14]. Moreover, high concentrations (100 μg/ml) of Con A, block the capping reaction induced by anti-surface immunoglobulin. This inhibition is not observed with succinyl-Con A.

There have been several attempts to correlate the receptor redistribution, seen by fluorescence and transmission microscopy with possible changes in membrane 'core' structure. Thus, Loor [90] reports that exposure of lymphocytes to lectin levels that cause capping leads to aggregation of the intramembranous particles revealed by freeze-etch electron microscopy. He interprets this finding to indicate a close association between intramembranous particles and lectin receptors. However, 'capping' and particle redistribution need not reflect migration of identical membrane components. Moreover, no aggregation of the intramembranous particles of lymphocyte plasma membranes has been observed in the 'capping' produced by Con A and anti-immunoglobulin [88,89].

Guerin et al. [33] in their comparison of ME_2, a Con A agglutinable clone of MOPC 173 murine plasmocytoma, and a non-agglutinable clone ME_2C^r, find that agglutination of ME_2 by 100 μg Con A/ml produces clustering of intramembranous particles, whereas exposure of the Con A resistant cells to the same lectin concentration does not change the distribution of the intramembranous particles. Guerin et al. [33] are very cautious in interpreting their data. Indeed, the meaning of their results cannot be easily assessed and the relationship of the results to membrane transposition phenomena is obscure: *First*, Con A does not induce 'capping' in lymphocytes at the high concentrations employed [14] and Guerin et al. [33] do not measure capping in their system. *Second*, Con A is toxic to many cells at levels >5 μg/ml, [14]. Clustering of intramembranous particles is often observed in damaged cells and Guerin et al. [33] did not report on the viability of ME_2 cells upon Con A treatment. Other studies, however, show that 25 μg Con A/ml, arrests cell division of ME_2 [91]. *Third*, the authors related their morphological data to their binding studies, but used unpurified Con A for the former and highly-purified, acetylated lectin for the latter. We clearly require further coordinated experimentation, correlating membrane transposition phenomena, lectin binding and possible variations of membrane 'core' structure.

Another issue to be considered is the likelihood that selection for Con A resistance produces multiple pleiotypic membrane changes. Thus, Wright, [92] has isolated lines of Chinese hamster ovary cells which are resistant to Con A-toxicity even though the variant and wild-type cells bind equivalent amounts of lectin. Significantly, however, the Con A-resistant cells are much more sensitive to a whole series of reagents, some of which are membrane-active (e.g., phenylethylalcohol, ouabain) and some of which (e.g., testosterone, progesterone) do not act primarily on the plasma membrane. Similar observations are reported for variants of the MOPC 173 plasmocytoma [36].

One must also appreciate that the membranes of diverse cells and of different regions of the same cell may exhibit disparate lectin responses. This is illustrated by the studies of Nicolson and Yanagimachi [93] on the interaction of rabbit spermatozoa with purified, ferritin-labelled castor bean agglutinin. These authors show convincingly that the lectin receptors in the postacrosomal regions of spermatozoa can cluster following lectin binding. However, the plasma membrane receptors in the acrosomal and tail-regions do not redistribute following lectin binding. This could be due to restraints intrinsic to the membrane, or to unusual degrees of lectin-receptor cross linking in the tail- and acrosomal-regions [94].

The most popular explanation of the lectin-induced redistribution of receptors is that it represents the diffusion of membrane proteins in a 'sea of fluid lipid'

[95] as suggested by the 'fluid mosaic model' of membranes [96]. However, this model rests entirely on facts which do not apply very well to plasma membranes of animal cells, lymphocytes in particular. We treat this matter in Chapters 1, 8 and 11; here it is sufficient to say that substantial micromorphologic evidence [88,89,97] indicates that the redistribution phenomena represent movement of receptors *on the external lymphocyte surfaces, not within the membrane core.* (This could come about if the diverse receptors are bound at external membrane surfaces in thermodynamic association–dissociation equilibria, which favor the membrane bound state. Binding of lectin, etc., would perturb this equilibrium leading to migration of unbound receptors to the site of initial binding, as in a typical nucleation phenomenon. Bivalency of the ligands would restrict the number of nucleation sites and might account for 'capping'. This hypothesis is fully consistent with the data in [88,89,97].)

10.7. Lectin toxicity

It has long been known that certain lectins, particularly those of the castor bean, are highly toxic. (However, at least in the case of Con A, toxicity does not necessarily correlate with lectin binding or lectin-induced agglutination [36].) Rather recently, Onozaki et al. [98] have demonstrated that this toxicity can also be produced by Sepharose-immobilized lectin, and that it can be blocked by galactose, the hapten-sugar. This experimentation suggests that an initial step in lectin-induced toxicity is the binding of the lectin to carbohydrate receptors on the plasma membranes of the target cells. This possibility has been studied more extensively by Kornfeld et al. [99]. Their work involved the use of L1210 mouse leukemic lymphocytes as target cells and highly purified ^{125}I-labelled castor bean lectins (mol. wt. 60,000 and mol. wt. 180,000).

It was found that human erythrocytes, human lymphocytes and L1210 leukemia cells bind 2.24×10^6, 4.06×10^6 and 2.42×10^6 molecules of 60,000 mol. wt. lectin per cell, respectively with association constants of 1.6, 3.0 and $3.7 \times 10^6 \cdot M^{-1}$, respectively. The values for the high-molecular weight lectin were 1.12×10^6, 3.35×10^6 and 3.53×10^6, binding sites per human erythrocyte, human lymphocyte, and L1210 leukemia cell respectively and the corresponding association constants were 12.5, 9.7, and $9.1 \times 10^6 \cdot M^{-1}$. Binding curves indicated a single type of receptor site. However, there were significantly more receptors for the low-molecular weight lectin on erythrocytes than for the high-molecular weight agglutinin, whereas the number of sites on lymphocytes appeared to be the same. The high-molecular weight lectin bound with greater affinity than the low-molecular weight molecule.

Incubation of the castor bean lectins with L1210 cells produces marked inhibition of protein- and DNA-synthesis. It is computed that 50% inhibition of DNA synthesis occurs when only 0.3% of the available binding sites on the cell membrane for the low-molecular weight species are occupied with lectin. Binding of the low-molecular weight lectin is rapid and (half maximal after about 7 min), but inhibition of protein synthesis could not be detected for the first 30 min. Blockage of DNA synthesis occurs somewhat later and inhibition of RNA synthesis appeared after about 1.5 hr. As reported previously, the toxic effects could be inhibited by hapten sugars.

The data of Kornfeld et al. [99] are not inconsistent with those of Onozaki et al. [98]. However, the former authors point out that their data do not prove that lectin-induced toxicity occurs simply by the binding of the lectin to the plasma membrane. They point out that the experiments using castor bean lectin bound covalently to large Sepharose beads, are not conclusive because very large amounts of immobilized lectins had to be added to the cells to cause toxicity and that dissociation of small amounts of soluble lectin (as it is known to occur with other immobilized lectins) might account for the observed effects.

A reasonable alternative is that, after binding to the plasma membrane, the lectin enters the cell by pinocytosis, or another mechanism and then proceeds to exert its toxic action intracellularly. The toxic process may be analogous to that suggested for the castor bean toxin Ricin, namely, inhibition of protein synthesis by prevention of the completion or release of nascent peptide chains [100]. The suggestion that, in order to effect its toxic effects, the lectin must first bind and then be translocated intracellularly, would certainly fit the lag between binding and metabolic inhibition observed by Kornfeld et al. [99].

10.8. Enhanced lectin-induced agglutination of certain neoplastic cells

10.8.1. Introduction

The increased susceptibility of many oncogenically transformed cells to agglutination by lectins such as Con A, WGA, SBA, lentil agglutinin and castor bean agglutinin is now an established fact [5]. However, despite intense research activity, the phenomenon remains inadequately quantified, over-interpreted and poorly understood. Particularly important, and often neglected, is the fact that lectin receptors are 1000 times more prevalent on most cells than are specific antigenic sites which allow cell agglutination in the presence of specific antibody. A preponderance of work has focussed on various rodent fibroblast lines and transformants thereof produced by oncogenic DNA viruses. The agglutination data obtained with these systems have been summarized by Lis and Sharon in Table 3 of their review [2]. Lis and Sharon express the diverse data in terms of a 'selectivity factor', that is, the ratio (minimal lectin concentration required for agglutination of normal cells per concentration required for agglutination of transformed cells).

The selectivity factor for Con A using normal rodent fibroblasts and SV40 or polyoma transformed variants ranges from 3–5 [101] to <15 [29,102,103].

The situation is complicated even at this level. Thus, Shoham et al. [104] find that Con A bound to transformed cells can cause their death or inhibit their replication – in contrast to the well known mitogenic action of this substance bound to lymphocytes. However, Burger and Noonan [105] claim that combination of the receptors on transformed cells with 'monovalent concanavalin A' restores normal 'contact-inhibitory' properties to these cells. They relate this observation to an apparent 'transient escape from growth control' produced in normal cell cultures by gentle protease treatment.

An important aspect of the lectin reactivity of cells is that mild protease treatment of normal cultured fibroblasts enhances agglutinability to the levels found in transformed cells [106, 107]. For example, treatment of contact inhibited 3T3 mouse fibroblasts with pronase, trypsin, ficin, papain, subtilisin and trypsin (less than 10 μg protease per ml) enhances agglutinability for about 5 hr after the cells are removed from the substrate using EDTA. Such protease treatment is not sufficient in itself to detach the cells from the culture substrate. Shortly after treatment, the cells begin to synthesize DNA and then go through another round of cell division. Thereafter, they return to their resting state, cease to divide and lose their agglutinability.

To exclude the possibility that the added proteases are pinocytosed and act internally, Burger [108] repeated the experiments, using trypsin or pronase coupled to Sephadex beads (5 μm diameter: ^{3}H-labelled). 3T3 mouse fibroblasts in confluence were exposed to protease-carrying beads for 10–15 min. Uncoupled beads were used as controls. After rinsing, less than one bead per 200 cells remained in the culture dishes. Exposure of cells to the protease-carrying beads induced another cycle of cell division, just as did the soluble proteases. Uncoupled beads had no effect. The growth stimulation was not due to release of proteases from the beads during incubation period. Growth activation by proteases has also been observed in the case of chick embryo fibroblasts [109]. It is important to note that protease-treatment of normal cultured fibroblasts does not change the amount of concanavalin A bound to these cells [29,102,110].

Burger and Noonan [107] argue that, if exposure of the surface binding sites reflects a change which prevents 'contact inhibition', then covering these receptors with fragments of concanavalin A might restore 'contact inhibition'. They claim that they have proven this connection. However, their data lack conviction and the mechanism and significance of 'contact inhibition' remain uncertain (Chapter 8).

Inbar et al. [111] have studied the effects of *temperature* on the Con A agglutinability of various transformed cell lines (SV40, polyoma, RNA viruses, chemicals) and their parental cells. They find that Con A agglutinates transformed cells at 24°C but not at 4°C. Further, transformed cells, as well as normal cells treated with trypsin, agglutinate at both 24°C and 4°C with low Con A concentrations (5 μg/ml). The authors find that the same number of Con A molecules bind to normal and transformed cells at both temperatures. They argue that the site for Con A on the surface membrane contains two activities, a component *A* that binds Con A and a component, *B*, that determines agglutination. *A* is not temperature sensitive and is active in normal and transformed cells, whereas *B*, which is temperature sensitive, is active only in transformed cells. *B* can be activated by trypsin, and the increased activity per cell allows agglutination at 4°C with a low concentration of Con A.

Inbar et al. [111] find that agglutination of transformed cells by wheat germ agglutinin, which binds to *N*-acetyl-D-glucosamine-like sites, and by soybean agglutinin, which binds to *N*-acetyl-D-galactosamine-like sites, are not temperature sensitive. Thus, the temperature-sensitive component, *A*, appears 'specific' for Con A. Neoplastic transformation of normal cells, which results in agglutinability by Con A, is associated with the activation of a specific temperature-sensitive activity on the surface membrane. Inbar et al. suggest that the observed temperature dependence may be *metabolic*, possibly involving a membrane enzyme.

Noonan and Burger [112] also observe a temperature dependence of Con A-induced agglutination of 3T3 fibroblasts, their polyoma transformants and trypsinized 3T3 cells. They find that transformed cells and trypsin-treated normal cells do not agglutinate at 0°C, although the amount of lectin bound at 0°C suffices to produce agglutination when the cells are shifted to >15°C; that is, both transformed cells and trypsin-treated normal cells agglutinate at 15°C. However, 'mild' glutaraldehyde fixation of the cell surface inhibits agglutination, but not agglutinin binding. Noonan and Burger [112] conclude that Con A-mediated cell agglutination requires free movement of the lectin receptor sites *within* the plane of the cell surface.

The selectivity factor of WGA, comparing normal rodent fibroblasts and their DNA viral transformants ranges between ~20 and 100 [29,113]. As in the case of Con A, the WGA selectivity factor of trypsinized normal mouse fibroblasts approaches that of transformed cells.

Two purified lectins from *castor bean* more readily agglutinate SV40 transformants of 3T3 mouse fibroblasts and polyoma transformants of BHK hamster fibroblasts, than they agglutinate parental cells [114]. For example, normal 3T3 and BHK cells show no agglutination at a concentration of 1 μg/ml of the 120,000 mol. wt. lectin and 1/2 maximal agglutination at ~5 μg/ml. In contrast, SV40 transformed cells showed greater agglutination, and this was 1/2 maximal between 0.1 and 0.5 μg/ml and maximal at 0.1 μg/ml. Polyoma transformed cells behave similarly, and the agglutination behavior of trypsinized normal 3T3 fibroblasts is at that of polyoma transformed cells. The selectivity factor is ~20 for the SV40-3T3 cells and ~10 for the polyoma-transformed BHK21 cells.

Sela et al. [115] find that SBA agglutinates the cells of mouse, rat and human cell lines which have been transformed by viral carcinogens, but not *hamster* cells transformed by viral or non-viral carcinogens. Parental cells are not aggregated by this agglutinin but they are rendered agglutinable after short incubation with trypsin or pronase. However, transformed hamster cells become agglutinable only after prolonged treatment with pronase. Agglutination by SBA is specifically inhibited by *N*-acetyl-D-galactosamine.

The data are interpreted to indicate that receptor sites containing *N*-acetyl-D-glucosamine exist in a cryptic form in normal cells, are exposed in transformed mouse, rat and human cells, but become less accessible in transformed hamster cells. Receptor sites for soybean agglutinin differ from the receptors for wheat germ agglutinin and concanavalin A, since the latter receptors become exposed upon transformation of *all* lines tested. In normal hamster cells, the receptors for all three agglutinins become exposed after incubation with trypsin, but the exposure of *N*-acetyl-D-galactosamine-like sites requires long enzyme treatment. The results are taken to suggest a difference in the location of different carbohydrate-containing sites in the surface membrane.

Chick embryo fibroblasts infected by Rous sarcoma virus, an oncogenic RNA virus, become more agglutinable by both Con A and WGA [116]. The differences between normal and transformed cells are more pronounced with WGA. This observation is contrary to that of Moore and Temin [117] and Burger and Martin [118] and suggests that the discrepancy is due to interference by the large amounts of acid mucopolysaccharide secreted by the chick cells. In their hands, hyaluronidase treatment removes this interference and makes the transformed cells highly agglutinable by Con A and WGA.

Of interest is Burger and Martin's [118] finding that cells infected with a temperature sensitive mutant of Rous sarcoma virus are more agglutinable when grown at the permissive (transforming) temperature (36 °C) than at the non-transforming temperature (41 °C), although the mutant grows equally well at both temperatures. The reasons for this finding are obscure, particularly in view of the ready agglutinability of many cells infected with non-oncogenic RNA viruses.

10.8.2. Possible involvement of adenylate cyclase

Treatment of fibroblasts transformed by DNA viruses with dibutyryl-cAMP and theophylline alters the characteristics of these cells, particularly their morphologic appearance and their high agglutinability by Con A [119–126]. Also, dibutyryl-cAMP and theophylline prevent the enhanced agglutinability that occurs when confluent cultures of 3T3 cells are treated with trypsin [126].

These observations must be related to the findings of Cuatrecasas and Tell [86], who show that in *adipocytes*, high concentrations of Con A activate membrane adenylate cyclase activity, and thus produce an elevation of cytoplasmic cAMP. WGA behaves similarly; in fact, both of these lectins act like insulin. cAMP has been implicated in the control of surface topology, through its stabilization of microtubules. Accordingly, concanavalin A, depending on its dose, may influence the surface characteristics of diverse cells by its influence on the activity of membrane adenyl late cyclase.

10.9. Lectin binding by normal and neoplastic cells

The enhanced agglutinability of transformed cells was originally attributed to increased lectin binding, but this inference is now in dispute. By radiolabelling diverse lectins with ^{125}I, and ^{3}H (acetylation) or ^{14}C (succinylation), their binding to various cells can be directly quantified. As shown in Tables 10.2 and 10.3, numerous cell types bear Con A receptors and, under the conditions generally used for binding (incubation at $> 20\,^{\circ}$C) one cannot discern clearly distinguishable differences in the saturation binding of this lectin to various cultured fibroblasts and their viral transformants [22,24,26,27,29,102,110,111,127–129]. Also, while human lymphoblasts in long-term culture are agglutinated far more readily by Con A than normal lymphocytes [35], these cells actually bind somewhat less ^{125}I-labelled Con A per unit surface area than normal lymphocytes or lymphoblasts (at 25 °C).

Moreover, under usual conditions, the number of Con A molecules bound to normal hamster cells does not appear to change between interphase and mitosis [130] and is approximately the same as the number of Con A molecules bound by chemically transformed (dimethylnitrosamine) hamster fibroblasts. Lectins other than Con A also do not appear to bind preferentially to transformed cells; for WGA the values range between $3–4 \times 10^{7}$ molecules/cell.

We virtually lack any data on lectin binding kinetics and equilibria. However, the experiments of Shoham and Sachs [130] suggest similar binding rates for normal, chemically-transformed and polyoma-transformed cells, whether in interphase or mitosis. However, the binding rates appear very slow, when compared with those observed in adipocytes [22]. This may be due to penetration of the lectin into the cells under the conditions employed.

Unfortunately, most binding studies do not discriminate between binding at the cell surface and total cellular uptake. Most workers have ignored this matter and have measured binding at $>20\,^{\circ}$C, where pinocytosis, etc., occur readily. This represents a crucial issue because entirely different processes are involved. Only surface binding can be related to agglutinability differences between normal and neoplastic cells. Indeed, Noonan and Burger [31,112] show very clearly that various transformed cells *bind 3–5 times as much [^{3}H]acetyl-Con A* than normal cells at 0°C, *but not at 22°C*. This pattern holds for BALB/c 3T3 cells and their polyoma or SV40 transformants and for H615 ts 3T3 cells bearing a temperature sensitive SV40 genome (transformed when grown at 32°C; normal when grown at 38°C).

The data, illustrated in Table 10.4, hold also for [^{14}C]succinyl-Con A, and the other cell lines tested. For example, the H615 3T3 cells grown at 32 °C (transformed) bind four times as much Con A, at 0°C as cells grown at 38°C (non-transformed). In all cases, no significant differences in binding could be detected at 22°C. The authors conclude very reasonably that the bulk of Con A taken up at temperatures >20°C has very little to do with the cell surface. This conclusion is supported by the fact that 1 mM α-methyl-D-mannopyranoside cannot reverse Con A binding as efficiently at 22°C as at 0°C. For example, at 0°C about 60% bound [^{3}H]acetyl-Con A can be eluted; at 22°C, the value is only 25%.

TABLE 10.4

Binding of [^{3}H]acetyl-Con A to normal and polyoma-transformed 3T3 fibroblasts at 0°C and 22°C[1]

	0°C		22°C	
	3T3	Py 3T3	3T3	Py 3T3
Molecules/cell	6.6×10^5	1.3×10^6	4.1×10^7	2.7×10^7
Molecules/nm^2 surface	3.0×10^2	1.7×10^3	1.8×10^4	1.8×10^4

[1] From [112].

The temperature effects also extend to protease-treated cells and cells in mitosis. Thus, at 0°C protease-treated cells bind 5 times as much Con A as untreated controls, whereas little binding difference can be detected at 22°C. Also, contrary to the arguments of Shoham and Sachs [130], who worked at 37°C, mitotic cells bind about 5 times as much Con A as interphase cells. Finally the lectin concentrations required to give half-maximal binding suggest that normal and transformed cells possess rather similar Con A affinities at 0°C. To sum, Noonan and Burger [112] clearly demonstrate a significant difference between the cell surfaces of normal and transformed cells, in terms of Con A binding.

Nicholson and Lacorbiere [131] have measured the reaction of *castor bean lectin* with normal BALB/c 3T3 fibroblasts and SV40 transformed variants thereof. They measured both cell agglutinability and the saturation binding of ^{125}I-labelled lectin. They observed only slight differences in the agglutinability of 3T3 cells whether sparsely growing or touching, but agglutination induced by castor bean lectin decreased after contact. Binding studies, performed at 4°C and 10 min (to avoid pinocytosis of the castor bean lectin) suggest that there are approximately 10^7 binding sites per confluent 3T3 or SV3T3 cell. In the case of the transformed cells, binding did not vary with cell density but 3T3 cells exhibit 2.5 times the number of binding sites at confluence cells as under conditions of sparse growth. On the basis of surface area, the density of receptor sites on the transformed cells is approximately five times as great as that on rapidly growing, non-contacting 3T3 cells.

10.10. The topological distribution of lectin receptors on the surfaces of normal and neoplastic cells

10.10.1. General Comments

The distribution of Con A receptors on cell surfaces can be evaluated by fluorescence microscopy, using fluorescein-labelled Con A [132,133] and by electron microscopy using ferritin-labelled Con A [44], peroxidase-labelled lectin [134] and localization of bound Con A by its reaction with hemocyanin [135,136]. In the ferritin technique, the cells are treated with a preformed, purified ferritin-Con A complex (mol. wt. of apoferritin ~450,000). The peroxidase method uses a different principle: horseradish peroxidase contains 18% carbohydrate, which can react with Con A. To use peroxidase for the localization of Con A receptors, the enzyme is added to Con A-treated cells after removal of excess lectin. If the bound lectin bears free valences for the carbohydrate on peroxidase, the enzyme is bound and can then be detected *enzymatically* by electron microscopic cytochemistry. The same principle applies to the use of hemocyanin. This is also added after Con A-binding and removal of excess lectin. Remaining free valences on membrane-associated Con A might then react with the sugars on hemocyanin and bind this molecule to the membrane also. Hemocyanin can then be detected by virtue of its *size* alone. However, the steric factors involved in the secondary binding of peroxidase (mol. wt. ~40,000) and hemocyanin (mol. wt. ~0.5–9 × 10^6; diam. 350 Å) clearly differ drastically.

Each of the high resolution labelling methods introduces certain ambiguities, which remain to be resolved. *First*, both the ferritin- and the peroxidase-techniques permit visualization of only selected regions of a cell surface or of only limited portions of a cell population. As employed by Nicolson and Singer [49], the ferritin technique allows visualization of relatively large surface areas, but only a *few cells* in a population are sampled. When thin-section electron microscopy is employed (e.g. [134,137]), only *limited* surface regions of a *few* cells can be sampled without prohibitive effort. These sampling problems are avoided in the hemocyanin method, where whole surfaces of many cells are visualized.

Second, basic principles of thermodynamics require that ferritin–Con A and native Con A exhibit different membrane binding equilibria and kinetics. However, no data exist on this point. We do not know the reaction rate of ferritin–Con A, the number of ferritin–Con A binding sites for a given cell or the affinities of these sites for ferritin–Con A vs. ^{125}I-labelled Con A. We are also not justified in assuming that ferritin–Con A is equivalent to Con A merely on the basis of its role in cell agglutination, because lectin binding and cell agglutinability are not related in a simple way. *Third*, the peroxidase technique and hemocyanin approach create other difficulties. Both methods depend upon the secondary binding of the marker molecules to membrane bound Con A via saccharide chains. But hapten *sugars* generally *dissociate* Con A complexes. It is unlikely that glycoproteins bearing hapten sugars are ineffective competitors for membranes. Indeed, in the case of WGA, ovomucoid readily dissociates the adipocyte–lectin complex. In any event, one would expect both peroxidase and hemocyanin to dissociate some Con A from the membranes in the form of peroxidase–Con A or hemocyanin–Con A complexes (possibly leading to a redistribution of bound Con A). Conceivably, secondary associations of Con A with membranes are sufficiently strong to minimize this possibility. Indeed, a recent study [138] shows that horseradish peroxidase does not dissociate concanavalin A bound to 3T3 cells; that is, the membrane sites have a higher affinity for the lectin than the peroxidase. On the other hand, the study also shows that peroxidase staining does not necessarily mirror concanavalin A-binding accurately. Indeed, the possibility exists that the peroxidase and the hemocyanin techniques may selectively demonstrate *high-affinity* membrane receptors.

10.10.2. Fluorescence measurements

Inbar et al. [139] have compared normal lymphocytes from rat lymph nodes and mouse spleens with lymphoma cells from a Moloney virus-induced mouse ascites lymphoma. They measured Con A binding using ^{3}H-labelled lectin, and evaluated receptor mobility, using Con A coupled to *fluorescein*. They also studied fluorescein conjugated WGA.

Their data show that lymphoma cells bind more Con A/cell than (but the same amount per mg of cell protein as) normal lymphocytes. Using fluorescein–Con A (F–Con A) they demonstrate cap formation in 30% of normal lymphocytes, but apparently none at all in lymphoma cells. They argue that the formation of caps shows that Con A sites can move on the surface membrane and that the greater capping in normal lymphocytes indicates *higher* receptor mobility in normal lymphocytes. Fixation with glutaraldehyde or formaldehyde inhibits capping, clustering on Con A sites and cell agglutination by Con A, but does not change Con A binding. The data are interpreted to indicate a different fluidity of normal and neoplastic plasma membranes (Chapter 1).

10.10.3. Electron microscopy

10.10.3.1. Concanavalin A

The introduction of the listed electronmicroscopic marker techniques has revealed another feature of the different lectin reactivity of normal cultured fibroblasts and various neoplastic transformants thereof. As first demonstrated by Nicolson [140], reaction of BALB/c 3T3 cells and their SV40 transformants with Con A–ferritin produces a uniform distribution of labelled lectin on the surface of the former and a clustering of deposited label on the latter (Fig. 10.1). In view of evidence suggesting the 3T3 cells and their SV40 transformants bear identical numbers of Con A receptors, Nicolson [140] concluded that the important difference between the two cell types, as revealed by their different agglutinability, was a function of the *localized* concentration of the receptors of transformants. However, Nicolson did not show whether binding of ferritin–Con A was as complete as that of native Con A, that is, whether the visualized

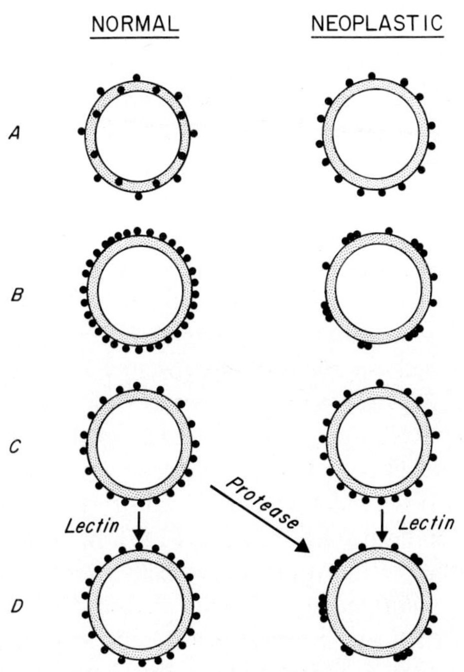

Fig. 10.1. Possible mechanisms involved in the different lectin reactivities of normal and neoplastic cells. *A* Uncovering of 'cryptic' sites; *B* altered intrinsic surface distribution; *C–D* altered induced distribution.

deposits were representative of all Con A binding sites. Indeed, we question whether full saturation of receptor sites was achieved for the following reasons: counts of ferritin deposits on normal 3T3 surfaces, using several fields of Fig. 1 in Nicolson [140] and also of Fig. 1a in Nicolson [141] show 1.3×10^3 and 9×10^2 particles per μm^2 respectively. This is less than one tenth of the receptor density typical for fibroblasts [130] and 1/100 that computed for diverse lymphoid cells! One must, therefore, suspect that either (a) some of the binding, reflected by use of labelled lectins, is due to *intracellular* receptors (e.g., via pinocytosis), or (b) not all the Con A receptors are revealed by ferritin–Con A, and the sites not seen represent low-affinity and/or sterically-protected loci.

Nicolson [140] comments that the density of ferritin–Con A deposits on the transformant surfaces is ~3.5 times that on normal cells. Since there is no difference between these cells in the binding of free Con A, the data suggest either (a) a change in affinity and/or accessibility, in addition to clustering, or (b) that significantly greater surface binding of Con A is masked by intracellular lectin accumulation in experiments using radiolabelled lectins.

Many studies show that mild protease treatment of normal cultured fibroblasts enhances their agglutinability by Con A and other lectins to an extent resembling that of their SV40 or Py transformants. However, mild trypsinization does not appreciably alter lectin binding to normal fibroblasts at >20°C [29,102,110, 111,129]. Using the ferritin–Con A technique, Nicolson [141] shows that light trypsinization of 3T3 cells, followed by treatment with ferritin–Con A, leads to enhanced agglutination of the cells and a clustered distribution of the visualized receptors. However, as noted before, the density of visualized receptors is considerably below that excepted from ^{125}I-labelled Con A binding studies. Nicolson suggests that (a) normally the lectin receptor sites are randomly distributed (b) protease treatment causes clustering by localized changes in the charge distribution of the cell surfaces and (c) clustering of receptor sites favors agglutination (Fig. 10.1).

Treatment of various normal fibroblasts with Con A followed by horseradish peroxidase and thin-section electron microscopy reveals a continuous surface layer of bound Con A [142–144]. The surfaces of many polyoma-transformed hamster fibroblasts have an appearance similar to that of normal cells, but in a significant number of cells, the Con A deposition is patchy and irregular [142]. Similar patterns are observed with SV40 transformed hamster cells [145] and Adeno 12-transformed hamster fibroblasts [143]. In all of these cases, a statistically significant number of the transformed cells showed a patchy deposition of Con A. However, no *difference* in labelling with Con A-peroxidase could be detected between normal and SV40 transformed *rat*-fibroblasts [144,145].

The concept that Con A induces the redistribution of at least some of its receptor sites in transformed and trypsinized 3T3 cells receives support from experiments in which bound Con A is reacted with hemocyanin and the reaction is visualized by shadowcast replication electron microscopy [136]. The data indicate that the inherent distribution of the bound Con A visualized by this technique is identical in 3T3 cells, trypsinized 3T3 cells and SV40 transformants. However, treatment of unfixed *transformants* or *trypsinized* normal cells with Con A and hemocyanin at 37°C or at 4°C with subsequent warming to 37°C, for 10 min, produces clustering. This is not observed with normal 3T3 cells or prefixed transformed cells (Table 10.5). The clustering of receptors found in

TABLE 10.5

Distribution of bound Con A on the surfaces of 3T3 Cells, SV40 transformants and trypsinized 3T3 cells as revealed by hemocyanin binding[1]

Cells	Labelling temperature[2]	Incubation temperature[3]	Distribution
3T3	4°C	4°C	Dispersed
	4°C	37°C	Dispersed
SV3T3	4°C	4°C	Dispersed
	4°C	37°C	Clustered
	37°C	4°C	Clustered
SV3T3-prefixed	37°C	37°C	Dispersed
SV3T3-trypsinized	4°C	4°C	Dispersed
	37°C	37°C	Clustered

[1] From Rosenblith et al. [136].
[2] Treated with Con A (10 min) washed, treated with hemocyanin (10 min).
[3] Postincubated 10 min.

protease-treated and transformed cells can thus be attributed directly to a crosslinking action of the lectin acting as a polyvalent ligand. The authors agree with Nicolson [141] that, once clusters form, these may promote agglutination. These experiments appear quite convincing in terms of the Con A molecules visualized. However, the data allow no conclusions as to the proportion of Con A receptors involved in the clustering process.

Nicolson [146] has used fluorescein-labelled Con A and fluorescein-labelled anti-Con A to monitor the effect of temperature on the distribution of Con A receptors on the surfaces of 3T3 cells and SV40 transformants thereof by fluorescence microscopy. His data support those of Rosenblith et al. [136]: Con A induces a redistribution of its binding sites at 37°C in the transformants, but not in normal cells.

Most investigators thus observe distinct differences between normal and transformed fibroblasts in terms of the distribution of bound Con A *after* reaction with the lectin at physiologic temperatures. All now agree that the distribution of Con A receptors *prior to* reaction with the lectin is uniform, dispersed and not appreciably different in normal, transformed and trypsinized normal cells. However, there are data indicating that lectin-induced clustering is not an invariant characteristic of neoplastic conversion. Thus, Smith and Revel [135] detect no differences in Con A receptor distribution between normal and SV40 transformed mouse fibroblasts (3T3) using the hemocyanin technique to localize bound Con A electron microscopically. Furthermore, DePetris et al. [137] report clustering of Con A–ferritin binding sites not only in transformed or trypsinized fibroblasts, but also in *normal* 3T3 cells (clone A31) and secondary mouse embryo fibroblasts. Indeed, these workers find equivalent densities of Con A–ferritin on the apposed surfaces of agglutinated transformed cells as on the free surfaces of normal cells. Using the Con A peroxidase method, Torpier and Montagnier [147] could also find no reliable correlation between receptor topology and transformation state. Furthermore, Walig et al. [148], using the Con A–peroxidase method, could not discern any differences between the surface topologies of a line of SV40-transformed rat cells, normal mouse 3T3 cells, and their somatic hybrids: all yielded a somewhat 'patchy' distribution.

The last listed data, as well as the reports of Bretton et al. [148] and Garrido et al. [144] on rat cells, indicate either that the clustering phenomenon is less general than previously assumed, or that it involves experimental variables as yet unrecognized. A clue to the discrepancies comes from the recent studies of Ukena et al. [149]. It appears that sampling inadequacies may be one source of the difficulties. Again, visualizing Con A–hemocyanin complexes by shadow-cast replication, Ukena et al. [149] confirm prior results [136] of an *intrinsically* dispersed distribution of visualizable Con A on both normal and SV40-transformed 3T3 fibroblasts, and temperature-dependent, Con A-induced clustering on the surfaces of the transformants. Significantly, however, they clearly demonstrate that clustering of Con A *does not occur on pseudopodia and other 'peripheral' membrane domains.* This distinction might not be readily discerned by thin section approaches.

Ukena et al. [149] make another important observation: colchicine, cytochalasin B and 2-deoxyglucose do not alter the intrinsic, dispersed distribution of the bound Con A visualized by the hemocyanin technique in the case of both normal and transformed 3T3 cells. However, *colchicine* induces a 'cap-like' association of clusters near the center of the cell, whereas cytochalasin B and 2-deoxyglucose lead to the appearance of Con A clusters over the *entire* cell surface, including pseudopodia! These drug effects can be eliminated by prefixing. As noted elsewhere, the precise roles of both colchicine and cytochalasin B remain to be defined. Colchicine at concentrations well above the 10^{-6} M used here may exert a primary membrane effect [150]. Also, colchicine inhibits the Con A-mediated agglutination of SV40-transformed 3T3 cells [151] and the gross surface distortions (not clustering of Con A receptors) induced by colchicine and related drugs occur in both normal and transformed cells. Cytochalasin B may also exert a primary membrane action at low concentrations (Chapter 7), but at the high levels ($\sim 5 \times 10^{-4}$ M) employed by Ukena et al. [149], both membrane and cytoplasmic responses could occur. Thus either substance might produce the responses observed in transformed cells acting directly on membranes or by influencing cytoplasmic microfilaments.

The action of 2-deoxy-D-glucose appears to be even more obscure. This sugar is transported by facilitated diffusion, phosphorylated but not readily metabolized (Chapter 7). It cannot act as an energy source, but shares some function with cytochalasin B, which inhibits its transport very efficiently.

Possibly these pleiotypic responses relate to the involvement of membrane adenylate cyclase. Indeed, dibutyryl-cAMP and theophylline prevent the concanavalin A-induced clustering of bound lectin on cultured 3T3 cells which usually occurs after brief trypsinization. In addition, these two reagents eliminate the concanavalin A-induced clustering of plasma membrane-bound concanavalin A which is typically found in SV40-transformed 3T3 fibroblasts [152]. Moreover, addition of concanavalin A to variant SV40-3T3 cells which are resistant to concanavalin A toxicity does not produce clustering of concanavalin A receptors, as detected by the hemocyanin technique. However, clusters do form when the concanavalin A-resistant cells are pre-treated with colchicine and dibutyryl-cAMP, as well as with theophylline.

10.10.3.2. Wheat germ agglutinin

Experimentation on Con A receptor topography has been extended to the distribution of WGA binding sites [144]. In these studies WGA, precoupled to peroxidase, was employed as a marker. As in the case of Con A-peroxidase, normal fibroblasts gave continuous, uniform surface labelling, whereas SV40-transformed *hamster* fibroblasts exhibited a patchy deposition of WGA. As in the case of Con A, one could not distinguish *normal rat* fibroblasts from SV40 or adeno-12 transformed cells according to WGA-peroxidase binding. The authors point out that the appearance of discontinuous labelling of transformed hamster cells required about 40 min in the case of Con A and 20 min in the case of WGA-peroxidase. This is consistent with data already presented, showing that the discontinuities reflect a *lectin-induced* membrane rearrangement. Clearly, despite their differences in specificity, Con A and WGA seem to report related surface peculiarities in transformed hamster cells.

10.11. Significance of the lectin reactivity of neoplastic cells

The enhanced lectin reactivity of transformed cultured fibroblasts appears to be a rather general phenomenon, closely related to many other characteristics of *in vitro* transformed cells. But what does this phenomenon signify?

First, can one consider the process peculiar to the transformed condition? The answer is definitely no. Thus Benjamin and Burger [113] observe enhanced agglutinability in cultured fibroblasts after *permissive* infection with polyoma virus. Also, Zarling and Tevethia [103] show very clearly that the *non-oncogenic* infection of rabbit kidney cells by vaccinia virus induces enhanced agglutinability soon after virus penetration into the cells. The process requires an active viral DNA and early protein synthesis. It does not require synthesis of viral DNA.

Similar data are reported for herpes simplex virus [153]. The surface change producing enhanced agglutinability is shown to be an early, virus-determined function, not a non-specific effect.

Indeed, infection of a variety of culture lines with diverse, non-transforming, enveloped RNA viruses (myxoviruses, arborviruses and rhabdoviruses) makes the cells agglutinable by Con A, usually within <12 hr after infection [154]. Not all cells react identically upon infection with a given virus. Thus the arborviruses do not induce agglutinability of chick embryo fibroblasts but do so of BHK cells, even though both cell lines produce equivalent, large amounts of virus. All the enveloped viruses are strongly agglutinated by Con A, a fact implying that the host membrane receptor enters the viral envelopes.

Studies on chick embryo and BHK fibroblasts infected with Newcastle disease virus indicate that these do become agglutinable by WGA as well as Con A [155,156]. The agglutination can be inhibited by specific hapten sugars. Again, the virus particles isolated from these cells can also be agglutinated by WGA and Con A.

Both influenza and vesicular stomatitis viruses bear sugar residues on their surfaces, which are capable of interacting with Con A, wheat germ agglutinin and castor bean lectin. This interaction can result in the agglutination of the viruses by these lectins. The amount of Con A or castor bean lectin bound to influenza particles suggests that the number of lectin receptors on a viral surface closely approximates the number of spike glycoproteins visualized by electron microscopy. Studies of the localization of ferritin–lectin conjugates on the viral membranes are also consistent with the notion that the glycoproteins are the receptors for the lectins [157].

As is true for other, non-oncogenic systems, infection of hamster BHK fibroblasts with influenza virus or vesicular stomatitis virus dramatically increases agglutination of these cells by Con A. However, Penhoet et al. [157] as well as Reeve et al. [158] show that this enhanced agglutinability does not arise from an increase in the number of Con A binding sites. Also, in both the influenza virus-infection and the vesicular stomatitis virus-infection the primary increase in lectin-mediated agglutinability occurs before significant levels of viral glycoproteins synthesis occur, suggesting involvement of a rather early event of the viral infectious cycle. Accordingly, Reeve and Poste [159] suggest that the effect is due to an increase in mobility of Con A receptors after *infection*, similar to that found after oncogenic transformation.

Penhoet et al. [157] find that, in contrast to the Con A results, the enhanced cell agglutination by castor bean lectin during influenza virus infection, is accompanied by 5- to 7-fold increase in the lectin binding sites, a fourth of which occurs very shortly after infection, when there is no significant synthesis of viral proteins. This early increase in castor bean agglutinability and castor bean lectin-binding is not blocked by inhibitors of protein synthesis, whereas later changes are. The implication is that there is an alteration of pre-existing membrane components during this time and that this modification is similar to the modification of the Con A receptors. It is possible that the neuraminidase activity of the viruses is involved here and it has been shown that treatment of BHK cells with soluble purified neuraminidases causes an increased agglutination by castor bean lectin as well as an increase in the number of castor bean binding sites.

Not unexpectedly, Con A also agglutinates tumorigenic murine RNA tumor viruses [160] and Con A receptors can be detected on the viral envelopes by the peroxidase technique.

It is interesting that Con A treatment of chick embryo fibroblasts prevents assembly or release of fowl plague virus and inhibits fusion of BHK cells by a paramyxovirus [161]. These effects could not be related to any Con A-induced metabolic impairment (although this matter was not rigorously studied) and have been attributed to a 'membrane-rigidifying' action of the lectin.

Second, how general is the phenomenon? Virtually all data on enhanced lectin agglutinability come from studies of cultured fibroblasts; however, even here the situation is complicated. Thus the experiments of Salzberg and Green [162] indicate that the genetic information carried by RNA *sarcoma* virus causes enhanced Con A agglutinability, whereas that in RNA *leukemia* viruses does not; that is, neoplastic transformation does not always produce enhanced Con A agglutinability. The only extensive information on normal, non-cultured cells concerns erythrocytes, adipocytes and lymphocytes. The first are 'end cells' and have no neoplastic equivalent. To our knowledge the lectin reactivity of neoplastic fat cells has not been investigated. As far as lymphocytes are concerned, these cells normally are easily agglutinated by various lectins and exhibit a set of elaborate metabolic reactions in response to mitogenic lectins. Unfortunately, we lack significant information about epithelial cells and the cells of spontaneous tumors. (Kapeller and Doljansky [166] report that cells from rat liver, both normal and regenerating, are agglutinated by Con A but not WGA. However they used only a single, high lectin concentration. Embryonic neural retina cells are readily agglutinated by 10 μg/ml Con A [164].) All in all, without further information, one cannot assume that all (or even most) tumor cells exhibit lectin reactivities different from those of their precursor cells.

Third, can one identify unique or unusual metabolic consequences of lectin binding to transformed cells (e.g., those recognized in lymphoid or fat cells)? Here again we lack much needed information.

As treated in detail in Chapter 7, Inbar et al. [165] argue that Con A exerts a different steric interference on sugar (or amino acid) transport in normal and neoplastic cells. They reason that the Con A receptors lie *closer* to the transport sites for D-glucose and D-galactose in tumor cells than in normal cells. As pointed in Chapter 7, it is also conceivable that the lectin inhibits the shuttling *rate* of the carrier, that is, its rate of oscillation between the two membrane surfaces.

Recent experimentation suggests involvement of a different metabolic mechanism, that is, the activity of a cell surface galactosyltransferase ([1-*O*-α-D-galactosyl-myso-inositol: raffinose galactosyltransferase [166]). Podolsky et al. [64] show a positive correlation between the Con A agglutinability of various cells and their ability to transfer galactose from UDP-galactose to fetuin (Table 10.6). However, although normal hamster BHK fibroblasts, as well as their polyoma transformants show appreciable galactosyl transferase activity, Con A agglutinates only the transformants.

It is noteworthy that Con A treatment significantly impaired galactosyl-

TABLE 10.6

Surface galactosyl transferring ability and Con A agglutinability of various cell types[1]

Cell Type	Galactosyl transferase activity[2]	Con A agglutinability[3]
Rat erythrocytes		
Strain 1	0	±
Strain 2	9,820	4+
Human erythrocyte ghosts	0	±
Rabbit erythrocyte ghosts	6,120	4+
Ascites tumor cells		
TA_3-H_a	2,410	+
TA_3-S_t	17,210	3+

1 From Podolsky et al. [164].
2 CPM UDP-[^{3}H]galactose transferred to fetuin per hr (37 °C).
3 500 μg/ml Con A.

transferase but did not affect other glycosyltransferases. Moreover, the galactosyltransferase, activity of galactosyltransferase-positive rat erythrocyte ghosts could be selectively purified by adsorption on to immobilized Con A and elution with α-methyl-D-mannopyranoside. Also, the purified enzyme could be precipitated by Con A, using the Ouchterlony technique. Finally, Podolsky et al. [64] show that human erythrocytes, which do not normally agglutinate in the presence of Con A, adsorb the enzyme purified from galactosyltransferase-positive rat erythrocytes and then become Con A agglutinable. Significantly, other glycoproteins extracted from rat erythrocytes lacked this capability.

The fact that BHK cells do not agglutinate in the presence of Con A, whereas polyoma transformed BHK cells do, even though both cells bear galactosyltransferase, shows that the presence of this surface enzyme is not in itself sufficient to confer Con A agglutinability. However, the data suggest that agglutinability occurs preferentially in cells bearing transferase(s). Two possibilities must be considered: (a) the presence of the enzyme is essential for Con A agglutination, or (b) agglutinable and non-agglutinable cells bear qualitatively different enzymes. In any event, the ability of purified galactosyltransferase to confer Con A agglutinability on otherwise nonagglutinable cells cannot be easily reconciled with those topographic hypotheses of lectin agglutinability which suggest that agglutinable and nonagglutinable cells bear identical receptors in clustered and unclustered sites respectively. On the other hand, the data are not inconsistent with the argument detailed elsewhere that Con A can induce the redistribution of lectin receptors in agglutinable cells and that agglutination depends upon this redistribution.

The precise role of the galactosyltransferase appears obscure. Is it merely a protein with Con A receptors that adsorbs to the membranes through unrelated processes? Or is its enzymatic function essential in agglutination, possibly as suggested in Roseman's enzymatic hypothesis of cell recognition (Chapter 8; [166])?

Vlodavsky et al. [16] have attempted to correlate the agglutinability or normal hamster fibroblasts and their SV40 and polyoma transformants with cellular ATP content. They find that transformed cells contain $\sim 1.3 \times 10^{-8}$ moles ATP/mg cell protein at low density and $\sim 1 \times 10^{-9}$ moles at high density in the absence of glucose. Con A agglutinability varies inversely, that is, it increases with diminishing ATP content. ATP depletion by diverse metabolic inhibitors also increases agglutinability. The enhanced agglutinability is not attributable to binding differences: SV40-transformed cells bind equivalent amounts of Con A whether ATP depleted or not. However, the agglutinability enhancement can be blocked by glutaraldehyde fixation. In contrast to transformed cells, ATP depletion of normal cells by metabolic inhibitors (both respiratory and glycolytic) enhances agglutination only in the case of cells near confluence.

The authors argue that agglutination requires clustering of surface receptors and that this can be prevented by an ATP-requiring process which is abnormally regulated in neoplastic cells. This conclusion appears consistent with Loor's view [46] that lymphocyte agglutination by lectins cannot be considered a passive process. However, neither view fits the concept of a 'fluid' membrane.

Fourth, can one establish a meaningful physical correlate for the lectin reactivity of neoplastic cells? This question has been posed by Inbar et al. [168] and Schinitzky et al. [169] who have attempted to employ fluorescein-labelled Con A (F–Con A) to probe the mobility of Con A receptors of normal and neoplastic cells (Chapter 1).

The authors claim that their data reflect differences in receptor mobility and that the difference is greater in neoplastic fibroblasts than normal cells and lower in neoplastic lymphocytes than normal lymphoid cells. They interpret the data to indicate major changes of membrane 'fluidity' with neoplastic conversion. As shown in Chapter 1, these conclusions appear premature.

Fifth, what might be the biological signifiance of the lectin reactivity of tumor cells? Is it related to the apparent 'unmasking' of certain glycolipids in tumor cells (Chapter 4). How does it relate to the altered agglutinability by basic copolymers of certain viral cells that have been virally converted? (A 1 : 1 ornithine: valine and 1 : 1 arginine: leucine copolymers induce specific aggregation of SV40-transformants [170].)

How does lectin-induced agglutinability relate to 'contact inhibition'? It has been suggested that the lectin reactivity of various neoplastic cells correlates with the escape from density-dependent growth control *in vitro*, that is, loss of 'contact inhibition' (cf. [171]). Moreover, this *in vitro* phenomenon is frequently associated with the malignant process *in vivo* despite the dubious relationship between 'contact inhibition' and malignancy (Chapter 8) and the fact that enhanced lectin agglutinability is not necessarily related even to *in vivo* oncogenicity. For example, no clear relationship has been established between enhanced Con A agglutinability *in vitro* and *in vivo* malignancy as measured by continued tumor growth and host death [172]. This is shown in Table 10.7, which reveals an only occasional correlation between enhanced agglutinability and tumorigenicity in several mouse systems.

Many highly agglutinable cells produce only transient neoplasms: that is, the tumors regress within a few days. Also, one highly malignant tumor, 2°KMSVT, agglutinates poorly and another non-malignant tumor does not agglutinate readily. Sakiyama and Robbins [173] also observe no correlation between the lectin-induced agglutinability of various transformed NIL hamster fibroblasts, and *in vivo* tumorigenicity.

Finally, Glynn et al. [174] show that pronase treatment of density inhibited 3T3 mouse fibroblasts markedly increases these cells' agglutinability by Con A, without inducing cell division, even though the treated cells remained responsive to stimulation by fresh serum and cortisol. Thus, surface changes producing

TABLE 10.7

Con A agglutinability, tumorigenicity and malignancy of several BALB/c mouse cell lines[1,2]

Cell line[3]	Agglutinability[4]	Tumorigenicity[5]	Malignancy[2]
A31	>1,000	−	
PBC	600	−	
SVT2	50	+	Regression
KMSV	100	+	Death
2°KMSVT	600	+	Death
3T12	50	+	Regression
3T12T	50	+	Death
STA31	100	+	Regression
STA31 clone 8	>1,000	+	Regression

1 From [173].
2 Malignancy defined as producing host death.
3 A31 is a clone of BALB/c 3T3 cells. PBC are primary BALB/c fibroblasts. SVT2 is a SV40 transformant of BALB/c 3T3 and KMSV is a murine sarcoma virus transformant. 2°KMSVT are from metastatic KMSV tumors. 3T12 is a spontaneous BALB/c 3T3 transformant. 3T12T is taken from tumors induced by injecting 3T12 into mice. STA31 clones 5 and 8 derive from a spontaneous transformant of A31.
4 Con A agglutination is in μg/ml at 1/2 maximal agglutination.
5 Response to s.c. infection of 2×10^6 cells.

increased susceptibility to Con A stimulation are not sufficient to bring about cell division and loss of density-dependent growth control. These data contradict hypotheses that relate the malignant process to the action of cathepsins on cell surfaces.

In searching for mechanisms explaining the enhanced agglutinability of transformed cells, Ben-Bassat et al. [175] have explored the possible basis for the enhanced agglutinability of transformed cells by Con A. Studying a large series of transformed cell lines (by viruses or chemicals) and their parental lines, they conclude that transformation causes (a) exposure of cryptic sites; (b) concentration of exposed sites by *decreasing cell size*; and (c) *rearrangement of sites* (clustering, without changes of cell size). Although the data are consistent with major membrane rearrangements, details remain obscure.

Noonan and Burger [112] propose a different *comprehensive concept* that attempts to explain the enhanced lectin agglutinability of transformed cells. They make two basic assumptions: (a) plasma membranes comprise a two-dimensional, fluid lipid continuum, within which lectin receptors normally exhibit free lateral mobility; (b) agglutination requires a localized clustering of lectin receptors. They do not consider the possibility that agglutination may be an active process requiring metabolic energy [46,167]. Noonan and Burger [112] argue as follows:

First, one binding site of a multivalent Con A molecule associates with one of the uniformly distributed Con A receptors on the surface of a transformed cell.

Second, the same Con A molecule, through another valence, binds to a second receptor when this randomly collides with the lectin, thereby initiating clustering of receptor sites.

Since Con A receptors can normally undergo free lateral translation within the fluid membrane, any process which would decrease 'fluidity' would hinder the mobility of the Con A receptors within the membrane plane and thereby inhibit agglutination. This sensitivity to 'fluidity' could explain the temperature sensitivity of the agglutination reaction, the 15 °C 'transition temperature' for agglutination, and the effect of glutaraldehyde on agglutination.

According to Noonan and Burger [112] once receptor clustering has begun, it continues because of the interaction of the multivalent lectin molecule with multivalent receptors until all available sites are stably crosslinked by agglutinin molecules. According to the model, trypsinization of normal cells might release the Con A receptor sites and permit them to move in the plane of the membrane. Noonan and Burger [112] suggest that the Con A binds to the uniformly distributed, mobile receptor sites on the surfaces of transformed and trypsin-treated normal cells and, because of its multivalency, aggregates the cells. Agglutination then proceeds by cross linking between *unoccupied* receptors on adjacent cells. How this might come about remains obscure.

Inbar et al. [168] argue as do Noonan and Burger [112] that Con A receptors move randomly in a 'fluid membrane'. Interaction with Con A alters receptor distribution. Depending on the membrane 'fluidity', the receptor–lectin complexes may form clusters, as in lymphoma cells, or may concentrate at one pole of the cell, as in normal lymphocytes. The altered distribution of Con A receptors after trypsinization treatment is also attributed to an increased mobility of receptors of Con A.

The authors suggest that the 'mobility' of bound Con A might be used as a probe for the mobility of the Con A receptor. They argue that in the case of cells that are in suspension *in vivo*, for example, normal lymphocytes and lymphoma cells, malignant transformation produces decreased mobility of Con A receptors. In the case of cells such as fibroblasts, which form a solid tissue, the transformation of normal into malignant cells produces increased mobility of receptors in the cell membrane.

This hypothesis, although less explicit than the proposal of Noonan and Burger [112], differs little in essence. Unfortunately, a careful consideration of the notion of 'membrane fluidity' shows this to be a rather imprecise concept, unsubstantiated by discriminating experiments (Chapter 1). Furthermore, in proposing that the membrane of transformed cells differ in 'fluidity', neither Inbar et al. [168] nor Noonan and Burger [112] propose any mechanism that would explain such differences. Moreover, data from numerous experiments are incompatible with the model and recent experiments, provide specific evidence for the complexity of the situation. Thus, mouse L-cells show two critical temperatures for Con A binding and lectin-mediated hemadsorption. Below the higher critical temperature (14–18 °C) binding and hemadsorption decrease 2-fold and 10-fold respectively. This inhibition rapidly reverses upon reheating to > 18 °C. These effects might be explained by changes in lipid state, but there is a second critical temperature (5–7 °C); chilling below this produces a further decrease in lectin binding and hemadsorption. However, reheating to > 20 °C after chilling below this second critical temperature does not produce rapid recovery of hemadsorption, although lectin binding is adequate. It is suggested that disruption of cytoplasmic microfilaments occurs at low temperature [176].

The possible role of lectin receptors in tissue organization is addressed very directly by Steinberg and Gepner [164]. These investigators prepared a 'univalent' Con A by chymotrypsin digestion of the native lectin. This Con A derivative lacks agglutinating capacity, but retains native binding specificity. The authors then measured whether high concentrations (20 μg/ml) of the 'univalent' lectin could affect various adhesion dependent processes of embryonic chick tissues (i.e. cell aggregation, cell sorting, tissue spreading). They found that the only detectable effect of exposure to univalent lectin was retardation of cell rearrangements. There was no detectable influence on specific sorting and reordering of dispersed mixtures of embryonic cells. The authors concluded that the Con A receptors do not participate importantly in specific cellular interactions in the system employed and question the argument of Burger and Noonan [105] concerning the role of lectin receptors in specific growth control.

In conclusion, while the modes of lectin reactivity with both normal and neoplastic cells are fascinating, they are also extremely complex and, for this reason the field is replete with controversy and contradiction. Although the extensive experimentation in the area shows that tumor cells often deviate in multiple ways from normal in terms of lectin reactivity, this research has not yet advanced our basic understanding of tumor cell membrane biology. Most troublesome probably are the complex binding characteristics of lectins and real progress in this area will require clear discrimination between binding to specific membrane molecules and adsorption processes that can be reversed by altering lectin conformation with 'hapten' sugars.

References

[1] Sharon, N.S. and Lis, H.L., *Science*, *177*, 949 (1972).
[2] Lis, H.L. and Sharon, N.S., *Annu. Rev. Biochem.*, *42*, 541 (1973).
[3] Aub, J.C., Tieslau, C. and Lankester, A., *Proc. Natl. Acad. Sci. U.S.*, *50*, 613 (1963).
[4] Weber, T.H., Aro, H. and Nordman, C.T., *Biochim. Biophys. Acta*, *263*, 94 (1972).

[5] Rapin, A. and Burger, M.M., *Advan. Cancer Res.*, *20*, 1 (1974).
[6] Nicolson, G.L., Blaustein, J. and Etzler, M.E., *Biochemistry*, *13*, 196 (1974).
[7] Podder, S.K., Surolia, A. and Bachhawat, B.K., *Eur. J. Biochem.*, *44*, 151 (1974).
[8] McKenzie, G.H., Sawyer, W.H. and Nichol, L.W., *Biochim. Biophys. Acta*, *263*, 283 (1972).
[9] McKenzie, G.H. and Sawyer, W.H., *J. Biol. Chem.*, *248*, 549 (1973).
[10] Yariv, J., Kalb, A.J. and Levitzki, A., *Biochim. Biophys. Acta*, *165*, 303 (1968).
[11] Edelman, G.M., Cunningham, B.A., Reeke, G.N., Jr., Becker, J.W., Wadal, M.J. and Wang, J.L., *Proc. Natl. Acad. Sci. U.S.*, *69*, 2580 (1972).
[12] Pflumm, M.N., Wang, J.L. and Edelman, G.M., *J. Biol. Chem.*, *246*, 4369 (1971).
[13] Hassing, G.S. and Goldstein, I.J., *Eur. J. Biochem.*, *16*, 549 (1970).
[14] Gunther, G.R., Wang, J.L., Yahara, I., Cunningham, B.A. and Edelman, G.M., *Proc. Natl. Acad. Sci U.S.*, *70*, 1012 (1973).
[15] Poretz, R.D. and Goldstein, I.J., *Biochemistry*, *9*, 2890 (1970).
[16] Bessler, W., Shafer, J.A. and Goldstein, I.J., *J. Biol. Chem.*, *249*, 2819 (1974).
[17] Nagata, Y. and Burger, M.M., *J. Biol. Chem.*, *247*, 2248 (1972).
[18] Nagata, Y. and Burger, M.M., *J. Biol. Chem.*, *249*, 3116 (1974).
[19] Rice, R.H. and Etzler, M.E., *Biochem. Biophys. Res. Commun.*, *59*, 414 (1974).
[20] Allen, A.K., Neuberger, A. and Sharon, N., *Biochem. J.*, *131*, 155 (1973).
[21] LeVine, D., Kaplan, M.J. and Greenway, P.J., *Biochem. J.*, *129*, 847 (1972).
[22] Cuatrecasas, P.C., *Biochemistry*, *12*, 1312 (1973).
[23] Gray, R.D. and Glew, R.H., *J. Biol. Chem.*, *248*, 7547 (1973).
[24] Krug, U., Hollenberg, M.D. and Cuatrecasas, P., *Biochem. Biophys. Res. Commun.*, *52*, 305 (1973).
[25] Betel, I. and van den Berg, K.J., *Eur. J. Biochem.*, *30*, 571 (1972).
[26] Edelman, G.M. and Millette, C.F., *Proc. Natl. Acad. Sci. U.S.*, *68*, 2436 (1971).
[27] Brown, J.C., *Biochem. Biophys. Res. Commun.*, *51*, 686 (1973).
[28] Hunt, R.C., Bullis, C.M. and Brown, J.C., *Biochemistry*, *14*, 109 (1975).
[29] Ozanne, B. and Sambrook, J., *Nature, New Biol.*, *232*, 156 (1971).
[30] Hunt, R.C. and Brown, J.C., *Biochemistry*, *13*, 22 (1974).
[31] Noonan, K.D. and Burger, M.M., *J. Biol. Chem.*, *248*, 4286 (1973).
[32] Gottlieb, C., Skinner, A.M. and Kornfeld, S., *Proc.. Natl. Acad. Sci. U.S.*, *71*, 1078 (1974).
[33] Guerin, C., Zachowski, A., Prigent, B., Paraf, A., Dunia, I., Diawara, M.-A. and Bennedetti, E.L., *Proc. Natl. Acad. Sci. U.S.*, *71*, 114 (1974).
[34] Kaneko, I., Satoh, H. and Ukita, T., *Biochem. Biophys. Res. Commun.*, *48*, 1504 (1972).
[35] DeSalle, L., Munakata, N., Pauli, R.M. and Strauss, B.S., *Cancer Res.*, *32*, 2463 (1972).
[36] Zachåwski, A., Migliore-Samour, D., Paraf, A. and Jollés, P., *FEBS Letters*, *52*, 57 (1975).
[37] Steinemann, A. and Stryer, L., *Biochemistry*, *12*, 1499 (1973).
[38] Changeux, J.P., Thiery, J., Tung, Y. and Kittel, C., *Proc. Natl. Acad. Sci. U.S.*, *57*, 335 (1967).
[39] Wyman, J. in *Summetry and Function of Biological Systems at the Macromolecular level*, A. Eugström and B. Straudberg, eds., John Wiley, New York, 1969, p. 267.
[40] Burger, M.M., *Fed. Proc.*, *32*, 91 (1973).
[41] Kornfeld, R., Keller, J., Baenziger, J. and Kornfeld, S., *J. Biol. Chem.*, *246*, 3259 (1971).
[42] Levitzki, A., *J. Theor. Biol.*, *44*, 367 (1974).
[43] Schmidt-Ullrich, R., Wallach, D.F.H. and Ferber, E., *Biochim. Biophys. Acta*, *356*, 288 (1974).
[44] Schmidt-Ullrich, R., Wallach, D.F.H. and Hendricks, J., *Biochim. Biophys. Acta*, *382*, 295 (1975).
[45] Steck, T.L. and Wallach, D.F.H., *Biochim. Biophys. Acta*, *97*, 510 (1965).
[46] Loor, F.L., *Exptl. Cell Res.*, *82*, 415 (1973).
[47] Nicolson, G.L., Lacorbiere, M. and Delmonte, P., *Exptl. Cell Res.*, *71*, 468 (1972).
[48] Hirano, H.H., Parkhouse, B., Nicolson, G., Lennox, E.S. and Singer, S.J., *Proc. Natl. Acad. Sci. U.S.*, *69*, 2945 (1972).
[49] Nicolson, G.L. and Singer, S.J., *Proc. Natl. Acad. Sci. U.S.*, *68*, 942 (1971).
[50] Nicolson, G.L. and Singer, S.J., *J. Cell Biol.*, *60*, 236 (1974).
[51] Wallach, D.F.H., Krantz, B., Ferber, E. and Fischer, H., *FEBS Letters*, *21*, 29 (1972).
[52] Zachowski, A. and Paraf, A., *Biochem. Biophys. Res. Commun.*, *45*, 787 (1974).
[53] Winqvist, L., Eriksson, L.C. and Dallner, G., *FEBS Letters*, *42*, 27 (1974).

[54] Barat, N. and Avrameas, S., *Exptl. Cell Res.*, *76*, 451 (1973).
[55] Huet, C.H. and Bernhard, W., *Int. J. Cancer*, *13*, 227 (1974).
[56] Allan, D. Auger, J. and Crumpton, M.J., *Nature, New Biol.*, *236*, 23 (1972).
[57] Hayman, M.J., Crumpton, M.J., *Biochem. Biophys. Res. Commun.*, *47*, 923 (1972).
[58] Wray, V.P. and Walborg, E.F., Jr., *Cancer Res.*, *31*, 2072 (1971).
[59] Hourani, B.T., Chace, N.M. and Pincus, J.H., *Biochim. Biophys. Acta*, *328*, 520 (1973).
[60] Jansons, V.K. and Burger, M.M., *Biochim. Biophys. Acta*, *291*, 127 (1973).
[61] Jansons, V.K., Sakamoto, C.K. and Burger, M.M., *Biochim. Biophys. Acta*, *291*, 136 (1973).
[62] Akedo, H., Mori, Y., Tanigaki, Y., Shinkai, K. and Morita, K., *Biochim. Biophys. Acta*, *271*, 378 (1972).
[63] Smith, D.F., Neri, G. and Walborg, E.F., Jr., *Biochemistry*, *12*, 2111 (1973).
[64] Podolsky, D.K., Weiser, M.M., LaMont, J.T. and Isselbacher, K.J., *Proc. Natl. Acad. Sci. U.S.*, *71*, 904 (1974).
[65] Klenk, H.-D., Rott, R. and Becht, H., *Virology*, *47*, 579 (1972).
[66] Greaves, M.F. and Bauminger, S., *Nature, New Biol.*, *235*, 67 (1972).
[67] Andersson, J., Edelman, G.M., Möller, G. and Sjöberg, O., *Eur. J. Immunol.*, *2*, 233 (1972).
[68] Hulser, D.F. and Peters, J.H., *Eur. J. Immunol.*, *1*, 494 (1971).
[69] Quastel, M.R. and Kaplan, J.G., *Exp. Cell Res.*, *63*, 230 (1970).
[70] Kay, J.E., *Exp. Cell Res.*, *71*, 245 (1972).
[71] Asherson, G.L., Davey, M.J. and Govelford, P.J., *J. Physiol. (London)*, *260*, 32 P (1970).
[72] Allwood, G., Asherson, G.L., Davey, M.J. and Goodford, P.J., *Immunology (London)*, *21*, 509 (1971).
[73] Whitney, R.B. and Sutherland, R.M., *Cell. Immunol.*, *5*, 137 (1972).
[74] Peters, J.H. and Hausen, P., *Eur. J. Biochem.*, *19*, 509 (1971).
[75] Peters, J.H. and Hausen, P., *Eur. J. Biochem.*, *19*, 502 (1971).
[76] Cross, M.E. and Ord, M.G., *Biochem. J.*, *124*, 241 (1971).
[77] van den Berg, K.J. and Betel, I., *Exptl. Cell Res.*, *76*, 63 (1973).
[78] Mendelsohn, J., Skinner, A. and Kornfeld, S., *J. Clin. Invest.*, *50*, 818 (1971).
[79] Quastel, M.R. and Kaplan, J.G., *Exp. Cell Res.*, *62*, 407 (1970).
[80] Alford, R.H., *J. Immunol.*, *104*, 698 (1970).
[81] Lucas, D.O., Shohet, S.B. and Merler, E., *J. Immunol.*, *106*, 768 (1971).
[82] Fisher, D.B. and Mueller, G.C., *Biochim. Biophys. Acta*, *176*, 316 (1969).
[83] Resch, K., Ferber, E., Odenthal, J. and Fischer, H., *Eur. J. Immunol.*, *1*, 162 (1971).
[84] Resch, K., Gelfand, E.W., Hansen, K. and Ferber, E., *Eur. J. Immunol.*, *2*, 598 (1972).
[85] Hokin, L.E., *Ann. N.Y. Acad. Sci. U.S.*, *165*, 695 (1969).
[86] Cuatrecasas, P. and Tell, G.P.E., *Proc. Natl. Acad. Sci. U.S.*, *70*, 485 (1973).
[87] Aoki, T., Hammerling, V., DeHarven, E., Boyse, E. and Old, L.J., *J. Exp. Med.*, *130*, 979 (1969).
[88] Karnovsky, M.J., Unanue, E.R. and Leventhal, M., *J. Exp. Med.*, *136*, 907 (1972).
[89] Karnovsky, M.J., Unanue, E.R., *Fed. Proc.*, *32*, 55 (1973).
[90] Loor, F.L., *Eur. J. Immunol.*, *3*, 112 (1973).
[91] Guerin, C., Prigent, B., Moyne, M.A. and Paraf, A., *Bull. Cancer*, *59*, 367 (1972).
[92] Wright, J.A., *J. Cell Biol.*, *56*, 666 (1973).
[93] Nicolson, G.L. and Yanagimachi, R.Y., *Science*, *184*, 1294 (1974).
[94] Yahara, I. and Edelman, G.M., *Proc. Natl. Acad. Sci. U.S.*, *69*, 608 (1972).
[95] Edidin, M. in *Transplantation Antigens*, Kahan and Reisfild, eds., 1972, pp. 125–140.
[96] Singer, S.J. and Nicolson, G., *Science*, *175*, 720 (1972).
[97] Unanue, E.R., Ault, K.A. and Karnovsky, M.J., *J. Exp. Med.*, *139*, 295 (1974).
[98] Onozaki, K., Tomita, M. Sakurai, I. and Ukita, T., *Biochem. Biophys. Res. Commun.*, *48*, 783 (1972).
[99] Kornfeld, S., Eider, W. and Gregory, W. in *Control and Proliferation in Animal Cells*, B. Clarkson and R. Baserga, eds., Cold Spring Harbor Laboratory, 1974, p. 435.
[100] Olsnes, S. and Pihl, A., *FEBS Letters*, *20*, 327 (1972).
[101] Tomita, M., Kurokawa, T. and Osawa, T., *Gann.*, *63*, 269 (1972).
[102] Cline, M.J. and Livingston, D.C., *Nature, New Biol.*, *232*, 155 (1971).
[103] Zarling, J.M. and Tevethia, S.S., *Virology*, *45*, 313 (1971).

[104] Shoham, J., Inbar, M. and Sachs, L., *Nature*, *225*, 1244 (1970).
[105] Burger, M.M. and Noonan, K.D., *Nature*, *228*, 512 (1970).
[106] Burger, M.M., *Proc. Natl. Acad. Sci. U.S.*, *62*, 994 (1969).
[107] Burger, M.M., *Nature*, *227*, 170 (1970).
[108] Burger, M.M. in *Membrane Research*, C. Fred Fox, ed., Academic Press, New York, 1972, pp. 231-237.
[109] Sefton, B.N. and Rubin, H., *Nature*, *227*, 843 (1970).
[110] Arndt-Jovin, D.J. and Berg., P., *J. Virol.*, *8*, 716 (1971).
[111] Inbar, M., Ben-Bassat, H. and Sachs, L., *Proc. Natl. Acad. Sci. U.S.*, *68*, 2748 (1971).
[112] Noonan, K.D. and Burger, M.M., *J. Cell Biol.*, *59*, 134 (1973).
[113] Benjamin, T.L. and Burger, M.M., *Proc. Natl. Acad. Sci. U.S.*, *67*, 929 (1970).
[114] Nicolson, G.L. and Blaustein, J., *Biochim. Biophys. Acta*, *266*, 543 (1972).
[115] Sela, B.A., Lis, H., Sharon, N. and Sachs, L., *J. Membrane Biol.*, *3*, 267 (1970).
[116] Kapeller, M. and Doljanski, F., *Nature, New Biol.*, *235*, 184 (1972).
[117] Moore, E.G. and Temin, H.M., *Nature*, *231*, 117 (1971).
[118] Burger, M.M. and Martin, G.S., *Nature, New Biol.*, *237*, 9 (1972).
[119] Hsie, A.W., Jones, C., Puck, T.T., *Proc. Natl. Acad. Sci. U.S.*, *68*, 1648 (1971).
[120] Johnson, G.S., Friedman, R.M., Pastan, I., *Proc. Natl. Acad. Sci. U.S.*, *68*, 425 (1971).
[121] Sheppard, J.R., *Proc. Natl. Acad. Sci. U.S.*, *68*, 1316 (1971).
[122] Goggins, J.F., Johnson, G.S. and Pastan, I., *J. Biol. Chem.*, *247*, 5759 (1972).
[123] Johnson, G.S. and Pastan, I., *Nature, New Biol.*, *237*, 267 (1972).
[124] Johnson, G.S. and Pastan, I., *Nature, New Biol.*, *236*, 247 (1972).
[125] Kurth, R. and Bauer, H., *Nature, New Biol.*, *243*, 243 (1973).
[126] Roberts, R.M., Walker, A. and Cetorelli, J.J., *Nature, New Biol.*, *244*, 86 (1973).
[127] Burger, M.M., Bombik, B.M., Breckenridge, B.M. and Sheppard, J.R., *Nature, New Biol.*, *239*, 161 (1972).
[128] Inbar, M., Ben-Bassat, H. and Sachs, L., *Nature, New Biol.*, *236*, 3 (1972).
[129] Sela, B.A., Lis, H., Sharon, N. and Sachs, L., *Biochim. Biophys. Acta*, *249*, 564 (1971).
[130] Shoham, J. and Sachs, L., *Exptl. Cell Res.*, *85*, 8 (1974).
[131] Nicolson, G.L. and Lacorbiere, M., *Proc. Natl. Acad. Sci. U.S.*, *70*, 1672 (1973).
[132] Shoham, J. and Sachs, L., *Proc. Natl. Acad. Sci. U.S.*, *69*, 2479 (1972).
[133] Inbar, M., Ben-Bassat, H. and Sachs, L., *Int. J. Cancer*, *12*, 93 (1972).
[134] Bernhard, W. and Arameas, S., *Exptl. Cell Res.*, *64*, 232 (1971).
[135] Smith, S.B., Revel, J.P., *Develop. Biol.*, *27*, 434 (1972).
[136] Rosenblith, J.Z., Ukena, T.E., Yin, H.H., Berlin, R.D. and Karnovsky, M.J., *Proc. Natl. Acad. Sci. U.S.*, *70*, 1625 (1973).
[137] dePetris, S., Raff, M.C. and Mallucci, L., *Nature, New Biol.*, *244*, 275 (1973).
[138] Collard, J.G. and Temmink, J.H.M., *Exptl. Cell Res.*, *86*, 81 (1974).
[139] Inbar, M., Ben-Bassat, H. and Sachs, L., *Int. J. Cancer*, *12*, 93 (1973).
[140] Nicolson, G.L., *Nature, New Biol.*, *233*, 244 (1971).
[141] Nicolson, G.L., *Nature, New Biol.*, *239*, 193 (1972).
[142] Martinez-Palomo, A., Wicker, R. and Bernhard, W., *Int. J. Cancer*, *9*, 676 (1972).
[143] Rowlatt, C.H., Wicker, R. and Bernhard, W., *Int. J. Cancer*, *11*, 314 (1973).
[144] Garrido, J., Burglen, M.-J., Samlyk, D., Wicker, R. and Bernhard, W., *Cancer Res.*, *34*, 230 (1974).
[145] Bretton, R., Wicker, R. and Bernhard, W., *Int. J. Cancer*, *10*, 397 (1972).
[146] Nicolson, G.L., *Nature, New Biol.*, *243*, 218 (1973).
[147] Torpier, G. and Montagnier, L., *Int. J. Cancer*, *11*, 604 (1973).
[148] Walig, C., Walboomers, J.M.M. and van der Noordaa, J., *J. Cell Biol.*, *61*, 553 (1974).
[149] Ukena, T.E., Borysenko, J.Z., Karnovsky, M.J. and Berlin, R.D., *J. Cell Biol.*, *61*, 70 (1974).
[150] Wunderlich, F., Müller, R. and Speth, V., *Science*, *182*, 1136 (1973).
[151] Yin, H.H., Ukena, T.E. and Berlin, R.D., *Science*, *178*, 867 (1972).
[152] Ukena, T.E., Borysenko, J.Z., Black, P.R., Karnovsky, M.J. and Berlin, R.D., *New England J. Med.*, *292*, 515 (1975).
[153] Tevethia, S.S., Lowry, S., Rawls, W.E., Melnick, J.L. and MacMillan, Y., *J. Gen. Virol.*, *15*, 93 (1972).

[154] Becht, H., Rott, R. and Klenk, H.D., *J. Gen. Virol.*, *14*, 1 (1972).
[155] Poste, G., Reeve, P., Alexander, D.J. and Terry, G., *J. Gen. Virol.*, *17*, 81 (1972).
[156] Poste, G., Reeve, P., *Nature, New Biol.*, *237*, 113 (1972).
[157] Penhoet, E., Olson, C., Carlson, S., Lacorbiere, M. and Nicolson, G.L., *Biochemistry*, *13*, 3561 (1974).
[158] Reeve, P., Poste, G. and Alexander, D. in *Negative Strand Viruses*, B.W.J. Mahy and R. Barry, eds., Academic Press, New York, in press.
[159] Poste, G. and Reeve, P., *Nature*, *247*, 469 (1974).
[160] Calafat, J. and Hageman, P.C., *J. Genl. Virol.*, *14*, 103 (1972).
[161] Rott, R., Becht, H., Klenck, H.-D. and Scholtissek, C., *Z. Naturforsch*, *27*, 227 (1972).
[162] Salzberg, S. and Green, M., *Nature, New Biol.*, *240*, 116 (1972).
[163] Kapeller, M. and Doljanski, F., *Nature, New Biol.*, *235*, 184 (1972).
[164] Steinberg, M.S. and Gepner, I.A., *Nature, New Biol.*, *241*, 249 (1973).
[165] Inbar, M., Ben-Bassat, H. and Sachs, L., *J. Memb. Biol.*, *6*, 195 (1971).
[166] Roseman, S., *Chem. Phys. Lipids*, *5*, 270 (1970).
[167] Vlodavsky, I., Inbar, M. and Sachs, L., *Proc. Natl. Acad. Sci. U.S.*, *70*, 1780 (1973).
[168] Inbar, M., Shinitzky, M. and Sachs, L., *J. Mol. Biol.*, *81*, 245 (1973).
[169] Shinitzky, M., Inbar, M. and Sachs, L., *FEBS Letters*, *34*, 247 (1973).
[170] Duskin, D., Katachalski, E. and Sachs, L., *Proc. Natl. Acad. Sci. U.S.*, *67*, 185 (1970).
[171] Burger, M.M., *Fed. Proc.*, *32*, 91 (1973).
[172] van Nest, G.A. and Grimes, W.J., *Cancer Res.*, *34*, 1408 (1974).
[173] Sakiyama, H. and Robbins, P.W., *Fed. Proc.*, *32*, 86 (1973).
[174] Glynn, R.D., Trash, C.R. and Cunningham, D.D., *Proc. Natl. Acad. Sci. U.S.*, *70*, 2676 (1973).
[175] Ben-Bassat, H., Inbar, M. and Sachs, L., *J. Memb. Biol.*, *6*, 183 (1971).
[176] Rittenhouse, H.G. and Fox, C.F., *Biochem. Biophys. Res. Commun.*, *57*, 323 (1974).

11 Epilogue

Unifying principles and general prospects

11.1. Introduction

A most frustrating feature of neoplastic deviation is its *polymorphism*. This appears in anomalies that clearly relate to membrane functions and in defects that have no obvious connection to membranes. Several major hypotheses – ranging in emphasis from the genetic through the molecular-biological, biochemical, biophysical to the physiological – have been offered to account for tumor cell polymorphism and related functional anomalies. We have treated these concepts in earlier sections, but will reintroduce them here, before presenting a different point of view.

The *protovirus* hypothesis of Temin [1] includes a gene-diversification – gene-amplification mechanism, which is not supplied by other genetic concepts such as the 'oncogene hypothesis' [2].

The *molecular correlation concept* of Weber (Chapter 6; [3]) states that the polymorphic properties of tumor cells derive from an ordered expression of genes coding for unregulated isozymes, creating an unresponsive cellular system with a competitive advantage over responsive systems. This concept is a coherent summation of a pattern common to many tumors, but it provides no mechanisms.

The *membron hypothesis* of Pitot et al. (Chapter 6; [4]) focuses on translational defects in tumors and accounts for tumor-cell enzyme polymorphism in terms of defective stabilization, by *endoplasmic reticulum membranes*, of the mRNA templates of diverse enzymes. While increasing evidence points to an important role of membranes in determining messenger lifetime, the membron hypothesis does not deal with possible membrane anomalies that could lead to altered membrane–polysome or membrane–mRNA associations.

The *Warburg hypothesis* (Chapter 6; [5]), in principle, assigns the critical deficiency in neoplasia to the *mitochondrial* membranes. Although never disproven, nor adequately tested by modern approaches, Warburg's hypothesis has been relegated to Limbo for many years. Nevertheless, an abundance of evidence points to extensive polymorphic anomalies in tumor mitochondria, which can, in turn, account for many facets of tumor cell behavior.

Several recent hypotheses attribute tumor polymorphism to deficiencies in cyclic nucleotide metabolism (Chapter 6). Of these proposals, that of Weber [3] is phrased in terms of isozyme shifts, but Tomkins and associates [6] implicate the *plasma membrane* at the receptor and/or cyclase level and Pastan et al. [7] specifically invoke altered membrane adenylate cyclase activity secondary to a modification of critical membrane protein, lipid or carbohydrate components. Deficient control of cyclic nucleotide metabolism can account for much of tumor polymorphism. However, if the critical restriction point is in the plasma membrane, we remain ignorant as to its primary deficiencies in neoplastic cells.

It is logical to focus on the properties common to most neoplasms and a most prominent common feature is the apparently inevitable alteration of cellular membranes. This alteration, however, is again polymorphic. Indeed, whenever tested, one encounters an extraordinary diversity of often seemingly unrelated membrane defects, involving multiple membrane systems (Fig. 11.1). This is true even within a single clone of transformed cells, derived from a single clone of parental cells through the action of a single oncogenic agent. Since these facts

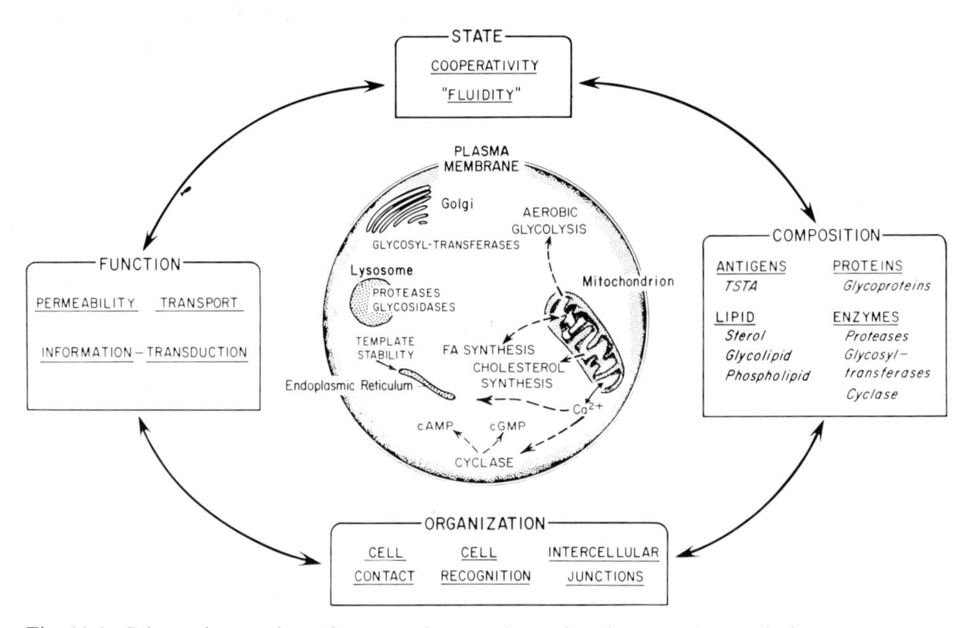

Fig. 11.1. Schematic overview of some major membrane involvements in neoplasia.

cannot be satisfactorily explained by any molecular-biologic concept yet proposed, one may reasonably enquire whether some fundamental properties of membranes might account for the membrane polymorphism (or general polymorphism) of tumors. We have previously speculated on this topic [8–10] proceeding from the fact that biomembranes constitute quasi-solid-state system, and the proposal of Changeux et al. [11] that important functional elements of biomembranes lie in ordered lattice systems whose behavior could be altered in a linked cooperative fashion by one or more of the following mechanisms:

(a) Insertion of a *new subunit* (mutation; viral gene *product*).
(b) Change in the *steady state concentration* (even deletion) of native, membrane ligands (mutation; gene product or viral).
(c) Emergence of an isomeric membrane ligand with abnormal binding affinity (mutation; viral gene product).

We have speculated that oncogenic agents might alter the concerted behavior of biomembranes and might thereby modify a diversity of membrane functions and membrane-related processes, accounting for the membrane- and general-polymorphism seen in neoplastic cells. We have suggested, for example, that the insertion into membranes of an abnormal protein (or lipid) could directly alter membrane antigenicity, modify transport and permeability, change binding of a variety of substances and functionally modify otherwise normal, membrane-associated enzymes. In addition, deviant membrane function might produce defects which show no immediately obvious relationship to membranes.

We will reevaluate this proposal in the light of major recent advances in membrane research, but will first present its theoretical foundations.

11.2. Models of cooperative membrane lattices

11.2.1. General description

Changeux et al. [11–13] have proposed that biomembranes consist of (or contain) 'two dimensional crystalline lattices', made up of ordered collections of repeating, globular lipoprotein subunits. These subunits, or 'protomers' might comprise a 'single polypeptide chain or several different protein subunits associated with lipids in characteristic amounts'. It is further assumed that:

(a) each 'protomer' can reversibly shift between several structural states;
(b) the state of each 'protomer', and its freedom to change state, depends upon association with surrounding protomers; that is, each protomer is subject to 'lattice constraint';
(c) each 'protomer' bears at least one receptor for each ligand capable of associating specifically with the membrane;
(d) the affinity for these ligands changes with the structural state of the protomers.

Changeux et al. [11–13] use the Bragg-Williams (molecular field) approximation [14] to evaluate the simple case of a two-dimensional lattice composed of subunits that can exist in either state S or state R. The expected fraction of subunits in state S is then $\langle s \rangle$ and the fraction in state R is $\langle r \rangle$.

In their original proposals, Changeux et al. [11–13] considered membranes composed of lattices essentially infinite in terms of molecular dimensions, as well as systems where interactions are limited to nearest neighbors. However, it is not essential to invoke long range interactions (i.e. non-nearest neighbor; >100 nm) to explain certain cooperative membrane phenomena. Thus Blumenthal [15] shows that, for membrane transport processes, an amorphous two-dimensional system with 'sparsely distributed oligomeric transport sites' can yield similar cooperative behavior as the 'infinite lattice' model, earlier suggested by Blumenthal et al. [16]. We will therefore, restrict our presentation to the nearest-neighbor case, and will later comment on the possibility of long range interactions.

Employing the notation of Changeux et al. [11], if ε is the energy required to promote one protomer from state S to state R, when all other protomers are in state S, the isomerization constant, l, for the reaction will be

$$l = \exp\,[\varepsilon/kT] \tag{11.1}$$

where k is the Boltzmann constant and T is in degrees Kelvin.

However, in a system where the S → R transition depends on the interaction between *nearest* neighbors, the promotion energy will depend on z, the number of nearest neighbors, η, the mean energy of interaction between nearest neighbors, $[\eta = (\varepsilon_{SS} - 2\varepsilon_{SR} + \varepsilon_{RR})]$, multiplied by $\langle r \rangle$, the expectation fraction of protomers already in the R-state. ε_{SS}, ε_{RR} and ε_{SR} are the interaction energies between S and S, R and R, and S and R protomers.

The isomerization constant, l', for this case is

$$l' = exp\left[\frac{\varepsilon}{kT} - \frac{z}{kT}(\varepsilon_{SS} - 2\varepsilon_{SR} + \varepsilon_{RR})\,\langle r \rangle\right] \tag{11.2}$$

$$l' = \langle s \rangle / \langle r \rangle = l \cdot \Lambda^{\langle r \rangle} \tag{11.3}$$

where Λ, the *cooperativity parameter*, equals

$$\Lambda = \exp\left[-\frac{z \cdot \eta}{kT}\right] \tag{11.4}$$

Accordingly, the cooperativity parameter varies with *temperature*, T, the number of nearest neighbors, z, and the nearest neighbor interaction, η.

The *binding* of a ligand, f, to a single protomer depends on the following equilibrium constants, l', $K_R = [f]\,\langle r_0 \rangle / \langle r_1 \rangle$ and $K_S = [f]\,\langle s_0 \rangle / \langle s_1 \rangle$, where $\langle s_0 \rangle$ and $\langle r_0 \rangle$ correspond to the expectation values of the unliganded protomers,

S_0 and R_0; $\langle s_1 \rangle$ and $\langle r_1 \rangle$ refer to the corresponding fraction of the liganded species, S_1 and R_1 and $c = K_R/K_S$.

Changeux et al. [11] have developed the following equations for the state function $\langle r \rangle$ and the binding function, $\langle y \rangle$

$$\langle r \rangle = \frac{1 + [f]/K_R}{1 + [f]/K_R + l\Lambda^{\langle r \rangle}(1 + c[f]/K_R)} \tag{11.5}$$

$$\langle y \rangle = \frac{[f]/K_R(1 + cl\Lambda^{\langle r \rangle})}{1 + [f]/K_R + l\Lambda^{\langle r \rangle}(1 + c[f]/K_R)} \tag{11.6}$$

These equations have been evaluated numerically for different values of Λ, $[f]$ and K_R [11]. The computations are conveniently presented as plots of $\langle r \rangle$ and $\langle y \rangle$ vs. $[f]/K_R$ and the principal patterns for $\langle r \rangle$ are summarized schematically in Fig. 11.2.

Without subunit interaction $\eta = 0$, $\Lambda = 1$ and the plots of $\langle r \rangle$ vs. $[f]/K_R$ constitute hyperbolas (Fig. 11.2). With increasing, *positively*-cooperative subunit interaction, (where thc transition of one protomer facilitates that of other protomers; $2\,\varepsilon_{SR} > \varepsilon_{SS} + \varepsilon_{RR}$), Λ decreases and the $\langle r \rangle$ curve assumes the S-shape characteristic of cooperative interactions (Fig. 11.2). At a critical value $\Lambda_c = e^{-4} = 0.018$, [i.e. $\cdot (z \cdot \eta)/kT = -4$], independent of l, K_S and K_R], the $\langle r \rangle$ curve become discontinuous (Fig. 11.2); that is, at a critical value of $[f]/K_R$, an abrupt change of state occurs. This constitutes a phase transition.

A similar sequence is found for the $\langle y \rangle$ curves; when $\Lambda = 1$, $\langle y \rangle$ varies hyperbolically with $[f]/K_R$. At Λ_c, the curve shows an abrupt jump in $\langle y \rangle$ at a critical value of $[f]/K_R$ and at intermediate values of Λ the curve is S-shaped (Fig. 11.2).

The 'concerted exclusive' model employed by Changeux et al. [11–13] is simpler than the more general 'sequential cooperative' or 'induced fit' model [17] but leads to many similar conclusions. However, the 'concerted exclusive' model allows both S- and R-states to exist *without* ligand binding. Therefore, $\langle r \rangle$ and $\langle y \rangle$ can vary separately as a function of $[f]/K_R$.

This independent course of $\langle r \rangle$ and $\langle y \rangle$ (Fig. 11.2) means that, at critical concentration of f, small changes of state can effect the binding or unloading of large amounts of ligand or, conversely, small changes in ligand binding can cause large changes of state, that is, the system becomes 'amplifying'.

When Λ is small, that is, when cooperativity is high, amplification can change very sharply with $[f]/K_R$ or may even approach infinity at critical values of $[f]/K_R$. *Then a very small change of $[f]$, causing only a minute change in the proportion of liganded subunits, can induce a phase transition.*

Hill [18] has extended the hypothesis of Changeux et al. [11–13] to include the possibility that cooperative changes in membrane state occur through a change of membrane potential or electric field, at a given temperature and ligand concentration $[f]$, or through a simultaneous change in field and $[f]$. This re-

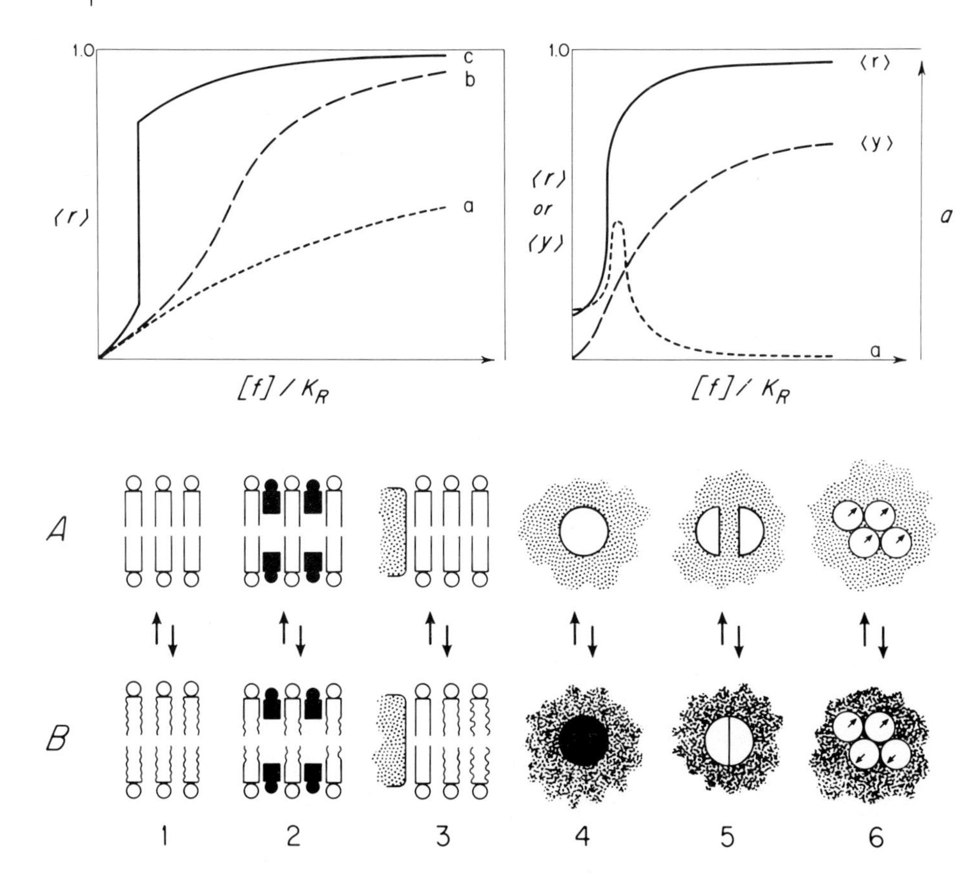

Fig. 11.2. Schematic presentation of the membrane lattice hypothesis. *Top left*: Change of 'state function' $<r>$, [11–13] with $[f]K_R$ (i.e. with $[f]$ at constant K_R, or K_R at constant $[f]$). Curve a: non-cooperative; curve b: intermediate cooperativity; curve c: high cooperativity $(z\eta/kT > -4)$ with phase transition at critical value of $[f]/K_R$ (discontinuous curve). *Top right:* Change of 'state-function' $<r>$, binding function, $<y>$, and amplification, a [12–13], for a moderately cooperative lattice. *Bottom:* Some possible membrane transitions ($A \rightleftharpoons B$).

1. Cooperative gel → liquid-crystal transition in a bilayer of pure phospholipid.
2. Poorly-cooperative transition in a cholesterol: phospholipid system. The sterol immobilizes acyl-chain segments nearest the membrane-water interfaces.
3. Influence of penetrating protein (stippled) on lipid transition. There is a gradient of 'fluidity' (transition-cooperativity) extending laterally from the penetrating protein. The boundary layer lipid immediately adjacent to the protein remains immobilized. Protein is shown to span membrane, but similar effect can be expected if protein penetrates into only 1/2 membrane.
4. Protein conformational transition (empty ⇄ full circle) with associated modification of lipid-boundary layer (light ⇄ dark stippling).
5. Monomer ⇄ multimer transition with associated modification of lipid-boundary layer (stippling).
6. Orientational transition in a polymeric membrane protein with associated modification of lipid-boundary layer (stippling).

1–3 viewed perpendicular to membrane plane; 4–6 represent hypothetical sections through apolar membrane core, parallel to membrane plane.

quires introduction of terms for the polarizabilities α_S and α_R of protomers in states S and R respectively and for the strength, E, of the electric field. Then

$$l'' = exp\{1/kT[\varepsilon - 2(\varepsilon_{SS} - 2\varepsilon_{SR} + \varepsilon_{RR})\langle r\rangle + \tfrac{1}{2}(\alpha_R - \alpha_S)E^2]\} \tag{11.7}$$

or

$$l'' = l \cdot \Lambda^{\langle r\rangle} \cdot exp\left[\tfrac{1}{2}(\alpha_R - \alpha_S)E^2\right] \tag{11.8}$$

Hill's numerical evaluations show that a cooperative membrane response can be induced by field changes. Moreover, as before, at $z\eta/kT = -4$ (or $z\eta/kT > -4$), $\langle r\rangle$ will change discontinuously at a critical field strength; that is, *changes* in *field strength* can *induce phase transitions in a lattice system.*

Up to now, we have considered situations where a change in the overall structure of a coherent membrane assembly follows a transition between different *discrete structural* states of the molecules making up the assembly.

However, Oosawa et al. (Chapter 8; [19–20]), show that the overall structure of a membrane can also be modified by a *reorientation* of constituent molecules. Such reorientation could result from slight changes at the *interaction sites* between membrane protomers and can follow a graded or discontinuous course (phase transition). A list of S$\rightleftharpoons$R transitions (Fig. 11.2) should therefore include changes of protomer *orientation* about an axis perpendicular to the membrane plane. As pointed out in Chapter 8, the model of Oosawa et al. [19, 20] has particular significance because the orientation-distribution of membrane components parallel to the membrane plane can provide for 'surface coding'.

11.2.2. Structural homologies in membranes

Animal cell membranes constitute multi-component, multi-functional, lipid–protein assemblies. For diverse membrane lipids and proteins to form even spatially limited, coherent, ordered two dimensional systems, requires that the *different membrane components interact through common homologous structures.*

The major membrane lipids are amphipathic, and their structural homology reside in the *apolar* moieties. These are essentially identical for the functionally very diverse class of sphingolipids, that is, sphingomyelin, neutral- and acidic-glycolipids. The phospholipids of animal cell membranes contain a larger diversity of acyl chains. All can be accommodated into a bilayer arrays, but some of these may be fluid under physiological conditions of temperature, pH and ionic environment and others not. Moreover, these arrays might include segregated, crystalline packets within a liquid-crystalline 'sea', or *vice versa.* For example, at temperatures where both dipalmitoyllecithin and dipalmitoylphosphatidic acid are fluid in the absence of Ca^{2+}, addition of Ca^{2+} can induce segregation of crystalline phosphatidic acid-domains from liquid-crystalline–lecithin regions [21]. As one might expect, a transition zone or *boundary layer*, of intermediate structure links the two structural phases, except in some systems containing cholesterol; this acts as a 'structure' buffer (Chapter 5) and tends to merge lipids that might otherwise segregate, into a coherent system of 'intermediate fluidity'.

As documented in Chapter 1, amphiphatic segregation of apolar and polar residues also typifies the architecture of *intrinsic* (rather than membrane-associated) *membrane proteins**. Indeed, hydrophobically-mediated structural homology constitutes an architectural device very common in soluble proteins. As noted, all normally-functioning vertebrate hemoglobins possess essentially identical secondary, tertiary and quaternary structures, although only ~9 residues per ~140-residue chain are invariant (including heme-linked histidines and proteins). However, only very conservative substitutions are permitted in the apolar core and the conformational segregation of apolar residues is maintained in the helical segments.

It is logical therefore, to propose that the structural homology between lipid subunits in biomembranes resides in the hydrophobic moieties that are specialized for intra-membranous processes. The polar hydrophilic entities at the membrane surfaces, in contrast, are specialized to participate in the multiplicity of events involving water-soluble substrates, regulators and products within the aqueous compartments.

As we suggested earlier [25] one can expect that *protein* segments inserted into apolar lipid domains would also associate with pure *lipid* regions through *boundary layers* of intermediate structure. The existence of such boundary layers has now been well documented [26–35]. However, such boundary layers cannot be assumed to be fixed, static domains and one can expect that their structure (composition) will depend on (a) the lipid affinity of the protein and (b) the state of the bulk lipid. Indeed, we have recently presented direct spectroscopic evidence that the state of membrane lipids can influence the function of membrane lipids, at least in boundary domains [31, 34, 35]. In any event, in terms of the present hypothesis it is logical to define the interacting units of a membrane lattice – the protomers – as the *intramembranous segments of membrane protein associated with their boundary layer lipid* (Fig. 11.3).

Extensive evidence indicates that biomembranes are structurally and func-

---

* Very recent data [22–24] indicate an arrangement even more complex than those discussed in Chapter 1, for yeast mitochondrial cytochrome oxidase. This oligomeric protein is composed of 7 dissimilar polypeptide chains of which three (mol. = 40,000, 33,000 and 22,000) are coded by mitochondrial DNA and synthesized within the mitochondria, and 4 subunits (mol. = 14,500, 12,700, 12,700 and 4,600) are coded by nuclear DNA and synthesized in the cytoplasm. The peptide chains synthesized by the mitochondria posess a much more hydrophobic character than the cytoplasmically-produced subunits. Moreover, a series of labelling experiments convincingly demonstrate that the cytochrome oxidase complex spans the membrane, that the two biggest, mitochondrially-synthesized subunits are largely sequestered in the membrane core and that the cytoplasmically synthesized chains, which are readily solubilized in aqueous media, lie at the membrane water interfaces.

tionally asymmetric [36,37]. Of the peptide chains penetrating into the apolar core, only few actually extend through the full membrane width. The remainder of the proteins, as well as certain phospholipids appear to be asymmetrically disposed. It is possible therefore that the 'two halves' of many membrane domains may change state independently and in a different way. Lateral movement of two 'half-membranes' relative to each other may also occur in some regions.

11.2.3. Possible ranges of coupling between membrane proteins

The cooperative lattice hypothesis of Changeux et al. [11–13] argues for a 'hierarchy in membrane cooperative effects'; that is, it allows for both *short-range* interactions *within* protein–lipid protomers and *long-range* interactions between protomers. In contrast, the 'fluid mosaic' model of membrane structure [38] implies propagation of lipid-phase changes over large distances. We have shown above and elsewhere (Chapter 1; [33]) that biomembranes assuredly contain lipid domains which are fluid under physiological conditions, and that such domains can undergo reversible disorder–order transitions. However, it is also clear (Chapter 1) that in most membrane systems of animal cells, such domains must be focal and of small dimensions (<1000 Å).

A number of factors may influence the possible range of interactions between membrane proteins. Of these, the high mass ratios of protein to lipid in animal cell membranes appears important. This ratio ranges between 1.5 and 2.5, and although only a portion of membrane protein is 'intrinsic' and presumably only portions of this 'penetrate', the concentration of protein in a hypothetical lipid continuum must be rather high. Indeed, X-ray estimates of the apolar volume occupied by protein in various membranes range from 30–50 % [39–40], implying that one third to one half of the cross-sectional area in the apolar cores of membranes might be protein. If one assumes the proteins to be cylinders, with mol. wt. ~60,000, partial specific volume ~0.73 and a cylinder radius of 25 Å, distributed uniformly and normal to the membrane plane, one arrives at distances between cylinder surfaces of 36–20 Å. The latter estimate is just enough to allow for a single layer of boundary lipid around each cylinder*. A membrane in which half the apolar volume is protein, would thus approach a continuous lipoprotein lattice and would not account for the known presence of fluid lipid domains capable of cooperative phase changes. However, a 30 % apolar protein volume would leave about 40 % of the apolar membrane volume for fluid lipid, the remaining 30 % constituting boundary-layer lipid. Still, even in this case,

---

* A radius of 20 Å would yield an intercylinder distance of 18 Å, which is probably not sufficient for a single layer of lipid around the protein. The intramembranous particles constituting proteins (lipoproteins) that span the full membrane width, take up <5% of the apolar membrane volume (Chapter 1).

the mean distances between the boundary layers of penetrating proteins are likely to be <30 Å, that is, less than the diameter of the hypothetical lipoprotein subunit.

Taking the latter case as a reasonable approximation, the presence or absence of coupling would depend on (a) the degree to which the position of the proteins relative to each other is stabilized by 'membrane-associated proteins' (e.g., microfilaments, spectrin; Chapter 1) and (b) the nature of the intermediate lipid between the boundary layers. The status of the intermediate lipid would depend very much on the presence of cholesterol. This sterol, through its tendency to form 'clusters', might act as a 'spacer', preventing close approach of protomers; its action as a 'fluidity' buffer would also tend to reduce the interaction between protomers. Acidic phosphatides (phosphatidic acid, phosphatidylserine, diphosphatidylglycerol, inositol phosphatides), may also be important because of the ability of Ca^{2+} to cause phase-separation in bilayers containing such phospholipids. Ca^{2+} might thus act to regulate the degree of coupling between membrane proteins.

On the whole, our careful assessment in Chapter 1 and here of interdisciplinary data on membrane structure, forces us to the counter-conventional view that a 'lattice mosaic' hypothesis is a more realistic representation of membrane structure than the fluid mosaic model [38]. In particular, we must conclude the following:

(a) Biomembranes in general, constitute condensed lipid protein mosaics, whose penetrating protein components (whether mono- or multi-meric) are associated with relatively tightly-bound lipid, arranged in a boundary layer whose composition and properties depend in large part on the overall properties of the proteins.
(b) Lipid that is fluid at physiological temperatures is restricted to microdomains (Chapter 1).
(c) Lateral contact, or coupling, between identical or non-identical protein-lipid protomers could be through 'fusion' of boundary layers, allowing for 'long range' cooperative effects. Contact might also be mediated via more fluid domains; these would tend to 'buffer' out or dilute out structural changes occurring in individual subunits and would reduce the range of coupling between protomers. One can suspect that both types of lateral contact may exist in a given membrane and that diverse membranes will differ in lateral contact relationships of subunits (Fig. 11.3)*.
(d) Rapid lateral translocation of *single* protomers within the apolar cores of native membranes is not probable – in contrast to movement of proteins asso-

* Close coupling is also required by a variety of well-studied physiological processes such as the linkage between alkali cation and amino acid transport, and between hormone receptors and adenylate cyclase.

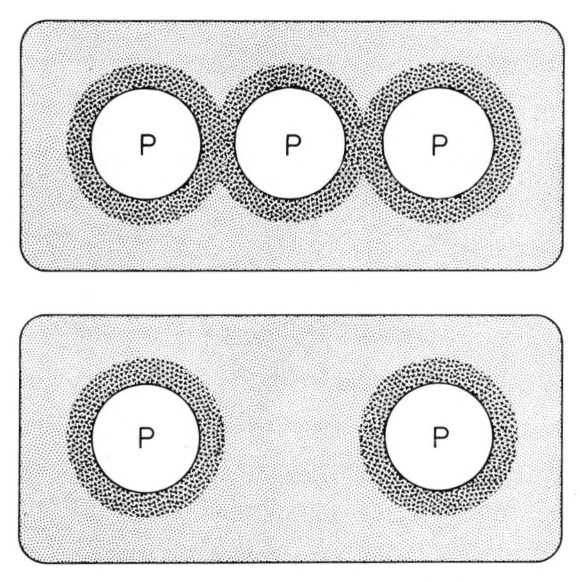

Fig. 11.3. Schematic representation of a section through a hypothetical membrane 'core', showing penetrating membrane proteins, P, with surrounding lipid boundary layers. The proteins are shown as cylinders perpendicular to the membrane plane and the section is parallel to the membrane plane. *Top*: Merged boundary layers ('coupled' protomers). *Bottom*: Protomers separated by 'fluid'-lipid.

ciated with membrane surfaces (Chapter 1). This allows for stable surface-coding (Chapter 8) patterns.

(e) Limited lateral motion of one half-membrane relative to the other half-membrane is feasible.
(f) Rotational motion of protomers about an axis normal to the membrane plane is feasible.
(g) Limited motion of protomers perpendicular to the membrane plane is feasible (Chapter 1).
(h) Rapid metabolic modifications occur (e.g., Chapter 2).

11.3. Cooperative phase changes in membranes

In his lattice model, Changeux [13] 'explicity assumes that the transition of the *protomer* is highly cooperative with respect to its constitutive *amino acid* or *phospholipid* elements', but that the unique features of membrane cooperativity derive from phase changes involving the 'statistical mechanical ensemble of protomers'. He introduces the phrase 'hierarchy in membrane cooperative effects' to refer to the hypothicated multiplicity of concerted interactions.

We lack substantial information as to cooperative processes initiated in membranes at the amino acid or polypeptide level, but we do understand the phase changes of membrane lipids rather well. As already discussed (Chapter 3–6), the gel → liquid-crystal phase transition of phospho- or glycolipid bilayers is a cooperative process in which the hydrocarbon chains of the membrane lipids switch from an ordered to a disordered state, with concurrent lateral expansion of the bilayer and a decrease in its thickness and density. The phase transition of membrane lipids is positively cooperative, that is, with cooperative units probably involving 50 to 200 molecules. Lipids can be shifted from the gel phase to the liquid-crystal phase (a) by *raising temperature* or (b) by *altering the ionic composition of the aqueous environment*, without change of temperature. The latter, a *gel → liquid-crystal transition at constant temperature, is the important phenomenon in the membranes of animal cells.*

The significance of cooperative lipid phase transition derives from two factors: *First*, as reviewed by Raison [42], an abundance of evidence indicates that numerous membrane functions (e.g. enzyme activity; membrane transport) vary with the state of the bulk lipid. The influence of lipid phase is generally on reaction-rate and for many membrane-functions this is greater when the lipid is in the liquid-crystal phase. However, much more drastic effects have been observed. Thus Zakim and Vassey [43] show that UDP-glucuronyltransferase from liver microsomes can be competitively inhibited by UDP-glucose at temperatures below but not above 16 °C, the lipid transition temperature for these membranes. Moreover, UDP-*N*-acetylglucosamine only activates the enzyme at temperatures $>16\,^{\circ}C$. In this instance, therefore, the membrane *lipid state controls both the substrate specificity and regulatory behavior* of a membrane-enzyme. Of perhaps even greater importance is the fact that the activation of *adenylate cyclase*, a most critical regulatory enzyme, whether by hormones [44,45] or by choleragen [46,47] depends sensitively upon the state and/or composition of plasma membrane lipid.

In any event, the fact that the function of membrane enzymes can change quantitatively or qualitatively when a lipid phase transition is induced by a temperature step, implies that the same functional alterations would occur if the phase transition were triggered by *ionic* changes.

11.3.1. The role of membrane ligands

According to the cooperative lattice hypotheses, if the transition between two membrane states is intrinsically cooperative, and if the two states have different binding affinities for a ligand, f, a small change in $[f]$ can, in certain concentration regions, produce a state transition. The role of $[f]$ is therefore of of major biological significance. For example, f might be an ion (H^+, Ca^{2+} or Mg^{2+}), a small organic molecule (cAMP and cGMP), a peptide hormone (glucagon and insulin), a macromolecule such as choleragen, or a supramolecular assembly.

The concentration of f may be subject to cellular genetic and enzymatic control, as well as to physiological regulatory mechanisms. Moreover, the concentration of f may be normally maintained at a 'poising' level, so that minor changes in $[f]$, due to some biologic event, membrane-related or not, may effect large changes in membrane state. We lack solid information about possible ligands that might induce a change of membrane state except for H^+, Ca^{2+} and Mg^{2+}, which are known to induce *isothermal* phase transitions in model lipid membranes.

The ionic conditions that can induce cooperative gel → liquid-crystal phase transitions in membrane lipids have only recently received attention [48–52]*. Within the physiologic range of pH only phosphatidic acid, phosphatidylethanolamine, phosphatidylserine and di- or tri-phosphoinositide can be expected to respond to changes of pH. In contrast, phosphatidic acid, phosphatidylserine, mono- and di-phosphatidylglycerol and inositol phosphatides can be expected to respond to variations in the divalent cation concentration of the aqueous milieu*.

Concerning pH, Träuble and Eibl [48] show that, in the case of phosphatidic acid, a very small increase of pH in the pH region of 7–9, can induce a gel → liquid-crystal transition at constant temperature. A smaller and reverse effect occurs with phosphatidylserine (titration of the serine amino-residue).

Concerning the role of divalent cations, Papahadjopoulos et al. [49] have demonstrated that the thermotropic gel → liquid-crystal transition of 7 mM phosphatidylserine was reduced in intensity in the presence of 0.5 moles Ca^{2+}/mole lipid and was abolished at a Ca^{2+}/lipid ratio = 5.0. Jacobson and Papahadjopoulos [50] have shown that beef brain phosphatidylserine bilayers in 100 mM sodium chloride, exhibit a phase transition that centers at about 5 °C in the absence of divalent cations (pH 7.4), but *shifts* to 18 °C in the presence of 5×10^{-4} Ca^{2+}, with a concurrent decrease of intensity. 10^{-3} M Ca^{2+} completely abolishes the transition, whereas $1–5 \times 10^{-3}$ M Mg^{2+} merely induces the shift of transition temperature. Ca^{2+} and Mg^{2+} influence the gel → liquid-crystal transitions also of phosphatidylglycerol bilayers; the transition temperature increases by ~12.5 °C in the presence of 10 mM Mg^{2+} but is abolished by the same concentration of Ca^{2+}.

These results are compatible with those of Verkleij et al. [51]. These authors show that phosphatidylglycerol (di-C12-phosphatidyl glycerol) dispersed in 100 mM NaCl at neutral pH has a gel → liquid-crystal transition near 0 °C. 1 mM Ca^{2+} (Ca^{2+}/phosphatidylglycerol = 1/100; M/M) raises the transition temperature to about 10 °C. A further rise in the Ca^{2+}/phosphatidylglycerol ratio to

* Ionic strength is precisely regulated in the intact cell of intact organism. The variables of potential physiological significance are therefore, pH, the concentrations of Ca^{2+} and Mg^{2+}, and possibly of organic bases.

1:4 causes only a small additional increase in the transition temperature, but when the ratio is greater than 1:2 the transition temperature jumps dramatically to 23 °C.

Jacobson and Papahadjopoulos [50] have also demonstrated that the addition of Ca^{2+} to mixed bilayers of phosphatidylserine (66%) and lecithin (34%) (100 mM NaCl; pH 7.4) causes isothermal, lateral *segregation* of the neutral lecithin molecules from the acidic phosphatidylserine molecules. A similar process was observed with mixtures of lecithin and phosphatidic acid, in which case phase segregation occurred only at pH 8 and not at pH 6; this indicates a requirement for two negative charges per molecule, favoring formation of 'linear polymers'. The data indicate that, at low levels both Ca^{2+} and Mg^{2+} exert their effects by charge neutralisation. However the response to high levels of Ca^{2+}, but not Mg^{2+}, suggest formation of specific, possibly polymeric complexes.

In examining the possible roles of Ca^{2+} or Mg^{2+} in triggering isothermal lipid phase transitions it is necessary to consider the physiological concentrations of these cations in various cellular spaces. Thus, the Ca^{2+} concentration in the extracellular- and presumably intra-mitochondrial spaces exceeds 10^{-3} M, but lies below 10^{-6} M in the cytoplasmic space. In contrast, the Mg^{2+} level $\geqq 10^{-3}$ M in both intra- and extracellular spaces. Ca^{2+} thus appears potentially capable of triggering phase changes in membranes bordering the cytoplasmic space. Important in this context is regulatory action of mitochondria on cytoplasmic Ca^{2+} (Chapter 6), the dramatic capacity of cAMP to induce Ca^{2+}-efflux from mitochondria [52] and the information indicating that the phosphatidylserine of some plasma membranes is asymmetrically disposed and faces the cytoplasmic space [36]*. Indeed, Rasmussen and associates [53,54] and Borle [52] argue that many of the effects mediated by cAMP are in fact *due to the role of cAMP in the regulation of intracellular levels of* Ca^{2+}.

11.3.2 Factors modifying lattice cooperativity

Cooperativity in polymeric proteins is complex indeed and the participating mechanisms vary markedly between proteins. The same can be expected for multimeric membrane proteins. In contrast, at least at this time, the factors determining the cooperativity of lipid-state transitions in lipid or lipid-protein systems appear better known; any element influencing the capacity of lipid acyl

* It is common to consider lipid phase transitions to involve both 'halves' of a bilayer membrane concurrently. This view is probably correct for membranes consisting of only one species of phosphatide. However, it is possible to manufacture stable model phosphatide bilayers that are asymmetric with regard to ionizable headgroups. Moreover, most biomembranes are assuredly asymmetric in terms of both protein and lipid disposition. Finally, many penetrating proteins probably do not span their membranes. It is therefore possible, and even probable, that a lipid- or lipid-protein-state transition may occur in one but not the other half of a biomembrane.

chains to switch from an all-(or multi)-*trans*-configuration to a multi-*gauche* configuration, alters the cooperativity of lipid state transitions. Three intramembranous 'cooperativity modifiers' must be considered, namely (a) unsaturated acyl chains, (b) protein and (c) cholesterol (or analogous sterols).

Concerning (a), unsaturation patterns in membrane lipids are complex and we lack substantial information as to the distribution of unsaturated acyl chains parallel to the membrane plane.

Concerning (b), recent experimentation by Grant and McConnell [55] on dimyristoyllecithin bilayers, into which the membrane-penetrating erythrocyte *M*, *N*-glycoprotein 'glycophorin', had been incorporated ($\sim$120 moles lipid per mole glycoprotein; wt. ratio lipid: protein 2:1–3:1), clearly document a lesser cooperativity of the thermotropic gel → liquid-crystal transition than with the pure lipid. The author data also suggest that a boundary layer of lecithin surrounds the protein and one can assume that there is a gradient of decreasing cooperativity extending from the bulk lipid to the protein-lipid interface (Fig. 2). Our own Raman data on intact erythrocyte- and lymphoid cell-plasma membranes (33, 56) support this contention.

Concerning (c), an abundance of evidence [33, 56, 57, 58] indicates that cholesterol reduces the cooperativity of gel → liquid-crystal transitions in phospholipid-cholesterol systems. This mechanism can be readily explained: Cholesterol associates with phospho- and/or glycolipid bilayers, with its OH residue at the membrane-water interface and its hydrophobic moiety in tight association with the acyl-chain segments in the outer half of the bilayer. In this way, the sterol diminishes, but does not abolish, cooperative interactions between acyl chain segments located 'deeper' in the membrane core (Fig. 11.2)*.

11.4. Interpretation of membrane polymorphism in tumors in terms of the membrane lattice hypothesis

11.4.1 Relevant predictions of the hypothesis

In the first presentation of their lattice hypothesis, Changeux et al. [11]** point out that a 'membrane phase transition may become *permanent* as a result of the insertion of a protomer of a different unit size, the structural modifications of

* Kimelberg and Papahadjopoulos [59] show that the activity of a Na^+, K^+-activated ATPase, incorporated into phosphatide bilayers is inhibited when cholesterol is also incorporated into the bilayers. This effect varied with the proportions of cholesterol; at cholesterol–phospholipid ratios of 1 : 1 (M/M) inhibition was complete in the case of bilayers made of saturated phospholipids, whereas it was only partial in the case of phospholipids with unsaturated fatty acid chains. Since cholesterol does not primarily interact with membrane proteins, it is thought to affect the activity of the enzyme by reducing its molecular motion within the membrane.

** Ref. 11, Footnote 24.

some of the native protomers by gain or loss of constitutive subunits, etc.'. It also follows from Eq. 11.4, that the 'number of protomers required to be altered in this process does not have to be high...'.

One should therefore inquire into the possibility that, in neoplasia, the overall properties of membrane ensembles, as well as the characteristics of their diverse functional entities are altered by one or more of the following processes:

(1) Insertion of a *new* protein or lipid.
(2) *Alteration* of existing proteins or lipids (e.g., by proteases or lipases).
(3) Change in the *proportions* of *phospholipids.*
(4) Change in the *proportions* of *glycolipids.*
(5) Change in the *proportion* of *cholesterol.* According to the lattice hypothesis, a similar effect may arise with:
(6) Change in the *steady-state of membrane ligands.*

Items 1–4 could involve the cooperativity, Λ, of the ensemble (Eq. 11.4, Fig. 11.2) and/or the affinities, K_R and K_S, of the membrane for ligand f (Eqs. 5 and 6; Fig. 11.2). Item 5 can be considered to relate primarily to Λ, and item 6 concerns $[f]$ (Eqs. 5 and 6; Fig. 11.2). Far-reaching effects due to changes in $[f]$ could only be expected at critical values of $[f]/K_R$ (Fig. 11.2).

11.4.2. 'New' membrane components

We now consider the possibility that the membrane alterations in neoplasia derive from the insertion of *new* material into the cellular membranes of tumor cells. This new material could be *lipid* in nature, but, although various 'neolipids' have been proposed over the years, these substances have proven to be either normal lipids present in unusual amounts or lipids synthesized by embryonic, but not adult cells, for example, Forssman glycolipid (Chapter 4). The status in the present context, of lacto-*N*-neotetraosylceramide, the unique ceramide tetrasaccharide recently discovered in polyoma transformed hamster cells [60, 61] remains to be established. However, an extensive body of *immunological* evidence indicates that *new proteins*, that is, *tumor specific transplantation antigens* (TSTA), do occur in at least the plasma membranes of tumor cells.*

Although the appearance of TSTA on tumor cells is an enormously important facet of tumor biology, we have not treated the subject before because one still knows extremely little about these antigens at the molecular level. However, since evidence is strong that TSTA are proteins and that they are membrane components, they are clearly candidates for the 'new' subunits hypothesized in

---

* TSTA must be distinguished from tumor associated antigens. Recently, Bystryn et al. [62] have purified the tumor-associated antigen of the B15 melanoma of C57 BL/6J mice. This is a glycoprotein with a molecular weight of a range of 150,000 to 200,000, which is not destroyed by papain or trypsin treatment and is also present in normal fibroblasts of C57 BL/6J mice. The glycoprotein is shed (Chapter 2) into culture media by both normal and malignant cells.

item 1. We will therefore briefly summarize current information on the biochemical properties of TSTA's.

TSTA can be inferred to constitute new plasma membrane components primarily because many of the immune reaction used to detect such antigens take place at the cell surface. However, there has been no rigorous study on the subcellular distribution of these transplantation antigens, as there have been of percertain histocompatibility antigens [63]; we therefore do not now know whether TSTA do not also occur on *intracellular* membranes.

Some have employed subcellular fractionations to localize the TSTA of chemically-induced tumor cells and all available data suggest that the antigens are membrane associated [64–65]. The only extensive studies are those of Baldwin and associates on the amino-azo-dye-induced D23 rat hepatoma carried in isogeneic rats [64–68]. It was found by use of routine differential centrifugation and rate-zonal centrifugation on sucrose density gradients, that the D23 TSTA is normally non-soluble and that it is associated with both large and small subcellular fragments, with the greatest specific activity in a fraction sedimenting 'between' nuclei and mitochondria. Unfortunately, the fractionation procedures employed do not exclude a rather wide distribution of plasma membrane fragments, and the question of whether the D23 TSTA is actually also present in subcellular membranes, remains unresolved.

The protein nature of the D23 TSTA is attested by the fact that it can be released in soluble form by limited proteolytic digestion of crude membrane preparations with papain [66–68]. However, a recent report by Baldwin et al. [68] shows that the D23 TSTA can also be released in water-soluble form by β-glucosidase digestion of isolated membranes; this process could be specifically inhibited by β-D-glucose. Release of TSTA was also blocked by iodoacetate and *p*-chloromercuriphenylsulfonate under conditions that did not alter β-glucosidase activity. The authors interpret their data to indicate that some intrinsic protease activity is involved in the β-glucosidase-mediated TSTA release.

The soluble antigen could be considerably purified by DEAE ion-exchange chromatography but the purified material was still highly heterogeneous by SDS-PAGE. The soluble material retained the capacity to elicit tumor-specific antibody production in syngeneic rats.

The TSTA of various tumor cells can also be released in soluble form by extracting homogenates with 3M KCl. The mechanisms involved are not clear; chaotropic effects have been invoked [69], as well as protease activation [70].

Meltzer et al. [71] have isolated and fractionated the TSTA of a diethylnitrosamine-induced hepatoma carried in inbred guinea pigs. Using delayed hypersensitivity as an immunologic detection system, they found that 15–40 % of the TSTA activity could be extracted with 3 M KCl. The extracted material could be precipitated with 2 M $(NH_4)_2SO_4$ and, upon molecular sieving on Sephadex, migrated in the 75,000 to 150,000 molecular weight region.

The same extraction procedure has been employed successfully with several methylcholanthrene induced mouse sarcomas [72] and the Gross surface antigens from Gross-virus-infected AKR mouse lymphomas [73]. The Gross antigens isolated by 3 M KCl extraction from intact cells and purified by $(NH_4)_2SO_4$ precipitation, when fractionated by molecular sieving, eluted as a broad peak ranging between 44,000 and 200,000 in molecular weight, with maximum activity at about 70,000 molecular weight. The yield was 47 %. The Gross antigen was inactivated by heating at 56 °C for 30 min and by trypsinization.

The TSTA induced by SV40 virus has also been isolated in soluble form, by extraction of homogenates from SV40-transformed cells with 3 M KCl [74] or by papain treatment of homogenates (or crude membrane preparations) [74, 75]. Drapkin et al. [75] have shown that crude membrane preparations isolated from SV40-transformed Balb/c cell lines could protect syngeneic mice against challenge by SV40 tumors. The yield of SV40 TSTA in the membrane fraction was about 50 %. Papain treatment of the membranes yielded a soluble material which, upon molecular sieving on Sephadex, showed a molecular weight of about 50,000. The yield of the antigen – which was also protected against tumor challenge – was about 5 %. The SV40 TSTA is not 'shed' (Chapter 2) in tissue culture [74] and is damaged by trypsinization [76].

The TSTA induced by the Rauscher leukemia virus can also be isolated by papain digestion [77]. This material showed a molecular weight of 50,000 to 600,000 by molecular sieving, and was also distinctly immunogenic.

To sum, TSTA are new membrane proteins. Whether they exist solely in the plasma membrane has not been determined, but can be tested. Their functional and structural roles remain to be established. However, since TSTA can now be extracted in soluble form and since techniques are now available for the insertion of isolated membrane proteins into lipid systems, the possible impact of the incorporation of TSTA into membranes may *be accessible* to experimentation*,**.

---

* In this context, the recent experiments of Uno et al. [78] on a yeast mutant lacking cytochrome oxidase, are extremely pertinent. In this mutant a cytochrome oxidase sub-unit coded by *mitochondrial* DNA and synthesized in the mitochondria, cannot be incorporated due to the synthesis of a *cytoplasmic* protein, coded by a mutant *nuclear* gene. It remains to be established how this new protein functions and it is possible that it is a variant of one of the cytochrome oxidase sub-units normally synthesized in the cytoplasm.

** Plasma membrane neo-antigens are not restricted to neoplastic transformation and/or viral infection. They are also induced in the membranes of mature erythrocytes during the intra-erythrocytic cycle of *malarial* infections [79] which also engender polymorphic membrane deficiencies [80]. Since erythrocytes are 'end cells' with well-studied, easily-isolated membranes, since the intra-erythrocytic cycle of *Plasmodia* is rapid and often highly synchronized and since one can readily raise antisera specific for the membrane antigen(s) induced by a given strain of *Plasmodium*, the study of membrane anomalies in malaria may offer some insights into the plasma membrane deviations in neoplasia.

11.4.3. Altered membrane protomers

Recent experimentation (Chapter 2) shows that many transformed cells lack one or more high molecular weight plasma membrane glycoproteins, possibly because of the abnormal release of proteases that act upon the external surface of the membrane. The proteins involved appear to be *membrane-associated*, rather than intrinsic, and thus are not likely to participate in the hypothetical interactions that would functionally couple diverse membrane protomers. However, it is conceivable that the release of proteases and glycosidases might modify or eliminate plasma membrane binding sites for extracellular membrane ligands; glycosidases could affect both glycoprotein and glycolipid receptors.

11.4.4. Altered proportions of membrane phospholipid

In Chapter 3 we have reviewed the limited but convincing experimentation documenting important deviations of membrane phospholipid composition in some tumors. The most impressive anomalies are found in mitochondria and include (a) anomalously high sphingomyelin levels and (b) discrepant proportions of acidic phosphatides, diphosphatidyl glycerol and phosphatidyl-serine in particular. The proportions of these phosphatides can also be abnormal in endoplasmic reticulum and plasma membrane.

The consequences of the phospholipid anomalies, detailed in Chapter 3, could be at (a) the level of protomers (boundary-layer lipid) or (b) the level of hypothetical lipid domains separating diverse protomers. Thus, boundary layers in which much of the neutral phospholipid is sphingomyelin would be differently 'coupled' from layers where the bulk of neutral phospholipid is lecithin or phosphatidylethanolamine. Moreover, a change in the proportion of charged lipids in boundary layers could critically modify the structure of these layers under given ionic conditions and could in essence produce a change in K_R. Finally, charged phosphatides such as diphosphatidylglycerol and phosphatidylserine will respond very differently from neutral phospholipids to subtle changes in $[Ca^{2+}]$, for example, throughout the 'hierarchy in cooperative effects'.

Phospholipid anomalies such as have been observed in some tumor membranes (Chapter 3) thus have very profound implications for tumor membrane polymorphism.

11.4.5. Altered glycolipid composition

As detailed in Chapter 4, tumors often exhibit an anomalous glycolipid composition. However, these anomalies involve only the sugar moieties extending into the aqueous compartments, not the apolar segments. For this reason, and because glycolipids are minor membrane components by mass, the glycolipid changes observed in transformed cells are not expected to modify membrane cooperativity.

11.4.6. Altered proportions of membrane cholesterol

In Chapter 5 we have summarized evidence showing that cholesterol biosynthesis is often inadequately regulated in tumors and that the cytoplasmic membranes of tumor cells may exhibit abnormally high cholesterol/phospholipid ratios.

Available information indicates that cholesterol does not associate tightly with membrane proteins, that is, the sterol is probably not a constitutive element of membrane protein–lipid protomers, but forms part of the lipid domains separating protomers. An increased proportion of cholesterol would reduce coupling between protomers by its 'damping' action, and would tend to reduce the cooperativity of potential phase transition. The tendency of the sterol to 'cluster' under certain conditions, would also tend to reduce the interaction between protomers. The high cholesterol levels reported for tumor mitochondria and endoplasmic reticulum might thus signify an inadequate responsiveness of these membranes to regulatory stimuli.

It would appear imperative, therefore, to rigorously study the basis and consequences of anomalous cholesterol metabolism in tumor cells. Many important questions can easily be answered, for example: Is defective control of cholesterol biosynthesis *generally* associated with abnormally high cholesterol/phospholipid ratios in cytoplasmic membranes? If so, can normal cholesterol/phospholipid ratios be restored by use of agents such as 25-OH cholesterol? What are the functional consequences of high cholesterol/phospholipid ratios in mitochondria and endoplasmic reticulum.

11.4.7. Altered concentration of membrane ligands

At critical values of $[f]/K_R$ small changes in $[f]$ can theoretically bring about large shifts in membrane state. A phase transition could, in effect, be 'made permanent' through a permanent change in the steady state value of $[f]$. But what is f? Ca^{2+} is a possible candidate, but we know nothing about intracellular Ca^{2+} levels in neoplastic vs. normal cells. cAMP and cGMP are also obvious possible candidates and their levels in neoplastic cells are commonly defective. At this writing we still know only little about the role of cyclic nucleotides on biomembranes, but this information gap will assuredly be soon filled.

11.5. Membrane alterations in neoplasia; relevance to malignancy

The ability to *invade* and *metastasize* constitute the lethal properties of *malignant* neoplasia and are the two clinical features that distinguish malignant neoplasia-cancer from neoplasia in general. A non-invasive, non-metastasizing neoplasm, even if rapidly growing, is clinically benign not malignant. However, once metastasis has occurred, the disease cannot usually be eradicated by elimination of the 'primary' tumor. In terms of clinical management, any maneuver that eliminates the invasive and metastatic potential of a tumor reduces the problem of therapy to one of *local control.* This one can now achieve by surgery and ionizing radiation. It is therefore appropriate to summarize our information on membrane anomalies that might account for malignant behavior. For neoplastic cells to form an invasive, metastatic tumor, they need to overcome the disadvantage of their unique antigenicity, that is, they must avoid or escape from the host immune defenses. They must also possess a significant *competitive advantage* over host cells.

An enhanced capacity to transport essential metabolites could clearly provide an important competitive advantage to a cell and, as described in Chapter 7, many virally transformed cells show enhanced sugar and/or amino acid transport *in vitro.* However, the possible significance of these observations must be evaluated quantitatively in terms of the *in vivo* situation. Specifically, tissue cells obtain essential nutrients by diffusion from capillaries. The concentration of a nutrient at a point in a tissue diffusion field depends on the rate of blood flow in the capillaries, the distance between capillaries, the diffusion coefficient of the intracellular material for the nutrient, the distance from the capillary wall, the distance from the arterial end of the capillary and cell metabolism. It is well established that, in a given tissue geometry, cells most remote from the capillary walls and from the arterial end of the capillary are least favored in terms of nutrient supply. It is also clear that the harmonious coexistence of cells in a tissue society requires either that one cell type does not accumulate essential nutrients more efficiently than another, or that the discrepant behavior is in some way compensated, for example, by *intercellular coupling* (Chapter 9). Barring such compensation, competitive elimination of normal cells would follow the emergence in a diffusion field of a cell clone which can accumulate a scarce essential metabolite in an abnormally acquisitive manner.

Whether the alterations in *sugar* transport observed *in vitro* can confer a significant physiological competitive advantage could be tested by an *in vitro* simulation of the *in vivo* situation. That is, can cells with enhanced sugar transport co-cultivated with otherwise equivalent, but 'unacquisitive' cells starve out normal cells at low glucose concentrations? However, considering the *in vivo* situation, the normal concentration of glucose in the plasma is $\sim$5 mM, considerably above the K_M values for transport observed in both normal and transformed cells (Chapter 6). The alterations of sugar transport described thus do not appear to provide an obvious selective advantage for the tumor cells.

Reported increases in the V_{max} values for amino acid transport may have greater *in vivo* significance than modifications of sugar transport, because the plasma concentrations of essential amino acids (0.02–0.2 mM) lie well below the K_M values for amino acid transport into cells. The amino acid concentrations in the diffusion field outside the capillaries must be still lower. Tissue cells must, therefore, generally operate far from saturation of their amino acid transport systems and small increases in V_{max} might provide a critical competitive advantage. This possibility could also be tested by an *in vitro* simulation of the *in vivo*, situation.

As already mentioned, *intercellular coupling* might protect a cell ensemble against the deleterious consequences, which, in an uncoupled system, might derive from an unusual capacity of some cells for the accumulation of essential nutrients. However, intercellular coupling depends critically on the mitochondrial control of intracellular Ca^{2+} levels, which may be impaired in many neoplastic cells, either through a mitochondrial deficiency or through an abnormal cyclic nucleotide metabolism (Chapter 6). Moreover, intercellular coupling can also be impaired by the action of proteases such as are released in abnormal amounts from neoplastic cells (Chapter 7).

The abnormal release of proteases and other hydrolases from neoplastic cells has profound implications that extend beyond the possible effects of such enzymes on intercellular coupling, particularly since the phenomenon correlates very well with clinically defined malignancy. Thus the release of these enzymes could lead to (a) polymorphic alteration of overall membrane behavior through a change in membrane cooperativity; (b) destruction of connective tissue allowing invasion by the tumor cells; (c) destruction of hormone receptors, releasing the cells from physiological controls (d) destruction of TSTA, creating a 'bland' surface character that allows the tumors to escape from immune surveillance; (e) interference with cell-mediated, or humorally-mediated immune reactions and (f) release of cells from growth control (as can be achieved by protease treatment of cultured fibroblasts). The propensity of tumor cells to release proteases and other hydrolases would thus appear to constitute a property that is extremely relevant to malignant behavior.

The release of lysosomal proteases may be particularly significant since many malignant tumors exhibit high lactic acid production, probably generating abnormally low local pH values (Chapter 6) that favor activation of lysosomal cathepsins. But an anomalously low pericellular pH might have other consequences relevant to malignant behavior. Thus, the acid 'boundary layer' expected to surround highly glycolysing tumors could alter the activities of pH-dependent *surface* enzymes (Chapter 8) as well as the surface charge characteristics influencing cell-contact. The aberrant contact behavior of tumor cells may therefore derive not from plasma membrane abnormalities but from high, localized acid production.

Clearly the issue of malignancy is a central one in tumor biology. It can be clarified by carefully balanced efforts that look not only at one membrane or one enzyme system but whole cells and cell communities.

References

[1] Temin, H.M., *Perspect. Biol. Med.*, *14*, 11 (1971).
[2] Huebner, R.J. and Todaro, G.J., *Proc. Natl. Acad. Sci.U.S.*, *64*, 1087 (1969).
[3] Weber, G. in *The Role of Cyclic Nucleotides in Carcinogenesis*, J. Schultz and H.G. Gratzner, eds., Academic Press, New York, 1973, p. 57.
[4] Pitot, H.C., Shires, T.K., Moyer, G. and Garrett, C.T., in *The Molecular Biology of Cancer*, H. Busch, ed., Academic Press, New York, 1974, p. 523.
[5] Warburg, O., *The Metabolism of Tumors*, Arnold Constable, London, 1930.
[6] Hershko, A., Mamont, P., Shields, R. and Tomkins, G.M., *Nature, New Biol.*, *232*, 206 (1971).
[7] Pastan, I., Anderson, W.B., Chapman, R.A., Willingham, M.L., Russell, T.R. and Johnson, G.S. in *Control of Proliferation in Animal Cells*, Clarkson, B., and Baserga, R., eds., Cold Spring Harbor Laboratory, New York, 1974, p. 563.
[8] Wallach, D.F.H., *Proc. Natl. Acad. Sci. U.S.*, *61*, 868 (1968).
[9] Wallach, D.F.H., *New Engl. J. Med.*, *280*, 761 (1969).
[10] Wallach, D.F.H., *Curr. Top. Microbiol. and Immunol.*, *47*, 152 (1969).
[11] Changeux, J.-P., Thiery, J., Tung, Y. and Kittel, *Proc. Natl. Acad. Sci. U.S.*, *57*, 335 (1967).
[12] Changeux, J.-P. and Thiery, J., in *Regulatory Functions of Biological Membranes*, J. Järnefelt, ed., Elsevier, Amsterdam, 1968, p. 115.
[13] Changeux, J.-P. in *Symmetry and Function of Biological Systems at the Macromolecular Level*, Engström, A. and Strandberg, B., eds., Wiley Interscience, New York, 1969, p. 235.
[14] Hill, T., *Introduction to Statistical Thermodynamics*, Addision-Wesley, New York, 2nd ed., 1960.
[15] Blumenthal, R., *J. Theoret. Biol.*, *49*, 219 (1975).
[16] Blumenthal, R., Changeux, J.-P. and Leféver, R., *J. Memb. Biol.*, *2*, 351 (1970).
[17] Koshland, D.E., Jr., Nemethy, G. and Filmer, D., *Biochemistry*, *5*, 365 (1966).
[18] Hill, T., *Proc. Natl. Acad. Sci. U.S.*, *58*, 111 (1967).
[19] Oosawa, F., Maruyama, M. and Fujime, S., *J. Theoret. Biol.*, *36*, 203 (1972).
[20] Maruyama, M. and Dosawa, F., *J. Theoret. Biol.*, *49*, 249 (1975).
[21] Jacobson, K. and Paphadjopoulos, D., *Biochemistry*, *14*, 153 (1975).
[22] Poynton, R.O. and Schatz, G., *J. Biol. Chem.*, *250*, 752 (1975).
[23] Poynton, R.O. and Schatz, G., *J. Biol. Chem.*, *250*, 762 (1975).
[24] Eytan, G.D. and Schatz, G., *J. Biol. Chem.*, *250*, 767 (1975).
[25] Wallach, D.F.H. and Zahler, H.P., *Proc. Natl. Acad. Sci. U.S.*, *56*, 1552 (1966).
[26] Jost, P.C., Griffith, O.H., Capaldi, R.A. and Vanderkooi, G., *Proc. Natl. Acad. Sci. U.S.*, *70*, 480 (1973).
[27] Jost, P.C., Griffith, O.H., Capaldi, R.A. and Vanderkooi, G., *Biochim. Biophys. Acta*, *311*, 141 (1973).
[28] Träuble, H. and Overath, P., *Biochim. Biophys. Acta*, *307*, 491 (1973).
[29] Overath, P. and Träuble, H., *Biochemistry*, *12*, 2625 (1973).
[30] Stier, A. and Sackmann, H., *Biochim. Biophys. Acta*, *311*, 400 (1973).
[31] Dehlinger, P., Jost, P. and Griffith, O.H., *Proc. Natl. Acad. Sci. U.S.*, *71*, 2280 (1974).
[32] Wallach, D.F.H., Verma, S.P., Weidekamm, E. and Bieri, V., *Biochim. Biophys. Acta*, *356*, 68 (1974).
[33] Verma, S.P. and Wallach, D.F.H., *Biochim. Biophys. Acta*, *382*, 73 (1975).

[34] Wallach, D.F.H., Bieri, V., Verma, S.P. and Schmidt-Ullrich, R., *Proc. N.Y. Acad. Sci.*, 1975, in press.
[35] Bieri, V. and Wallach, D.F.H., *Biochim. Biophys. Acta*, *406*, 415 (1975).
[36] Zwaal, R.F.A., Roelofsen, B. and Colley, C.M., *Biochim. Biophys. Acta*, *300*, 159 (1973).
[37] Wallach, D.F.H., *Biochim. Biophys. Acta*, *265*, 61 (1972).
[38] Singer, S.J. and Nicolson, G., *Science*, *175*, 720 (1972).
[39] Engelman, D.M., *Chem. Phys. Lipid*, *8*, 298 (1972).
[40] Finean, J.B., *Chem. Phys. Lipids*, *8*, 279 (1972).
[41] Blaurock, A.E., *Chem. Phys. Lipids*, *8*, 285 (1972).
[42] Raison, J.K., *Bioenergetics*, *4*, 285 (1973).
[43] Zakim, D. and Vessey, D.A., *J. Biol. Chem.*, *250*, 343 (1975).
[44] Birnbaumer, L., Pohl, S.L. and Rodbell, M., *J. Biol. Chem.*, *246*, 1857 (1971).
[45] Pohl, S.L., Krans, H.M., Kozyreff, V., Birnbaumer, L. and Rodbell, M., *J. Biol. Chem.*, *256*, 4447 (1971).
[46] Bennett, V., O'Keefe, E. and Cuatrecasas, P., *Proc. Natl. Acad. Sci. U.S.*, *72*, 33 (1975).
[47] Siegel, M.I. and Cuatrecasas, P. in *Cellular Membranes and Tumor Cell Behavior*, Walborg, E.F., ed., Williams and Wilkins, Baltimore, in press, 1975.
[48] Träuble, H. and Eibl, H., *Proc. Natl. Acad. Sci. U.S.*, *71*, 214 (1974).
[49] Papahadjopoulos, D., Poste, G. and Shaeffer, B.E., *Biochim. Biophys. Acta*, *323*, 23 (1973).
[50] Jacobson, K. and Papahadjopoulos, D., *Biochemistry*, *14*, 152 (1975).
[51] Verkleij, A.J., DeKruijff, B., Ververgaert, P.H.J., Th., Tocanne, J.F. and Van Deenen, L.L.M., *Biochim. Biophys. Acta*, *339*, 432 (1974).
[52] Borle, A.B., *J. Memb. Biol.*, *16*, 221 (1974).
[53] Rasmussen, H., *Science*, *170*, 404 (1970).
[54] Rasmussen, H., Goodman, D.B.P., Tenenhouse, A., *CRC Crit. Rev. Biochem.*, *1*, 95 (1972).
[55] Grant, C.W.M. and McConnell, H.M., *Proc. Natl. Acad. Sci. U.S.*, *71*, 4653 (1974).
[56] Chapman, D. in *Biological Membranes*, D. Chapman and D.F.H. Wallach, eds., Academic Press, London, 1973, Vol. 2, p. 91.
[57] Lippert, J.L. and Peticolas, W.L., *Proc. Natl. Acad. Sci. U.S.*, *68*, 1572 (1971).
[58] Verma, S.P., Wallach, D.F.H. and Schmidt-Ullrich, R., *Biochim. Biophys. Acta*, *394*, 633 (1975).
[59] Kimelberg, H.K. and Papahadjopoulos, D., *J. Biol. Chem.*, *249*, 1071 (1974).
[60] Gahmberg, C.F. and Hakomori, S.-I., *Biochem. Biophys. Res. Commun.*, *59*, 283 (1974).
[61] Gahmberg, C.F. and Hakomori, S.-I., *J. Biol. Chem.* *250*, 2438, 2447 (1975).
[62] Bustryn, J.-C., Schenkein, I., Baur, S. and Uhr, J.W., *J. Natl. Cancer Inst.*, *52*, 1263 (1974).
[63] Ozer, J.H. and Wallach, D.F.H., *Transplantation*, *5*, 652 (1967).
[64] Baldwin, R.W., Glaves, D., Harris, J.R. and Price, M.R., *Transplantation Proc.*, *3*, 1189 (1972).
[65] Baldwin, R.W. and Glaves, D., *Clin. Exp. Immunol.*, *11*, 51 (1973).
[66] Baldwin, R.W., Harris, J.R. and Price, M.R., *Int. J. of Cancer*, *11*, 385 (1973).
[67] Harris, J.R., Price, M.R. and Baldwin, R.W., *Biochim. Biophys. Acta*, *311*, 600 (1973).
[68] Baldwin, R.W., Bowen, J.G. and Price, M.R., *Biochim. Biophys. Acta*, *367*, 47 (1974).
[69] Reisfeld, R.A. and Kahan, B.D., *Fed. Proc.*, *29*, 2034 (1970).
[70] Mann, D.L., *Transplantation*, *14*, 398 (1972).
[71] Meltzer, S.M., Leonard, E.J., Rapp, H.J. and Borsos, T., *J. Natl. Cancer Inst.*, *47*, 703 (1971).
[72] Brannen, G.E., Adams, J.S. and Santos, G.W., *J. Natl. Cancer Inst.*, *53*, 165 (1974).
[73] Brandchaft, P.B. and Boone, C.W., *J. Natl. Cancer Inst.*, *53*, 1079 (1974).
[74] Blasecki, J.W. and Tevethia, S.S., *J. Immunol.*, *110*, 590 (1973).
[75] Drapkin, M.S., Aella, E. and Law, L.L., *J. Natl. Cancer Inst.*, *52*, 259 (1974).
[76] Ting, C.-C., Ortaldo, J.R. and Herbermann, R.B., *J. Natl. Cancer Inst.*, *815* (1974).
[77] Law, L.L. and Aella, E., *Nature*, *243*, 83 (1973).
[78] Uno, B.-I., Fink, G. and Schatz, G., *J. Biol. Chem.*, *250*, 775 (1975).
[79] Miller, L. and Wallach, D.F.H. to be published.
[80] Wallach, D.F.H. in *Biological Membranes*, Chapman, D. and Wallach, D.F.H., eds., Academic Press, London, 1973, p. 254.

Addendum

Research on the membrane biology of tumor cells continues to move very rapidly. Thus, during the interval between the submission of the manuscript for this book and return of the galleys to the printers, some 70 or so relevant papers have come to my attention. These are presented briefly here but cannot be treated in depth.

Membrane structure and organization

Several authors have examined normal and virally-transformed cells for differences in membrane freeze-fracture topology. Torpier et al. [1] report a significant increase in the density of intramembranous particles of BHK21 hamster fibroblasts that have been transformed by polyoma or hamster sarcoma viruses, as well as of chick embryo fibroblasts infected by a thermosensitive Rous sarcoma virus at temperatures both permissive and non-permissive for transformation. In contrast, no such differences were observed by Da Silva and Martinez-Palomo [2] in a comparison between normal Balb/c or Swiss 3T3 fibroblasts and their SV40 or murine sarcoma transformants. Da Silva and Martinez-Palomo also observed no modification of freeze-fracture topology following concanavalin A treatment of transformed cells and argue that macromolecules can move *on* the membrane surface without necessarily changing the structure of the membrane core. A study comparing normal hamster lymphocytes with a line (GD 248) of SV40 transformed lymphocytes also failed to reveal significant differences in the numbers and distribution of intramembranous particles [3]. No thermotropic response was detected in the plasma membranes of either normal or neoplastic cells but particle free areas appeared in the nuclear membranes of normal cells below 15°C and in the membranes of GD 248 cells only below 10°C, that is, the transformed cells differ from normals in nuclear membrane thermotropism [3].

The ^{1}H and ^{13}C nuclear magnetic resonance spectra of particulate fractions from normal and SV40 transformed fibroblasts have been interpreted to indicate increased 'fluidity' of the membrane lipids in transformed cells [4]. This does not concur with Gafney's comparison of 3T3 cells and SV40 or Py transformants thereof [5], showing no significant differences in 'inherent lipid flexibility' between normal and transformed cells by spin labelling techniques. It is difficult to evaluate these two spectroscopic studies because only signals from whole cells (or their homogenates) were recorded, preventing recognition of possible differences between membrane classes.

Membrane proteins

The role of membranes in protein synthesis, the metabolism of membrane proteins and differences between the membrane proteins of normal and neoplastic cells are all topics attracting increasing attention.

Thus Lande et al. [6] demonstrate a direct association of mRNA and endoplasmic reticulum membranes in WI38 fibroblasts. This association does not involve attachment of ribosomes or nascent polypeptides with the membranes. Also, Lodish and Small [7], extending their work on the biosynthesis of membrane proteins by rabbit reticulocytes, show that the two proteins under study are confined to the cytoplasmic surface of the plasma membrane and that 20–40 amino acids are proteolytically split off one of the proteins after its incorporation into the membrane. In a related study, Wreschner et al. [8] demonstrate progressive 'differentiation' of membrane protein patterns with reticulocyte maturation.

The mechanisms of membrane glycoprotein synthesis have been further elucidated and it is documented that membrane preparations from hen oviduct catalyze the transfer of mannose from GDP-mannose via a mannose-containing oligosaccharide-lipid to glycoprotein [9]. Indeed the data indicate that the oligosaccharide chain is transferred '*en bloc*' from the oligosaccharide-lipid to the protein. A different study, using a crude plasma membrane preparation from HeLa cells [10] suggests that non-glycosylated and glycosylated membrane proteins are inserted into plasma membranes sequentially (proteins more rapidly than glycoproteins) or independently.

Concerning membrane protein changes associated with neoplastic transformation, Isaka et al. [11] have examined the membrane proteins of chick embryo fibroblasts transformed by diverse strains of avian sarcoma virus.

All transformed cells showed appearance of a polypeptide with apparent molecular weight of 90,000, increase of polypeptide in the range of 79,000 daltons, decrease of a polypeptide with an apparent molecular weight near 50,000 and marked depletion of a protein with molecular weight near 200,000. Experiments with viral mutants temperature-sensitive for transformation suggest that the appearance of the 90,000 dalton protein and decrease of the 50,000 dalton protein are closely correlated with neoplastic transformation. In this connection it is significant that the amount of *actin* associated with membrane isolated from cultured chick embryo fibroblasts, decreases by 30–50% after neoplastic transformation by Rous sarcoma virus [12]. This phenomenon is not due to a decrease in cellular actin but to an altered association of this protein with membranes. Use of mutants temperature-sensitive for transformation indicates that the altered actin-membrane association is not due to viral infection but to transformation. Finally, Shin et al. [13] have purified plasma membrane fractions from normal, lactating rat mammary gland and from the R3230AC mammary tumor. The membranes from normal and neoplastic cells did not differ markedly in their complement of non-glycosylated proteins (judging by dodecyl sulfate polyacrylamide electrophoresis). However a membrane fraction from tumor cells was depleted in a glycoprotein (apparent mol. wt. 110,000) that is prominent in normal materials. The tumor membranes also contained a glycoprotein (apparent mol. wt. 70,000), lacking in normal membranes.

Phospholipids

In membrane-containing systems the inhibition of acetyl-CoA carboxylase activity by long chain acyl-CoA molecules, depends on the capacity of the membranes to accommodate long chain acyl-CoA [14]. This in turn relates to ionic strength, ionic composition and the intrinsic charge characteristics of the membranes. Accordingly, one might suspect that modifications in the proportion of charged phospholipids in cytoplasmic membranes (subsequent to neoplastic transformation) could interfere with regulation of fatty acid biosynthesis (as does occur in tumors). The local level and proportion of divalent cations could also play a role. For example, it has been shown that phosphatidylglycerol, in the presence of Mg^{2+}, exhibits super cooling and formation of metastable gel phases; these do not occur in the presence of Ca^{2+} [15]. The steady-state level of long chain acyl-CoA will also depend on the transport of carnitine and long chain acyl-carnitines across the inner mitochondrial membrane. The translocation of these intermediates by exchange diffusion has been recently characterized [16].

Glycolipids

The amount of Forssman glycolipid (G1–5) of two clones of NIL hamster fibroblasts increases with cell density in monolayer culture but not in spinner culture [17]. However, NIL cells transformed by hamster sarcoma virus showed no density dependent rise in Gl–5 synthesis even in monolayer culture.

Progress continues in the purification of Golgi membranes rich in glycosyl transferase activities [18] and the lipid composition of Golgi membranes has been thoroughly characterized [19].

Cholesterol

Sterols, such as 25-OH cholesterol, that inhibit sterol biosynthesis by L cells *in vitro* (by suppressing the activity of 3-hydroxy-3-methyl glutaryl-CoA reductase) appear to require specific binding to saturable cellular receptors [20], whereas the uptake of cholesterol proceeds in a different manner. In this connection, Brown et al. [21] have isolated a line of mutant human fibroblasts deficient in a plasma membrane receptor for low-density lipoproteins (LDL). Lack of this receptor leads to an overproduction of cholesterol due to the inability of LDL to normally suppress the activity of 3-hydroxy-3-methylgulateryl-CoA reductase. One should consider the possibility that such receptor deficiencies might occur also in neoplastia cells, since the overproduction of cholesterol by leukemic L_2C lymphocytes cannot be suppressed by *in vivo* cholesterol feeding or *in vitro* exposure to cholesterol phospholipid liposomes, but is suppressed by small amounts of 25-OH cholesterol or a nitroxide derivative of androstane [22]. Some inhibitory sterols also enhance esterification of endogenous and exogenous cholesterol (by membrane-bound fatty acyl-CoA: cholesteryl acyl transferase) in human fibroblasts [23]. This suggests coordinated reciprocal regulation of cholesterol synthesis and cholesterol esterification.

Blastogenic stimulation of mouse lymphocytes with phytohemagglutinin stimulates cholesterol biosynthesis within 4 hr, reaching a maximum within 24 hr [24]. Stimulation of the lymphocytes could be blocked by inhibiting cholesterol biosynthesis with 25-OH cholesterol, before sterol synthesis had become maximal. Thereafter, the inhibitors could not block phytohemagglutinin induced activation, indicating that cholesterol synthesis is required for the successful initiation of blastogenic stimulation. On the other hand, artificial enrichment of lymphoid cells by equilibration with cholesterol-enriched phospholipid liposomes, interferes with blastogenic stimulation by concanavalin A, but does not interfere with lectin binding [25].

Hamsters bearing tumors induced by simian virus 40 exhibit an elevation of all serum lipids, including cholesterol, in a manner characteristic of the tumor strain. These anomalies persist even after surgical excision of the tumors and are ascribed to a tumor-induced inhibition of host pre-betalipoprotein catabolism [25].

Enzymes and enzyme regulation

Numerous experimenters continue in efforts to evaluate the significance of divergent cyclic nucleotide levels in normal vs. neoplastic cells. Thus, Willingham and Pastan [27] argue that the greater tendency of neoplastic cells to form surface microvilli relates to their lower cAMP content and also to their anomalous lectin-induced agglutinability. Another study [28] reports that various viral transformants of mouse 3T3 fibroblasts contain less cAMP/cell than parental cells. Moreover, several SV40 transformed cell lines that are temperature-sensitive with regard to transformation exhibited lower cAMP levels than normal 3T3 cells, but no temperature sensitivity of cAMP content. This suggests that expression of the transformed phenotype does not necessarily correlate with cAMP level. Related experiments [29] confirm other studies, reporting that mouse 3T3 cells neoplastically transformed by SV40 exhibit about half and twice the cAMP and cGMP levels, respectively, of normal parental cells. This experimentation further shows that density-dependent inhibition of growth in 3T3 cells, as well as four revertant lines, derived from SV40 transformants, correlates well with reduction of cGMP, but not cAMP. These studies also suggest that cell density restriction on proliferation differs from that due to serum restriction in its effect on cyclic nucleotide metabolism. A report relevant to the problem of malignancy [30] shows that membrane isolates of metastasizing rat mammary carcinomas exhibit a threefold enhancement in the complement of low K_m adenosine-3,5-cyclic monophosphate phosphodiesterase activity relative to the high K_M enzyme, whereas in other subcellular fractions the proportions of high and low K_M activity enzymes were similar. In all cases the diesterase activities were 1.3–2.0-fold higher in metastasizing cells. The specific adenylate cyclase activity of a hepatocyte membrane fraction from Yoshida ascites hepatoma AH130 is reported to be lower than in liver plasma membrane [31]. In contrast the basal- or fluoride-stimulated adenylate cyclase activity in a plasma membrane isolate of Morris hepatoma 5123 (cCh), a tumor with a low cAMP content, [32] is reported similar to that of membranes from normal liver [33], whereas the response of the tumor membrane enzyme to glucagon was less than that of membranes from normal hepatocytes.

While most investigations suggest that cyclic nucleotides are critical regulators of the cell cycle and cell metabolism, the recent elegant studies of Coffino et al. [33] on mutants of the S49 mouse lymphoma, indicate the periodic fluctuations of cAMP levels are not *essential* for the complete cell cycle.

Adenylate cyclase is of course critical in the cAMP levels of all cells, normal or transformed, but the enzyme is not well understood. Considerable progress in this direction has been achieved by Neer [34], who shows that the enzyme activity of rat renal medulla exists in two forms. The predominant species has a mol. wt. of about 159,000 in Triton X-100; the minor form has a mol. wt. of 38,000. The data also suggest that only a small proportion (~ 5%) of the adenylate cyclase molecule is associated with other membrane components through apolar associations.

Kasărov and Friedman [35] report that the Na^+-K^+-activated adenosine triphosphatase activities in homogenates of several lines of neoplastically-transformed rat and mouse fibroblasts were 4–5 times greater than those in non-neoplastic parental lines. A similar pattern was observed in comparisons of normal and SV40 or Rous sarcoma-transformed BHK fibroblasts during logarithmic growth [36], but not in a comparison of rat hepatocytes and Yoshida ascites hepatoma [37]. K^+ uptake, measured as $^{86}Rb^+$ accumulation, was also enhanced in the transformed cells and Arrhenius plots of their Na^+-K^+-activated ATPase showed a discontinuity at 24°C not found in 3T3 cells [36]. This is taken to indicate a different state of phospholipids in the transformed cells. Such a difference might well escape bulk spectroscopic measurements such as in [5].

Permeability

Fibroblasts neoplastically transformed by chemical carcinogens *in vitro* show enhanced uptake of 2-deoxyglucose [38], or 2-deoxyglucose and 2-aminoisobutyric acid [39]. Malignant hepatocytes from a chemically induced hepatoma, also accumulate 2-deoxyglucose more effectively than non-neoplastic cells in culture [40]. Kinetic analyses [39,40] indicate an increase in V_{max} and no change in apparent K_M. 2-Deoxy-D-glucose is incorporated into the glycolipids of both normal fibroblasts and cells transformed by diverse oncogenic viruses [41]. However, the 2-deoxy-D-glucose-glycolipid complement of the transformants differs from that of the normal cells, and experimentation with viral mutants that are temperature-sensitive for transformation indicate that this difference represents an expression of the transformed phenotype.

Adriamycin and bleomycin, both antibiotics used in cancer chemotherapy act far more effectively at 42–43°C, than at 37–41°C [42]. This 'synergism' between antibiotic action and hyperthermia, 'thermochemotherapy', appears to involve a sharp modification of plasma membrane permeability to the antibiotics, as the temperature is raised from 41–43°C. The basis for the phenomenon may be the membrane protein thermotropism at 41–42°C, described in Chapter 1.

Cell contact and cell recognition

An extensive bibliography on the relationship between cell surface properties of normal and neoplastic fibroblasts and their growth characteristics has been published by Pardee [43]. This review also deals with the action of proteases on cell surfaces, which continues to be studied intensively. Thus, Christman et al. [44], continuing their experiments on the tumorigenicity of mouse melanoma $B_5$59, document a close correlation between the presence of plasminogen activator and malignant behavior. However, while the presence of plasminogen activator often correlates with the depletion of certain trypsin-sensitive external membrane proteins, and there are suggestions that the impaired growth control of neoplastic cells *in vitro* may derive from proteolytic surface modifications, a study [45] on the action of thrombin, a potent mitogen [46], shows that the mitogenic action of this agent on chick embryo fibroblasts is not accompanied by deletion of the trypsin-sensitive membrane-associated proteins. The removal of these proteins is thus not necessary to induce cell proliferation. Studies with low concentrations of other peptidases further indicate that protease-mediated removal of the trypsin-sensitive proteins is not *per se* sufficient to induce cell division.

Cell coupling

The mechanisms involved in the formation of permeable intercellular junctions continue subject to aggressive investigation. Elegant experiments documenting the role of cytoplasmic Ca^{2+} activity are presented in [47] and the importance of cell contact mode, junctional conductance and junctional insulation are described in [47,48]. A detailed microphysiological–micromorphological correlation of junction formation in Novikoff ascites hepatoma is given in [50].

Lectin reactivity

Numerous investigators continue to focus their attention on the different interactions of diverse lectins with normal and neoplastic cells, despite the extreme complexities of lectin-binding properties already discussed. Important progress has been achieved through (a) detailed structural analyses of concanavalin A [51–54] and (b) recognition that thymocytes possess high- and low-affinity binding sites for concanavalin A, both of which can be blocked by certain hapten sugars [55]. High-affinity binding is positively cooperative and is mediated through a specific membrane glycoprotein. The low-affinity binding sites are 5–10 times more prevalent but do not involve specific membrane macromolecules. Lectin-induced agglutination, toxicity and possibly 'capping' probably involve low-affinity binding. The different agglutinability of both normal and SV40 transformed 3T3 cells propagated in culture media varying in fatty acid composition [56] may well relate to low-affinity adsorption processes. In any case, the occurrence on cell surfaces of both receptor-specific binding and non-specific adsorption, both blocked by hapten sugars [55], complicates interpretations of the interaction of diverse cell types with various lectins under saturation conditions. This includes observation described in the main text, as well as experiments published more recently (e.g. [57–64]).

If conclusions about the 'fluidity' of membranes, as inferred from cell–lectin interactions, were originally suspect (because of the complex and poorly understood properties of lectins), recent data, using 1,6-diphenyl-1,3,5-hexatriene (DPH) to monitor membrane microviscosity following cell–lectin interaction [65–66] are even more open to criticism. The reason for this is that the signals emitted by probes such as DPH, depend strongly on the lipid distribution within cells under given conditions. Measurements on intact cells will yield signals that depend on (a) the distribution of the probe among the various membrane compartments and (b) the proportion of these compartments in a cell. If transformed cells have, for example, a higher content of low-cholesterol membranes (e.g. nuclear envelope) than normal cells, and DPH partitions into these membranes, one might get the impression of 'increased fluidity'. Also, small oil droplets, as occur in many tumor cells, will cause highly 'fluid' signals. Moreover, recent experimentation on glutaraldehyde-fixed normal and polyoma transformed hamster fibroblasts, show that such cells coagglutinate equally well with human erythrocytes as unfixed cells [67]. This suggests that agglutinability is not a simple function of membrane 'fluidity'.

Epilogue

Increasing evidence supports the proposal presented elsewhere in this volume that membranes are segregated into domains of different composition and behavior. This appears the case not only for plasma membranes, but also cytoplasmic membranes and the membrane envelope of Newcastle Disease virus [68]. Penetrating proteins within such domains are surrounded by lipid boundary layers, whose detailed composition (structure) depends upon the character of the hydrophobic aspect of the penetrating proteins [69]. Moreover, lipid phase transitions can regulate the activity of mitochondrial enzymes, such as β-hydroxybutyrate dehydrogenase [69]. Also, studies on lipid vesicles bearing incorporated calcium transport protein [70], support our conclusion that, in general, cholesterol is excluded from the lipid boundary layer surrounding penetrating proteins and that penetrating proteins impose a lateral segregation of lipid classes. Studies on artificial lipid bilayer membranes composed of binary phosphatidyl choline mixtures further indicate that even without protein, membranes can exhibit segregation of fluid phases parallel to the membrane plane, due to 'fluid-fluid immiscibility' [71]. Transverse phase separation, to give an asymmetrical bilayer, is also possible.

References

[1] Torpier, G., Montagnier, L., Biquard, J. M. and Vigier, P., *Proc. Natl. Acad. Sci. U.S.*, *72*, 1695 (1975).
[2] Pinto Da Silva, P. and Martinez-Palomo, A., *Proc. Natl. Acad. Sci. U.S.*, *72*, 572 (1975).
[3] Wallach, D. F. H., Schmidt-Ullrich, R., Alroy, J. and Weinstein, R., *Biochim. Biophys. Acta*, submitted.
[4] Nicolau, Cl., Dietrich, W., Steiner, M. R., Steiner, S. and Melnick, J. L., *Biochim. Biophys. Acta.*, *382*, 311 (1975).
[5] Gaffney, B. J., *Proc. Natl. Acad. Sci. U.S.*, *72*, 664 (1975).
[6] Laude, M. A., Adesnik, M., Sumida, M., Tashir, Y. and Sabatini, D. D., *J. Cell Biol.*, *65*, 513 (1975).
[7] Lodish, H. F. and Small, B., *J. Cell Biol.*, *65*, 51 (1975).
[8] Wreschner, D., Foglizzo, R. and Herzberg, M., *FEBS Letters*, *52*, 255 (1975).
[9] Lucas, J. J., Waechter, C. J. and Lennarz, W. J., *J. Biol. Chem.*, *250*, 1992 (1975).
[10] Atkinson, P. H., *J. Biol. Chem.*, *250*, 2123 (1975).
[11] Isaka, T., Yoshida, M., Owada, M. and Toyoshima, K., *Virology 65*, 226 (1975).
[12] Wickus, G., Gruenstein, E., Robbins, P. W. and Rich, A., *Proc. Natl. Acad. Sci. U.S.*, *72*, 746 (1975).
[13] Shin, B. C., Ebner, K. E., Hudson, B. A. and Carraway, K. L., *Cancer Res. 35*, 1135 (1975).
[14] Sumper, M., *Eur. J. Biochem. 49*, 469 (1974).
[15] Ververgaert, P. H., J. Th., de Kruyff, B., Verkleij, A. J., Tocanne, J. F. and van Deenen, L. L. M., *Chem. Phys. Lipids*, *13*, 97 (1975).
[16] Pande, S. V., *Proc. Natl. Acad. Sci. U.S. 72*, 883 (1975).
[17] Sakiyama, H. and Terasima, T., *Cancer Res.*, *35*, 1723 (1975).
[18] Mitranic, M. M., Sturgess, J. M. and Moscarello, M. A., *J. Memb. Biol.*, *19*, 397 (1974).
[19] Zambrano, F., Fleischer, S. and Fleischer, B., *Biochim. Biophys. Acta*, *380*, 357 (1975).
[20] Kandutsch, A. A. and Chen, H. W., *J. Cell. Physiol.*, *85*, 415 (1975).
[21] Brown, M. S., Brannan, P. G., Bohmfalk, H. A., Brunschede, G. Y., Dana, S. E., Helgeson, J. and Goldstein, J. L., *J. Cell. Physiol.*, *85*, 425 (1975).
[22] Philippot, J., Wallach, D. F. H. and Cooper, A., *Biochim. Biophys. Acta.*, *406*, 161 (1975).
[23] Brown, M. S., Dana, S. E. and Goldstein, J. L., *J. Biol. Chem.*, *250*, 4025 (1975).
[24] Chen, H. W., Heiniger, H. J. and Kandutsch, A. A., *Proc. Natl. Acad. Sci. U.S.*, *72*, 1950 (1975).
[25] Alderson, J. C. E. and Green, C., *FEBS Letters*, *52*, 208 (1975).
[26] Cox, R. and Gökcen, M., *J. Natl. Cancer Inst.*, *54*, 379 (1975).
[27] Willingham, M. and Pastan, I., *Proc. Natl. Acad. Sci. U.S.*, *72*, 1263 (1975).
[28] Burstin, S. J., Renger, H. C. and Basilico, C., *J. Cell. Physiol.*, *84*, 69 (1975).
[29] Moens, W., Vokaer, A. and Kram, R., *Proc. Natl. Acad. Sci. U.S.*, *72*, 1063 (1975).
[30] Chatterjee, S. K. and Kim, U., *J. Natl. Cancer Inst.*, *54*, 181 (1975).
[31] Hickie, R. A., Walker, C. M. and Croll, G. A., *Biochem, Biophys. Res. Commun.*, *59*, 167 (1974).
[32] Hickie, R. A., Jan, S. H. and Datta, A., *Cancer Res.*, *35*, 596 (1975).
[33] Coffino, P., Gray, J. W. and Tomkins, G. M., *Proc. Natl. Acad. Sci. U.S.*, *72*, 878 (1975).
[34] Neer, E. J., *J. Biol. Chem.*, *249*, 6527 (1974).
[35] Kasărov, L. B. and Friedman, H., *Cancer Res. 34*, 1862 (1974).
[36] Kimelberg, H. K. and Mayhew, E., *J. Biol. Chem.*, *250*, 100 (1975).

[37] Réthy, A., Trevisani, A., Manservigi, R. and Tomasi, V., *J. Memb. Biol.*, *20*, 99 (1975).
[38] Oshiro, Y. and DiPaolo, J. A., *J. Cell. Physiol.*, *83*, 193 (1974).
[39] Kiroki, T. and Yamakawa, S., *Int. J. Cancer*, *13*, 240 (1974).
[40] Siddiqui, M. and Iype, P. T., *Int. J. Cancer*, *15*, 773 (1975).
[41] Steiner, M. R., Somers, K. and Steiner, S., *Biochem. Biophys. Res. Commun.*, *61*, 795 (1975).
[42] Hahn, G. M., Braun, J. and Har-Kedar, I., *Proc. Natl. Acad. Sci. U.S.*, *72*, 937 (1975).
[43] Pardee, A. B., *Biochim. Biophys. Acta*, *417*, 153 (1975).
[44] Christman, J. K., Silagi, S., Newcomb, E. W., Silverstein, S. C. and Acs, G., *Proc. Natl. Acad, Sci. U.S.*, *72*, 47 (1975).
[45] Teng, N. N. H. and Chen, L. B., *Proc. Natl. Acad, Sci. U.S.*, *72*, 413 (1975).
[46] Chen, L. B. and Buchanan, J. M., *Proc. Natl. Acad. Sci. U.S.*, *72*, 131 (1975).
[47] Rose, B. and Loewenstein, W. R., *Nature*, *254*, 250 (1975).
[48] Ito, S., Sato, E. and Loewenstein, W. R., *J. Memb. Biol.*, *19*, 305 (1974).
[49] Ito, S., Sato, E. and Loewenstein, W. R., *J. Memb. Biol.*, *19*, 339 (1974).
[50] Johnson, R., Hammer, M., Sheridan, J. and Revel, J. P., *Proc. Natl. Acad. Sci. U.S.*, *71*, 4536 (1974).
[51] Wang, J. L., Cunningham, B. A., Waxdal, M. J., and Edelman, G. M., *J. Biol. Chem.*, *250*, 1490 (1975).
[52] Cunningham, B. A., Wang, J. L., Waxdal, M. J. and Edelman, G. M., *J. Biol. Chem.*, *250*, 1503 (1975).
[53] Becker, J. W., Reeke, G. N., Jr., Wang, J. L., Cunningham, B. A. and Edelman, G. M., *J. Biol. Chem.*, *250*, 1513 (1975).
[54] Reeke, G. N., Jr., Becker, J. W. and Edelman, G.M., *J. Biol. Chem.*, *250*, 1525 (1975).
[55] Schmidt-Ullrich, R., Wallach, D. F. H. and Hendricks, J., *Biochim. Biophys. Acta*, submitted.
[56] Horwitz, A. F., Hatten, M. E. and Burger, M. M., *Proc. Natl. Acad. Sci., U.S.*, *71*, 3115 (1974).
[57] Penhoet, E., Olson, C., Carlson, Lacorbière, M. and Nicholson, G. L., *Biochemistry*, *13*, 3561 (1974).
[58] Ben-Bassat, H. and Goldblum, N., *Proc. Natl. Acad. Sci. U.S.*, *72*, 1046 (1975).
[59] Chauvenet, A. R. and Scott, D. W., *J. Immunol.*, *114*, 470 (1975).
[60] De Petris, S., *J. Cell. Biol.*, *65*, 123 (1975).
[61] Rutishauser, U. and Sachs, L., *J. Cell. Biol.*, *65*, 247 (1975).
[62] Grimaldi, J. J. and Sykes, B. D., *J. Biol. Chem.*, *250*, 1618 (1975).
[63] Killion, J. J. and Kollmorgen, G. M., *Nature*, *254*, 247 (1975).
[64] Marikovsky, Y., Inbar, M., Danon, D. and Sachs, L., *Exp. Cell. Res.*, *89*, 359 (1974).
[65] Inbar, M. and Shinitzky, M., *Eur. J. Immunol.*, *5*, 166 (1975).
[66] Toyoshima, S. and Osawa, T., *J. Biol. Chem.*, *250*, 1655 (1975).
[67] Gibson, D. A., Marquardt, M. D. and Gordon, J. A., *Science*, *189*, 45 (1975).
[68] Wisnieski, B. J., Parkes, J. G., Huang, Y. O. and Fox, C. F., *Proc. Natl. Acad. Sci. U.S.*, *71*, 4381 (1974).
[69] Houslay, M. D., Warren, G. B., Birdsall, N. J. M. and Metcalfe, J. C., *FEBS Letters*, *51*, 146 (1975).
[70] Warren, G. B., Houslay, M. D., Metcalfe, J. C. and Birdsall, N. J. M., *Nature*, *255*, 684 (1975).
[71] Wu, S. H. W. and McConnell, H. M., *Biochemistry*, *14*, 847 (1975).

Index